T0254678

Modern Birkhäuser Classics

Many of the original research and survey monographs, as well as textbooks, in pure and applied mathematics published by Birkhäuser in recent decades have been groundbreaking and have come to be regarded as foundational to the subject. Through the MBC Series, a select number of these modern classics, entirely uncorrected, are being re-released in paperback (and as eBooks) to ensure that these treasures remain accessible to new generations of students, scholars, and researchers.

Methods of Algebraic Geometry in Control Theory: Part II

Multivariable Linear Systems and Projective Algebraic Geometry

Peter Falb

Reprint of the 1999 Edition

Peter Falb
Division of Applied Mathematics
Brown University
Providence, Rhode Island, USA

ISSN 2197-1803 ISSN 2197-1811 (electronic)
Modern Birkhäuser Classics
ISBN 978-3-319-96573-4 ISBN 978-3-319-96574-1 (eBook)
https://doi.org/10.1007/978-3-319-96574-1

Library of Congress Control Number: 2018951899

Printed on acid-free paper

This book is published under the imprint Birkhäuser, www.birkhauser-science.com by the registered company Springer Nature Switzerland AG part of Springer Nature.
The registered company address is: Gewerbestrasse 11, 6330 Cham, Switzerland

Peter Falb

Methods of Algebraic Geometry in Control Theory: Part II

Multivariable Linear Systems and Projective Algebraic Geometry

Springer Science+Business Media, LLC

Peter Falb
Division of Applied Mathematics
Brown University
Providence, RI 02912

Library of Congress Cataloging-in-Publication Data

Falb, Peter L.

Methods of algebraic geometry in control theory / Peter Falb.

p. cm.—

Includes bibliographical references

Contents: v. 1. Scalar linear systems and affine algebraic geometry.

ISBN 978-1-4612-7194-9 ISBN 978-1-4612-1564-6 (eBook)

DOI 10.1007/978-1-4612-1564-6

1. Control theory. 2. Geometry, Algebraic. I. Title. II. Series.

QA402.3.F34 1990

629.8'—dc20 90-223 CIP

AMS Subject Classifications: 14-01, 14L17, 14M15, 14N05, 93A25, 93B27, 93C35

Printed on acid-free paper.

Originally published by Birkhäuser Boston in 1999
Softcover reprint of the hardcover 1st edition 1999

ISBN 978-1-4612-7194-9

Typeset in LaTeX by TeXniques, Inc., Cambridge, MA.

9 8 7 6 5 4 3 2 1

Contents

Preface

"Control theory represents an attempt to codify, in mathematical terms, the principles and techniques used in the analysis and design of control systems. Algebraic geometry may, in an elementary way, be viewed as the study of the structure and properties of the solutions of systems of algebraic equations. The aim of this book is to provide access to the methods of algebraic geometry for engineers and applied scientists through the motivated context of control theory".*

The development which culminated with this volume began over twenty-five years ago with a series of lectures at the control group of the Lund Institute of Technology in Sweden. I have sought throughout to strive for clarity, often using constructive methods and giving several proofs of a particular result as well as many examples.

The first volume dealt with the simplest control systems (i.e., single input, single output linear time-invariant systems) and with the simplest algebraic geometry (i.e., affine algebraic geometry). While this is quite satisfactory and natural for scalar systems, the study of multi-input, multi-output linear time-invariant control systems requires projective algebraic geometry.

Thus, this second volume deals with multi-variable linear systems and projective algebraic geometry. The results are deeper and less transparent, but are also quite essential to an understanding of linear control theory. A review of

*From the Preface to Part I.

the scalar theory is included along with a brief summary of affine algebraic geometry (Appendix E).

Acknowledgements. Although he did not play a direct role in this work, I should like to express my deep appreciation for the inspiration provided by my very dear friend, the late George Zames.

There are a great many friends, colleagues, teachers, and students to whom considerable thanks are due, but I should like especially to express my appreciation to Karl Astrom of the Lund Institute of Technology for my original involvement in the precursor lectures and to the Laboratory for Information and Decision Systems at M.I.T. and its Director, Sanjoy K. Mitter, for the use of a quiet office (without which this book would never have been written). Thanks are also due to Elizabeth Loew for the excellent computer preparation of the manuscript.

Finally, I dedicate this work to my (long suffering) dear wife, Karen.

Peter Falb
Cambridge, 1999

Introduction

We recall from Part I that "the overall goal of these notes is to provide an introduction to the ideas of algebraic geometry in the motivated context of system theory." We shall suppose familiarity with the development in Part I and we adopt the same general approach and conventions. (See also Appendix E.)

Part II deals with multivariable (i.e., several input, several output) linear systems and projective algebraic geometry. This represents, in essence, the second stage of both system theory and algebraic geometry. The results extend the material in Part I, but, as we shall see, the extension is not entirely straightforward.

We begin with a brief review of the scalar theory and extend it directly to systems with either a single input or a single output. We recall and introduce seven representations, namely: (1) a strictly proper rational meromorphic vector; (2) a vector of coprime polynomials; (3) a block (with vector blocks) Hankel matrix of finite rank; (4) a triple of matrices; (5) a curve in a projective space; (6) a pair of matrices modulo equivalence (the observability, controllability pair); and, (7) a causal $k[z]$-module homomorphism of power series. The representations (1), (2), (3), (4) extend those developed in Part I. The representations (5) and (6) were sketched in Chapter I.23 and the representation (7) is developed here. We treat the case of a single input $p \times 1$ with $p \geq 1$, system here. We develop the entire theory (with the exception of the pole placement theorem) including the appropriate transition theorems and the geometric quotient theorem. We also introduce some notation for indices and the idea of a p-partition of n. The single input theory essentially mimics the scalar theory of Part I.

In order to motivate the general theory, we treat a number of examples of two or three input, two output systems in detail. We observe that the *degree* of such a system need *not* be the length of the shortest recurrence, that a coprime matrix representation involves equivalence under the unimodular group; and, that the curve in a projective space actually lies in a Grassmannian. We give an explicit algebraic structure to $\mathrm{Rat}(n, 2, 2)$ and $\mathrm{Rat}(n, 3, 2)$ involving a set of degree conditions and some Grassmann equations. Critical is a natural embedding into a Grassmannian. Next we examine the Laurent map and develop an explicit algebraic structure for $\mathrm{Hank}(2, 2, 2)$ and $\mathrm{Hank}(2, 3, 2)$. We also "prove" a geometric quotient theorem for these low order cases. It should be clear from this critical Chapter 2 that the general extension of the theory is not straightforward.

Chapters 3–8 are devoted to some basic concepts of system theory and projective algebraic geometry. The transfer and Hankel matrices are introduced using the module formulation ([F-3]) and the notion of a strictly proper rational meromorphic transfer matrix is defined. Block Hankel matrices are defined and the key structure lemma of Risannen (Lemma 3.42) is established. The system spaces $\mathbf{Rat}(n, m, p)$ and $\mathbf{Hank}(n, m, p)$ are defined and shown to depend on a finite number of parameters via an embedding in $\mathbb{A}_k^{n(mp+1)}$. Polynomial matrices are studied in Chapter 4. The unimodular group, the Hermite normal form and the concepts of divisibility (right or left) and coprimeness are developed. The representation $(\mathbf{P}(z), \mathbf{Q}(z))$ by coprime matrices is analyzed and minimal realizations defined. The *degree* and *Kronecker set* of a system are also introduced. The Transfer Lemma (Lemma 4.50) which is the transition theorem to the $(\mathbf{P}(z), \mathbf{Q}(z))$ representation is also established.

We begin the work on projective algebraic geometry with a brief account of projective space. Linear subspaces, dimension, homogeneous and affine coordinates, the notion of general position and the projective group, $\mathrm{PGL}(N)$, are developed. The *join* of subspaces is analyzed. The critical concept of a *projection* with center a linear subspace is also treated. In Chapter 6, we deal with some basic concepts revolving around the notion of a projective algebraic set. Graded rings, homogeneous ideals and the Hilbert polynomial are commutative algebra ideas we need. The (homogeneous) ideal of V, the Zariski topology and the homogeneous coordinate ring are defined and the projective Nullstellensatz is proved. Some significant differences with the affine case occur; notably, (i) the homogeneous coordinate ring is *not* an invariant for a projective variety (it depends on the embedding in projective space); and (ii) the "irrelevant" ideal in the Nullstellensatz. The concept of dimension for projective varieties is treated and several results established including a theorem (Theorem 6.57) on the dimension of components of an intersection and the fact that the dimension is the degree of the Hilbert polynomial. Regular functions, local rings and morphisms are examined in Chapter 7 and $\mathbb{P}^N$ is shown to be locally affine ($\mathbb{A}^N$) using the notion of a sheaf. A map $\psi\colon W \to V$ is a *morphism* if ψ is continuous (Zariski topology) and if $f \in o_W(U)$ (regular functions on U open in W) implies $\psi^*(f) = f \circ \psi \in o_V(\psi^{-1}(U))$ (regular functions on $\psi^{-1}(U)$ open in V). In contrast to the affine case where the concept is global, here it is local behavior which counts. A key result, Theorem 7.28, the *affine criterion* for morphisms, states that it is enough to check on an affine covering. We also consider the *local ring* of a *subvariety* and prove the crucial Nakayama lemma (Lemma 7.45). Finally, in this area of Part II, we develop in some detail the exterior algebra required to define $\mathrm{Gr}(p, N)$, the Grassman variety of p-dimensional subspaces of N-space. The so-called Plücker relations (Equation 8.59) define $\mathrm{Gr}(p, N)$ which is an irreducible projective variety of dimension $p(N - p)$.

The next portion of the book is devoted to the Laurent Isomorphism Theorem and the algebraic structure of $\mathrm{Rat}(n, m, p)$ and $\mathrm{Hank}(n, m, p)$ as well as some required material from projective algebraic geometry (Chapters 9–11). We first introduce the system matrix $M_{\mathbf{F}}(z)$ of an element $\mathbf{F}(z)$ of $\mathrm{Rat}(n, m, p)$. $M_{\mathbf{F}}(z)$ is a $(p + m) \times m$ matrix of rank m for *all* z and we let $\psi_{\mathbf{F},\boldsymbol{\alpha}}(z)$ denote, for $\boldsymbol{\alpha} = (i_1, \ldots, i_m)$, $1 \le i_1 < \cdots < i_m \le p + m$, the $m \times m$ minor with rows $i_1, \ldots, i_m$. Then $(\psi_{\mathbf{F},\boldsymbol{\alpha}}(z))$ gives an embedding of $\mathrm{Rat}(n, m, p)$ into $\mathrm{Gr}(m, p + m)$. "Homogenizing", we get a morphism $(\psi_{\mathbf{F},\boldsymbol{\alpha}}(x_0, x_1))$ of $\mathbb{P}_k^1$ into $\mathrm{Gr}(m, p + n)$ where the $\psi_{\mathbf{F},\boldsymbol{\alpha}}(x_0, x_1)$ are each a form of degree n. These give a variety $\mathbf{V}(n, m, p)$ which is an algebraic structure for $\mathrm{Rat}(n, m, p)$. Before developing a similar structure for $\mathrm{Hank}(n, m, p)$, we need some projective algebraic geometry results on products. The concept of a product for projective varieties is defined in two ways, namely: (i) via local (affine) products and patching; and, (ii) concretely by the Segre map which embeds $\mathbb{P}^N \times \mathbb{P}^M$ as a closed subvariety of $\mathbb{P}^{NM+M+N}$. The complex of ideas relating to completeness (a projective analog of compactness) and the Main Theorem of Elimination Theory (Theorem 10.16) is considered. The key result is that the image of a projective variety under a morphism is closed and hence, in contrast to the affine case, projections are closed maps. We, then, can study the structure of $\mathrm{Hank}(n, m, p)$ and prove the Laurent Isomorphism Theorem (Theorem 11.30) that $\mathrm{Rat}(n, m, p)$ and $\mathrm{Hank}(n, m, p)$ are isomorphic. We show that the Hankel variety is defined by four sets of equations which are: (a) Block symmetry; (b) Hankel structure; (c) Dependence; and, (d) "Grassmann conditions" (a necessary internal consistency).

The subsequent portion of the book deals with the state space representation and concludes with the Geometric Quotient Theorem (Theorem 16.28), (Chapters 12–16). We begin with some projective algebraic geometry. In par-

particular, we introduce the family of hypersurface sections and prove a geometric form of the Noether-Normalization Theorem (Theorem 12.7) using the Main Theorem of Elimination Theory (see [M-5]). Incidence varieties, e.g., $\mathcal{L}_p(V) = \{L_p\colon L_p \cap V \neq \emptyset,\ L_p$ a p-dimensional linear subspace$\}$, are studied and we show that $\deg V$ is the number of points of intersection of V with a general L_{N-r}, $r = \dim V$. In the key Chapter 13, we treat the state space representation. We prove the existence of state space realizations of a Hankel matrix $\mathbf{H}$ using an argument of Clark ([C-1]) which also gives a nice local coordinatization of $\mathrm{Hank}(n,m,p)$. Moreover, we get $\dim \mathrm{Hank}(n,m,p) = n(m+p)$ and the normality of $\mathrm{Hank}(n,m,p)$. The controllability and observability maps are defined and the space $S^n_{m,p}$ of minimal linear systems introduced. Equivalence under the action of $G = \mathrm{GL}(n,k)$ leads to the State Space Isomorphism Theorem (Theorem 13.42). We next treat three representations of $S^n_{m,p}$; namely, (1) The A-diagonal Δ_A in $\mathfrak{o}^n_p \times \mathfrak{c}^n_m$ (observable $\times$ controllable) with $\Delta_A = \{(A_1, C) \times (A_2, B)\colon A_1 = A_2\}$; (2) the set of $(\mathbf{Z}, \mathbf{Y}, A)$ in $M_*((n+1)p, n) \times M_*(n, (n+1)m) \times \mathbb{A}^{n^2}$; and (3) the variety $\mathbf{V}_A(n,m,p)$ in $M_*((n+1)p, n) \times M_*(n, (n+1)m) \times \mathbb{A}^{n^2}$ defined by $Z^i = Z^1 X^{i-1}$, $Y_j = X^{j-1} Y_1$. All are G-isomorphic to $S^n_{m,p}$. We define a natural map into $\mathrm{Hank}(n,m,p)$ and show that the image is defined by Hankel type equations: (a) Block symmetry; (b) Hankel structure; (c) Dependence; and (d) "Grassmann" conditions. A version of the Lemma of Risannen (Lemma 13.66) shows the power of (a), (b), and (c). The critical example 13.68 shows the necessity of the conditions (d). Before proving the Geometric Quotient Theorem we need some results on fibers of morphisms. In Chapter 14, we develop the results analogous to those of Chapter I.18 which give conditions for an open map. We also show that if ψ is a dominant morphism, then the number of points in $\psi^{-1}(\psi(x))$ is $[k(X) : k(Y)]_s$ (separable degree) and introduce the concept of a separable morphism (used in studying feedback). Tangents, differentials, and simple subvarieties are studied in Chapter 15. The (Zariski) tangent space is defined and the notion of a simple point is related to independent differentials (in Proposition 15.29). The differential, $d\psi$, of a morphism $\psi\colon V \to W$ is defined as a map of tangent spaces. A morphism is *smooth* at a point ξ, if ξ is simple and $d\psi$ is surjective at ξ. We prove (Theorem 15.31) that if a morphism is dominant and separable, then it is smooth on an open set; and, conversely, if ψ is smooth at ξ, then it is dominant and separable. We are now ready to prove the Geometric Quotient Theorem (Theorem 16.28). We show that $(\mathrm{Hank}(n,m,p), \psi_A)$ is such a quotient (using an analog of the classical proof of Part I) and that $\mathrm{Hank}(n,m,p)$ is a nonsingular variety of dimension $n(m+p)$. We next show that in the multivariate case continuous canonical forms do not exist using ideas of Hazewinkel ([H-4]). To do so, we develop the notion of a principal G-bundle. We have (Theorem 16.41): a continuous canonical form for the action of G on V exists if and only if V is a trivial principal G-bundle. An example of Hazewinkel on $S^2_{2,2}$ which is universal via an embedding shows the nonexistence of a continuous canonical form. In effect, we show (Theorem 16.52) that a continuous canonical form exists if and only if either $m = 1$ or $p = 1$.

The final portion of the book (Chapters 17–20) deals with feedback. We begin with some projective algebraic geometry. If V is a normal variety, then a *divisor*, D, on V is a sum $\Sigma n_i \Gamma_i$, Γ_i a closed, irreducible subvariety of codimension 1. Call D *equivalent* to D_1 if $D - D_1 = (h)$, h a rational function. The Picard Group, $\mathrm{Pic}(V)$, is $\mathrm{Div}(V)$ modulo this equivalence. We introduce the complete linear system of divisors, $|D| = \{(f) + D\colon f = 0 \text{ or } (f) + D \geq 0\}$, show that $\dim_k |D| < \infty$, and call any projective linear subspace of $|D|$, a linear system of divisors. We relate this to the poles of a system. In Chapter 18, we give an analysis of intersections so that we can examine $\psi_{\mathbf{F}}(x_0, x_1) \cap H_N$, $H_N = V(Y_N)$ (the system poles). We show that (as divisors) $\deg(\gamma_{\mathbf{F}} \cdot H_N) = \deg \gamma_{\mathbf{F}} \cdot \deg H_N = \deg \gamma_{\mathbf{F}} = \deg \mathbf{F}$ where $\gamma_{\mathbf{F}}$ is the curve $\psi_{\mathbf{F}}(x_0, x_1)$. We then prove (a generalization of) Bezout's Theorem (Theorem 18.17) which (loosely) says: $(\deg V_1)(\deg V_2) = \Sigma m(Z_j; V_1 \cap V_2, \mathbb{P}^N) Z_j$, Z_j the components of $V_1 \cap V_2$, and $m(Z_j; V_1 \cap V_2, \mathbb{P}^N)$ an appropriate (albeit complicated) notion of multiplicity. State feedback is the subject of Chapter 19. We introduce the state feedback group and describe its structure. We prove Heymann's Lemma (Lemma 19.13) which leads to the coefficient assignment theorem. We use Theorem 15.31 (on smooth morphisms) to show that the map $\varphi_{A,B}(g, K, \alpha) = g(A + B\alpha^{-1}K)g^{-1}$ is dominant if and only if (A, B) is controllable (Proposition 19.17). The Hermite indices are *not invariant* under state feedback while the Kronecker indices *are invariant* under state feedback. We examine system structure under the action of state feedback and see that the stabilizer is in "block stripe" form ([W-1], Wang and Davison). Finally, we exhibit a quotient but not a geometric quotient. The last chapter is devoted to output feedback and is quite geometric. We introduce the output feedback group and describe its structure. Using ideas of Wang ([W-2], [W-3]), we view the pole placement map as a projection $\Lambda_{\mathbf{F}}\colon \mathrm{Gr}(m, m+p) - E_{\mathbf{F}} \to \mathbb{P}^n$ where $E_{\mathbf{F}}$ is a linear variety determined by $\mathbf{F}$. The system $\mathbf{F}$ is *assignable* if $\Lambda_{\mathbf{F}}$ is surjective. We show that $mp \geq n$ is necessary for assignability and that if $mp \geq n$ then assignability is generic. However, $mp \geq n$ is not sufficient for global assignability. We carefully consider the case $mp = n$ and introduce the concept of degeneracy. If $mp = n$ and $\mathbf{F}$ is not degenerate, then F is assignable. We also sketch the frequency domain version of the theory.

We reiterate that we do not strive for the greatest generality and that we tend to use constructive methods with a view towards applications. As in Part I, the exercises are an integral part of the account and are often used in the main body of the text. We have, of course, used many sources and we acknowledge their considerable contribution (e.g., [A-2], [B-6], [B-9], [D-2], [H-2], [H-3], [K-5], [M-1], [M-2], [M-5], [W-4], [Z-3], etc.) even if explicit reference to them is not made at a particular point in the text.

Conventions

All rings are commutative with an identity element 1 and a ring homomorphism maps 1 into 1. Neither an integral domain nor a field is the zero ring (i.e., $0 \neq 1$) and consequently, a prime (or maximal) ideal is necessarily a proper ideal. The notation $A \subset B$ means A is contained in B and A may equal B while the notation $A < B$ means A is contained in B but A is not equal to B. For sets, the notation $A - B$ means the complement of B in A. References to Part I take the form Theorem I.14.20 (i.e., Theorem 14.20 of Part I) and references to Part II take the form Theorem 7.28. Generally, matrices are in boldface as $\mathbf{M}$ and $\mathbf{M}^i(\mathbf{M}_j)$ represent the ith row (jth column of $\mathbf{M}$). If $\mathbf{M}$ is $r \times s$ and $t \leq \min(r, s)$, then $\det[\mathbf{M}^{i_1,\dots,i_t}_{j_1,\dots,j_t}]$ is the determinant of the $t \times t$-minor from the $i_1, \dots, i_t$ rows and $j_1, \dots, j_t$ columns.

1
Scalar Input or Scalar Output Systems

Let us suppose, for ease of exposition, that our field is simply the complex numbers $\mathbb{C}$ in this section (although all that we do is valid for any algebraically closed field k). We recall that there were four basic representations of a scalar input-scalar output linear system introduced in Part I. These representations were:

(1) a proper rational meromorphic function $f(z)$;

(2) a pair of relatively prime (coprime) polynomials $(p(z), q(z))$;

(3) a Hankel matrix H of finite rank; and,

(4) a triple (A, b, c) in $\mathbb{A}^{n^2+2n}$.

In addition, we saw in Chapter 23 of Part I that the system could be represented by:

(5) a map $\psi_f\colon \mathbb{P}^1_{\mathbb{C}} \to \mathbb{P}^1_{\mathbb{C}}$ given by $\psi_f(x_0, x_1) = (x_0^n p(x_1/x_0),\ x_0^n q(x_1/x_0))$; and,

(6) a point $x_f = (\mathbf{Z}^1, \mathbf{Y}_1)$ in $M_*(n+1, n)_\alpha \times M_*(n, n+1)_\beta$ or, modulo equivalence, as a point in $\mathrm{Gr}(n, n+1) \times \mathrm{Gr}(n, n+1)$.

There is one more representation which we now describe.

Let $\mathbb{C}[[1/z]]$ be the ring of formal power series in $1/z$ with coefficients in $\mathbb{C}$ (see [Z-3] or Appendix A). Since this ring is an integral domain, it has a quotient field $\mathbb{C}((1/z))$. The set $\{1, 1/z, \dots\}$ is a multiplicatively closed set in

P. Falb, *Methods of Algebraic Geometry in Control Theory: Part II*,
Modern Birkhäuser Classics, https://doi.org/10.1007/978-3-319-96574-1_1

$\mathbb{C}[[1/z]]$ and it is easy to see that $\mathbb{C}((1/z)) = \mathbb{C}[[1/z]]_{(1/z)}$. In other words, if $w \in \mathbb{C}((1/z))$, then $w = z^t \hat{w}$ with $\hat{w} \in \mathbb{C}[[1/z]]$. More formally, we have:

Lemma 1.1 *Let A be a ring and X be an indeterminate. Then $a_0 + a_1 X + \cdots$ is a unit in $A[[X]]$ if and only if a_0 is a unit in A.*

Proof. If a_0 is a unit in A, then there is an $\alpha_0 \in A$ with $a_0 \alpha_0 = 1$. Let us define $\alpha_1, \alpha_2, \ldots, \alpha_n, \ldots$ by

$$\begin{aligned} \alpha_1 &= -\alpha_0 a_1 \alpha_0, \quad \alpha_2 = -\alpha_0 a_1 \alpha_1 - \alpha_0 a_2 \alpha_0, \ldots \\ \alpha_n &= -\alpha_0 a_1 \alpha_{n-1} - \cdots - \alpha_0 a_n \alpha_0, \ldots . \end{aligned}$$

Then

$$(a_0 + a_1 X + \cdots)(\alpha_0 + \alpha_1 X + \cdots) = \sum_{n=0}^{\infty} \left(\sum_{i+j=n} a_i \alpha_j \right) X^n.$$

But $\sum_{i+j=n} a_i \alpha_j = a_0 \alpha_n + a_1 \alpha_{n-1} + \cdots + a_n \alpha_0 = 0$ for $n > 0$ since $a_0 \alpha_n = -(a_0 \alpha_0) a_1 \alpha_{n-1} - \cdots - (a_0 \alpha_0) a_n \alpha_0$. Conversely, if $(a_0 + a_1 X + \cdots)(\beta_0 + \beta_1 X + \cdots) = 1$, then $a_0 \beta_0 = 1$ and a_0 is a unit in A.

Corollary 1.2 *If K is a field, then $K((X)) = K[[X]]_X$.*

Proof. Let $b_0 + b_1 X + \cdots / a_0 + a_1 X + \cdots$ be an element of $K((X))$ and let ν be the order of $a_0 + a_1 X + \cdots$ so that $a_0 + a_1 X + \cdots = X^\nu (a_\nu + a_{\nu+1} X + \cdots)$ with $a_\nu \neq 0$. Then $a_\nu + a_{\nu+1} X + \cdots$ has an inverse $\alpha(X) = \alpha_\nu + \alpha_{\nu+1} X + \cdots$ and it follows that $b_0 + b_1 X + \cdots / X^\nu (q_\nu + q_{\nu+1} X + \cdots) = (b_0 + b_1 X + \cdots)\alpha(X)/X^\nu$.

We let $\hat{S} = \mathbb{C}[[1/z]]$, $R = \mathbb{C}[z]$, and $\hat{S}_+$ be the ideal generated by $1/z$ in $\hat{S}$ (i.e., $\hat{S}_+$ is the set of power series of positive order). Then $\hat{S}_{1/z} = \hat{S}_{z^{-1}} = \mathbb{C}((1/z))$ and we have the following proposition.

Proposition 1.3 $\hat{S}_{z^{-1}} = R \oplus \hat{S}_+$ *as $\mathbb{C}$-vector spaces.*

Proof. If $w \in \hat{S}_{z^{-1}}$, then $w = z^t \hat{w}$ with $\hat{w} \in \hat{S}$ and $t \geq 0$. But $\hat{w} = a_0 + a_1 z^{-1} + \cdots + a_t z^{-t} + \sum_{j=1}^{\infty} b_j z^{-t} z^{-j}$ so that $z^t \hat{w} = a_0 z^t + \cdots + a_t + \sum_{j=1}^{\infty} b_j z^{-j} \in R + \hat{S}_+$. If $w \in R \cap \hat{S}_+$, then $w = a_0 + a_1 z + \cdots + a_t z^t = a_1' z^{-1} + a_2' z^{-2} + \cdots$ and $z^{-t} w = a_t + a_{t-1} z^{-1} + \cdots + a_0 z^{-t} = a_1' z^{-1-t} + a_2' z^{-t-2} + \cdots$. Thus, $a_0 = 0$, $a_1 = 0, \ldots, a_t = 0$ and $w = 0$.

Since R is a subring of $\hat{S}_{z^{-1}}$, $\hat{S}_{z^{-1}}$ is an R-module and there is an exact sequence of R-modules

$$0 \longrightarrow R \xrightarrow{i} \hat{S}_{z^{-1}} \xrightarrow{\pi} \hat{S}_{z^{-1}}/R \longrightarrow 0 \tag{1.4}$$

where i is the natural injection and π is the natural projection. We define an R-module structure on $\hat{S}_+$ by setting

$$z \cdot \sum_{j=1}^{\infty} a_j z^{-j} = \sum_{j=1}^{\infty} a_{j+1} z^{-j} \tag{1.5}$$

(i.e., the left-shift). Then $\hat{S}_{z^{-1}}/R$ and $\hat{S}_+$ are isomorphic R-modules. We call an R-module homomorphism $\varphi\colon \mathbb{C}((1/z)) \to \mathbb{C}((1/z))$ a *system* if

$$\varphi(\hat{S}) \subset \hat{S}_+ \tag{1.6}$$

(this is equivalent to causality). Now, how does this abstract idea relate to our previous representations of a scalar linear system?

Definition 1.7 Let $\Omega_c = \{\omega\colon \mathbb{Z} \to \mathbb{C} \text{ with } \omega(n) = 0 \text{ for } n \leq -N_\omega,\ N_\omega \geq 0\}$ be the *set of admissible sequences.* If $\omega \in \Omega_c$, then the *transform of* ω, $\hat{\omega}(z)$, is the element of $\hat{S}_{z^{-1}}$ given by

$$\hat{\omega}(z) = \sum_{n=-\infty}^{\infty} \omega(n) z^{-n} = \sum_{-N_\omega}^{\infty} \omega(n) z^{-n} = z^{N_\omega}\left(\sum_{j=0}^{\infty} \omega(j - N_\omega) z^{-j}\right) \tag{1.8}$$

(cf. [F-3]). If $\omega(n)$ is any sequence, then the sequence $\{(\sigma\omega)(n)\}$ with

$$(\sigma\omega)(n) = \omega(n+1) \tag{1.9}$$

is the *left-shift of* ω.

If $\omega \in \Omega_c$, then $N_{\sigma\omega} = N_\omega + 1$ and

$$\begin{aligned}(\widehat{\sigma\omega})(z) &= z^{N_\omega+1}\left(\sum_{j=0}^{\infty} (\sigma\omega)(j - N_\omega - 1) z^{-j}\right) \\ &= z \cdot z^{N_\omega}\left(\sum_{j=0}^{\infty} \omega(j - N_\omega) z^{-j}\right) = z \cdot \hat{\omega}(z)\end{aligned}$$

so that the left-shift transforms to multiplication by z. Let $\{h_j\}$, $\{u_j\}$ be any sequences and let $\{y_n\}$ be the formal *convolution* of $\{h_j\}$ and $\{u_j\}$, i.e., $y_n = \sum_{j=-\infty}^{\infty} h_{n-j} u_j = \sum_{k=-\infty}^{\infty} h_k u_{n-k}$ (note these sums have no real meaning as yet). We write $\{y_n\} = \{h_j\} * \{u_j\}$.

Proposition 1.10 *If $\{h_j\}$ and $\{u_j\}$ are in Ω_c, then $\{y_n\} = \{h_j\} * \{u_j\}$ is a well-defined element of Ω_c and $\hat{y}(z) = \hat{h}(z)\hat{u}(z)$.*

Proof. If $k \leq -N_h$, then $h_k = 0$ and if $k \geq n + N_u$, then $n - k \leq -N_u$ and $u_{n-k} = 0$. It follows that

$$y_n = \sum_{k=-N_h+1}^{n+N_u-1} h_k u_{n-k} \tag{1.11}$$

is a *finite* sum for every n. In other words, $\{y_n\} = \{h_j\} * \{u_j\}$ is well-defined. If $n \leq -(N_n + N_u)$ then $n + N_u \leq -N_h$ and $y_n = 0$ by (1.11) so that $\{y_n\} \in \Omega_c$. The final assertion is a well-known straightforward calculation ([F-3] or Exercise 1).

Now let (A, b, c) be a state space representation of $f(z)$ and let $h_j = cA^{j-1}b$ for $j = 1, \ldots, h_j = 0$ for $j \leq 0$. Then $\{h_j\} \in \Omega_c$ and if $\{u_j\} \in \Omega_c$, we have

$$y_n = \sum_{k=1}^{n+N_u-1} h_k u_{n-k} = \sum_{k=1}^{n+N_u-1} (cA^{k-1}b) u_{n-k} \tag{1.12}$$

and

$$\hat{y}(z) = \left(\sum_{k=1}^{\infty} (cA^{k-1}b) z^{-k} \right) \hat{u}(z) \tag{1.13}$$

so that the transfer function $f(z) = \sum cA^{k-1}bz^{-k} = \hat{h}(z)$. Thus, we are led to the following representation of a scalar linear system:

(7) a $\mathbb{C}[z]$-module homomorphism $\varphi_f \colon \mathbb{C}((1/z)) \to \mathbb{C}((1/z))$ such that $\varphi_f(\mathbb{C}[[1/z]]) \subset \mathbb{C}_+[[1/z]]$ ([F-3]).

We note that every such homomorphism is given by multiplication by an $f(z)$ in $\mathbb{C}_+[[1/z]]$.

Now let us assume that our system has $m = 1$ input but $p \geq 1$ outputs.* We assert that the extension of the results of Part I to this case is quite straightforward as we shall show.

Consider the representation (1). We replace the single rational meromorphic function $f(z)$ by a $p \times 1$ vector $\mathbf{f}(z)$ of proper rational meromorphic functions $f^i(z) = p^i(z)/q^i(z)$ with p^i, q^i relatively prime. Let $q(z)$ be the least common multiple of the q^i, $i = 1, \ldots, p$. Then

$$q(z) = s^i(z) q^i(z), \quad f^i(z) = s^i(z) p^i(z)/q(z) = P^i(z)/q(z)$$

where $P^i(z) = s^i(z) p^i(z)$, $i = 1, \ldots, p$.

*The case of $m \geq 1$ inputs and $p = 1$ output is entirely analogous.

Proposition 1.14 *$P^1, \dots, P^p$, q are coprime in the sense that there are $x_1(z), \dots, x_p(z)$, $y(z)$ such that*

$$x_1 P^1 + \cdots + x_p P^p + yq = 1 \tag{1.15}$$

identically in z.

Proof. Let $y = y_1 + \cdots + y_p$ are write (1.15) in the form

$$x_1 p^1 s^1 + \cdots + x_p p^p s^p + y_1 q^1 s^1 + \cdots + y_p q^p s^p = 1 \tag{1.16}$$

or, equivalently, in the form

$$(x_1 p^1 + y_1 q^1) s^1 + \cdots + (x_p p^p + y_p q^p) s^p = 1. \tag{1.17}$$

Since p^i, q^i are relatively prime, it will be enough to show that the s^i are coprime. Since $\mathbb{C}[z]$ is a UFD, let $\pi_1, \dots, \pi_\nu$ be the primes which divide some q^i. Then $q^i = \pi_1^{\alpha_{1i}} \pi_2^{\alpha_{2i}} \cdots \pi_\nu^{\alpha_{\nu i}}$ and

$$q = \pi_1^{\max \alpha_{1i}} \pi_2^{\max \alpha_{2i}} \cdots \pi_\nu^{\max \alpha_{\nu i}}. \tag{1.18}$$

It follows that

$$s^i = \pi_1^{\max\limits_j \alpha_{1j} - \alpha_{1i}} \pi_2^{\max\limits_j \alpha_{2j} - \alpha_{2i}} \cdots \pi_\nu^{\max\limits_j \alpha_{\nu j} - \alpha_{\nu i}} \tag{1.19}$$

for $i = 1, \dots, p$. But for any of the primes π_j, there is by (1.19) an $s^{i(j)}$ such that π_j does not divide $s^{i(j)}$. Thus, the s^i are coprime.

If $\mathbf{P}(z) = (P^1(z), \dots, P^p(z))$, then $\mathbf{f}(z) = \mathbf{P}(z)/q(z)$ with $\mathbf{P}(z)$, $q(z)$ coprime. Thus the representation (1) is essentially the same for the single-input, multi-output case as for the scalar case. Similarly, the polynomial representation (2) is replaced by a set of $p+1$ coprime polynomials $(P^1(z), \dots, P^p(z), q(z))$.

If $\mathbf{f}(z) = \sum\limits_{i=1}^{\infty} \mathbf{H}_i z^{-i}$ where each $\mathbf{H}_i$ is $p \times 1$, then we can consider the Hankel matrix $H_\mathbf{f}$ given by

$$H_\mathbf{f} = \begin{pmatrix} \mathbf{H}_1 & \mathbf{H}_2 & \cdots \\ \mathbf{H}_2 & \mathbf{H}_3 & \cdots \\ \mathbf{H}_3 & \cdot & \cdots \\ \vdots & \vdots & \end{pmatrix} = (\mathbf{H}_{i+j-1})_{i,j=1}^{\infty}$$

where each $\mathbf{H}_{i+j-1}$ is $p \times 1$. We observe that if $\mathbf{f}(z) = \mathbf{P}(z)/q(z)$ is rational with $q(z)$ of degree n, then there is a recurrence relation

$$a_0 \mathbf{H}_j + a_1 \mathbf{H}_{j+1} + \cdots + a_{n-1} \mathbf{H}_{j+n-1} + \mathbf{H}_{j+n} = 0 \tag{1.20}$$

for $j = 1, \ldots$ where $q(z) = a_0 + a_1 z + \cdots + a_{n-1}z^{n-1} + z^n$. Thus, $\rho(H_{\mathfrak{f}}) = \operatorname{rank} H_{\mathfrak{f}} \leq n$. If we say that a polynomial $r(z) = r_0 + r_1 z + \cdots + r_\nu z^\nu$ *annihilates the sequence* $\{\mathbf{H}_j\}$ if

$$r_0 \mathbf{H}_j + r_1 \mathbf{H}_{j+1} + \cdots + r_\nu \mathbf{H}_{j+\nu} = 0 \tag{1.21}$$

for $j = 1, \ldots$ and if we let $\mathfrak{a}(\{\mathbf{H}_j\}) = \{r(z)\colon r(z) \text{ annihilates } \{\mathbf{H}_j\}\}$, then (i) $\mathfrak{a}(\{\mathbf{H}_j\})$ is an ideal in $\mathbb{C}[z]$; (ii) $\mathfrak{a}(\{\mathbf{H}_j\}) \neq (0)$ if and only if $\{\mathbf{H}_j\}$ is recurrent; and, (iii) if $\mathfrak{a}(\{\mathbf{H}_j\}) \neq (0)$, then there is a monic $q(z)$ of lowest degree with $\mathfrak{a}(\{\mathbf{H}_j\}) = (q(z))$ since $\mathbb{C}[z]$ is a principal ideal domain. Clearly, if $H_{\mathfrak{f}}$ has finite rank $n_{\mathfrak{f}} = n$, then there is a recurrence relation of the form (1.20) and $\mathfrak{a}(\{\mathbf{H}_j\}) = (q_{\mathfrak{f}}(z)) \neq (0)$. We have:

Proposition 1.22 *Let $\rho(H_{\mathfrak{f}}) = n < \infty$. Then there are an $n \times n$ A, an $n \times 1$ b, and a $p \times n$ C such that $\mathbf{H}_j = CA^{j-1}b$, $j = 1, 2, \ldots$. Hence, $n \geq$* degree $q_{\mathfrak{f}}$.

Proof. ([C-1]) Let $\nu = (\nu_1, \ldots, \nu_n)$ be such that the rows H_∞^ν span the row space of $H = H_{\mathfrak{f}}$. Then there is a unique $p \times n$ C_ν such that

$$H_\infty^{\mathbf{p}} = C_\nu H_\infty^\nu \tag{1.23}$$

where $\mathbf{p} = (1, \ldots, p)$, or, in other words,

$$\mathbf{H}_j = C_\nu (h_j^{\nu_i}) \tag{1.24}$$

$i = 1, \ldots, n$, $j = 1, \ldots$. Let b_ν be the $n \times 1$ vector given by

$$b_\nu = (h_1^{\nu_i}) \tag{1.25}$$

so that $\mathbf{H}_1 = C_\nu b_\nu$. Let $\nu + p = (\nu_1 + p, \ldots, \nu_n + p)$. Since H_∞^ν span the row space of H, there is a (unique) $n \times n$ A_ν such that $H_\infty^{\nu+p} = A_\nu H_\infty^\nu$ or, equivalently, such that

$$(h_j^{\nu_i + p}) = A_\nu (h_j^{\nu_i}) \tag{1.26}$$

for $i = 1, \ldots, n$, $j = 1, 2, \ldots$. But H is a Hankel matrix so that

$$(h_j^{\nu_i + p}) = (h_{j+1}^{\nu_i}) \tag{1.27}$$

for $i = 1, \ldots, n$, $j = 1, 2, \ldots$. It follows that

$$\mathbf{H}_j = C_\nu A_\nu^{j-1} b_\nu \tag{1.28}$$

for $j = 1, \ldots,$ and $n \geq$ degree $q_{\mathfrak{f}}$ by the Cayley-Hamilton theorem.

Proposition 1.29 *If $\rho(H_{\mathfrak{f}}) = n$, then* degree $q_{\mathfrak{f}} = n$.

Proof. In view of Proposition 1.22, it is enough to show that $\tilde{n} = \text{degree } q_{\mathbf{f}} \geq n$. Let $q_{\mathbf{f}}(z) = a_0 + a_1 z + \cdots + a_{\tilde{n}-1} z^{\tilde{n}-1} + z^{\tilde{n}}$ and set

$$A = \begin{bmatrix} 0 & 0 & \cdots & 0 & -a_0 \\ 1 & 0 & & 0 & -a_1 \\ 0 & 1 & & \vdots & \\ \cdot & \cdot & & 1 & \\ \cdot & \cdot & \cdots & 0 & -a_{\tilde{n}-1} \end{bmatrix}, \quad b = \epsilon_1 = \begin{bmatrix} 1 \\ 0 \\ \vdots \\ 0 \end{bmatrix}$$

$$C = \begin{bmatrix} \mathbf{H}_1 & \cdots & \mathbf{H}_{\tilde{n}} \end{bmatrix}$$

so that

$$A^{j-1} b = \epsilon_j, \quad C A^{j-1} b = \mathbf{H}_j \tag{1.30}$$

for $j = 1, \ldots, \tilde{n}$, and

$$C A^{\tilde{n}+t-1} b = -a_{\tilde{n}+t-2} C A^{\tilde{n}+t-2} b \cdots - a_0 C A^{t-1} b \tag{1.31}$$

for $t = 1, \ldots$. It follows that $\mathbf{H}_j = C A^{j-1} b$ for *all* j and again, by Cayley-Hamilton, $\tilde{n} > \operatorname{rank} H_{\mathbf{f}} = n$.

We have, in effect, shown that, just as in the scalar case, $\rho(H_{\mathbf{f}})$ is the same as the degree of the denominator $q_{\mathbf{f}}(z)$ of $\mathbf{f}(z) = \mathbf{P}(z)/q_{\mathbf{f}}(z)$ which is the same as the degree of the generator of $\mathfrak{a}(\{\mathbf{H}_j\})$ which is the same as the length of the shortest recurrence for $\{\mathbf{H}_j\}$. This is, as we shall see, not true in general for multivariable systems.

We call the rank of $H_{\mathbf{f}}$, the *degree of* $\mathbf{f}$, and if $H = (\mathbf{H}_{i+j-1})_{i,j=1}^{\infty}$ is a (block) Hankel matrix, then we say that H has *degree* n if H has rank n.

Definition 1.32 Let $\mathbf{Rat}(n; p, 1) = \{\mathbf{f}(z)\colon \mathbf{f}(z)$ is a strictly proper rational meromorphic $p \times 1$ vector of degree $n\}$ and let $\mathbf{Hank}(n; p, 1) = \{\mathfrak{h}_{\infty}^{\infty}\colon \mathfrak{h}_{\infty}^{\infty}$ a block Hankel matrix of $p \times 1$ blocks with $\rho(\mathfrak{h}_{\infty}^{\infty}) = n\}$.

We observe that $\mathbf{f}(z) = \mathbf{P}(z)/q(z)$ is an element of $\mathbf{Rat}(n; p, 1)$ if and only if there are $(b_0^1, \ldots, b_{n-1}^1, \ldots, b_0^p, \ldots, b_{n-1}^p, a_0, \ldots, a_{n-1})$ such that the polynomials

$$\begin{aligned} p^j(z) &= b_0^j + b_1^j z + \cdots + b_{n-1}^j z^{n-1}, \qquad j = 1, \ldots, p \\ q(z) &= a_0 + a_1 z + \cdots + a_{n-1} z^{n-1} + z^n \end{aligned} \tag{1.33}$$

are coprime. Similarly, we note that $\mathcal{H}_{\infty}^{\infty} = (\mathbf{H}_{i+j-1})_{i,j=1}^{\infty}$ is an element of $\mathbf{Hank}(n; p, 1)$ if and only if there are $\tilde{\mathbf{H}}_1, \ldots, \tilde{\mathbf{H}}_n \in \mathbb{A}_{\mathbb{C}}^p$ and $a_0, \ldots, a_{n-1}$ in $\mathbb{C}$ such that $\mathbf{H}_i = \tilde{\mathbf{H}}_i$, $i = 1, \ldots, n$, $\mathbf{H}_{j+n} = -a_0 \mathbf{H}_j - a_1 \mathbf{H}_{j+1} \cdots - a_{n-1} \mathbf{H}_{n-1+j}$, $j = 1, 2, \ldots$, and, setting $\tilde{\mathbf{H}}_{j+n} = \mathbf{H}_{j+n}$, if $\rho(\mathfrak{h}(\tilde{\mathbf{H}}_j)) = n$.

Definition 1.34 Let $\operatorname{Rat}(n; p, 1) = \{(b_0^1, \ldots, b_{n-1}^1, \ldots, b_0^p, \ldots, b_{n-1}^p, a_0, \ldots, a_{n-1}) \in \mathbb{A}_{\mathbb{C}}^{np+n}\colon p^j(z),\ q(z)$ given by (1.33) are coprime$\}$ and let

$\text{Hank}(n; p, 1) = \{(\mathbf{H}_1, \ldots, \mathbf{H}_n, a_0, \ldots, a_{n-1}) \in \mathbb{A}_{\mathbb{C}}^{np+n}$: if $\mathbf{H}_{j+n} = -a_0\mathbf{H}_j - \cdots - a_{n-1}\mathbf{H}_{j+n-1}$, then $\rho((\mathbf{H}_{i+j-1})_{i,j=1}^{\infty}) = n\}$.

We have noted that $\mathbf{Rat}(n; p, 1)$ and $\text{Rat}(n; p, 1)$ can be identified and that $\mathbf{Hank}(n; p, 1)$ and $\text{Hank}(n; p, 1)$ can also be identified. We view $\text{Rat}(n; p, 1)$ and $\text{Hank}(n; p, 1)$ as contained in $\mathbb{A}_{\mathbb{C}}^{np+n}$ and we shall show that these are quasi-affine varieties. If $(\mathbf{H}_{i+j-1})_{i,j=1}^{\infty}$ is a $p \times 1$ block Hankel matrix, then it is an element of $\mathbf{Hank}(n; p, 1)$ if and only if

$$\begin{bmatrix} \mathbf{H}_1 & \cdots & \mathbf{H}_n \\ \vdots & & \vdots \\ \mathbf{H}_n & & \mathbf{H}_{2n-1} \end{bmatrix} \begin{bmatrix} x_0 \\ x_1 \\ \vdots \\ x_{n-1} \end{bmatrix} = \begin{bmatrix} \mathbf{H}_{n+1} \\ \vdots \\ \mathbf{H}_{2n} \end{bmatrix} \tag{1.35}$$

has a *unique* solution $x_j = -a_j$, $j = 0, \ldots, n-1$ such that $\mathfrak{h}_n^n = (\mathbf{H}_{i+j-1})_{i,j=1}^{n}$ has rank n. If we consider the set of $(\mathbf{H}_1, \ldots, \mathbf{H}_{2n}) \in \mathbb{A}_{\mathbb{C}}^{2np}$ such that (1.35) holds and $\rho(\mathfrak{h}_n^n) = n$, then this set is a quasi-affine variety X in $\mathbb{A}_{\mathbb{C}}^{2np}$. Define a map $\psi \colon X \to \mathbb{A}_{\mathbb{C}}^{np+n}$ by setting

$$\psi(\mathbf{H}_1, \ldots, \mathbf{H}_{2n}) = (\mathbf{H}_1, \ldots, \mathbf{H}_n, x_0, \ldots, x_{n-1}) \tag{1.36}$$

with the x_i determined by 1.35. ψ is a well-defined morphism. In fact, $\psi(X) = \text{Hank}(n; p, 1)$ and ψ is an isomorphism so that $\text{Hank}(n; p, 1)$ is a quasi-affine variety.

Example 1.37 Let $p = 2$, $n = 1$ so that

$$\mathbf{H}_1 = \begin{bmatrix} h_1^1 \\ h_1^2 \end{bmatrix}, \quad \mathbf{H}_2 = \begin{bmatrix} h_2^1 \\ h_2^2 \end{bmatrix}, \quad x_0\mathbf{H}_1 = \mathbf{H}_2$$

and

$$\psi(\mathbf{H}_1, \mathbf{H}_2) = (\mathbf{H}_1, x_0).$$

Since $\rho(\mathbf{H}_1) = 1$, either $h_1^1 \neq 0$ or $h_1^2 \neq 0$. Let $U_1 = (\mathbb{A}_{\mathbb{C}}^4)_{x_1}$, $U_2 = (\mathbb{A}_{\mathbb{C}}^4)_{x_2}$, $V_1 = (\mathbb{A}_{\mathbb{C}}^3)_{x_1}$, and $V_2 = (\mathbb{A}_{\mathbb{C}}^3)_{x_2}$ be principal open sets. Then $\psi(U_i) \subset V_i$ and $\psi^*(\mathbb{C}[V_i]) \subset \mathbb{C}[U_i]$ on $\text{Hank}(1; 2, 1)$ since $x_0 \circ \psi = h_2^1/h_1^1$ on U_1 and $x_0 \circ \psi = h_2^2/h_1^2$ on U_2. Moreover, since $x_0\mathbf{H}_1 = \mathbf{H}_2$, these agree on $U_1 \cap U_2$. Alternatively, observe that $\mathbf{H}_1'\mathbf{H}_1 = (h_1^1)^2 + (h_1^2)^2 = (h_1^1 + ih_2^1)(h_1^1 - ih_1^2)$ and that if

$$g = \begin{bmatrix} 1 & i \\ 1 & -i \end{bmatrix} \Big/ \sqrt{2}$$

then

$$\begin{aligned}\tilde{\mathbf{H}}_1 &= g\mathbf{H}_1 = \begin{bmatrix} h_1^1 + ih_1^2 \\ h_1^1 - ih_1^2 \end{bmatrix} \Big/ \sqrt{2} \\ \tilde{\mathbf{H}}_1'\tilde{\mathbf{H}}_1 &= (h_1^1)^2 - (h_2^1)^2\end{aligned}$$

so that either $\det \mathbf{H}_1'\mathbf{H}_1 \neq 0$ or $\det \tilde{\mathbf{H}}_1'\tilde{\mathbf{H}}_1' \neq 0$ and we can use a similar approach. If either determinant is 0, then the outputs are dependent so that the system is "really" a scalar system. The example represents a slight chink in the affine armor as the map ψ is defined locally and pieced together (which is the typical projective situation).

If $\mathbf{f}(z) = (P^j(z)/q(z)) \in \mathrm{Rat}(n; p, 1)$ and if we set

$$\mathbf{P}_i = \begin{bmatrix} b_i^1 \\ b_i^2 \\ \vdots \\ b_i^p \end{bmatrix}, \quad i = 0, 1, \ldots, n-1$$

then $\mathbf{f}(z) = (\mathbf{P}_0 + \mathbf{P}_1 z + \cdots + \mathbf{P}_{n-1} z^{n-1})/q(z)$ and we can define a map $L\colon \mathrm{Rat}(n; p, 1) \to \mathrm{Hank}(n; p, 1)$ by

$$L(\mathbf{P}_0, \mathbf{P}_1, \ldots, \mathbf{P}_{n-1}, a_0, \ldots, a_{n-1}) = (\mathbf{H}_1, \ldots, \mathbf{H}_n, a_0, \ldots, a_{n-1}) \tag{1.38}$$

where

$$\begin{aligned}[\mathbf{H}_1 \ \cdots \ \mathbf{H}_n] &= [\mathbf{P}_0 \ \cdots \ \mathbf{P}_{n-1}]\tilde{X}(a_0, \ldots, a_{n-1})^{-1} \\ \tilde{X}(a_0, \ldots, a_{n-1}) &= \begin{bmatrix} a_1 & a_2 & \cdots & 1 \\ \vdots & & & \vdots \\ a_{n-1} & 1 & & 0 \\ 1 & 0 & \cdots & 0 \end{bmatrix}\end{aligned}$$

i.e.,

$$\begin{aligned}\mathbf{P}_0 &= a_1\mathbf{H}_1 + a_2\mathbf{H}_2 + \cdots + \mathbf{H}_n \\ &\vdots \\ \mathbf{P}_{n-1} &= \mathbf{H}_{n-1}\end{aligned}$$

L is clearly an isomorphism of $\mathbb{A}_{\mathbb{C}}^{np+n}$ onto $\mathbb{A}_{\mathbb{C}}^{np+n}$. The mapping $\mathbf{L} = \psi^{-1} \circ L$ is thus a bijective mapping of $\mathrm{Rat}(n; p, 1)$ onto $\mathrm{Hank}(n; p, 1)$ with inverse $\mathbf{L}^{-1} = L^{-1} \circ \psi$. Thus, to "prove" the Laurent isomorphism in this context, we need only show that $\mathrm{Rat}(n; p, 1)$ is a quasi-affine variety.

Example 1.39 Let $p(z) = b_0 + b_1 z$, $q(z) = a_0 + a_1 z + z^2$. Then (I.3.15) p, q are relatively prime if and only if $\text{Res}(p,q) = b_0^2 - a_1 b_0 b_1 + a_0 b_1^2 \neq 0$. Let $\mathbf{s} = (s_1, s_2)$, $\mathbf{t} = (t_1, t_2)$ with the s_i and t_i indeterminates. If $\mathbf{g}(z) = (p(z), q(z))$, then $\mathbf{s} \cdot \mathbf{g} = s_1 p + s_2 q$, $\mathbf{t} \cdot \mathbf{g} = t_1 p + t_2 q$. We let $\rho(\mathbf{s} \cdot \mathbf{g}, \mathbf{t} \cdot \mathbf{g}) = \psi(s_1, s_2, t_1, t_2, b_0, b_1, a_0, a_1) = \text{Res}(\mathbf{s} \cdot \mathbf{g}, \mathbf{t} \cdot \mathbf{g}) = \text{Res}(s_1 b_0 + s_2 a_0 + (s_1 b_1 + s_2 a_1)z + s_2 z^2, t_1 b_0 + t_2 a_0 + (t_1 b_1 + t_2 a_1)z + t_2 z^2) = (t_1 s_2 - s_1 t_2)^2 (b_0^2 - a_1 b_0 b_1 + a_0 b_1^2) = (t_1 s_1 - s_1 t_2)^2 \text{Res}(p, q)$. If $d(z)$ is a non-constant factor of p, q, then $d(z)$ is a common factor of $\mathbf{s} \cdot \mathbf{g}$, $\mathbf{t} \cdot \mathbf{g}$ which implies that $\rho(\mathbf{s} \cdot \mathbf{g}, \mathbf{t} \cdot \mathbf{g}) \equiv 0$ in $\mathbf{s}, \mathbf{t}$ and hence that $\text{Res}(p,q) = 0$. On the other hand, if $\text{Res}(p,q) \neq 0$, then $\rho(\mathbf{s} \cdot \mathbf{g}, \mathbf{t} \cdot \mathbf{g}) \not\equiv 0$ in $\mathbf{s}, \mathbf{t}$ and p, q are relatively prime.

Bearing the example in mind, we consider $(p^1(z), \dots, p^p(z), q(z))$ as polynomials in z with coefficients $\boldsymbol{\xi} = (b_0^1, \dots, b_{n-1}^1, \dots, b_{n-1}^p, a_0, \dots, a_{n-1})$ and we let $\mathbf{s} = (s_1, \dots, s_p, s_{p+1})$, $\mathbf{t} = (t_1, \dots, t_p, t_{p+1})$ with the s_i, t_i as indeterminates. If $\mathbf{g}(z) = (p^1(z), \dots, p^p(z), q(z))$ then $\mathbf{s} \cdot \mathbf{g} = s_1 p^1 + \cdots + s_p p^p + s_{p+1} q$, $\mathbf{t} \cdot \mathbf{g} = t_1 p^1 + \cdots + t_p p^p + t_{p+1} q$ and

$$\text{Res}(\mathbf{s} \cdot \mathbf{g}, \mathbf{t} \cdot \mathbf{g}) = \Sigma R_{\alpha\beta}(\boldsymbol{\xi}) \mathbf{s}^\alpha \mathbf{t}^\beta = \psi(\boldsymbol{\xi}, \mathbf{s}, \mathbf{t}) \tag{1.40}$$

where $\alpha = (i_1, \dots, i_{p+1})$, $\beta = (j_1, \dots, j_{p+1})$, etc. Suppose $d(z)$ is a common factor of $p^1, \dots, p^p, q$, then $d(z)$ is a common factor of $\mathbf{s} \cdot \mathbf{g}$ and $\mathbf{t} \cdot \mathbf{g}$ for all $\mathbf{s}, \mathbf{t}$. Hence, $\psi(\boldsymbol{\xi}, \mathbf{s}, \mathbf{t}) \equiv 0$ in $\mathbf{s}, \mathbf{t}$ so that $R_{\alpha\beta}(\boldsymbol{\xi}) = 0$ for all α, β, i.e., $\boldsymbol{\xi} \in \bigcap\limits_{\alpha,\beta} V(R_{\alpha\beta})$. On the other hand, if $R_{\alpha\beta}(\boldsymbol{\xi}) \neq 0$ for some α, β, then $\psi(\boldsymbol{\xi}, \mathbf{s}, \mathbf{t})$ is not identically zero and $p^1, \dots, p^p, q$ are coprime. Thus, $p^1, \dots, p^p, q$ are coprime if and only if $\boldsymbol{\xi} \notin \bigcap\limits_{\alpha,\beta} V(R_{\alpha\beta})$. It follows that $\text{Rat}(n; p, 1)$ is a quasi-affine variety.

We could also develop this part of the theory using realization maps as in I.10. For instance, we define morphisms $\mathfrak{R}_f$, $\mathfrak{R}_x$, $\mathfrak{R}_h$ as follows:

(1) $\mathfrak{R}_f \colon \mathbb{A}_{\mathbb{C}}^{n^2+n(p+1)} \to \mathbb{A}_{\mathbb{C}}^{n(p+1)}$ is given by

$$\begin{aligned}\mathfrak{R}_f(\mathbf{X}, Y, \mathbf{Z}) = (&\mathbf{P}_0(\mathbf{X}, Y, \mathbf{Z}), \dots, \mathbf{P}_{n-1}(\mathbf{X}, Y, \mathbf{Z}), \\ &-\chi_n(\mathbf{X}), \dots, -\chi_1(\mathbf{X}))\end{aligned}$$

where

$$\mathbf{Z} \operatorname{adj}(zI - \mathbf{X}) Y = \sum_{j=0}^{n-1} \mathbf{P}_j(\mathbf{X}, Y, \mathbf{Z}) z^j$$

$$\det(zI - \mathbf{X}) = z^n - \chi_1(\mathbf{X}) z^{n-1} - \cdots - \chi_n(\mathbf{X})$$

$\mathfrak{R}_f$ is called the *transfer matrix realization map*;

(2) $\mathfrak{R}_\chi \colon \mathbb{A}_{\mathbb{C}}^{n^2+n(p+1)} \to \mathbb{A}_{\mathbb{C}}^{n(p+1)}$ is given by

$$\mathfrak{R}_\chi(\mathbf{X}, Y, \mathbf{Z}) = (\mathbf{Z}Y, \dots, \mathbf{Z}\mathbf{X}^{n-1}Y, \quad -\chi_1(\mathbf{X}), \dots, -\chi_n(\mathbf{X}))$$

$\mathfrak{R}_\chi$ is called the *characteristic function realization map*;

(3) $\mathfrak{R}_h\colon \mathbb{A}_{\mathbf{C}}^{n^2+n(p+1)} \to \mathbb{A}_{\mathbf{C}}^{2np}$ is given by

$$\mathfrak{R}_h(\mathbf{X}, Y, \mathbf{Z}) = (\mathbf{Z}Y, \dots, \mathbf{Z}\mathbf{X}^{n-1}Y, \dots, \mathbf{Z}\mathbf{X}^{2n-1}Y)$$

$\mathfrak{R}_h$ is called the *Hankel matrix realization map.*

$\mathfrak{R}_f, \mathfrak{R}_x$ are surjective but $\mathfrak{R}_h$ is not. We can also define the connecting morphisms ψ_χ, ψ_h and ψ_f. If $(\mathbf{U}_1, \dots, \mathbf{U}_n, x_{n+1}, \dots, x_{2n}), (\mathbf{V}_1, \dots, \mathbf{V}_n, y_{n+1}, \dots, y_{2n})$ are coordinates on $\mathbb{A}_{\mathbf{C}}^{n(p+1)}$ and $(\mathbf{H}_1, \dots, \mathbf{H}_{2n})$ are coordinates on $\mathbb{A}_{\mathbf{C}}^{2np}$, then

(a) $\psi_\chi\colon \mathbb{A}^{n(p+1)} \to \mathbb{A}^{n(p+1)}$ is given by

$$\begin{aligned}
\mathbf{V}_r &= \sum_{j=1}^{n-r} x_{2n+1-j-r}\mathbf{U}_j + \mathbf{U}_{n-r+1}, \quad r = 1, \dots, n-1 \\
\mathbf{V}_n &= \mathbf{U}_1 \\
y_{n+j} &= x_{2n+1-j}, \quad j = 1, \dots, n.
\end{aligned}$$

(b) $\psi_h\colon \mathbb{A}^{n(p+1)} \to \mathbb{A}^{2n}$ is given by

$$\begin{aligned}
\mathbf{H}_i &= \mathbf{U}_i, \quad i = 1, \dots, n \\
\mathbf{H}_{n+j} &= -\sum_{i=1}^{n} x_{2n+1-i}\mathbf{H}_{i+j-1}(x), \quad j = 1, \dots, n.
\end{aligned}$$

For example,

$$\mathbf{H}_{n+1} = -\sum_{i=1}^{n} x_{2n+1-i}\mathbf{U}_i.$$

(c) $\psi_f\colon \mathbb{A}^{n(p+1)} \to \mathbb{A}^{2n}$ is given by

$$\mathbf{H}_j = \tilde{C}(x)\tilde{A}^{j-1}(x)\tilde{b}, \quad j = 1, \dots, 2n$$

where

$$\begin{aligned}
\tilde{A}(x) &= \begin{bmatrix} 0 & 0 & \cdots & 0 \\ 0 & 1 & \cdots & 0 \\ \vdots & & & \vdots \\ 0 & 0 & & 1 \\ -x_{n+1} & -x_{n+2} & & -x_{2n} \end{bmatrix}, \\
\tilde{C}(x) &= [\mathbf{U}_1 \cdots \mathbf{U}_n], \quad \tilde{b} = \epsilon^n = \begin{bmatrix} 0 \\ \vdots \\ 1 \end{bmatrix}.
\end{aligned}$$

We observe that ψ_χ is an isomorphism as ψ_χ^{-1} is given by

$$\begin{aligned} \mathbf{U}_1 &= \mathbf{V}_n \\ \mathbf{U}_r &= \mathbf{V}_{n-r+1} - \sum_{j=1}^{r-1} y_{2n+1+j-r}\mathbf{U}_j(y), \quad r = 2, \ldots, n \\ x_{n+j} &= y_{2n+1-j}, \quad j = 1, \ldots, n. \end{aligned}$$

For example, $\mathbf{U}_2 = \mathbf{V}_{n-1} - y_{2n}\mathbf{V}_n$. In addition,

Proposition 1.41 *The realization maps $\mathfrak{R}_f$, $\mathfrak{R}_x$, $\mathfrak{R}_h$ and the associated morphisms satisfy the relation*

$$\begin{aligned} \psi_\chi \circ \mathfrak{R}_\chi &= \mathfrak{R}_f \\ \psi_h \circ \mathfrak{R}_\chi &= \mathfrak{R}_h \\ \psi_f \circ \mathfrak{R}_f &= \mathfrak{R}_h \\ \psi_h &= \psi_f \circ \psi_\chi \end{aligned}$$

(cf.: I.10.14).

We leave it to the reader to complete this aspect of the theory.

The state space representation (4) is given by a triple (A, b, C) in $\mathbb{A}^{n^2+n(p+1)}$ with C $p \times n$. The *controllability map* $\mathbf{Y}$ is given by

$$\mathbf{Y}(A, b, C) = \begin{bmatrix} b & Ab & \cdots & A^{n-1}b \end{bmatrix} \tag{1.42}$$

and the *observability map* $\mathbf{Z}$ is given by

$$\mathbf{Z}(A, b, C) = \begin{bmatrix} C \\ CA \\ \vdots \\ CA^{n-1} \end{bmatrix} \tag{1.43}$$

so that $\mathbf{Z}(A, b, C)$ is $np \times n$. The system is *controllable* if $\mathbf{Y}$ has rank n; the system is *observable* if $\mathbf{Z}$ has rank n; and the system is *minimal* if it is both controllable and observable, or, equivalently, if

$$\mathbf{H}(A, b, C) = \mathbf{Z}(A, b, C)\mathbf{Y}(A, b, C) = (CA^{i+j-2}b)_{i,j=1}$$

has full rank n. We let $S_{1,p}^n = \{x = (A, b, C)\colon x \text{ is minimal}\}$ be the set of *linear systems of degree n with p outputs and one input.* If $G = \mathrm{GL}(n, \mathbb{C})$, then G acts on $\mathbb{A}^{n^2+n(p+1)}$ via

$$g \cdot (A, b, C) = (gAg^{-1}, gb, Cg^{-1}) \tag{1.44}$$

or, on $\mathbb{C}[X,Y,Z]$ (where $X = (X^i_j)^n_{i,j=1}$, $Y = (Y^j)^n_{j=1}$, $Z = (Z^r_s)^{p,n}_{r=1,s=1}$) via

$$g \cdot (X,Y,Z) = (gXg^{-1}, gY, Zg^{-1}) \tag{1.45}$$

and it is clear that controllability, observability and minimality are preserved under the action of G. If we let $\gamma(X,Y,Z) = \det \mathbf{Y}(X,Y,Z)$, then the principal affine open set $U_c = (\mathbb{A}^{n^2+n(p+1)})_\gamma$ is the set of controllable systems and is G-invariant. If

$$X^c(X) = \begin{bmatrix} 0 & 0 & 0 & -\chi_n(x) \\ 1 & 0 & 0 & -\chi_{n-1}(x) \\ 0 & 1 & & \\ \vdots & & \vdots & \vdots \\ 0 & & 1 & -\chi_1(X) \end{bmatrix} \tag{1.46}$$

then

$$\begin{aligned} \mathbf{Y}(X,Y,Z)X^c(X) &= X\mathbf{Y}(X,Y,Z) \\ Z\mathbf{Y}(X,Y,Z) &= [ZY\ ZXY \cdots ZX^{n-1}Y] \\ \mathbf{Y}(X,Y,Z)\epsilon^1 &= Y \end{aligned} \tag{1.47}$$

where

$$\epsilon^1 = \begin{bmatrix} 1 \\ 0 \\ \vdots \\ 0 \end{bmatrix}.$$

If $x = (A,b,C) \in (\mathbb{A}^{n^2+n(p+1)})_\gamma = U_c$, then $g \cdot x = x^c = (A^c,\ \epsilon^1,\ [cA^{j-1}b]_{j=1})$ where $A^c = X^c(A)$ and $g = \mathbf{Y}(A,b,C)^{-1}$.

Proposition 1.48 *If $x \sim x_1 \bmod G$, then $x^c = x_1^c$ and conversely, if x, x_1 are controllable and $x^c = x_1^c$, then $x \sim x_1 \bmod G$.*

Proof. If $x \sim x_1 \bmod G$, then $g \cdot x = x_1$ for some $g \in G$ and $\chi_i(A) = \chi_i(A_1)$ for all i so that $X^c(A) = X^c(A_1)$. Moreover, $CA^{j-1}b = (Cg^{-1})(gAg^{-1})^{j-1}gb = C_1A_1^{j-1}b_1$ for all j. Conversely, if $x = (A,b,C)$, $x_1 = (A_1,b_1,C_1)$ are controllable and $x^c = x_1^c$, then $\chi_i(A) = \chi_i(A_1)$ for $i = 1,\ldots,n$ and

$$[Cb\ CAb \cdots CA^{n-1}b] = [C_1b_1 \ \cdots\ C_1A_1^{n-1}b_1].$$

If $g_x = [b\ Ab\ \cdots\ A^{n-1}b]^{-1}$, $g_{x_1} = [b_1\ A_1b_1\ \cdots\ A_1^{n-1}b_1]^{-1}$, then $g_x \cdot x = x^c = x_1^c = g_{x_1} \cdot x_1$ and so $(g_{x_1}^{-1}g_x) \cdot x = x_1$ (i.e., $x \sim x_1 \bmod G$).

Proposition 1.49 *If $x = (A,b,C)$ is an element of $S^n_{1,p}$, then $H_x = (CA^{i+j-1}b)^\infty_{i,j=1} \in$* **Hank**$(n;p,1)$ *and $\mathfrak{R}_h(x) \in$* Hank$(n;p,1)$. *Conversely, if*

$H \in \text{Hank}(n; p, 1)$, then there is an x in $S^n_{1,p}$ with $\mathfrak{R}_h(x) = H = H(x)$. In other words, $\mathfrak{R}_h(S^n_{1,p})$ is $\text{Hank}(n; p, 1)$. (Similarly for $\text{Rat}(n; p, 1)$ and $\text{Char}(n; p, 1)$).

Proof. If $x = (A, b, C) \in S^n_{1,p}$, then $\mathbf{Y}(A, b, C)$ is nonsingular and $\mathbf{Z}(A, b, C)$ has rank n so that $\mathbf{H}(A, b, C) = \mathbf{Z}(A, b, C)\mathbf{Y}(A, b, C)$ has rank n. In other words, $\mathfrak{R}_h(x)$ is in $\text{Hank}(n; p, 1)$. Conversely, if $H \in \text{Hank}(n; p, 1)$, then, letting $x = (A, b, C)$ where

$$A = \begin{bmatrix} 0 & 0 & \cdots & 0 & -a_0 \\ 1 & 0 & & 0 & -a_1 \\ 0 & 1 & & 0 & \\ \vdots & \vdots & & \vdots & \vdots \\ 0 & 0 & & 1 & -a_n \end{bmatrix}, \quad C = \begin{bmatrix} \mathbf{H}_1 & \cdots & \mathbf{H}_n \end{bmatrix}, \quad b = \epsilon^1$$

and $q_H(z) = a_0 + a_1 z + \cdots + a_{n-1} z^{n-1} + z^n$, we have $H = H_x$. Since $\mathbf{Y}(A, b, C) = I$ and $\mathbf{H}(A, b, C) = (H_x)^n_n = \mathbf{Z}(A, b, C)$ has rank n, $x \in S^n_{1,p}$.

In effect, Proposition 1.48 states that there is a canonical form for equivalence and Proposition 1.49 is the appropriate version of the Theorem I.10.18. Let us rewrite $\mathbf{Z}(A, b, C)$ in the (so-called Hermite) form as

$$\hat{\mathbf{Z}}(A, b, C) = \begin{bmatrix} C^1 \\ \vdots \\ C^1 A^{n-1} \\ C^2 \\ \vdots \\ C^p A^{n-1} \end{bmatrix} \tag{1.50}$$

where $C^1, \ldots, C^p$ are the rows of C. Clearly, $\mathbf{Z}$ and $\hat{\mathbf{Z}}$ have the same rank. Let

$$\mathbf{e} = (e^1, \ldots, e^p) \tag{1.51}$$

where the e^i are integers with $0 \le e^i \le n$ and $\sum_{i=1}^{p} e^i = n$. We call $\mathbf{e}$ a p-partition of n. Let $\alpha_{\mathbf{e}} = (1 \cdots e^1 \; n+1 \cdots n+e^2 \cdots n(p-1)+e^p)$ so that $\alpha_{\mathbf{e}}$ is a selection of n rows of $\hat{\mathbf{Z}}$ (if some $e^i = 0$, then the corresponding rows are omitted). We let $\omega_{\mathbf{e}}(X, Y, Z) = \det \hat{\mathbf{Z}}_{\alpha_{\mathbf{e}}}(X, Y, Z)$. Then (A, b, C) is observable if and only if $\omega_{\mathbf{e}}(A, b, C) \neq 0$ for some $\mathbf{e}$. In other words, $\bigcup_{\mathbf{e}} (\mathbb{A}^{n^2+n(p+1)})_{\omega_{\mathbf{e}}} = U_0$ is the set of observable systems; it is open and G-invariant. Thus, $S^n_{1,p} = U_0 \cap U_c$ is open in $\mathbb{A}^{n^2+n(p+1)}$ (therefore irreducible and dense) and G-invariant and x^c is a canonical form (global!) for the action of G on $S^n_{1,p}$.

Proposition 1.52 *If $x \in S^n_{p,1}$, then $\dim \mathcal{O}_G(x) = n^2$ and hence, orbits in $S^n_{1,p}$ are closed.*

Proof. Since $\mathcal{O}_G(x) \simeq G/S_G(x)$ where $S_G(x) = \{g \in G\colon g \cdot x = x\}$ is the stabilizer of x, it follows that for $g \in S_G(x)$, $\mathbf{Y}(x) = \mathbf{Y}(g \cdot x) = g\mathbf{Y}(x)$. Since $\mathbf{Y}(x)$ is nonsingular, $S_G(x) = \{I\}$.

If we let $\operatorname{Char}(n;p,1) = \mathfrak{R}_\chi(S^n_{1,p})$, then $\operatorname{Char}(n;p,1) = \psi_\chi^{-1} \circ \mathfrak{R}_f(S^n_{1,p}) = \psi_\chi^{-1}(\operatorname{Rat}(n;p,1))$ (by Proposition 1.49) is an irreducible quasi-affine subvariety of $\mathbb{A}^{n(p+1)}$. Since $\operatorname{Char}(n;p,1)$ is open in $\mathbb{A}^{n(p+1)}$, $\dim \operatorname{Char}(n;p,1) = n(p+1)$ (Part I) and $\operatorname{Char}(n;p,1)$ is normal (I.16.34). We consider$(\operatorname{Char}(n;p,1), \mathfrak{R}_\chi)$.

Proposition 1.53 (a) $\mathfrak{R}_\chi(g \cdot x) = \mathfrak{R}_\chi(x)$ *for* $g \in G$ *and* $x \in \mathbb{A}^{n^2+n(p+1)}$ *(i.e., $\mathfrak{R}_\chi$ is G-invariant)*; (b) *if* $\xi \in \operatorname{Char}(n;p,1)$, *then* $\mathfrak{R}_\chi^{-1}(\xi) = O_G(x)$ *for some* $x \in S^n_{p,1}$*; and,* (c) $\mathfrak{R}_\chi\colon S^n_{p,1} \to \operatorname{Char}(n;p,1)$ *is an open map.*

Proof. (a) is clear. As for (b), let $\xi = (\mathbf{H}_1, \dots, \mathbf{H}_n, -\chi_1, \dots, -\chi_n) \in \operatorname{Char}(n;p,1)$, then there is an $x \in S^n_{1,p}$ with $\mathfrak{R}_\chi(x) = \xi$. If $x_1 \in O_G(x)$, then $\mathfrak{R}_\chi(x) = \mathfrak{R}_\chi(x_1)$ by (a) and $O_G(x) \subset \mathfrak{R}_\chi^{-1}(\xi)$. On the other hand, if $\mathfrak{R}_\chi(x_2) = \xi$, then $x_2^c = (A^c(x), \epsilon^1, [\mathbf{H}_1 \ \cdots \ \mathbf{H}_n]) = x^c$ and $x_2 \sim x^c \bmod G$ by Proposition 1.48. Hence, $\mathfrak{R}_\chi^{-1}(\xi) = O_G(x)$. In view of (b) and Proposition 1.52,

$$\dim \mathfrak{R}_\chi^{-1}(\xi) = n^2 = n^2 + n(p+1) - n(p+1) = \dim S^n_{1,p} - \dim \operatorname{Char}(n;p,1)$$

and $\mathfrak{R}_\chi$ is open by Proposition I.18.22.

Let ψ_i, ψ_i be given by

$$\begin{aligned} \psi_i(X,Y,Z) &= ZX^{i-1}Y, \quad i = 1,\dots,n \\ \psi_i(X,Y,Z) &= -\chi_{i-n}(X), \quad i = n+1,\dots,2n \end{aligned} \tag{1.54}$$

and consider the regular map $\psi\colon \mathbb{A}^{n^2+n(p+1)} \to \mathbb{A}^{n(p+1)}$ with $\psi = (\psi_1, \dots, \psi_n, \psi_{n+1}, \dots, \psi_{2n})$ (i.e., $\psi = \mathfrak{R}_\chi$ in coordinates). It is clear that $S_0 = \mathbb{C}[\psi_1, \dots, \psi_{2n}] \subset k[X,Y,Z]^G = R^G = S$.

Proposition 1.55 $S_0 = S$.

Proof. Let $f(X,Y,Z)$ be an element of R and let $f_\gamma(X,Y,Z) = f(X^c(X), \epsilon^1, [ZX^{j-1}Y]_{j=1}^n)$. If f is in $S = R^G$, then $f = f_\gamma$ on $(\mathbb{A}^{n^2+n(p+1)})_\gamma$ and so, $f = f_\gamma$. But $f_\gamma \in S_0$.

Now consider $\mathbb{A}^{n(p+1)}$ with coordinates $(\mathbf{H}_1, \dots, \mathbf{H}_n, x_1, \dots, x_n)$ and let

$$\begin{aligned} \phi_j(\mathbf{H}_1, \dots, \mathbf{H}_n, x_1, \dots, x_n) &= \mathbf{H}_j, & j &= 1,\dots,n \\ \phi_{n+j}(\mathbf{H}_1, \dots, \mathbf{H}_n, x_1, \dots, x_n) &= x_1\phi_j + \cdots + x_n\phi_{n+j-1}, & j &= 1,\dots,n. \end{aligned} \tag{1.56}$$

If $\tilde{H}$ is the $p \times 1$ block Hankel matrix with

$$\tilde{H} = \begin{bmatrix} \phi_1 & \cdots & \phi_n \\ \phi_2 & & \phi_{n+1} \\ \vdots & & \vdots \\ \phi_n & & \phi_{2n-1} \end{bmatrix}$$

and if θ_ν are the $n \times n$ minors of $\tilde{H}$ (which is $np \times n$), then $V_\nu = (\mathbb{A}^{n(p+1)})_{\theta_\nu}$ are open in $\mathbb{A}^{n(p+1)}$. Let $U_\nu = (\mathbb{A}^{n^2+n(p+1)})_{\theta_\nu \circ \mathbf{H}}$ where $\mathbf{H}(X,Y,Z) = \mathbf{Z}(X,Y,Z)\mathbf{Y}(X,Y,Z)$. Then $\psi(U_\nu) \subset V_\nu$ (actually equal) for all ν and also, $\psi^*(\mathbb{C}[V_\nu]) \subset \mathbb{C}[U_\nu]$. Moreover, $\mathbb{C}[V_\nu] = \mathbb{C}[\boldsymbol{\psi}_1, \dots, \boldsymbol{\psi}_n, \psi_1, \dots, \psi_n]_{\theta_\nu}$ and $\mathbb{C}[U_\nu] = \mathbb{C}[S^n_{p,1}]_{\theta_\nu \circ \mathbf{H}} = \mathbb{C}[X,Y,Z]_{\theta_\nu \circ \mathbf{H}}$. Since U_ν is G-invariant and $\theta_\nu \circ \mathbf{H}$ is G-invariant, to show that $\mathbb{C}[V_\nu]$ (or better $\psi^*(\mathbb{C}[V_\nu])$) is $\mathbb{C}[U_\nu]^G$), it is enough to show that $R^G = S = S_0$ which is Proposition 1.55. In other words, we have:

Proposition 1.56 $\mathbb{C}[S^n_{p,1}]^G \simeq \mathbb{C}[\mathrm{Char}(n;p,1)]$.

Theorem 1.57 $(\mathrm{Char}(n;p,1), \mathfrak{R}_\chi)$ *is a geometric quotient for the action of* G *on* $S^n_{p,1}$.

We note that since $m = 1$, there is a global canonical form which we can exploit. When both m and p are greater than 1, this is impossible.

So far we have shown that the theory for representations (1), (2), (3) and (4) is essentially the same as in the single input-single output case.

Now suppose that $\mathbf{f}(z) = \mathbf{P}(z)/q(z)$ with

$$\begin{aligned} p^j(z) &= b_0^j + b_1^j z + \cdots + b_{n-1}^j z^{n-1}, \quad j = 1, \dots, p \\ q(z) &= a_0 + a_1 z + \cdots + a_{n-1} z^{n-1} + z^n \end{aligned}$$

co-prime. Then we can define a map $\psi_{\mathbf{f}} \colon \mathbb{C} \to \mathbb{C}^{p+1}$ via

$$\psi_{\mathbf{f}}(z) = (p^1(z), \dots, p^p(z), q(z)). \tag{1.58}$$

If we let $\xi_\infty = (0,1)$ and $\mathbb{C} = \{(1,\xi) \colon \xi \in \mathbb{C}\}$, then we can extend $\psi_{\mathbf{f}}$ to a map $\mathbb{P}^1_{\mathbb{C}} \to \mathbb{P}^p_{\mathbb{C}}$ via

$$\begin{aligned} \psi_{\mathbf{f}}(x_0, x_1) = \; & (b_0^1 x_0^n + \cdots + b_{n-1} x_0 x_1^{n-1}, b_0^2 x_0^n + \cdots, \dots, \\ & a_0 x_0^n + \cdots + a_{n-1} x_0 x_1^{n-1} + x_1^n) \end{aligned} \tag{1.59}$$

(cf. I.23.8). Noting that $\psi_{\mathbf{f}}(\xi_\infty) = (0, \dots, 0, 1)$ and that

$$\psi_{\mathbf{f}}(1, \xi) = (p^1(\xi), \dots, p^p(\xi), q(\xi)),$$

we see that $\psi_{\mathbf{f}}$ does indeed map $\mathbb{P}^1_{\mathbb{C}} \to \mathbb{P}^p_{\mathbb{C}}$. Moreover, the "poles" of $\mathbf{f}(z)$ are the points $(1, \xi)$ where $\psi_{\mathbf{f}}(1, \xi) = (p^1(\xi), \ldots, p^p(\xi), 0)$ i.e., the points where $\psi_{\mathbf{f}}(\mathbb{P}^1_{\mathbb{C}})$ meets the hyperplane $Y_p = 0$ (the coordinates on $\mathbb{P}^p_{\mathbb{C}}$ are $(Y_0, \ldots, Y_p)$). So the representation (5) is the "same" as in the scalar case.

Now let us examine the representation (7) by module homomorphisms.

Definition 1.60 Let $\Omega^r_c = \{\omega\colon \mathbb{Z} \to \mathbb{C}^r \text{ with } \omega(n) = 0 \text{ for } n \leq -N_\omega, N_\omega \geq 0\}$. If $\omega \in \Omega^r_c$, then the *transform of* ω, $\hat{\omega}(z)$, is the element of $\hat{S}_{z^{-1}} \otimes_{\mathbb{C}} \mathbb{C}^r = \mathbb{C}^r((1/z))$ given by

$$\hat{\omega}(z) = \sum_{n=-\infty}^{\infty} \omega(n) z^{-n} = z^{N_\omega}\left(\sum_{j=0}^{\infty} \omega(j - N_\omega) z^{-j}\right) \tag{1.61}$$

(cf. [F-3]).

Just as before, the left-shift of ω is given by multiplication by z. In addition, we have:

Proposition 1.62 *If* $H = \{\mathbf{H}_j\} \in \Omega^p_c$ *and* $u = \{u_j\} \in \Omega^1_c = \Omega_c$, *then* $\{y_n\} = \{\mathbf{H}_j\} * \{u_j\}$ *is a well-defined element of* Ω^p_c *with* $\hat{y}(z) = \hat{H}(z)\hat{u}(z)$.

The proof is exactly the same as the proof of Proposition 1.10.

If $x = (A, b, C)$ is a state representation of $\mathbf{f}(z)$, then $H_x = \{\mathbf{H}_j\} \in \Omega^p_c$ where $\mathbf{H}_j = CA^{j-1}b$ for $j = 1, \ldots$ and $\mathbf{H}_j = 0$ for $j \leq 0$. If $u = \{u_j\} \in \Omega^1_c$, then

$$\begin{aligned} y_n &= \sum_{k=1}^{n+N_u-1} \mathbf{H}_k u_{n-k} = \sum_{k=1}^{n+N_u-1} (CA^{k-1}b) u_{n-k} \\ \hat{y}(z) &= \left(\sum_{k=1}^{\infty} (CA^{k-1}b) z^{-k}\right) \hat{u}(z) \end{aligned} \tag{1.63}$$

so that the transfer matrix $\mathbf{f}(z) = \Sigma CA^{k-1}bz^{-k} = \hat{H}_x(z)$. Now $\mathrm{Hom}_{\mathbb{C}}(\mathbb{C}, \mathbb{C}^p)$ $(= \mathbb{C}^p)$ is a $\mathbb{C}$-vector space and so $\hat{S}_{z^{-1}} \otimes_{\mathbb{C}} \mathrm{Hom}_{\mathbb{C}}(\mathbb{C}, \mathbb{C}^p) = \mathrm{Hom}_{\mathbb{C}}(\mathbb{C}, \mathbb{C}^p)((1/z))$ $(= \mathbb{C}^p((1/z)))$ is an $\hat{S}_{z^{-1}}$-module (a fortiori an $R = \mathbb{C}[z]$-module) with submodule (under left-shift) $\hat{S}_+ \otimes_{\mathbb{C}} \mathrm{Hom}_{\mathbb{C}}(\mathbb{C}, \mathbb{C}^p) = \mathrm{Hom}_{\mathbb{C}}(\mathbb{C}, \mathbb{C}^p)_+[[1/z]] = \mathbb{C}^p_+[[1/z]]$. We observe that $\hat{H}_x(z)$ is an element of $\mathrm{Hom}_{\mathbb{C}}(\mathbb{C}, \mathbb{C}^p)_+[[1/z]]$ and that the map $\psi_x = \psi_{H_x}\colon \mathbb{C}((1/z)) \to \mathbb{C}^p((1/z))$ given by

$$\psi_x(y) = \hat{H}_x \cdot y \tag{1.64}$$

is an $\hat{S}_{z^{-1}}$ (a fortiori an R) homomorphism of $\mathbb{C}((1/z))$ into $\mathbb{C}^p((1/z))$ such that

$$\psi_x(\hat{S}) \subset \mathbb{C}^p_+[[1/z]] \tag{1.65}$$

(i.e., ψ_x is causal). We have:

Proposition 1.66 *ψ is an $\hat{S}_{z^{-1}}$-module homomorphism of $\mathbb{C}((1/z))$ into $\mathbb{C}^p((1/z))$ if and only if there is a Ψ in* $\mathrm{Hom}_{\mathbb{C}}(\mathbb{C}, \mathbb{C}^p)((1/z))$ *with $\psi(y) = \Psi \cdot y$.*

Proof. Given such a $\Psi = z^s(\psi_0 + \psi_1 z^{-1} + \cdots)$ and $y = z^t(y_0 + y_1 z^{-1} + \cdots)$, then $\Psi \cdot y = z^{s+t} \sum_{n=0}^{\infty} \alpha_n z^{-n}$ where $\alpha_n = \sum_{i+j=n} \psi_i(y_j)$ and clearly, this is an $\hat{S}_{z^{-1}}$-homomorphism. Now suppose that ψ is given. Then 1 generates $\mathbb{C}((1/z))$ over $\hat{S}_{z^{-1}}$ and

$$\psi(1) = z^s(v_0 + v_1 z^{-1} + \cdots) \tag{1.67}$$

where the v_i are in $\mathbb{C}^p$. Let $\psi_i \in \mathrm{Hom}_{\mathbb{C}}(\mathbb{C}, \mathbb{C}^p)$ be given by $\psi_i(\alpha) = \alpha v_i$ (i.e., $\psi_i(1) = v_i$). Then $\Psi = z^s(\psi_0 + \psi_1 z^{-1} + \cdots)$ is in $\mathrm{Hom}_{\mathbb{C}}(\mathbb{C}, \mathbb{C}^p)((1/z))$ and $\Psi \cdot 1 = \psi(1)$ so that $\Psi \cdot y = \psi(y)$.

Corollary 1.68 *If $\psi(\hat{S}) \subset \mathbb{C}^p_+[[1/z]]$, then Ψ is in* $\mathrm{Hom}_{\mathbb{C}}(\mathbb{C}, \mathbb{C}^p)_+[[1/z]]$.

Proof. Since $\psi(1) \in \mathbb{C}^p_+[[1/z]]$ and $\Psi \cdot 1 = \psi(1)$, the corollary follows.

Let ψ be a causal element of $\mathrm{Hom}_R(\mathbb{C}((1/z)), \mathbb{C}^p((1/z)))$ and let $i \colon R \to \mathbb{C}((1/z))$ and $\pi \colon \mathbb{C}^p((1/z)) \to \mathbb{C}^p((1/z))/R \simeq \mathbb{C}^p_+[[1/z]]$ be the natural injection and projection. Then $\Psi = \pi \circ \psi \circ i \colon R \to \mathbb{C}^p_+[[1/z]]$ is a R-homomorphism. Conversely, given a $\overline{\varphi}$ in $\mathrm{Hom}_R(R, \mathbb{C}^p_+[[1/z]])$, there is a *unique* causal φ in $\mathrm{Hom}_R(\mathbb{C}((1/z)), \mathbb{C}^p((1/z)))$ with $\overline{\varphi} = \pi \circ \varphi \circ i$ (φ is given by $\varphi\left(\sum_{-N \le j} y_j/z^j\right) = \sum_{-N \le j} \overline{\varphi}(y_j)/z^j$). Note that φ is also an $\hat{S}_{z^{-1}}$-homomorphism.

Proposition 1.69 *If $\psi, \tilde{\psi}$ are elements of* $\mathrm{Hom}_{\hat{S}_{z^{-1}}}(\mathbb{C}((1/z)), \mathbb{C}^p((1/z)))$, *then* (a) *$\psi = \tilde{\psi}$ on R implies $\psi = \tilde{\psi}$; and,* (b) *$\psi = \tilde{\psi}$ on $\mathbb{C}_+[[1/z]]$ implies $\psi = \tilde{\psi}$.*

Proof. (a) Since $\psi = \tilde{\psi}$ on R, $\psi(1) = \tilde{\psi}(1)$ and so $\psi = \tilde{\psi}$. (b) Since $\psi(1/z) = \tilde{\psi}(1/z)$, $\psi(1) = \psi(z \cdot 1/z) = z\psi(1/z) = z\tilde{\psi}(1/z) = \tilde{\psi}(z \cdot 1/z) = \tilde{\psi}(1)$.

Corollary 1.70 *If ψ is a causal element of* $\mathrm{Hom}_R(\mathbb{C}((1/z)), \mathbb{C}^p((1/z)))$, *then there is a Ψ in* $\mathrm{Hom}_{\mathbb{C}}(\mathbb{C}, \mathbb{C}^p)_+[[1/z]]$ *with $\psi(y) = \overline{\Psi} \cdot y$.*

Proof. Let $\Psi = \psi(1) = v_1 z^{-1} + \cdots$ and set $\tilde{\psi}(y) = \Psi \cdot y$. Then $\tilde{\psi}$ is an $\hat{S}_{z^{-1}}$-homomorphism (a fortiori an R-homomorphism) which is causal. Since $\psi = \tilde{\psi}$ on R, we have by 1.69, $\psi = \tilde{\psi}$.

Thus, we can again represent the system as a $\mathbb{C}[z]$-module homomorphism $\varphi_{\mathfrak{f}}$ of $\mathbb{C}((1/z))$ into $\mathbb{C}^p((1/z))$ such that $\varphi_{\mathfrak{f}}(\mathbb{C}[[1/z]]) \subset \mathbb{C}^p_+[[1/z]]$. In other words, the representation (7) is essentially the same as in the scalar case.

Now let us turn our attention to the analog of the representation (6). We let $\mathbf{Y}_1 : \mathbb{A}_{\mathbb{C}}^{n^2+n(p+1)} \to M(n, n+1)$ be given by

$$\mathbf{Y}_1(x) = \mathbf{Y}_1(A, b, C) = \begin{bmatrix} b \; Ab \; \cdots \; A^n b \end{bmatrix} = \begin{bmatrix} \mathbf{Y}(A, b, C) \; A^n b \end{bmatrix} \tag{1.71}$$

and we let

$$\beta(\tilde{M}) = \det\left[\tilde{M}\big|_{1\cdots n}^{1\cdots n}\right] \tag{1.72}$$

for $\tilde{M} \in M(n, n+1)$. Then, just as in Part I, we have:

Proposition 1.73 (i) $\mathbf{Y}_1(x) \in M_*(n, n+1)$ *if and only if* x *is controllable; and* (ii) $\mathbf{Y}_1(S_{p,1}^n) = (M(n, n+1))_\beta$.

The map $\mathbf{Z}^1 : \mathbb{A}_{\mathbb{C}}^{n^2+n(p+1)} \to M((n+1)p, n)$ that we use is given by

$$\mathbf{Z}^1(x) = \mathbf{Z}^1(A, b, C) = \begin{bmatrix} c^1 \\ \vdots \\ c^1 A^n \\ c^2 \\ \vdots \\ c^p A^n \end{bmatrix} \tag{1.74}$$

where the c^i, $i = 1, \ldots, p$ are the rows of C.

Proposition 1.75 $\mathbf{Z}^1(x) \in M_*((n+1)p, n)$ *if and only if* x *is observable.*

Proof. Obvious.

Example 1.76 Let $p = 2$, $n = 2$ and let $M \in M_*(6, 2)$ be given by

$$M = \begin{bmatrix} 0 \\ \epsilon^1 \\ 0 \\ 0 \\ \epsilon^2 \\ 0 \end{bmatrix}, \quad \epsilon^1 = (1\ 0), \quad \epsilon^2 = (0\ 1).$$

Then M is *not* in the range of $\mathbf{Z}^1$, i.e., $\mathbf{Z}^1(\mathbb{A}_{\mathbb{C}}^{n^2+n(p+1)}) < M_*(6, 2)$.

Example 1.77 Let $p = 2$, $n = 3$ and let $M \in M_*(8,3)$ be given by

$$M = \begin{bmatrix} \epsilon^1 \\ \epsilon^2 \\ 0 \\ 0 \\ \epsilon^3 \\ 0 \\ 0 \\ 0 \end{bmatrix}, \quad \epsilon^1 = (1\ 0\ 0),\ \ \epsilon^2 = (0\ 1\ 0),\ \ \epsilon^3 = (0\ 0\ 1).$$

If $c^1 = \epsilon^1$, $c^2 = \epsilon^3$ and

$$A = \begin{bmatrix} 0 & 1 & 0 \\ 0 & 0 & 0 \\ 0 & 0 & 0 \end{bmatrix}$$

then $\mathbf{Z}^1(A,b,C) = M$ for *any* b so that (A,b,C) is observable for *any* b. However, there does not exist a b such that (A,b) is controllable since $\begin{bmatrix} b & Ab & A^2b \end{bmatrix} = \begin{bmatrix} b & Ab & 0 \end{bmatrix}$ cannot have rank 3 (note $A^2 = 0$). In other words, for $p > 1$ and $m = 1$, even if (A,C) is observable, there may *not* be *any* b such that (A,b,C) is minimal.

When $p > 1$, we require a more complex notation for dealing with $M((n+1)p,n)$. If $M \in M((n+1)p,n)$, then $M = [M^i]$, $i = 1,\dots,p$ with $M^i \in M(n+1,n)$. Let $W^i(M) = \text{span}[M^i]$ be the row space of M^i and let $V^j(M) = \sum_{i=1}^{j} W^i(M)$ for $j = 1,\dots,p$. Set $\tau_0 = 0$, $\tau_j(M) = \dim V^j(M)$, $e^j(M) = \tau_j(M) - \tau_{j-1}(M)$, $j = 1,\dots,p$ and $\mathbf{e}(M) = (e^1(M),\dots,e^p(M))$. We note that $\tau_p(M) = e^1(M) + \cdots + e^p(M) = \text{rank}\, M \le n$ and that $0 \le e^j(M) \le n$. We also observe that

$$\mathbf{Z}^1(A,b,C) = [M^i(A,b,C)] \tag{1.78}$$

where $M^i(A,b,C) = [c^i A^j]_{j=0}^{n}$, $i = 1,\dots,p$. We label the rows of M as $(11 \cdots n+11\ 12 \cdots n+12 \cdots n+1p)$ so that $M^i = [\mathbf{m}^{ji}]_{j=1}^{n+1}$ and the $\mathbf{m}^{ji}$ are row n-vectors. We use multiindices $\boldsymbol{\alpha} = (\alpha_1,\dots,\alpha_n)$ where $\alpha_r = \ell_r j_r$, $\ell_r \in (1,\dots,n+1)$ and $j_r \in (1,\dots,p)$. Set

$$U_{\boldsymbol{\alpha}} = \{M\colon\ \det[M|_{1\cdots n}^{\alpha_1\cdots\alpha_n}] \ne 0\} \tag{1.79}$$

so that $U_{\boldsymbol{\alpha}} \subset M_*((n+1)p,n)$. If $\mathbf{e}(M) = (e^1(M),\dots,e^p(M))$ and $M \in M_*((n+1)p,n)$, then call $\boldsymbol{\alpha} = (\alpha_1,\dots,\alpha_n)$ *nice relative to* $\mathbf{e}(M)$ ([H-4]) if

$$\boldsymbol{\alpha} = (11 \cdots e^1 1\ 12 \cdots e^2 2 \cdots e^p p) \tag{1.80}$$

where the terms $\cdot j$ are omitted if $e^j(M) = 0$.

Proposition 1.81 *Let $S_p^n = \{x \in \mathbb{A}_{\mathbb{C}}^{n^2+n(p+1)} : x$ is observable$\}$. Then $S_p^n = \{(A, b, C) : \det \mathbf{Z}^1(A, b, C)\boldsymbol{\alpha} \neq 0$ for $\boldsymbol{\alpha}$ nice relative to $\mathbf{e}(\mathbf{Z}^1(A, b, C))\}$.*

Proof. If $\det \mathbf{Z}^1(A, b, C)\boldsymbol{\alpha} \neq 0$, then $x = (A, b, C)$ is clearly observable. On the other hand, if $x = (A, b, C)$ is observable, then $\mathbf{Z}^1(x) \in M_*((n+1)p, n)$. Since $W^i(\mathbf{Z}^i(x))A \subset W^i(\mathbf{Z}^1(x))$ as $c^i A^n \in \text{span}[c^i A^i]_{j=0}^{n-1}$ and $V^j(\mathbf{Z}^1(x))A \subset V^j(\mathbf{Z}^1(x))$, we can choose a basis for the row space of $\mathbf{Z}^1(x)$ with index $\boldsymbol{\alpha}$ nice relative to $\mathbf{e}(\mathbf{Z}^1(x))$.

Example 1.82 Let $n = 2$, $p = 2$ and let $M \in M_*(6, 2)$ be given by

$$M = \begin{bmatrix} \epsilon^1 \\ \epsilon^2 \\ \epsilon^1 + \epsilon^2 \\ \epsilon^2 \\ 0 \\ 0 \end{bmatrix} = \begin{bmatrix} 1 & 0 \\ 0 & 1 \\ 1 & 1 \\ 0 & 1 \\ 0 & 0 \\ 0 & 0 \end{bmatrix}.$$

Then for $\mathbf{e} = (1, 1)$, $\boldsymbol{\alpha} = (11\ 12) = (\alpha_1\ \alpha_2)$ is nice relative to $\mathbf{e}$ and $\det[M|_{1\ \ 2}^{\alpha_1\ \alpha_2}] \neq 0$ (note this $\mathbf{e}$ is *not* $\mathbf{e}(M) = (2, 0)$). However, M is not in the range of $\mathbf{Z}^1$ for: if $c^1 = \epsilon^1$, $c^1 A = \epsilon^2$ and $c^1 A^2 = \epsilon^1 + \epsilon^2 = (c^1 A)A = \epsilon^2 A$, then $c^2 = \epsilon^2$ and $c^2 A = \epsilon^2 A = 0$ is not possible. This means that $\mathbf{Z}^1(S_p^n) \cap U_{\boldsymbol{\alpha}_e} < U_{\boldsymbol{\alpha}_e}$ in general.

Let $\mathbf{e} = (e^1, \dots, e^p)$ be a p-partition of n so that $e^j \in (0, 1, \dots, n)$, $\sum_{j=1}^{p} e^j = n$. Then $\boldsymbol{\alpha}_e = (11 \cdots e^1 1\ 12 \cdots e^2 2 \cdots c^p p)$ is the (unique) nice multiindex relative to $\mathbf{e}$. We let

$$\omega_e(M) = \det\left[M|_{1 \cdots n}^{\alpha_e}\right] \tag{1.83}$$

and

$$U_e = U_{\boldsymbol{\alpha}_e} = (M((n+1)p, n))_{\omega_e} \tag{1.84}$$

so that we have:

Corollary 1.85 $\mathbf{Z}^1(S_{1,p}^n) \subset \mathbf{Z}^1(S_p^n) \subset \cup U_e$.

Let $\varphi \colon \mathbb{A}_{\mathbb{C}}^{n^2+n(p+1)} \to M((n+1)p, n) \times M(n, n+1)$ be given by

$$\varphi(x) = (\mathbf{Z}^1(x), \mathbf{Y}_1(x)) \tag{1.86}$$

for $x = (A, b, C)$ in $\mathbb{A}_{\mathbb{C}}^{n^2+n(p+1)}$. φ is clearly a morphism.

Corollary 1.87 *$\varphi(S^n_{1,p}) \subset (\cup U_e) \times M(n, n+1)_\beta$. Set $\tilde{S}^n_{1,p} = \varphi(S^n_{1,p})$. As a map of $S^n_{1,p}$ into $\tilde{S}^n_{1,p}$, φ is surjective by definition.*

Proposition 1.88 *φ is injective.*

Proof. Suppose that $\varphi(x) = \varphi(x_1)$ where $x = (A, b, C)$, $x_1 = (A_1, b_1, C_1)$ and x, x_1 are in $S^n_{1,p}$. Then $b = b_1$ and $C = C_1$ as $c^i = c^i_1$ for $i = 1, \ldots, p$ and also $\mathbf{Y}_1(A, b, C) = \mathbf{Y}_1(A_1, b_1, C_1) = \mathbf{Y}_1(A_1, b, C)$, $\mathbf{Z}^1(A, b, C) = \mathbf{Z}^1(A_1, b, C)$. From $\mathbf{Y}_1(A, b, C) = \mathbf{Y}_1(A_1, b, C)$, we have $A = A_1$ just as in I.19.43. Similarly, from $\mathbf{Z}^1(A, b, C) = \mathbf{Z}^1(A_1, b, C)$, we have $CA^k = CA^k_1$ for $k = 0, 1, \ldots$ and if $c^i A^j = c^i A^j_1$ is a basis $w^1, \ldots, w^n$, then $w^r A = w^r A_1$ for $r = 1, \ldots, n$ so that (again) $A = A_1$.

Assuming for the moment that $\tilde{S}^n_{1,p}$ is a (quasi-affine) variety, we have:

Proposition 1.89 *φ is an isomorphism.*

Proof. Since φ is bijective, we need only show that φ^{-1} is a morphism. If $\varphi(x) = (\mathbf{Z}^1(x), \mathbf{Y}(x)) \in \tilde{S}^n_{1,p}$, then $\varphi^{-1}(\mathbf{Z}^1(x), \mathbf{Y}_1(x)) = (A, b, C)$ where $b = (\mathbf{Y}_1(x))_1$, $C = (\mathbf{Z}^1(x))^{1, n+2, \ldots, (p-1)(n+1)+1}$ and $A = g^{-1} A^* g$ where $g = [\mathbf{Y}_1(x)_{1, \ldots, n}]^{-1}$, $\mathbf{Y}_1(x)_{n+1} = \sum_{j=1}^{n} a_j \mathbf{Y}_1(x)_j$, and

$$A^* = \begin{bmatrix} 0 & 0 & & 0 & a_1 \\ 1 & 0 & & 0 & a_2 \\ & 1 & & & \\ \vdots & \vdots & & \vdots & \vdots \\ 0 & & & 1 & a_n \end{bmatrix}.$$

All are regular as in I.19.44.

So in order to show that the representation (6) is the same as in the scalar case, we need only show that $\tilde{S}^n_{1,p}$ is a variety. We use an argument similar to that given in I.19.

Let $X = \mathbb{A}^n \times M((n+1)p, n+1)$ and view an element $\mathbf{M}$ of $M((n+1)p, n+1)$ as $[\mathbf{M}^i]_{i=1}^p$ where $\mathbf{M}^i$ is in $M(n+1, n+1)$. Let $\mathcal{L}_{n+1}$ be defined by the equations

$$\left.\begin{aligned} &\mathbf{M}^i_{rs} - \mathbf{M}^i_{sr} = 0, && r, s = 1, \ldots, n+1 \\ &\mathbf{M}^i_{rs+1} - \mathbf{M}^i_{r+1s} = 0, && r = 1, \ldots, s-1, s = 2, \ldots, n \end{aligned}\right\} \tag{1.90}$$

$$\mathbf{M}^i_{rn+1} - \sum_{\ell=1}^{n} x_\ell \mathbf{M}^i_{r\ell} = 0, \quad r = 1, \ldots, n+1 \tag{1.91}$$

where $i = 1, \ldots, p$ (this views the "block" Hankel matrix as a block matrix of scalar Hankel matrices).[†] Noting that $X \simeq \mathbb{A}^{n+(n+1)^2 p}$ and that for fixed i, the variety in $M(n+1, n+1)$ is defined by $\frac{n(n+1)}{2} + \frac{n(n-1)}{2} + (n+1) = n^2 + n + 1$ equations, we can readily see that $\dim \mathcal{L}_{n+1} = n + (n+1)^2 p - (n^2+n+1)p = n + np = n(p+1)$. Now define morphisms $\psi\colon M((n+1)p, n) \times M(n, n+1) \to M((n+1)p, n+1)$, $\tilde{\varphi}\colon \mathbb{A}^{n^2+n(p+1)} \to \mathbb{A}^n \times [M((n+1)p, n) \times M(n, n+1)]$, and $\tilde{\psi}\colon \mathbb{A}^n \times [M((n+1)p, n) \times M(n, n+1)] \to X$ via

$$\psi(Z, Y) = ZY \tag{1.92}$$

$$\tilde{\varphi}(x) = (\chi(x), \varphi(x)) \tag{1.93}$$

$$\tilde{\psi}(\mathbf{a}, Z, Y) = (\mathbf{a}, \psi(Z, Y)) = (\mathbf{a}, ZY) \tag{1.94}$$

where $x = (A, b, C)$, $\chi(x) = (\chi_i(A))$ and $\varphi(x) = (\mathbf{Z}^1(x), \mathbf{Y}_1(x))$. We observe that $\varphi(S^n_{1,p})$ is irreducible (since $S^n_{1,p}$ is) and that $\tilde{S}^n_{1,p} = \pi_2(\tilde{\varphi}(S^n_{1,p}))$ is therefore also irreducible where $\pi_2\colon \mathbb{A}^n \times [M((n+1)p, n) \times M(n, n+1)] \to M((n+1)p, n) \times M(n, n+1)$ is the projection. Also $\varphi(S^n_{1,p}) = \tilde{S}^n_{1,p} = \pi_2(\tilde{\varphi}(S^n_{1,p}))$. If $\mathcal{H} = (\tilde{\psi} \circ \tilde{\varphi})(S^n_{1,p})$, then $\mathcal{H}$ is irreducible and $\tilde{\psi}^{-1}(\mathcal{H}) = \tilde{\varphi}(S^n_{1,p})$ so that $\tilde{S}^n_{1,p} = \pi_2(\tilde{\psi}^{-1}(\mathcal{H}))$. We shall show that $\mathcal{H}$ is open in $\mathcal{L}_{n+1}$ and hence, that $\tilde{S}^n_{p,1}$ is a quasi affine variety.

Example 1.95 Let $p = 2$, $n = 2$ and let $x = (\mathbf{a}, \mathbf{M})$ with $\mathbf{a} = (a_1, a_2)$ and

$$\mathbf{M} = \begin{bmatrix} \mathbf{M}^1_{11} & \mathbf{M}^1_{12} & \mathbf{M}^1_{13} \\ \mathbf{M}^1_{21} & \mathbf{M}^1_{22} & \mathbf{M}^1_{23} \\ \mathbf{M}^1_{31} & \mathbf{M}^1_{32} & \mathbf{M}^1_{33} \\ \mathbf{M}^2_{11} & \mathbf{M}^2_{12} & \mathbf{M}^2_{13} \\ \mathbf{M}^2_{21} & \mathbf{M}^2_{22} & \mathbf{M}^2_{23} \\ \mathbf{M}^2_{31} & \mathbf{M}^2_{32} & \mathbf{M}^2_{33} \end{bmatrix}.$$

Then x is an element of $\mathcal{L}_{n+1}$ if

$$\begin{aligned} &\mathbf{M}^1_{12} - \mathbf{M}^1_{21} = 0, \quad \mathbf{M}^1_{13} - \mathbf{M}^1_{31} = 0, \quad \mathbf{M}^1_{23} - \mathbf{M}^1_{32} = 0 \\ &\mathbf{M}^1_{13} - \mathbf{M}^1_{22} = 0 \\ &\mathbf{M}^1_{13} - a_1\mathbf{M}^1_{11} - a_2\mathbf{M}^1_{12} = 0 \\ &\mathbf{M}^1_{23} - a_1\mathbf{M}^1_{21} - a_2\mathbf{M}^1_{22} = 0 \\ &\mathbf{M}^1_{33} - a_1\mathbf{M}^1_{31} - a_2\mathbf{M}^1_{32} = 0 \end{aligned}$$

and similarly for $\mathbf{M}^2_{rs}$.

[†]If L_{n+1} is the linear variety defined by 1.90 and $R = \mathbb{C}[L_{n+1}]$, $X_1 = \mathbb{A}^n \times L_{n+1}$, then $\mathbb{C}[L_{n+1} \times \mathbb{A}^n] = R[x_1, \ldots, x_n]$ and $\mathcal{L}_{n+1}$ is the subvariety of X_1 defined by the equations $\mathbf{m}^i_{rn+1} - \sum_{\ell=1}^{n} x_\ell \mathbf{m}^i_{r\ell} = 0$, $r = 1, \ldots, n+1$ in $R[x_1, \ldots, x_n]$ which are "linear" over R.

We note that (1.90) insures that $\mathbf{M}^i$ is an $n+1 \times n+1$ scalar Hankel matrix and that (1.91) insures that the rank of $\mathbf{M}$, $\rho(\mathbf{M}) \leq n$. We now write $x = (\mathbf{a}, \mathbf{H})$ for elements of $\mathcal{L}_{n+1}$ and we shall show that a basis for the row space of $\mathbf{H}$ can be taken from the "top" down.

Example 1.96 Let $p = 2$, $n = 2$ and $(\mathbf{a}, \mathbf{H})$ be an element of $\mathcal{L}_{n+1}$ with $\mathbf{a} = (a_1, a_2)$ and

$$\mathbf{H} = \begin{bmatrix} h_1^1 & h_2^1 & h_3^1 \\ h_2^1 & h_3^1 & h_4^1 \\ h_3^1 & h_4^1 & h_5^1 \\ h_1^2 & h_2^2 & h_3^2 \\ h_2^2 & h_3^2 & h_4^2 \\ h_3^2 & h_4^2 & h_5^2 \end{bmatrix}.$$

Note that $h_3^1 = a_1 h_1^1 + a_2 h_2^1$, $h_4^1 = a_1 h_2^1 + a_2 h_3^1$, $h_5^1 = a_1 h_3^1 + a_2 h_4^1$, $h_3^2 = a_1 h_1^2 + a_2 h_2^2$, $h_4^2 = a_1 h_2^2 + a_2 h_3^2$, $h_5^2 = a_1 h_3^2 + a_2 h_4^2$ so that $\rho(\mathbf{H}) \leq 2$. Suppose $\rho(\mathbf{H}) = 2$. If the first row $(h_1^1\ h_2^1\ h_3^1) = 0$, then clearly the second and third rows are zero and so, the fourth and fifth rows are independent ($\rho(\mathbf{H}) = 2$) and are a basis of the row space of $\mathbf{H}$. If $(h_1^1\ h_2^1\ h_3^1) \neq 0$ but $(h_2^1\ h_3^1\ h_4^1) = \alpha(h_1^1\ h_2^1\ h_3^1)$, then the equations (1.91) imply $(h_3^1\ h_4^1\ h_5^1)$ is also in the span of the first row. Since $\rho(\mathbf{H}) = 2$, we claim that $(h_1^2\ h_2^2\ h_3^2)$ is not in the span of the first row and hence, $(h_1^1\ h_2^1\ h_3^1)$, (h_1^2, h_2^2, h_3^2) are a basis of the row space of $\mathbf{H}$. Suppose this were false; then $h_1^2 = \beta h_1^1$, $h_2^2 = \beta h_2^1$, $h_3^2 = \beta h_3^1$ and $h_4^2 = a_1 h_2^2 + a_2 h_3^2 = \beta(a_1 h_2^1 + a_2 h_3^1) = \beta h_4^1$ so that $(h_2^2\ h_3^2\ h_4^2) = \beta(h_2^1\ h_3^1\ h_4^1) = \beta\alpha(h_1^1\ h_2^1\ h_3^1)$ which contradicts $\rho(\mathbf{H}) = 2$. Note also that (1.91) is critical for if (1.91) does not hold then

$$\begin{bmatrix} 0 & 0 & 0 \\ 0 & 0 & 1 \\ 0 & 1 & * \\ * & * & * \\ * & * & * \\ * & * & * \end{bmatrix}$$

provides a counterexample.

If $x = (\mathbf{a}, \mathbf{H}) \in \mathcal{L}_{n+1}$, then $\mathbf{e}(x) = \mathbf{e}(\mathbf{H}) = (e^1(\mathbf{H}), \ldots, e^p(\mathbf{H}))$ with $\sum_{j=1}^{n} e^j(\mathbf{H}) = \operatorname{rank} \mathbf{H} \leq n$ and $0 \leq e^j(\mathbf{H}) \leq n$. We let $\boldsymbol{\alpha}_e = \boldsymbol{\alpha}_e(\mathbf{H}) = (11 \cdots e^1 1$ $12 \cdots e^p p)$ be the (unique) nice multiindex relative to $\mathbf{e}(\mathbf{H})$. If $\mathbf{H}$ is of rank n, we let

$$\omega_e(\mathbf{H}) = \det[\mathbf{H}|_{1\cdots n}^{\boldsymbol{\alpha}_e}] \tag{1.97}$$

and we now prove a lemma which indicates some of the power of the structure of $\mathcal{L}_{n+1}$.

Lemma 1.98 (Risannen) *Let $x = (\mathbf{a}, \mathbf{H})$ be an element of $\mathcal{L}_{n+1}$ and let $\mathbf{e} = \mathbf{e}(\mathbf{H}) = (e^1, \dots, e^p)$. Then the rows $(\mathbf{H}^1)^1, \dots, (\mathbf{H}^1)^{e^1}$, $(\mathbf{H}^2)^1, \dots$, $(\mathbf{H}^2)^{e^2}, \dots$, $(\mathbf{H}^p)^{e^p}$ are a basis of the row space of $\mathbf{H}$ (where $\mathbf{H} = [\mathbf{H}^i]_{i=1}^p$ with $\mathbf{H}^i$ in $M(n+1, n+1)$).*

Proof. First consider the case $p = 1$ so that $\mathbf{H} = \mathbf{H}^1$, $\rho(\mathbf{H}^1) = r \leq n$ and we claim that $(\mathbf{H}^1)^1, \dots, (\mathbf{H}^1)^r$ are independent. If $(\mathbf{H}^1)^1 = \mathbf{0}$, then all rows are $\mathbf{0}$ by 1.90 and 1.91. If $r = n$, then again the claim is true in view of 1.91. So suppose that $(\mathbf{H}^1)^1, \dots, (\mathbf{H}^1)^s$ are independent $1 \leq s \leq r < n$ but

$$(\mathbf{H}^1)^{s+1} = \alpha_1 (\mathbf{H}^1)^1 + \cdots + \alpha_s (\mathbf{H}^1)^s. \tag{1.99}$$

We assert that then $(\mathbf{H}^1)^{s+2} \in \mathrm{span}[(\mathbf{H}^1)^1 \cdots (\mathbf{H}^1)^{s+1}] = \mathrm{span}[(\mathbf{H}^1)^1 \cdots (\mathbf{H}^1)^s]$. Now

$$(\mathbf{H}^1)^i = \left[h_i^1 \; h_{i+1}^1 \; \cdots \; h_{n+i}^1\right] \tag{1.100}$$

so that (1.99) gives

$$\begin{aligned} h_{s+1}^1 &= \alpha_1 h_1^1 + \cdots + \alpha_s h_s^1 \\ h_{s+1+j}^1 &= \alpha_1 h_{j+1}^1 + \cdots + \alpha_s h_{s+j}^1, \quad j = 1, \dots, n. \end{aligned}$$

Since $h_{n+i}^1 = \sum_{\ell=1}^{n} a_\ell h_{\ell+i-1}^1$, $i = 1, \dots, \mathbf{a} = (a_1, \dots, a_n)$ we have

$$\begin{aligned} h_{n+s+2}^1 &= \sum_{\ell=1}^{n} a_\ell h_{\ell+s+1}^1 \\ &= \sum_{\ell=1}^{n} a_\ell \left(\sum_{r=1}^{s} \alpha_r h_{\ell+r}^1 \right) \\ &= \sum_{r=1}^{s} \alpha_r \left(\sum_{\ell=1}^{n} a_\ell h_{\ell+r}^1 \right) \\ &= \sum_{r=1}^{s} \alpha_r h_{n+r+1}^1 \end{aligned}$$

and consequently,

$$h_{s+1+j}^1 = \alpha_j h_{j+1}^1 + \cdots + \alpha_s h_{s+j}^1, \quad j = 1, \dots, n+1.$$

In other words, $(\mathbf{H}^1)^{s+2} \in \mathrm{span}[(\mathbf{H}^1)^1, \dots, (\mathbf{H}^1)^{s+1}]$ and the claim for $p = 1$ holds. Next consider the case $p = 2$ so that $e^1 = \dim \mathrm{span}[\mathbf{H}^1]$ and $e^2 = r - e^1 =$

$\dim \text{span}[\mathbf{H}] - \dim \text{span}[\mathbf{H}^1]$. By the case $p = 1$, $(\mathbf{H}^1)^1, \ldots, (\mathbf{H}^1)^{e^1}$ are a basis of span$[\mathbf{H}^1]$. If $(\mathbf{H}^2)^1 \in \text{span}[\mathbf{H}^1]$, then, noting that

$$\begin{aligned}(\mathbf{H}^1)^i &= \begin{bmatrix} h_i^1 & \cdots & h_{n+i}^1 \end{bmatrix} \\ (\mathbf{H}^2)^i &= \begin{bmatrix} h_i^2 & \cdots & h_{n+i}^2 \end{bmatrix}\end{aligned}$$

we have

$$\begin{aligned} h_1^2 &= \alpha_1 h_1^1 + \cdots + \alpha_{n+1} h_{n+1}^1 \\ h_{j+1}^2 &= \alpha_1 h_{j+1}^1 + \cdots + \alpha_{n+1} h_{n+1+j}^1, \quad j = 1, \ldots, n \end{aligned} \tag{1.101}$$

and

$$h_{n+i}^2 = \sum_{\ell=1}^{n} a_\ell h_{\ell+i-1}^2$$

(as for h_{n+i}^1). Thus, for instance,

$$\begin{aligned} h_2^2 &= \alpha_1 h_2^1 + \cdots + \alpha_{n+1} h_{n+1+1}^1 \\ &\vdots \\ h_{n+1}^2 &= \alpha_1 h_{n+1}^1 + \cdots + \alpha_{n+1} h_{2n+1}^1 \end{aligned}$$

and

$$\begin{aligned} h_{n+2}^2 &= \sum_{\ell=1}^{n} a_\ell h_{\ell+1}^2 = \sum_{\ell=1}^{n} a_\ell \left(\sum_{r=1}^{n+1} \alpha_r h_{\ell+r}^1 \right) \\ &= \sum_{r=1}^{n+1} \alpha_r \left(\sum_{\ell=1}^{n} a_\ell h_{\ell+r}^1 \right) = \sum_{r=1}^{n+1} \alpha_r h_{n+1+r}^1 \end{aligned}$$

so that $(\mathbf{H}^2)^2 \in \text{span}[\mathbf{H}^1]$. By an entirely similar argument if $(\mathbf{H}^1)^1, \ldots, (\mathbf{H}^1)^{e^1}, (\mathbf{H}^2)^1, \ldots, (\mathbf{H}^2)^t$ are independent ($t \leq e^2$) but $(\mathbf{H}^2)^{t+1} \in \text{span}[(\mathbf{H}^1)^1, \ldots, (\mathbf{H}^2)^t]$, then $(\mathbf{H}^2)^{t+2} \in \text{span}[(\mathbf{H}^1)^1, \ldots, (\mathbf{H}^2)^t]$. Thus we are done for $p = 2$. The general case follows analogously by induction.

Corollary 1.102 *If $\rho(\mathbf{H}) = n$, then $\omega_e(\mathbf{H}) \neq 0$.*

Consider the quasi-affine variety $\mathcal{H}_1 = \mathcal{L}_{n+1} \cap (\mathbb{A}^n \times \bigcup_{\alpha_e} M_{*n}((n+1)p, n+1)_{\alpha_e}) = \bigcup_{\alpha_e} (\mathcal{L}_{n+1} \cap (\mathbb{A}^n \times U_e))$ where U_e is the principal affine open set where $\omega_e \neq 0$. Clearly $\mathcal{H}_1$ is open in $\mathcal{L}_{n+1}$.

Proposition 1.103 $\mathcal{H} = \mathcal{H}_1$.

Proof. Clearly $(\tilde{\psi} \circ \tilde{\varphi})(S_{p,1}^{n}) = \mathcal{H} \subset \mathcal{H}_1$. Conversely, if $(\mathbf{a}, \mathbf{H})$ is in $\mathcal{H}_1$, then

$$\mathbf{H}^i = \begin{bmatrix} h_1^i & \cdots & h_{n+1}^i \\ \vdots & & \vdots \\ h_{n+1}^i & \cdots & h_{2n+1}^i \end{bmatrix}$$

and

$$h_{n+j}^i = \sum_{\ell=1}^{n} a_\ell h_{\ell+j-1}^i, \quad j = 1, \ldots, n+1$$

for $i = 1, \ldots, p$. Since $\omega_e(\mathbf{H}) \neq 0$, there is by 1.49 a realization $x = (A, b, C)$ with $h_\ell^i = c^i A^{\ell-1} b$, and since $\mathbf{H}$ has rank n, $\chi_\ell(A) = a_\ell$ for $\ell = 1, \ldots, n$. Thus, $(\tilde{\psi} \circ \tilde{\varphi})(x) = (\mathbf{a}, \mathbf{H})$ and $\mathcal{H}_1 \subset \mathcal{H}$.

We have now completed the demonstration that the representation (6) is the "same" as in the scalar case. However, we note here an indication of the issues to come. In essence, we have shown that the results of Part I extend without undue difficulty to the case of a single input (or single output) and multiple outputs (or multiple inputs). What happens in the truly multivariable situation when both m *and* p are greater than one, will be the subject of the rest of this volume.

Exercises

(1) Show that $\hat{y}(z) = \hat{h}(z)\hat{u}(z)$ in Proposition 1.10.

(2) Show that the map ψ of (1.36) is an isomorphism between X and Hank$(n; p, 1)$.

(3) Develop the analogs of the results in I.10 for the realization maps and associated morphisms in the $p \geq 1$ case.

(4) Complete the proof of Lemma 1.98.

2

Two or Three Input, Two Output Systems: Some Examples

We saw in Chapter 1 that, for systems with either a single input ($m = 1$) or a single output ($p = 1$), the affine theory of Part I extends in an entirely satisfactory way. We now treat some examples with $m = 2$, $p = 2$ to indicate ways in which the scalar theory breaks down.

A system with two inputs ($m = 2$) and two outputs ($p = 2$) can be represented by a 2×2 matrix $\mathbf{F}(z)$. If each entry $\mathbf{F}_j^i(z)$ is a strictly proper rational function, then $\mathbf{F}(z)$ is a strictly proper rational transfer matrix. The effect of the system is represented by

$$\mathbf{y}(z) = \mathbf{F}(z)\mathbf{u}(z) \tag{2.1}$$

where $\mathbf{u}(z) \in M(2,1)$, $\mathbf{y}(z) \in M(2,1)$.

Example 2.2 Suppose that

$$\mathbf{F}(z) = \begin{bmatrix} 1/z & 0 \\ 0 & 1/z^2 \end{bmatrix}.$$

Let $\mathbf{P}(z) = I_2$, $\mathbf{Q}(z) = \mathrm{diag}[z\ z^2]$ so that $\mathbf{F}(z) = \mathbf{P}(z)\mathbf{Q}^{-1}(z)$. Since the rank of the matrix

$$\psi_{\mathbf{F}}(z) = \begin{bmatrix} \mathbf{P}(z) \\ \mathbf{Q}(z) \end{bmatrix} \tag{2.3}$$

in $M(4,2)$ is 2 (full rank) for all z, there are polynomial matrices $\mathbf{X}(z)$, $\mathbf{Y}(z)$ such that $\mathbf{X}(z)\mathbf{P}(z) + \mathbf{Y}(z)\mathbf{Q}(z) \equiv I_2$. Note that deg det$\mathbf{Q}(z) = 3$. If $\mathbf{F}_j^i(z) = p_j^i(z)/q_j^i(z)$ and $q(z) = \mathrm{lcm}(q_j^i(z))$ (least common multiple), then $q(z) = z^2$ has degree 2 and

$$\mathbf{F}(z) = (\mathbf{P}_0 + \mathbf{P}_1 z)/z^2 = \frac{\mathbf{H}_1}{z} + \frac{\mathbf{H}_2}{z^2} + \frac{\mathbf{H}_3}{z^2} + \cdots \tag{2.4}$$

P. Falb, *Methods of Algebraic Geometry in Control Theory: Part II*, Modern Birkhäuser Classics, https://doi.org/10.1007/978-3-319-96574-1_2

where

$$\mathbf{P}_0 = \begin{bmatrix} 0 & 0 \\ 0 & 1 \end{bmatrix}, \quad \mathbf{P}_1 = \begin{bmatrix} 1 & 0 \\ 0 & 0 \end{bmatrix}, \quad \mathbf{H}_1 = \mathbf{P}_1, \ \mathbf{H}_2 = \mathbf{P}_0$$

and

$$0 \cdot \mathbf{H}_j + 0 \cdot \mathbf{H}_{j+1} + 1 \cdot \mathbf{H}_{j+2} = 0$$

for $j = 1, 2, \ldots$. In other words, the polynomial $q(z)$ annihilates the sequence $\{\mathbf{H}_1, \mathbf{H}_2, \mathbf{H}_3, \ldots\}$. However, if $\mathbf{H}_\mathbf{F} = (\mathbf{H}^i_j)$ with $\mathbf{H}^i_j = \mathbf{H}_{i+j-1}$ is the Hankel matrix of $\mathbf{F}(z)$, then

$$\mathbf{H}_\mathbf{F} = \begin{bmatrix} 1 & 0 & 0 & 0 & \\ 0 & 0 & 0 & 1 & \\ & & & & \mathbf{0} \\ 0 & 0 & 0 & 0 & \\ 0 & 1 & 0 & 0 & \\ & \mathbf{0} & & & \mathbf{0} \end{bmatrix}$$

has rank 3. If we call the rank of $\mathbf{H}_\mathbf{F}$, the *degree of* $\mathbf{F}(z)$, then the degree of $\mathbf{F}(z)$ is *not* the same as the length of the shortest recurrence for $\{\mathbf{H}_j\}_{j=1}^{\infty}$. Since $\mathbf{F}(z)$ has degree 3, the "right" representation for $\mathbf{F}(z)$ is not (2.4) but rather is

$$\mathbf{F}(z) = (\tilde{\mathbf{P}}_0 + \tilde{\mathbf{P}}_1 z + \tilde{\mathbf{P}}_2 z^2)/z^3 \tag{2.5}$$

where

$$\tilde{\mathbf{P}}_0 = \begin{bmatrix} 0 & 0 \\ 0 & 0 \end{bmatrix}, \quad \tilde{\mathbf{P}}_1 = \begin{bmatrix} 0 & 0 \\ 0 & 1 \end{bmatrix}, \quad \tilde{\mathbf{P}}_2 = \begin{bmatrix} 1 & 0 \\ 0 & 0 \end{bmatrix}$$

and

$$\mathbf{H}_1 = \tilde{\mathbf{P}}_2, \quad \mathbf{H}_2 = \tilde{\mathbf{P}}_1, \quad \tilde{\mathbf{H}}_3 = \mathbf{P}_0.$$

Let

$$A = \begin{bmatrix} 0 & 0 & 0 \\ 0 & 0 & 1 \\ 0 & 0 & 0 \end{bmatrix}, \quad B = \begin{bmatrix} 1 & 0 \\ 0 & 0 \\ 0 & 1 \end{bmatrix}, \quad C = \begin{bmatrix} 1 & 0 & 0 \\ 0 & 1 & 0 \end{bmatrix}.$$

Then $\mathbf{F}(z) = C(zI - A)^{-1}B$. If we let

$$\mathbf{Y}(A, B, C) = [B \ AB \ A^2B] = \begin{bmatrix} 1 & 0 & 0 & 0 & 0 & 0 \\ 0 & 0 & 0 & 1 & 0 & 0 \\ 0 & 1 & 0 & 0 & 0 & 0 \end{bmatrix}$$

and

$$\mathbf{Z}(A,B,C) = \begin{bmatrix} C \\ CA \\ CA^2 \end{bmatrix} = \begin{bmatrix} 1 & 0 & 0 \\ 0 & 1 & 0 \\ 0 & 0 & 0 \\ 0 & 0 & 1 \\ 0 & 0 & 0 \\ 0 & 0 & 0 \end{bmatrix}$$

then $\rho(\mathbf{Y}(A,B,C)) = 3$, $\rho(\mathbf{Z}(A,B,C)) = 3$ and

$$\rho(\mathbf{H_F}) = \rho(\mathbf{Z}(A,B,C)\mathbf{Y}(A,B,C)) = 3.$$

In other words, $x = (A,B,C) \in \mathbb{A}^{9+6+6} = \mathbb{A}^{n^2+n(m+p)}$ $(n = 3, m = 2, p = 2)$ is a "minimal" realization of $\mathbf{F}(z)$. Let $\text{Rat}(3,2,2)$ be the set of all 2×2 rational (strictly proper) transfer matrices of degree 3 (i.e., $\mathbf{F}(z) \in \text{Rat}(3,2,2)$) and let $\text{Hank}(3,2,2)$ be the set of block Hankel matrices

$$\tilde{\mathbf{H}} = \begin{bmatrix} \mathbf{H}_1 & \mathbf{H}_2 & \mathbf{H}_3 & \mathbf{H}_4 \\ \mathbf{H}_2 & \mathbf{H}_3 & \mathbf{H}_4 & \mathbf{H}_5 \\ \mathbf{H}_3 & \mathbf{H}_4 & \mathbf{H}_5 & \mathbf{H}_6 \\ \mathbf{H}_4 & \mathbf{H}_5 & \mathbf{H}_6 & \mathbf{H}_7 \end{bmatrix} = (\tilde{\mathbf{H}}^i_j)$$

of rank 3 where $\tilde{\mathbf{H}}^i_j = \mathbf{H}_{i+j-1}$ is in $M(2,2)$. Note that $\tilde{\mathbf{H}} \in M(8,8) = M((3+1)\cdot 2, (3+1)\cdot 2) = M((n+1)p, (n+1)m)$. If $U(z)$ is any 2×2 matrix with polynomial entries such that $\det U(z) = \alpha \in k^*$ (i.e., is a non-zero constant), then (i) $\mathbf{F}(z) = [\mathbf{P}(z)U(z)] \cdot [\mathbf{Q}(z)U(z)]^{-1}$; (ii) the matrix $\psi_{\mathbf{F}}(z)U(z)$ in $M(4,2)$ has full rank for all z; and, (iii) there are polynomial matrices $\mathbf{X}_1(z)$ $(= U^{-1}(z)\mathbf{X}(z))$, $\mathbf{Y}_1(z)$ $(= U^{-1}(z)\mathbf{Y}(z))$ such that $\mathbf{X}_1(z)[\mathbf{P}(z)U(z)] + \mathbf{Y}_1(z)[\mathbf{Q}(z)U(z)] \equiv I_2$. For instance, if $\varphi(z)$ is any polynomial and if

$$U_\varphi(z) = \begin{bmatrix} 1 & \varphi(z) \\ 0 & 1 \end{bmatrix} \tag{2.6}$$

then

$$\mathbf{P}(z)U_\varphi(z) = \begin{bmatrix} 1 & \varphi(z) \\ 0 & 1 \end{bmatrix}, \quad \mathbf{Q}(z)U_\varphi(z) = \begin{bmatrix} z & z\varphi(z) \\ 0 & z^2 \end{bmatrix}.$$

So, even though $\mathbf{F}(z)$ has degree 3, the degree of the entries in a coprime representation $(\mathbf{P}(z), \mathbf{Q}(z))$ of $\mathbf{F}(z)$ can be arbitrarily large. For instance, if $\varphi(z) = z^t$, then $\mathbf{Q}(z)U_\varphi(z)$ has an entry of degree $t+1$. However, as we shall see, the embedding into $\text{Gr}(2,4) \subset$ the projective space $\mathbb{P}^5$, is not affected. In the case of a scalar system, the representation $\mathbf{f}(z) = \mathbf{P}(z)/q(z)$ with $\mathbf{P}^j(z)$, $q(z)$

coprime and $q(z)$ monic is unique. However, the corresponding representation $(\mathbf{P}(z), \mathbf{Q}(z))$, even with det $\mathbf{Q}(z)$ monic is certainly *not unique.* Let

$$\mathbf{M} = \begin{bmatrix} v^1 \\ v^2 \\ v^3 \\ v^4 \end{bmatrix} \tag{2.7}$$

be an element of $M_*(4,2)$ with $v^i \in M(1,2)$ and let $|v^i\ v^j|$ be the determinant of the 2×2 matrix with rows v^i, v^j. If $p_0 = |v^1\ v^2|$, $p_1 = |v^1\ v^3|$, $p_2 = |v^1\ v^4|$, $p_3 = |v^2\ v^3|$, $p_4 = |v^2\ v^4|$, and $p_5 = |v^3\ v^4|$, then

$$\mathcal{P}(\mathbf{M}) = (p_0(\mathbf{M}), \ldots, p_5(\mathbf{M})) \tag{2.8}$$

is a non-zero element of $\mathbb{A}_k^6$. If $\mathbf{M}, \mathbf{M}_1$ determine the same 2-dimensional subspace of k^4, then there is a g in $\text{GL}(2,k) = G$ with $\mathbf{M}g = \mathbf{M}_1$ and conversely. Clearly, $\mathcal{P}(\mathbf{M}_1) = \mathcal{P}(\mathbf{M})\det g$ and so, $\mathcal{P}(\mathbf{M})$ (modulo the equivalence $\mathbf{M} \equiv_G \mathbf{M}_1$ if $\mathbf{M}_1 = \mathbf{M}g$) is an element of the *projective space* $\mathbb{P}_k^5$ (see, I.23 or Chapter 5). If $\mathbf{F}(z) = \mathbf{P}_1(z)\mathbf{Q}_1^{-1}(z)$, then $\mathbf{P}_1(z)$, $\mathbf{Q}_1(z)$ will be (right) coprime if and only if 1 is the gcd (greatest common divisor) of the 2×2 minors of

$$\begin{bmatrix} \mathbf{P}_1(z) \\ \mathbf{Q}_1(z) \end{bmatrix} = \mathbf{M}_\mathbf{F}(z) \tag{2.9}$$

i.e., if and only if $\mathcal{P}(\mathbf{M}_\mathbf{F}(z)) \in \mathbb{P}_k^5$ for *all* z. Since $\mathcal{P}(\mathbf{M}_\mathbf{F}(z)) = (1, 0, z^2, -z, 0, z^3)$ and

$$\mathcal{P}\left(\begin{bmatrix} \mathbf{P}U \\ \mathbf{Q}U \end{bmatrix}\right) = (1, 0, z^2, -z, 0, z^3) \cdot \det U,$$

the point $\mathcal{P}(\psi_\mathbf{F}(z))$ in $\mathbb{P}_k^5$ is well-defined. If $(x_0, x_1) \in \mathbb{P}_k^1$, then we may define a map, also denoted by $\psi_\mathbf{F}$, of $\mathbb{P}_k^1 \to \mathbb{P}_k^5$ by

$$\psi_\mathbf{F}(x_0, x_1) = (x_0^3, 0, x_0 x_1^2, -x_0^2 x_1, 0, x_1^3). \tag{2.10}$$

For $x_0 \neq 0$, $(x_0, x_1) \equiv (1, z)$ and $\psi_\mathbf{F}(1, z) = \mathcal{P}(\psi_\mathbf{F}(z))$. If $\xi_\infty = (0, 1)$, then $\psi_\mathbf{F}(\xi_\infty) = (0,0,0,0,0,1)$. The *poles* of $\mathbf{F}(z)$ are the points $(1, z)$ where $\psi_\mathbf{F}(1, z)$ meets the hyperplane H_5 $(= \{\boldsymbol{\xi} \in \mathbb{P}_k^5 \colon \xi_5 = 0,\ \boldsymbol{\xi} = (\xi_0, \xi_1, \ldots, \xi_5)\})$. We note that $\psi_\mathbf{F}(\mathbb{P}_k^1) \not\subset H_5$, $\dim \psi_\mathbf{F}(\mathbb{P}_k^1) = 1$, $\dim H_5 = 4$, and that $\psi_\mathbf{F}(\mathbb{P}_k^1) \cap H_5 = \{(1,0,0,0,0,0)\}$ with "multiplicity" 3.

Example 2.11 Suppose that

$$\mathbf{F}(z) = \begin{bmatrix} 1/z+1 & 1/(z+1)(z-1) \\ 0 & z/(z+1)(z-1) \end{bmatrix} \tag{2.12}$$

and let

$$\mathbf{P}(z) = \begin{bmatrix} 1 & 1 \\ 0 & z \end{bmatrix}, \quad \mathbf{Q}(z) = \begin{bmatrix} z+1 & 0 \\ 0 & (z+1)(z-1) \end{bmatrix} \tag{2.13}$$

so that $\mathbf{F}(z) = \mathbf{P}(z)\mathbf{Q}^{-1}(z)$. Since the rank of

$$\psi_{\mathbf{F}}(z) = \begin{bmatrix} \mathbf{P}(z) \\ \mathbf{Q}(z) \end{bmatrix} = \begin{bmatrix} 1 & 1 \\ 0 & z \\ z+1 & 0 \\ 0 & (z+1)(z-1) \end{bmatrix} \tag{2.14}$$

is 2 for *all* z (as $\det \mathbf{Q}(z) \neq 0$ if $z \neq \pm 1$ and $\det \mathbf{P}(z) \neq 0$ when $z = \pm 1$), the matrices $\mathbf{P}(z)$, $\mathbf{Q}(z)$ are coprime. Let $z = x_1/x_0$. Then

$$\mathbf{F}(x_0, x_1) = \begin{bmatrix} x_0/x_1 + x_0 & x_0^2/(x_1+x_0)(x_1-x_0) \\ 0 & x_1x_0/(x_1+x_0)(x_1-x_0) \end{bmatrix} \tag{2.15}$$

$\mathbf{F}$ maps $k^2 \to k^2$ and the *graph of* $\mathbf{F}(x_0, x_1)$, $\mathrm{gr}_{\mathbf{F}}(x_0, x_1)$, is given by

$$\mathrm{gr}_{\mathbf{F}}(x_0, x_1) = \begin{bmatrix} u_1 \\ u_2 \\ \mathbf{F}(x_0, x_1)\begin{bmatrix} u_1 \\ u_2 \end{bmatrix} \end{bmatrix} = \begin{bmatrix} u_1 \\ u_2 \\ \frac{x_0}{x_1+x_0}u_1 + \frac{x_0^2}{x_1^2-x_0^2}u_2 \\ x_1x_0u_2/x_1^2 - x_0^2 \end{bmatrix}$$

so that if $\mathbf{u}, \mathbf{u}'$ span k^2, then $\mathrm{gr}_{\mathbf{F}}(x_0, x_1)$ spans a 2-dimensional subspace of k^4. More precisely, if $\mathbf{u} = (u_i)_{i=1,2}$, $\mathbf{u}' = (u_i')_{i=1,2}$ and $\mathrm{span}[\mathbf{u}, \mathbf{u}'] = k^2$, then $\mathrm{gr}_{\mathbf{F}}(x_0, x_1)(\mathbf{u})$, $\mathrm{gr}_{\mathbf{F}} = (x_0, x_1)(\mathbf{u}')$ span a 2-dimensional subspace of k^4 i.e., the matrix $[\mathrm{gr}_{\mathbf{F}}(x_0, x_1)(\mathbf{u})\ \mathrm{gr}_{\mathbf{F}}(x_0, x_1)(\mathbf{u}')]$ has rank 2 i.e., is an element of $M_*(4, 2)$. If $\mathbf{u} = \epsilon_1$, $\mathbf{u}' = \epsilon_2$, then

$$\begin{aligned} &[\mathrm{gr}_{\mathbf{F}}(x_0, x_1)(\epsilon_1)\ \mathrm{gr}_{\mathbf{F}}(x_0, x_1)(\epsilon_2)] \\ &= \begin{bmatrix} 1 & 0 \\ 0 & 1 \\ x_0/x_1 + x_0 & x_0^2/x_1^2 - x_0^2 \\ 0 & x_1x_0/x_1^2 - x_0^2 \end{bmatrix} = M_{\mathrm{gr}_{\mathbf{F}}}(x_0, x_1) \end{aligned}$$

and, "homogenizing,"

$$\begin{aligned} \mathcal{P}(M_{\mathrm{gr}_{\mathbf{F}}}(x_0, x_1)) = ((x_1 + x_0)(x_1^2 - x_0^2), x_0^2(x_1 + x_0), x_1x_0(x_1 + x_0), \\ - x_0(x_1^2 - x_0^2), 0, x_1x_0^2). \end{aligned} \tag{2.16}$$

Let $U_0 = \{(x_0, x_1): x_0 \neq 0\}$ and $U_1 = \{(x_0, x_1): x_1 \neq 0\}$. On U_0, let $z = x_1/x_0$ and on U_1, let $w = x_0/x_1$. If $\tilde{\psi}_{\mathbf{F}}(x_0, x_1) = \mathcal{P}(M_{\mathrm{gr}_{\mathbf{F}}}(x_0, x_1))$, then, on U_0

$$\tilde{\psi}_{\mathbf{F}}(1, z) = ((z+1)(z^2-1), z+1, z(z+1), -(z^2-1), 0, z) \tag{2.17}$$

and, on U_1,

$$\tilde{\psi}_{\mathbf{F}}(w, 1) = ((w+1)(1-w^2), w^2(w+1), w(w+1), w(w^2-1), 0, w^2)$$

so that $\tilde{\psi}_{\mathbf{F}}$ does map $\mathbb{P}_k^1$ into $\mathbb{P}_k^5$. Note that on $U_0 \cap U_1$, $(1, z) = z(1/z, 1) = (w, 1)$ (in the projective sense). Viewing the point as $(1, z)$ in U_0, $\tilde{\psi}_{\mathbf{F}}(1, z)$ is given by (2.17), and viewing the point as $(1/z, 1)$ in U_1, $\tilde{\psi}_{\mathbf{F}}(1/z, 1) = ((z+1)(z^2-1)/z^3$, $z+1/z^3$, $z+1/z^2$, $-(z^2-1)/z^3$, 0, $1/z^2) = 1/z^3((z+1)(z^2-1)$, $z+1$, $z(z+1)$, $-(z^2-1)$, 0, $z) = \tilde{\psi}_{\mathbf{F}}(1, z)$ (as points in $\mathbb{P}_k^5$ since $z \neq 0$ on $U_0 \cap U_1$). If, however, we consider $\psi_{\mathbf{F}}$ as in Example 2.2, then

$$\begin{aligned} \psi_{\mathbf{F}}(x_0, x_1) &= (x_1 x_0^2, -(x_1 + x_0)x_0^2, x_0(x_1^2 - x_0^2), \\ &\quad -x_1 x_0(x_1 + x_0), 0, (x_1 + x_0)(x_1^2 - x_0^2)) \end{aligned}$$

and $\psi_{\mathbf{F}}(\xi_\infty) = (0, 0, 0, 0, 0, 1)$. The poles of $\mathbf{F}(z)$ are the points $(1, z)$ where $\psi_{\mathbf{F}}(1, z)$ meets the hyperplane H_5. These are also the points where $\tilde{\psi}_{\mathbf{F}}(1, z)$ meets H_0 $(= \{\boldsymbol{\xi}: \xi_0 = 0, \boldsymbol{\xi} \in \mathbb{P}_k^5\})$. Here, $\psi_{\mathbf{F}}(\mathbb{P}_k^1) \cap H_5$ consists of the point $(-1, 0, 0, 0, 0, 0)$ with "multiplicity" 2 and the point $(1, -2, 0, -2, 0, 0)$ with "multiplicity" 1. This system also has degree 3.

Example 2.18 If $\mathbf{F}(z)$ is any element of $\mathrm{Rat}(3, 2, 2)$, then there are polynomial matrices $\mathbf{P}(z)$, $\mathbf{Q}(z)$ such that (i) $\mathbf{F}(z) = \mathbf{P}(z)\mathbf{Q}^{-1}(z)$; (ii) $\mathbf{P}(z)$, $\mathbf{Q}(z)$ are coprime; and, (iii) $\det \mathbf{Q}(z) = q(z) = z^3 + a_1 z^2 + a_2 z + a_3$ is monic of degree 3. Let $\mathrm{SO}(2, k[z]) = \{U(z): U(z)$ a 2×2 matrix with entries in $k[z]$ and $\det U(z) \equiv 1\}$ (e.g. $U_\varphi(z)$ of (2.6)). $\mathrm{SO}(2, k[z])$ is a group which acts on $(\mathbf{P}(z), \mathbf{Q}(z))$ via

$$U \cdot (\mathbf{P}(z), \mathbf{Q}(z)) = (\mathbf{P}(z)U, \mathbf{Q}(z)U). \tag{2.19}$$

If $\mathbf{P}(z)$, $\mathbf{Q}(z)$ are coprime with $\det \mathbf{Q}(z)$ monic of degree 3, then $\mathbf{P}(z)U$, $\mathbf{Q}(z)U$ are coprime, $\det(\mathbf{Q}(z)U)$ is monic of degree 3, and

$$\mathbf{F}(z) = \mathbf{P}(z)\mathbf{Q}^{-1}(z) = (\mathbf{P}(z)U)(\mathbf{Q}(z)U)^{-1}.$$

Suppose that $\mathbf{F}(z) = \mathbf{P}(z)\mathbf{Q}^{-1}(z) = \mathbf{P}_1(z)\mathbf{Q}_1^{-1}(z)$ where $\mathbf{P}_1(z)$, $\mathbf{Q}_1(z)$ are also coprime and $\det \mathbf{Q}_1(z)$ is monic of degree 3. Does there exist a U in $\mathrm{SO}(2, k[z])$ such that $\mathbf{P}_1(z) = \mathbf{P}(z)U$, $\mathbf{Q}_1(z) = \mathbf{Q}(z)U$? Since $\det \mathbf{Q}_1(z) = z^3 + \cdots$ and $\det \mathbf{Q}(z) = z^3 + \cdots$ and $\det \mathbf{Q}(z)U = \det \mathbf{Q}(z) \cdot \det U$, if there is *any unimodular* U (i.e., $\det U(z) \equiv a \in k^*$) with $\mathbf{P}_1(z) = \mathbf{P}(z)U$, $\mathbf{Q}_1(z) = \mathbf{Q}(z)U$, then U must be in $\mathrm{SO}(2; k[z])$. We *claim* that $U(z) = \mathbf{Q}^{-1}(z)\mathbf{Q}_1(z)$ is in fact unimodular. For this $U(z)$, $\mathbf{P}(z)U = \mathbf{P}(z)\mathbf{Q}^{-1}(z)Q_1(z) = \mathbf{F}(z)\mathbf{Q}_1(z) = \mathbf{P}_1(z)\mathbf{Q}_1^{-1}(z)\mathbf{Q}_1(z) = \mathbf{P}_1(z)$. Since $\mathbf{P}_1(z) = \mathbf{P}(z)U$ and $\mathbf{Q}_1(z) = \mathbf{Q}(z)U$, U is a common right divisor of $\mathbf{P}_1(z)$, $\mathbf{Q}_1(z)$. If U has entries in $k[z]$, then $I = U_1U$ for some U_1 with entries in $k[z]$ and U will be unimodular. Since $\mathbf{P}(z)$, $\mathbf{Q}(z)$ are coprime, there are $\mathbf{X}(z)$, $\mathbf{Y}(z)$ with $\mathbf{X}(z)\mathbf{P}(z) + \mathbf{Y}(z)\mathbf{Q}(z) \equiv I$ and, consequently,

$$\begin{aligned}\mathbf{X}(x)\mathbf{P}(x)[\mathrm{Adj}\ \mathbf{Q}(z)]\mathbf{Q}_1(z) + \mathbf{Y}(z)\mathbf{Q}(z)[\mathrm{Adj}\ \mathbf{Q}(z)]\mathbf{Q}_1(z)\\ = [\mathrm{Adj}\ \mathbf{Q}(z)]\mathbf{Q}_1(z).\end{aligned}$$

But $\mathbf{P}(z)\mathbf{Q}^{-1}(z) = \mathbf{P}(z)[\mathrm{Adj}\ \mathbf{Q}(z)]/\det \mathbf{Q}(z)$, $\mathbf{P}(z)\mathbf{Q}^{-1}(z) = \mathbf{P}_1(z)\mathbf{Q}_1^{-1}(z)$ and $\mathbf{Q}(z)[\mathrm{Adj}\ \mathbf{Q}(z)] = \det \mathbf{Q}(z)\cdot I$. It follows that $\mathbf{X}(z)\mathbf{P}_1(z) + \mathbf{Y}(z)\mathbf{Q}_1(z) = \mathbf{Q}^{-1}(z)\mathbf{Q}_1(z) = U$ is a polynomial matrix. For now, let $\mathcal{R} = \mathcal{R}(3,2,2) = \{(\mathbf{P}(z)\mathbf{Q}(z))\colon \mathbf{P}(z),\ \mathbf{Q}(z)$ coprime, $\mathbf{P}(z)\mathbf{Q}^{-1}(z) = \mathbf{F}(z) \in \mathrm{Rat}(3,2,2)$, $\det \mathbf{Q}(z)$ monic of degree $3\}$. Then $\mathrm{SO}(2, k[z])$ acts on $\mathcal{R}$ via (2.19) and the equivalence classes under this action are in 1–1 correspondence with the elements of $\mathrm{Rat}(3,2,2)$. Here, in contrast with the scalar case, we must deal with an equivalence in the so called polynomial or differential operator representation of a linear system. As sets, $\mathrm{Rat}(3,2,2) \simeq \mathcal{R}(3,2,2)/\,\mathrm{SO}(2,k[z])$. Define the *degree of a polynomial vector* as the highest degree of its entries. If $\mathbf{P}(z)\mathbf{Q}^{-1}(z)$ is strictly proper, then each column of $\mathbf{P}(z)$ has degree less than the degree of the corresponding column of $\mathbf{Q}(z)$. Unfortunately, the converse is not true. Let

$$\tilde{\mathbf{P}}(z) = \begin{bmatrix} 2z^2 & 1 \\ z^2+1 & 0 \end{bmatrix}, \quad \tilde{\mathbf{Q}}(z) = \begin{bmatrix} z^4 & z^2 \\ z^2 - z & 1 \end{bmatrix}. \tag{2.20}$$

Then $\deg(\tilde{\mathbf{P}}(z))_1 = 2$, $\deg(\tilde{\mathbf{P}}(z))_2 = 0$, $\deg(\tilde{\mathbf{Q}}(z))_1 = 4$, and $\deg(\tilde{\mathbf{Q}}(z))_2 = 2$ so that $\deg(\tilde{\mathbf{P}}(z))_i < \deg(\tilde{\mathbf{Q}}(z))_i$ for $i = 1, 2$. Moreover, $\det \tilde{\mathbf{Q}}(z) = z^3$ is monic of degree 3 and $\tilde{\mathbf{P}}(z)$, $\tilde{\mathbf{Q}}(z)$ are coprime. However,

$$\tilde{\mathbf{P}}(z)\tilde{\mathbf{Q}}^{-1}(z) = \begin{bmatrix} z^2 + 1/z^3 & -z \\ z^2 + 1/z^3 & -(z^2+1)/z \end{bmatrix}$$

is not even proper. This cannot occur in the scalar case.

Example 2.21 If $\mathbf{Q}(z)$ is any 2×2 polynomial matrix and $\partial_1 = \deg(\mathbf{Q}(z))_1$, $\partial_2 = \deg(\mathbf{Q}(z))_2$ are the column degrees of $\mathbf{Q}(z)$, then

$$\mathbf{Q}(z) = \Delta_c(Q)\,\mathrm{diag}\left[z^{\partial_1}, z^{\partial_2}\right] + \hat{\mathbf{Q}}(z) \tag{2.22}$$

where $\partial_i(\hat{\mathbf{Q}}(z)) = \deg(\hat{\mathbf{Q}}(z))_i < \partial_i$ for $i = 1, 2$. $\Delta_c(\mathbf{Q})$ is in $M(2,2,k)$ and is called the *column coefficient of* $\mathbf{Q}(z)$. We note that $\det \mathbf{Q}(z) = (\det \Delta_c(\mathbf{Q}))z^{\partial_1+\partial_2}$

+ lower degree terms. Since $\deg \det \mathbf{Q}(z) \leq \partial_1 + \partial_2$, we shall say that $\mathbf{Q}(z)$ is *column proper* ([W-4]) if $\deg \det \mathbf{Q}(z) = \partial_1 + \partial_2$ or, equivalently, if $\det \Delta_c(\mathbf{Q}) \neq 0$. For the $\tilde{\mathbf{Q}}(z)$ of (2.20),

$$\tilde{\mathbf{Q}}(z) = \begin{bmatrix} 1 & 1 \\ 0 & 0 \end{bmatrix} \begin{bmatrix} z^4 & 0 \\ 0 & z^2 \end{bmatrix} + \begin{bmatrix} 0 & 0 \\ z^2 - z & 1 \end{bmatrix}$$

and $\det \Delta_c(\tilde{\mathbf{Q}}) = 0$ so that $\tilde{\mathbf{Q}}(z)$ is not column proper. If $\mathbf{Q}(z)$ is column proper, then $\mathbf{F}(z) = \mathbf{P}(z)\mathbf{Q}^{-1}(z)$ is strictly proper if and only if $\partial_i(\mathbf{P}(z)) < \partial_i(\mathbf{Q}(z))$ for all i. Let $\mathcal{R}_c = \mathcal{R}_c(3,2,2) = \{(\mathbf{P}(z), \mathbf{Q}(z))\colon \mathbf{P}(z), \mathbf{Q}(z)$ are coprime, $\mathbf{Q}(z)$ is column proper, $\partial_i(\mathbf{P}(z)) < \partial_i(\mathbf{Q}(z))$, $\partial_1(\mathbf{Q}(z)) \geq \partial_2(\mathbf{Q}(z))$, and $\det \mathbf{Q}(z)$ monic of degree $3\}$. Let $S(\mathcal{R}_c)$ be the stabilizer subgroup of $\mathcal{R}_c$ in $\mathrm{SO}(2, k[z])$ i.e., $S(\mathcal{R}_c) = \{U \in \mathrm{SO}(2, k[z])\colon \mathcal{R}_c U = \mathcal{R}_c\}$. Then $S(\mathcal{R}_c)$ acts on $\mathcal{R}_c$. Since *any* $\mathbf{Q}(z)$ can be made column proper by applying elementary column operations (Chapter 4 or [W-4]) and since columns can be permuted, the equivalence classes of $\mathcal{R}_c \bmod S(\mathcal{R}_c)$ are in 1–1 correspondence with $\mathrm{Rat}(3,2,2)$. In other words, $\mathrm{Rat}(3,2,2) \simeq \mathcal{R}_c/S(\mathcal{R}_c)$ as sets. Since $\det \mathbf{Q}(z)$ is monic of degree 3, $\partial_1(\mathbf{Q}(z)) + \partial_2(\mathbf{Q}(z)) = 3$. Thus, $\mathcal{R}_c = \mathcal{R}_{c,3,0} \cup \mathcal{R}_{c,2,1}$ where $\mathcal{R}_{c,3,0} = \{(\mathbf{P}(z), \mathbf{Q}(z)) \in \mathcal{R}_c\colon \partial_1(\mathbf{Q}(z)) = 3, \partial_2(\mathbf{Q}(z)) = 0\}$ and $\mathcal{R}_{c,2,1} = \{(\mathbf{P}(z), \mathbf{Q}(z)) \in \mathcal{R}_c\colon \partial_1(\mathbf{Q}(z)) = 2, \partial_2(\mathbf{Q}(z)) = 1\}$. We *claim* that $\mathcal{R}_{c,3,0}$ and $\mathcal{R}_{c,2,1}$ are invariant under the action of $S(\mathcal{R}_c)$. To verify the claim, we prove the following (cf. Proposition 4.47):

Proposition 2.23 *If $\mathbf{Q}_1(z) = \mathbf{Q}_2(z)U$ with $\mathbf{Q}_1(z)$, $\mathbf{Q}_2(z)$ column proper and U unimodular, then $\{\partial_1(\mathbf{Q}_1(z)), \partial_2(\mathbf{Q}_1(z))\} = \{\partial_1(\mathbf{Q}_2(z)), \partial_2(\mathbf{Q}_2(z))\}$ as sets.*

Proof. Since $U = \mathbf{Q}_2^{-1}(z)\mathbf{Q}_1(z)$,

$$U = [\mathrm{Adj}\ \mathbf{Q}_2(z)]\mathbf{Q}_1(z)/\det \mathbf{Q}_2(z).$$

But $\deg[\mathrm{Adj}\ \mathbf{Q}_2(z)]^i_j \leq 3 - \partial_i(\mathbf{Q}_2(z))$ for $i, j = 1, 2$ and so,

$$u^i_j(z) = \sum_{\ell=1}^{2} [\mathrm{Adj}\ \mathbf{Q}_2(z)]^i_\ell (\mathbf{Q}_1(z))^\ell_j / \det \mathbf{Q}_2(z)$$

has degree $\leq \partial_j(\mathbf{Q}_1(z)) - \partial_i(\mathbf{Q}_2(z)) = \partial_j(\mathbf{Q}_1(z)) + 3 - \partial_i(\mathbf{Q}_2(z)) - 3$, $i, j = 1, 2$. But for $j = 1$ (say), not all $u^i_1(z)$ are zero (otherwise $\det U \equiv 0$) and $\partial_1(\mathbf{Q}_1(z)) - \partial_1(\mathbf{Q}_2(z)) \geq 0$ or, $\partial_1(\mathbf{Q}_1(z)) - \partial_2(\mathbf{Q}_2(z)) \geq 0$ (similarly for $j = 2$). Since $\partial_1(\mathbf{Q}_1(z)) + \partial_2(\mathbf{Q}_1(z)) = \partial_1(\mathbf{Q}_2(z)) + \partial_2(\mathbf{Q}_2(z))$, the result follows. Note also that if $(\mathbf{P}(z), \mathbf{Q}(z)) \in \mathcal{R}_{c,3,0}$, then $\det \mathbf{P}(z) \equiv 0$ and $\det \mathbf{F}(z) \equiv 0$. Thus, already the structure of $\mathrm{Rat}(3,2,2)$ is complex.

Example 2.2.4 ($\mathrm{Rat}(n,2,2)$) Let $\mathbf{F}(z) = \mathbf{P}(z)\mathbf{Q}^{-1}(z)$ be an element of $\mathrm{Rat}(n,2,2)$. Then, for $\mathbf{P}(z)$, $\mathbf{Q}(z)$ coprime, there is a map $\psi_{\mathbf{F}} : \mathbb{P}^1_k \to \mathbb{P}^5_k$ (defined via $\mathcal{P}(\mathbf{M}_{\mathbf{F}}(z))$) such that (1) the coordinates $\psi_{\mathbf{F},i}(x_0, x_1)$ of $\psi_{\mathbf{F}}(x_0, x_0)$ are each a form of degree n; (2) $\psi_{\mathbf{F}}(\mathbb{P}^1) \subset \mathrm{Gr}(2,4)$; and, (3) $\psi_{\mathbf{F}}(\xi_\infty) = \psi_{\mathbf{F}}(0,1) = (0,0,0,0,0,1)$. Moreover, the map $\mathbf{F}(z) \to \psi_{\mathbf{F}}$ is injective. If $(\mathbf{P}(z), \mathbf{Q}(z))$ are coprime with $\mathbf{Q}(z)$ column proper and $\det \mathbf{Q}(z)$ monic of degree n, then

$$\begin{array}{lcll} \psi_{\mathbf{F},0}(1,z) & = & \det \mathbf{P}(z) & \text{degree } \leq n-2 \\ \psi_{\mathbf{F},1}(1,z) & = & -\mathbf{P}^1(z)[\mathrm{Adj}\ \mathbf{Q}(z)]_2 & \text{degree } \leq n-1 \\ \psi_{\mathbf{F},2}(1,z) & = & \mathbf{P}^1(z)[\mathrm{Adj}\ \mathbf{Q}(z)]_1 & \text{degree } \leq n-1 \\ \psi_{\mathbf{F},3}(1,z) & = & -\mathbf{P}^2(z)[\mathrm{Adj}\ \mathbf{Q}(z)]_2 & \text{degree } \leq n-1 \\ \psi_{\mathbf{F},4}(1,z) & = & \mathbf{P}^2(z)[\mathrm{Adj}\ \mathbf{Q}(z)]_1 & \text{degree } \leq n-1 \\ \psi_{\mathbf{F},5}(1,z) & = & \det \mathbf{Q}(z) & \text{monic, degree } n \end{array} \tag{2.25}$$

and

$$\psi_{\mathbf{F},0}\psi_{\mathbf{F},5} - \psi_{\mathbf{F},1}\psi_{\mathbf{F},4} + \psi_{\mathbf{F},2}\psi_{\mathbf{F},3} = 0 \tag{2.26}$$

(equation for $\mathrm{Gr}(2,4)$). If we write

$$\psi_{\mathbf{F},i}(x_0,x_1) = \alpha_{i0}x_0^n + \cdots + \alpha_{in}x_1^n \tag{2.27}$$

for $i = 0,1,\ldots,5$, then (2.25) becomes

$$\begin{array}{rcll} \alpha_{in} & = & 0, & i = 0,1,\ldots,4 \\ \alpha_{0n-1} & = & 0, & \alpha_{5n} = 1 \neq 0. \end{array} \tag{2.28}$$

In view of (2.25) and (2.26), we have

(a) $\psi_{\mathbf{F},5}(1,z) = \det \mathbf{Q}(z)$

(b) $\mathbf{P}(z)[\mathrm{Adj}\ \mathbf{Q}(z)] = \begin{bmatrix} \psi_{\mathbf{F},2}(1,z) & -\psi_{\mathbf{F},1}(1,z) \\ \psi_{\mathbf{F},4}(1,z) & -\psi_{\mathbf{F},3}(1,z) \end{bmatrix}.$

or, equivalently,

(b)′ $\mathbf{P}(z)\mathbf{Q}^{-1}(z) = \mathbf{F}(z) = \begin{bmatrix} \psi_{\mathbf{F},2}(1,z)/\psi_{\mathbf{F},5}(1,z) & -\psi_{\mathbf{F},1}(1,z)/\psi_{\mathbf{F},5}(1,z) \\ \psi_{\mathbf{F},4}(1,z)/\psi_{\mathbf{F},5}(1,z) & -\psi_{\mathbf{F},3}(1,z)/\psi_{\mathbf{F},5}(1,z) \end{bmatrix}.$

Consider now the set of *all* regular maps $\psi\colon \mathbb{P}^1 \to \mathbb{P}^5$ such that $\psi(x_0,x_1) = (\psi_i(x_0,x_1))$ has a representation by forms of degree n i.e.,

$$\psi_i(x_0,x_1) = a_{i0}x_0^n + \cdots + a_{in}x_1^n \tag{2.29}$$

for $i = 0,1,\ldots,5$. We may view ψ as the point $\psi = (a_{00},\ldots,a_{0n},\ldots,a_{50},\ldots,a_{5n})$ in $\mathbb{P}_k^{6n+5}$ (or $\mathbb{A}_k^{6n+6}$). The degree conditions on ψ become

$$a_{in} = 0, \quad i = 0,1,\ldots,4, \quad a_{0n-1} = 0 \tag{2.30}$$

and so, if ψ satisfies (2.30), then ψ lies in a linear subspace of dimension $6n-1$ ($\simeq \mathbb{P}_k^{6n-1}$). If, in addition, ψ lies in $\mathrm{Gr}(2,4)$ and so satisfies a "Grassmann" equation which is homogeneous of degree $2n-2$, then there are $2n-1$ quadratic equations in the coefficients and ψ lies in a variety V of dimension $4n$. If, finally, ψ_5 is monic (or, better $a_{5n} \neq 0$), then ψ is an element of the quasi-projective variety $V_{a_{5n}}$ of dimension $n \cdot 4 = n \cdot (m+p)$! If $\mathbf{F}(z)$ is an element of $\text{Rat}(n, 2, 2)$, then $\psi_{\mathbf{F}}$ is an element of $V_{a_{5n}}$ and the map $\mathbf{F} \to \psi_{\mathbf{F}}$ is an injective map of $\text{Rat}(n, 2, 2)$ into $V_{a_{5n}}$. We *claim* it is surjective and consequently that $\text{Rat}(n, 2, 2)$ can be identified with the quasi-projective variety $V_{a_{5n}}$. Let $\psi = (\psi_0, \ldots, \psi_5)$ be an element of $V_{a_{5n}}$ and let

$$\tilde{\mathbf{F}}(z) = \begin{bmatrix} \psi_2/\psi_5 & -\psi_1/\psi_5 \\ \psi_4/\psi_5 & -\psi_3/\psi_5 \end{bmatrix}.$$

Note that $\det \tilde{\mathbf{F}}(z) = (\psi_1\psi_4 - \psi_2\psi_3)/\psi_5^2 = \psi_0\psi_5/\psi_5^2 = \psi_0/\psi_5$ by virtue of the Grassmann equation. Moreover, $\tilde{\mathbf{F}}(z)$ is a strictly proper 2×2 rational transfer matrix in view of the degree conditions (2.30). Let $\tilde{n}$ be the degree of $\tilde{\mathbf{F}}(z)$. There are $\tilde{\mathbf{P}}(z)$, $\tilde{\mathbf{Q}}(z)$ with $\tilde{\mathbf{F}}(z) = \tilde{\mathbf{P}}(z)\tilde{\mathbf{Q}}^{-1}(z)$, $\tilde{\mathbf{P}}(z)$, $\tilde{\mathbf{Q}}(z)$ coprime, $\tilde{\mathbf{Q}}(z)$ column proper and $\det \tilde{\mathbf{Q}}(z)$ monic of degree $\tilde{n}$. Let $\tilde{\varphi}_i = \psi_{\tilde{\mathbf{F}},i}$, $i = 0, 1, \ldots, 5$. Then

$$\tilde{\mathbf{F}}(z) = \begin{bmatrix} \tilde{\varphi}_2/\tilde{\varphi}_5 & -\tilde{\varphi}_1/\tilde{\varphi}_5 \\ \tilde{\varphi}_4/\tilde{\varphi}_5 & -\tilde{\varphi}_3/\tilde{\varphi}_5 \end{bmatrix}$$

and, hence,

$$\tilde{\varphi}_i/\tilde{\varphi}_5 = \psi_i/\psi_5, \quad i = 1, 2, 3, 4, 5.$$

Since $\psi_0/\psi_5 = \det \tilde{\mathbf{F}}(z) = \tilde{\varphi}_0/\tilde{\varphi}_5$ (by the Grassmann equation), we have

$$\tilde{\varphi}_i\psi_5 = \psi_i\tilde{\varphi}_5 \tag{2.31}$$

for $i = 0, 1, 2, 3, 4, 5$. Both $(\tilde{\varphi}_0, \ldots, \tilde{\varphi}_5)$ and $(\psi_0, \ldots, \psi_5)$ are relatively prime and so, there are $\alpha^i(z)$, $\beta^i(z)$ such that

$$1 = \sum_{i=0}^{5} \alpha^i(z)\tilde{\varphi}_i(1, z), \quad 1 = \sum_{i=0}^{5} \beta^i(z)\psi_i(1, z). \tag{2.32}$$

It follows that

$$\begin{aligned} \psi_5(1, z) &= \left(\sum_{i=0}^{5} \alpha^i(z)\psi_i(1, z) \right) \tilde{\varphi}_5(1, z) \\ \tilde{\varphi}_5(1, z) &= \left(\sum_{i=0}^{5} \beta^i(z)\tilde{\varphi}_i(1, z) \right) \psi_5(1, z) \end{aligned} \tag{2.33}$$

and hence that $\psi_5(1, z)$ and $\tilde{\varphi}_5(1, z)$ are associates. In other words, $\tilde{\varphi}_5(1, z) = \epsilon\psi_5(1, z)$ with ϵ a unit. Since $\tilde{\varphi}_5$, ψ_5 are monic, $\epsilon = 1$ and $\tilde{n} = n$. Thus,

$\tilde{\mathbf{F}}(z) \in \text{Rat}(n,2,2)$ and $\psi_{\tilde{\mathbf{F}}} = \psi$. This approach gives $\text{Rat}(n,2,2)$ an algebraic structure.

Example 2.34 ($\text{Rat}(n,3,2)$) Let $\mathbf{F}(z) = \mathbf{P}(z)\mathbf{Q}^{-1}(z)$ be an element of $\text{Rat}(n,3,2)$ i.e., $m = 3$ inputs, $p = 2$ outputs, with $\mathbf{P}(z)$, $\mathbf{Q}(z)$ coprime. Set

$$\mathbf{M}_{\mathbf{F}}(z) = \begin{bmatrix} \mathbf{P}(z) \\ \mathbf{Q}(z) \end{bmatrix} \tag{2.35}$$

and (dropping "z's")

$$\begin{aligned}
\psi_{\mathbf{F},0} &= \det[\mathbf{M}_{\mathbf{F}}\big|_{123}^{123}], & \psi_{\mathbf{F},1} &= \det[\mathbf{M}_{\mathbf{F}}\big|_{123}^{124}] \\
\psi_{\mathbf{F},2} &= \det[\mathbf{M}_{\mathbf{F}}\big|_{123}^{125}], & \psi_{\mathbf{F},3} &= \det[\mathbf{M}_{\mathbf{F}}\big|_{123}^{134}] \\
\psi_{\mathbf{F},4} &= \det[\mathbf{M}_{\mathbf{F}}\big|_{123}^{135}], & \psi_{\mathbf{F},5} &= \det[\mathbf{M}_{\mathbf{F}}\big|_{123}^{145}] \\
\psi_{\mathbf{F},6} &= \det[\mathbf{M}_{\mathbf{F}}\big|_{123}^{234}], & \psi_{\mathbf{F},7} &= \det[\mathbf{M}_{\mathbf{F}}\big|_{123}^{235}] \\
\psi_{\mathbf{F},8} &= \det[\mathbf{M}_{\mathbf{F}}\big|_{123}^{245}], & \psi_{\mathbf{F},9} &= \det[\mathbf{M}_{\mathbf{F}}\big|_{123}^{345}] = \det \mathbf{Q}(z)
\end{aligned} \tag{2.36}$$

then $\psi_F = (\psi_{\mathbf{F},0}, \ldots, \psi_{\mathbf{F},9})$ is a map of $\mathbb{P}_k^1 \to \mathbb{P}_k^9$ such that (when homogenized) (1) the $\psi_{\mathbf{F},i}$ are all forms of degree n; (2) $\psi_F(\mathbb{P}^1) \subset \text{Gr}(3,5)$; and (3) $\psi_F(\xi_\infty) = (0, \ldots, 0, 1)$. The map $\mathbf{F} \to \psi_F$ is again injective. If $(\mathbf{P}, \mathbf{Q})$ are coprime, $\mathbf{Q}$ is column proper with $\partial_1 \geq \partial_2 \geq \partial_3$, and $\det \mathbf{Q}$ is monic of degree n, then

$$\begin{aligned}
\deg \psi_{\mathbf{F},i} &\leq n-2 & i &= 0,1,2, \\
\deg \psi_{\mathbf{F},i} &\leq n-1 & i &= 3, \ldots, 8 \\
\deg \psi_{\mathbf{F},q} &= n
\end{aligned} \tag{2.37}$$

and $(\psi_{\mathbf{F},0}, \ldots, \psi_{\mathbf{F},9})$ are relatively prime. If X is any 2×3 matrix, let

$$\pi_1(X) = \det[X\big|_{12}^{12}], \quad \pi_2(X) = \det[X\big|_{13}^{12}], \quad \pi_3(X) = \det[X\big|_{23}^{12}].$$

Also, let

$$\text{Adj } \mathbf{Q} = \begin{pmatrix} Q_{11} & -Q_{21} & Q_{31} \\ -Q_{12} & Q_{22} & -Q_{32} \\ Q_{13} & -Q_{23} & Q_{33} \end{pmatrix} = ((-1)^{i+j} Q_{ji})$$

where the Q_{ji} are the cofactors of $\mathbf{Q}$. We have

$$\psi_{\mathbf{F},0} = \psi_{\mathbf{F},9}\pi_3(\mathbf{F}), \quad \psi_{\mathbf{F},1} = -\psi_{\mathbf{F},9}\pi_2(\mathbf{F}), \quad \psi_{\mathbf{F},2} = \psi_{\mathbf{F},9}\pi_1(\mathbf{F})$$

$$\begin{aligned}
\psi_{\mathbf{F},3} &= \mathbf{P}^1(\text{Adj } \mathbf{Q})_3 = \psi_{\mathbf{F},9}\mathbf{F}_3^1 & \psi_{\mathbf{F},6} &= \mathbf{P}^2(\text{Adj } \mathbf{Q})_3 = \psi_{\mathbf{F},9}\mathbf{F}_3^2 \\
\psi_{\mathbf{F},4} &= -\mathbf{P}^1(\text{Adj } \mathbf{Q})_2 = -\psi_{\mathbf{F},9}\mathbf{F}_2^1 & \psi_{\mathbf{F},7} &= -\mathbf{P}^2(\text{Adj } \mathbf{Q})_2 = \psi_{\mathbf{F},9}\mathbf{F}_2^2 \\
\psi_{\mathbf{F},5} &= \mathbf{P}^1(\text{Adj } \mathbf{Q})_2 = \psi_{\mathbf{F},9}\mathbf{F}_1^1 & \psi_{\mathbf{F},8} &= \mathbf{P}^2(\text{Adj } \mathbf{Q})_1 = \psi_{\mathbf{F},9}\mathbf{F}_1^2.
\end{aligned} \tag{2.38}$$

The $\psi_{\mathbf{F},i}$ satisfy the Grassmann equations

$$\begin{aligned}
\psi_{\mathbf{F},0}\psi_{\mathbf{F},5} - \psi_{\mathbf{F},1}\psi_{\mathbf{F},4} + \psi_{\mathbf{F},2}\psi_{\mathbf{F},3} &= 0 \\
\psi_{\mathbf{F},0}\psi_{\mathbf{F},8} - \psi_{\mathbf{F},1}\psi_{\mathbf{F},7} + \psi_{\mathbf{F},2}\psi_{\mathbf{F},6} &= 0 \\
\psi_{\mathbf{F},0}\psi_{\mathbf{F},9} - \psi_{\mathbf{F},3}\psi_{\mathbf{F},7} + \psi_{\mathbf{F},4}\psi_{\mathbf{F},6} &= 0 \\
\psi_{\mathbf{F},1}\psi_{\mathbf{F},9} - \psi_{\mathbf{F},3}\psi_{\mathbf{F},8} + \psi_{\mathbf{F},5}\psi_{\mathbf{F},6} &= 0 \\
\psi_{\mathbf{F},2}\psi_{\mathbf{F},9} - \psi_{\mathbf{F},4}\psi_{\mathbf{F},8} + \psi_{\mathbf{F},5}\psi_{\mathbf{F},7} &= 0
\end{aligned} \tag{2.39}$$

In addition, when we homogenize, $\psi_{\mathbf{F}}(\mathbb{P}^1) \subset \mathrm{Gr}(3,5)$ and $\psi_F(\xi_\infty) = (0,\dots,0,1)$. Again consider the set of *all* regular maps $\psi\colon \mathbb{P}^1_k \to \mathbb{P}^9_k$ such that $\psi(x_0,x_1) = (\psi_i(x_0,x_1))$ with the ψ_i forms of degree n. We view ψ as a point $\boldsymbol{\psi}$ in $\mathbb{P}^{10n+9}_k$ (or $\mathbb{A}^{10n+10}_k$). If ψ satisfies the degree conditions (2.37), then $\boldsymbol{\psi}$ lies in a linear subspace of dimension $10n-3$. If, in addition, $\boldsymbol{\psi}$ satisfies the Grassmann equations (2.39) and the condition that ψ_9 be monic (or better $a_{9n} \neq 0$), then $\boldsymbol{\psi}$ lies in a quasi-projective variety $V_{a_{9n}} \subset \mathbb{P}^{10n+9}$. If $\mathbf{F}(z) \in \mathrm{Rat}(n,3,2)$, then $\psi_F \in V_{a_{9n}}$ and the map $\mathbf{F} \to \psi_F$ is an injective map of $\mathrm{Rat}(n,3,2)$ into $V_{a_{9n}}$. We *claim* it is surjective so that $\mathrm{Rat}(n,3,2)$ can be identified with the quasi-projective variety $V_{a_{9n}}$. Let $\boldsymbol{\psi} = (\psi_0,\dots,\psi_9)$ be an element of $V_{a_{9n}}$ and let

$$\tilde{\mathbf{F}}(z) = \begin{bmatrix} \psi_5/\psi_9 & -\psi_4/\psi_9 & \psi_3/\psi_9 \\ \psi_8/\psi_9 & -\psi_7/\psi_9 & \psi_6/\psi_9 \end{bmatrix}. \tag{2.40}$$

Then $\tilde{\mathbf{F}}(z)$ is, by the degree conditions, a strictly proper rational transfer matrix of degree $\tilde{n}$. In view of the Grassmann equations

$$\psi_0 = \psi_9\pi_3(\tilde{\mathbf{F}}), \quad \psi_1 = -\psi_9\pi_2(\tilde{\mathbf{F}}), \quad \psi_3 = \psi_9\pi_1(\tilde{\mathbf{F}}) \tag{2.41}$$

e.g. $\pi_3(\tilde{\mathbf{F}}) = (-\psi_4\psi_6 + \psi_3\psi_7)/\psi_9^2 = \psi_0\psi_9/\psi_9^2 = \psi_0/\psi_9$. Since $\tilde{\mathbf{F}}(z)$ is a 2×3 strictly proper rational transfer matrix, there are $\tilde{\varphi}_0, \tilde{\varphi}_1, \dots, \tilde{\varphi}_9$ (degree $\tilde{n}$) satisfying the Grassmann equations, the condition at ξ_∞, the relatively prime condition, and coming from $(\tilde{\mathbf{P}}, \tilde{\mathbf{Q}})$. It follows that

$$\tilde{\varphi}_i\psi_9 = \psi_i\tilde{\varphi}_9 \tag{2.42}$$

for $i = 0,1,\dots,9$. Arguing exactly as in the previous example, we verify the claim that the map $\mathbf{F} \to \psi_F$ is surjective. So, the same approach gives an algebraic structure to $\mathrm{Rat}(n,3,2)$.

Example 2.43 Consider $\mathrm{Rat}(2,2,2)$ as a quasi-projective variety in $\mathbb{P}^{17}_k$ and let $\boldsymbol{\varphi} = (\varphi_0,\dots,\varphi_5)$ where

$$\begin{aligned}
\varphi_0(z) &= a_{00}, \quad \varphi_1(z) = b_{01} + b_{11}z, \quad \varphi_2(z) = b_{02} + b_{12}z \\
\varphi_3(z) &= b_{03} + b_{13}z, \quad \varphi_4(z) = b_{04} + b_{14}z, \\
\varphi_5(z) &= a_{05} + a_{15}z + z^2
\end{aligned} \tag{2.44}$$

be a typical element of Rat(2, 2, 2). Set

$$H_1 = H_1(\varphi) = \begin{bmatrix} b_{12} & -b_{11} \\ b_{14} & -b_{13} \end{bmatrix}, \quad B_0 = B_0(\varphi) = \begin{bmatrix} b_{02} & -b_{01} \\ b_{04} & -b_{03} \end{bmatrix}$$

and

$$H_2 = H_2(\varphi) = \begin{bmatrix} b_{02} - a_{15}b_{12} & -b_{01} + a_{15}b_{11} \\ b_{04} - a_{15}b_{14} & -b_{03} + a_{15}b_{13} \end{bmatrix} = B_0 - a_{15}H_1.$$

The Grassmann equation $\varphi_0\varphi_5 = \varphi_1\varphi_4 - \varphi_2\varphi_3$ becomes the three equations

$$\begin{aligned} a_{00} &= \det H_1 = b_{11}b_{14} - b_{12}b_{13} \\ a_{00}a_{05} &= \det B_0 = b_{01}b_{04} - b_{02}b_{03} \\ a_{00}a_{15} &= \det \begin{bmatrix} b_{01} & b_{12} \\ b_{03} & b_{14} \end{bmatrix} - \det \begin{bmatrix} b_{02} & b_{11} \\ b_{04} & b_{13} \end{bmatrix} \\ &= b_{01}b_{14} + b_{04}b_{11} - b_{03}b_{12} - b_{02}b_{13}. \end{aligned} \tag{2.45}$$

Let

$$H_{n+1} = H_{n+1}(\varphi) = -a_{05}H_{n-1} - a_{15}H_n \tag{2.46}$$

for $n = 2, \ldots$. Since

$$F_\varphi(z) = \begin{bmatrix} \varphi_2/\varphi_5 & -\varphi_1/\varphi_5 \\ \varphi_4/\varphi_5 & -\varphi_3/\varphi_5 \end{bmatrix} \tag{2.47}$$

we have $\varphi_5 F_\varphi(z) = (H_2 + a_{15}H_1) + H_1 z = B_0 + H_1 z$ and

$$F_\varphi(z) = \frac{H_1}{z} + \frac{H_2}{z^2} + \frac{H_3}{z^3}$$

so that the block Hankel matrix of $F_\varphi(z)$, $\mathbf{H}_\varphi$, is given by

$$\mathbf{H}_\varphi = \begin{bmatrix} H_1 & B_0 - a_{15}H_1 \\ B_0 - a_{15}H_1 & (-a_{05} + a_{15}^2)H_1 - a_{15}B_0 \end{bmatrix}. \tag{2.48}$$

We shall, provisionally, call the map L given by

$$L(\varphi) = (B_0(\varphi), H_1(\varphi), a_{05}(\varphi), a_{15}(\varphi)) \tag{2.49}$$

the *Laurent map*. We can reasonably view L as a morphism by virtue of its explicit form. If $\varphi \in$ Rat(2, 2, 2), then $\mathbf{H}_\varphi$ has rank 2 (Exercise 5) and consequently, so does the infinite block Hankel matrix

$$\mathcal{H}(\varphi) = (H_{i+j-1})_{i,j=1,\ldots,\infty}. \tag{2.50}$$

Since $\rho(\mathbf{H}_\varphi) = 2$, L maps Rat(2, 2, 2) into the elements (H_1, H_2, ξ_1, ξ_2) of $\mathbb{A}^{10}$ such that the matrix $\mathcal{H}(\varphi) = (H_{i+j-1})_{i,j=1,\ldots,\infty}$, $H_{n+1} = -\xi_1 H_{n-1} - \xi_2 H_n$ has rank 2. We wish to characterize these Hankel matrices in an appropriate algebraic way. In $\mathbb{P}_k^{17}$ with coordinates $(a_{00}, a_{10}, a_{20}, b_{01}, \ldots, a_{15}, a_{25})$, $\mathrm{Rat}(2,2,2)$ is the variety $V_{a_{25}}$ where V is defined by the equations

$$a_{10} = 0, \quad a_{20} = 0, \quad b_{2j} = 0, \quad j = 1,2,3,4 \tag{2.51}$$

$$\left.\begin{array}{rcl} a_{00}a_{25} & = & b_{11}b_{14} - b_{12}b_{13} \\ a_{00}a_{05} & = & b_{01}b_{04} - b_{02}b_{03} \\ a_{00}a_{15} & = & b_{01}b_{14} + b_{04}b_{11} - b_{03}b_{12} - b_{02}b_{13} \end{array}\right\} \tag{2.52}$$

which are homogeneous. Thus, $\dim \mathrm{Rat}(2,2,2) = 8$. Viewing φ as the point $(a_{00}, a_{05}, a_{15}, a_{25}, b_{01}, b_{11}, b_{02}, b_{12}, \ldots, b_{04}, b_{14})$ in $\mathbb{P}_k^{11}$, then $\mathrm{Rat}(2,2,2) = W_{a_{25}}$ where W is given by the equations (2.52). If Θ_3 is the hyperplane in $\mathbb{P}_k^{11}$ where $a_{25} = 0$, then $W_{a_{25}} \subset \mathbb{P}_k^{11} - \Theta_3$ and $\varphi \in W_{a_{25}}$ is of the form $(\xi_0, \xi_1, \xi_2, \lambda, b_{01}, b_{11}, \ldots, b_{14})$ with $\lambda \neq 0$ (eventually we take $\lambda = 1$). Consider the space $X = \mathbb{A}_k^1 \times \mathbb{P}^2 \times M(6,6)$ with coordinates $(\lambda, \xi_0, \xi_1, \xi_2, \mathbf{M}_j^i)$, $i,j = 1, \ldots, 3$, $\mathbf{M}_j^i \in M(2,2)$. Note here that $n = 2$, $(n+1)p = 6$, $(n+1)m = 6$. We define a map $\psi_L \colon \mathbb{P}_k^{11} \to X$ by setting

$$\psi_L(\varphi) = (\lambda, \xi_0, \xi_1, \xi_2, \mathbf{M}_j^i(\varphi)) \tag{2.53}$$

where the $\mathbf{M}_j^i(\varphi)$ are 2×2 $(p \times m)$ matrices with

$$\mathbf{M}_j^i(\varphi) = H_{i+j-1}(\varphi) \tag{2.54}$$

for $i, j = 1,2,3$ and

$$\begin{array}{rcl} H_1(\varphi) & = & B_1 = \begin{bmatrix} b_{12} & -b_{11} \\ b_{14} & -b_{13} \end{bmatrix} \\ H_2(\varphi) & = & B_0 - \xi_2 B_1, \quad B_0 = \begin{bmatrix} b_{02} & -b_{01} \\ b_{04} & -b_{03} \end{bmatrix} \\ H_{r+1}(\varphi) & = & -\xi_2 H_r(\varphi) - \xi_1 H_{r-1}(\varphi) \end{array}$$

for $r = 2, \ldots$. The map ψ_L is a morphism which is injective on $\mathrm{Rat}(2,2,2)$ since $F_\varphi(z) = \sum_{i=1}^{\infty} H_i(\varphi)/z^i$. We call ψ_L the *Laurent map.* Let $\mathcal{L}(3,2,2)$ $(= \mathcal{L}(n+1, m, p))$ be defined by the equations

$$(\mathbf{M}_j^i)_{rs} - (\mathbf{M}_i^j)_{rs} = 0, \quad i,j = 1, \ldots, n+1, \ r = 1, \ldots, p, \ s = 1, \ldots, m, \tag{2.55}$$

$$\begin{array}{l} (\mathbf{M}_{j+1}^i)_{rs} = (\mathbf{M}_j^{i+1})_{rs} = 0, \\ \qquad i = 1, \ldots, j-1, \ j = 2, \ldots, n, \ r = 1, \ldots, p, \ s = 1, \ldots, m, \end{array}$$

$$(\mathbf{M}^i_{n+1})_{rs} - \sum_{\ell=1}^{n} x_\ell (\mathbf{M}^i_\ell)_{rs} = 0, \quad i = 1, \dots, n+1, \; r = 1, \dots, p, \; s = 1, \dots, m \tag{2.56}$$

where here $n = 2$, $m = 2$, $p = 2$. $\mathcal{L}(3,2,2)$ has dimension $39 - 28 = 11$ $(= 1 + n + [(n+1)^2 - (n^2 + n + 1)]pm)$. Consider the Grassmann equations, where $|\ | = \det(\ )$,

$$\begin{aligned} x_0 &= |\mathbf{M}^1_1| \\ x_0 x_1 &= |\mathbf{M}^1_2 + x_2 \mathbf{M}^1_1| \\ x_0 x_2 &= -|(\mathbf{M}^1_1)_1 (\mathbf{M}^1_2)_2| + |(\mathbf{M}^1_1)_2 (\mathbf{M}^1_2)_1| \end{aligned} \tag{2.57}$$

$$\det[x_1 \mathbf{M}^1_1 + x_2 \mathbf{M}^1_2 + x_2 \mathbf{M}^1_2 z] = (x_1 + x_2 z + z^2) \cdot \lambda.$$

Let V_{Gr} be the zero set of (2.57) in X. Then $\psi_L(\mathrm{Rat}(2,2,2)) \subset \mathcal{L}(3,2,2) \cap V_{\mathrm{Gr}}$. Letting $\lambda = 1$, we have $\mathcal{L}(3,2,2) \cap V_{\mathrm{Gr}} \subset \mathbb{A}^3_k \times M_2(6,6)$ where $M_2(6,6) = \{\mathbf{M}\colon \mathbf{M}$ a 6×6 matrix of rank $\leq 2\}$. Let $\mathcal{H}(2,2,2) = \{(\boldsymbol{\xi}, (\mathbf{M}^i_j)) \in \mathcal{L}(3,2,2) \cap V_{\mathrm{Gr}}$, $\rho((\mathbf{M}^i_j)) = 2\}$. Call $\mathcal{H}(2,2,2)$ the set of Hankel matrices and write $\mathcal{H}(2,2,2) = \mathrm{Hank}(2,2,2)$. Then $\mathrm{Hank}(2,2,2)$ is a quasi-projective (normal) variety of dimension 8 and it can be shown that

$$\psi_L(\mathrm{Rat}(2,2,2)) = \mathrm{Hank}(2,2,2). \tag{2.58}$$

If $(\boldsymbol{\xi}, (\mathbf{M}^i_j))$ is an element of $\mathrm{Hank}(2,2,2)$ with $\boldsymbol{\xi} = (\xi_0, \xi_1, \xi_2)$ and if we let $H_1 = \mathbf{M}^1_1$, $H_2 = \mathbf{M}^1_2$ $(= \mathbf{M}^2_1)$, then ψ_L^{-1} is given by

$$\psi_L^{-1}(\boldsymbol{\xi}, (\mathbf{M}^i_j)) = (\varphi_0, \varphi_1, \dots, \varphi_5) = \boldsymbol{\varphi} \tag{2.59}$$

where

$$\begin{aligned} \varphi_0 &= \xi_0 = |H_1|, \quad \varphi_5 = \xi_1 + \xi_2 z + z^2 \\ \varphi_1 &= -(h^1_{2,2} + \xi_2 h^1_{1,2}) - h^1_{1,2} z \\ \varphi_2 &= (h^1_{2,1} + \xi_2 h^1_{1,1}) + h^1_{1,1} z \\ \varphi_3 &= -(h^2_{2,2} + \xi_2 h^2_{1,2}) - h^2_{1,2} z \\ \varphi_4 &= (h^2_{2,1} + \xi_2 h^2_{1,1}) + h^2_{1,1} z \end{aligned} \tag{2.60}$$

and $H_1 = (h^i_{1,j})$, $H_2 = (h^i_{2,j})$. Note that (2.57) ensures that $\varphi_0 \varphi_5 = \varphi_1 \varphi_4 - \varphi_2 \varphi_3$ so that $\boldsymbol{\varphi}$ lies in $\mathrm{Gr}(2,4)$. ψ_L^{-1} is a morphism and so ψ_L is an isomorphism between $\mathrm{Rat}(2,2,2)$ and $\mathrm{Hank}(2,2,2)$.

Example 2.61 Let $n = 2$, $m = 3$, $p = 2$ and consider $\mathrm{Rat}(2,3,2)$. We let $\boldsymbol{\psi} = (\psi_0, \psi_1, \dots, \psi_9)$ with

$$\begin{aligned} \psi_0 &= \xi_0, \quad \psi_1 = \eta_1, \quad \psi_2 = \eta_2 \\ \psi_j(z) &= b_{0j} + b_{1j} z, \quad j = 3, \dots, 8 \\ \psi_9(z) &= \xi_1 + \xi_2 z + z^2 \end{aligned} \tag{2.62}$$

be a typical element. Then

$$\begin{aligned} \psi_9(z)\mathbf{F}(z) &= (\xi_2 H_1 + H_2) + H_1 z \\ &= B_0 + H_1 z \end{aligned}$$

where

$$\begin{aligned} B_0 &= \begin{bmatrix} b_{05} & -b_{04} & b_{03} \\ b_{08} & -b_{07} & b_{06} \end{bmatrix} \\ H_1 &= \begin{bmatrix} b_{15} & -b_{14} & b_{13} \\ b_{18} & -b_{17} & b_{16} \end{bmatrix} \\ H_2 &= B_0 - \xi_2 H_1 \\ H_{\ell+1} &= -\xi_2 H_\ell - \xi_1 H_{\ell-1}, \quad \ell = 2, 3, \ldots . \end{aligned} \tag{2.63}$$

If we view ψ as the point $(\xi_0, \xi_1, \xi_2, \xi_3, \eta_1, \eta_2, b_{03}, b_{13}, \cdots, b_{08}, b_{18})$ in $\mathbb{P}^{17}$, then the Grassmann equations (i.e., those derived from the equations of $\mathrm{Gr}(3,5) \subset \mathbb{P}^9$) become

$$\left.\begin{aligned} \xi_0 b_{05} - \eta_1 b_{04} + \eta_2 b_{03} &= 0 \\ \xi_0 b_{15} - \eta_1 b_{14} + \eta_2 b_{13} &= 0 \\ \xi_0 b_{08} - \eta_1 b_{07} + \eta_2 b - 06 &= 0 \\ \xi_0 b_{18} - \eta_1 b_{17} + \eta_2 b_{16} &= 0 \end{aligned}\right\} \tag{2.64a}$$

$$\left.\begin{aligned} \xi_0\xi_3 - b_{13}b_{17} + b_{14}b_{16} &= 0 \\ \xi_0\xi_1 - b_{03}b_{07} + b_{04}b_{06} &= 0 \\ \xi_0\xi_2 - b_{07}b_{13} - b_{03}b_{17} + b_{06}b_{14} + b_{04}b_{16} &= 0 \end{aligned}\right\} \tag{2.64b}$$

$$\left.\begin{aligned} \eta_1\xi_3 - b_{13}b_{18} + b_{15}b_{16} &= 0 \\ \eta_1\xi_1 - b_{03}b_{08} + b_{05}b_{06} &= 0 \\ \eta_1\xi_2 - b_{08}b_{13} - b_{03}b_{18} + b_{05}b_{16} + b_{06}b_{15} &= 0 \end{aligned}\right\} \tag{2.64c}$$

$$\left.\begin{aligned} \eta_2\xi_3 - b_{14}b_{18} - b_{15}b_{17} &= 0 \\ \eta_2\xi_1 - b_{04}b_{08} + b_{05}b_{07} &= 0 \\ \eta_2\xi_2 - b_{08}b_{14} - b_{04}b_{18} + b_{05}b_{17} + b_{07}b_{15} &= 0 \end{aligned}\right\}. \tag{2.64d}$$

These are homogeneous quadrics. The equations (2.64) are not independent. For example, if $\xi_0 \neq 0$, then (2.64a) and (2.64b) suffice; or, if $\eta_1 \neq 0$, then (2.64a) and (2.64c) suffice; or, if $\eta_2 \neq 0$, then (2.64a) and (2.64d) suffice. We observe that, for this example, rather than having a single set of equations define $\mathrm{Rat}(2,3,2)$, we have different sets each working on an affine space with compatability on the overlap. This illustrates the projective nature of the example. Let $X = \mathbb{A}_k^3 \times \mathbb{P}_k^2 \times M(6,9)$ with coordinates $(\lambda_{12}, \lambda_{13}, \lambda_{23}, \alpha, \xi_1, \xi_2, (\mathbf{M}_j^i))$ $i = 1,2$, $j = 1,2,3$. $\dim X = 59$. Let $\mathcal{L}(3,3,2)$ be defined by the equations (2.55) and (2.56) with $n = 2$, $m = 3$, $p = 2$. $\mathcal{L}(3,3,2)$ has dimension $17 = \gamma(2,3,2) - 1$ where $\gamma(n,m,p) = \sum_{t=0}^{p}(n+1-t)\binom{p}{t}\binom{m}{t}$. Let us consider the equations

$$\begin{aligned} (\xi_1 + \xi_2 z + \alpha z^2)\lambda_{23} &= \det\left[(\xi_2\mathbf{M}_1^1 + \mathbf{M}_2^1) + \mathbf{M}_1^1 z\right|_{23}^{12}\right] \\ (\xi_1 + \xi_2 z + \alpha z^2)\lambda_{12} &= \det\left[(\xi_2\mathbf{M}_1^1 + \mathbf{M}_2^1) + \mathbf{M}_1^1 z\right|_{12}^{12}\right] \\ (\xi_1 + \xi_2 z + \alpha z^2)\lambda_{13} &= \det\left[(\xi_2\mathbf{M}_1^1 + \mathbf{M}_2^1) + \mathbf{M}_1^1 z\right|_{13}^{12}\right] \end{aligned}$$

and let V_{Gr} be the zero set of these equations. Then $\mathcal{H}(2,3,2) = \mathcal{L}(3,3,2) \cap V_{\text{Gr}}$ is a projective variety and

$$\text{Hank}(2,3,2) = \{(\boldsymbol{\lambda}, \boldsymbol{\xi}, (\mathbf{M}_j^i)) \in \mathcal{H}(2,3,2) : \rho((\mathbf{M}_j^i)_{i=1,2,j=1,2,3}) = 2\}$$

is a quasi-projective variety. Now, if $\boldsymbol{\psi} = (\psi_0, \ldots, \psi_9) \in \text{Rat}(2,3,2)$ is given by (2.62) and H_1, H_2, B_0, $H_{\ell+1}$ are given by (2.63), then

$$\begin{aligned} \psi_0 &= \xi_0 = \pi_3(H_1) = \det\left[H_1\big|_{23}^{12}\right] \\ \psi_1 &= \eta_1 = -\pi_2(H_1) = \det\left[H_1\big|_{13}^{12}\right] \\ \psi_2 &= \eta_2 = \pi_1(H_1) = \det\left[H_1\big|_{12}^{12}\right]. \end{aligned} \tag{2.65}$$

Since we may suppose $\alpha = 1$ in $\mathcal{L}(3,3,2) \cap V_{\text{Gr}}$, we can define a map ψ_L: $\text{Rat}(2,3,2) \to \text{Hank}(2,3,2)$ by

$$\psi_L(\boldsymbol{\psi}) = (\eta_2, \eta_1, \xi_0, 1, \xi_1, \xi_2, (H_{i+j-1})) \tag{2.66}$$

where ξ_0, η_1, η_2 are given by (2.65) and the H_ℓ, $\ell = 1, 2, \ldots$ by (2.63) and $\psi_9(z) = \xi_1 + \xi_2 z + z^2$. The map ψ_L is the Laurent map. We also have $\dim \text{Hank}(2,3,2) = 10 = n(m+p)$!

Example 2.67 Let $n = 2$, $m = 2$, $p = 2$ and consider $x = (A, B, C) \in \mathbb{A}_k^{12}$ $(= \mathbb{A}_k^{n^2+n(m+p)})$. Define a map $\boldsymbol{\chi}\colon \mathbb{A}_k^{12} \to \mathbb{A}_k^1 \times \mathbb{P}_k^2 \times M(6,6) = X$ by

$$\boldsymbol{\chi}(x) = \left(1, \det[CB], -\chi_2(A), -\chi_1(A), (CA^{i+j-2}B)_{i,j=1}^{n+1}\right) \tag{2.68}$$

where $A^2 - \chi_1(A)A - \chi_2(A)I = 0$. Clearly $\boldsymbol{\chi}$ is a morphism and $\boldsymbol{\chi}(x) \in \mathcal{L}(3,2,2)$ in view of the Cayley–Hamilton theorem. We *claim* that $\boldsymbol{\chi}(x) \in V_{\text{Gr}}$. First, $x_0 = \det[CB] = \det[CA^{1+1-2}B]$. Then $x_1 = -\chi_2(A) = \det A$ so that $x_0 x_1 = \det[CB](-\chi_2(A)) = \det[CB] \cdot \det A = \det C \cdot \det B \cdot \det A = \det[CAB] = \det[CAB - \chi_1(A)CB] = \det[CAB + x_2 CB]$ since $\chi_1(A) = \text{Tr } A$ and $\det[(A - \text{Tr } A \cdot I)A] = (\det A)^2$. Finally, $x_0 x_2 = \det[CB]x_2 = \det[CB](-\chi_1(A)) = \det C \cdot \det B(-\text{Tr } A)$ and $-\det[(CB)_1(CAB)_2] + \det[(CB)_2(CAB)_1] = \det C \cdot \{-\det[B_1(AB)_2] + \det[B_2(AB)_1]\} = \det C \cdot \{-\det[B_1 AB_2] + \det[B_2 AB_1]\} = \det C \cdot \det B \cdot (-\text{Tr } A)$ so that $x_0 x_2 = -\det[(CB)_1(CAB)_2] + \det[(CB)_2(CAB)_1]$. Let the controllability and observability matrices be given by

$$\mathbf{Y}(x) = \left[B \;\, AB \;\, A^2B\right], \quad \mathbf{Z}(x) = \begin{bmatrix} C \\ CA \\ CA^2 \end{bmatrix}$$

so that $\mathbf{Z}(x)\mathbf{Y}(x) = (CA^{i+j-2}B)$. If $\rho(\mathbf{Z}(x)\mathbf{Y}(x)) = 2$ then $\rho(\mathbf{Z}(x)) = 2$, $\rho(\mathbf{Y}(x)) = 2$ and $\boldsymbol{\chi}(x) \in \text{Hank}(2,2,2)$. We let $S_{2,2}^2 = \{x\colon \rho(\mathbf{Z}(x)) = 2,$

$\rho(\mathbf{Y}(x)) = 2\}$ be the set of *minimal linear systems of degree* 2. If $G = \mathrm{GL}(2,k)$, then G acts on $\mathbb{A}_k^{12}$ via

$$g \cdot x = (gAg^{-1}, gB, Cg^{-1}) \tag{2.69}$$

and $\chi(g \cdot x) = \chi(x)$ is invariant. We shall show that $(\mathrm{Hank}(2,2,2), \chi)$ is a geometric quotient of $S_{2,2}^2$ modulo G. Our approach will involve an imbedding of $S_{2,2}^2$ into $M_*(6,2) \times M_*(2,6) \times \mathbb{A}_k^3$ (here, $= M_*((n+1)p, n) \times M_*(n,(n+1)m) \times \mathbb{A}_k^{n+1}$). Consider the map ψ given by

$$\psi(A,B,C) = (\tilde{\mathbf{Z}}(A,B,C), \tilde{\mathbf{Y}}(A,B,C), \det[CB], -\chi_2(A), -\chi_1(A)) \tag{2.70}$$

where

$$\begin{aligned} \tilde{\mathbf{Y}}(A,B,C) &= \begin{bmatrix} B_1 & AB_1 & A^2B_1 & B_2 & AB_2 & A^2B_2 \end{bmatrix} \\ \tilde{\mathbf{Z}}(A,B,C) &= \begin{bmatrix} C^1 \\ C^1A \\ C^1A^2 \\ C^2 \\ C^2A \\ C^2A^2 \end{bmatrix}. \end{aligned} \tag{2.71}$$

Set $\tilde{\mathbf{H}}(A,B,C) = \tilde{\mathbf{Z}}(A,B,C)\tilde{\mathbf{Y}}(A,B,C)$. If $\mathbf{x} = (\tilde{\mathbf{Z}}, \tilde{\mathbf{Y}}, \xi_0, \xi_1, \xi_2) \in M_*(6,2) \times M_*(2,6) \times \mathbb{A}_k^3 = \mathbb{A}_k^{27}$ and if $\tilde{\mathbf{H}}(x) = \tilde{\mathbf{Z}}\tilde{\mathbf{Y}}$, then

$$\tilde{\mathbf{H}}(x) = (\tilde{\mathbf{Z}}^i \tilde{\mathbf{Y}}_j), \quad i = 1, \ldots, 6, \; j = 1, \ldots, 6$$

is given by

$$\begin{bmatrix} Z^1Y_1 & Z^1Y_2 & Z^1Y_3 & Z^1Y_4 & Z^1Y_5 & Z^1Y_6 \\ Z^2Y_1 & Z^2Y_2 & Z^2Y_3 & Z^2Y_4 & Z^2Y_5 & Z^2Y_6 \\ Z^3Y_1 & Z^3Y_2 & Z^3Y_3 & Z^3Y_4 & Z^3Y_5 & Z^3Y_6 \\ Z^4Y_1 & Z^4Y_2 & Z^4Y_3 & Z^4Y_4 & Z^4Y_5 & Z^4Y_6 \\ Z^5Y_1 & Z^5Y_2 & Z^5Y_3 & Z^5Y_4 & Z^5Y_5 & Z^5Y_6 \\ Z^6Y_1 & Z^6Y_2 & Z^6Y_3 & Z^6Y_4 & Z^6Y_5 & Z^6Y_6 \end{bmatrix}.$$

Suppose that $\mathbf{x}$ satisfies the "Hankel" equations and the "Grassmann" equations i.e.,

$$\begin{array}{ll} Z^3 = -\xi_2 Z^2 - \xi_1 Z^1 & Y_3 = -\xi_2 Y_2 - \xi_1 Y_1 \\ Z^6 = -\xi_2 Z^5 - \xi_1 Z^4 & Y_6 = -\xi_2 Y_5 - \xi_1 Y_4 \\ Z^1Y_2 = Z^2Y_1 & Z^1Y_5 = Z^5Y_4 \\ Z^4Y_2 = Z^5Y_1 & Z^4Y_5 = Z^5Y_4 \\ Z^1Y_3 = Z^2Y_2 & Z^1Y_6 = Z^2Y_5 \\ Z^4Y_3 = Z^5Y_2 & Z^4Y_6 = Z^5Y_5 \end{array} \tag{2.72}$$

$$\begin{aligned}
\xi_0 &= \det \begin{bmatrix} Z^1Y_1 & Z^1Y_4 \\ Z^4Y_1 & Z^4Y_4 \end{bmatrix} \\
\xi_0\xi_1 &= \det \left\{ \begin{bmatrix} Z^2Y_1 & Z^2Y_4 \\ Z^5Y_1 & Z^5Y_4 \end{bmatrix} + \xi_2 \begin{bmatrix} Z^1Y_1 & Z^1Y_4 \\ Z^4Y_1 & Z^4Y_4 \end{bmatrix} \right\} \\
\xi_0\xi_2 &= -\det \begin{bmatrix} Z^1Y_1 & Z^2Y_4 \\ Z^4Y_1 & Z^5Y_4 \end{bmatrix} + \det \begin{bmatrix} Z^1Y_4 & Z^2Y_1 \\ Z^4Y_4 & Z^5Y_1 \end{bmatrix}.
\end{aligned}$$

If $\mathbf{V}$ is the quasi-affine variety given by these equations (2.72) and the conditions $\rho(\tilde{\mathbf{Z}}) = 2$, $\rho(\tilde{\mathbf{Y}}) = 2$, then $\psi(A, B, C) \in \mathbf{V}$ for $(A, B, C) \in S_{2,2}^2$. Clearly ψ is a morphism. G acts on $\mathbf{V}$ by

$$g \cdot \mathbf{x} = (\tilde{\mathbf{Z}}g^{-1}, g\tilde{\mathbf{Y}}, \xi_0, \xi_1, \xi_2) \tag{2.73}$$

and ψ is a G-morphism (i.e., commutes with the action of G). We *claim* that $\psi\colon S_{2,2}^2 \to \mathbf{V}$ is injective. If $\psi(A, B, C) = \psi(A_1, B_1, C_1)$, then clearly $B = B_1$ and $C = C_1$. Moreover, if $v_1, \dots, v_n$ is a "basis" of $\tilde{\mathbf{Y}}$, then $v_1, \dots, v_n$ is also a "basis" of $\tilde{\mathbf{Y}}_1$ (here $n = 2$). It follows that $Av_i = A_1v_i$ and hence that $A = A_1$. We next *claim* that ψ is surjective. If $\tilde{\mathbf{Z}}\tilde{\mathbf{Y}} = \tilde{\mathbf{Z}}_1\tilde{\mathbf{Y}}_1$, then there is a $g \in G$ with $\tilde{\mathbf{Y}}_1 = g\tilde{\mathbf{Y}}$, $\tilde{\mathbf{Z}}_1 = \tilde{\mathbf{Z}}g^{-1}$. For instance, if $\tilde{\mathbf{Z}}^\alpha$, $\alpha = (\alpha_1, \alpha_2)$, are the first 2 independent rows of $\tilde{\mathbf{Z}}$ and if $\tilde{\mathbf{Y}}_\beta$, $\beta = (\beta_1, \beta_2)$ are the first 2 independent columns of $\tilde{\mathbf{Y}}$, then $\tilde{\mathbf{Z}}^\alpha\tilde{\mathbf{Y}}_\beta = \tilde{\mathbf{Z}}_1^\alpha\tilde{\mathbf{Y}}_{1\beta}$ and all are nonsingular. Then, $g = (\tilde{\mathbf{Z}}_1^\alpha)^{-1}(\tilde{\mathbf{Z}}^\alpha) = \tilde{\mathbf{Y}}_{1\beta}(\tilde{\mathbf{Y}}_\beta)^{-1}$ does the job. Let $\mathbf{x} \in \mathbf{V}$ so that $\rho(\tilde{\mathbf{H}}(\mathbf{x})) = 2$. Let $H(\mathbf{x}) = \mathbf{Z}\mathbf{Y}$ (extended)

$$\begin{aligned}
\mathbf{Y} &= \begin{bmatrix} Y_1 & Y_4 & Y_2 & Y_5 & Y_3 & Y_6 \end{bmatrix} \\
\mathbf{Z} &= \begin{bmatrix} Z^1 \\ Z^4 \\ Z^2 \\ Z^5 \\ Z^3 \\ Z^6 \end{bmatrix}
\end{aligned}$$

so that $\rho(H(\mathbf{x})) = 2$ ($H(\mathbf{x})$ is the "standard Hankel"). Let $\mathbf{e} = (e^1, e^2)$ be the first 2 independent rows of $H(\mathbf{x})$ and let

$$H^{\mathbf{e}} = \begin{bmatrix} H^{e^1} \\ H^{e^2} \end{bmatrix}.$$

Then there is a C_1 with

$$C_1 H^{\mathbf{e}} = \begin{bmatrix} H^1 \\ H^2 \end{bmatrix}.$$

Let B_1 be the 2×2 matrix consisting of the first 2 columns of $H^{\mathbf{e}}$. Note that

$$H(\mathbf{x}) = \begin{bmatrix} H_1 & H_2 & H_3 & \cdots \\ H_2 & H_3 & H_4 & \cdots \\ H_3 & H_4 & H_5 & \cdots \\ \vdots & \vdots & \vdots & \end{bmatrix}$$

where the H_i are 2×2, by virtue of the equations defining $\mathbf{V}$. Then $C_1 B_1 = H_1$. There is a (unique) 2×2 A_1 with

$$A_1 H^{\mathbf{e}} = \begin{bmatrix} H^{e^1+2} \\ H^{e^2+2} \end{bmatrix}$$

(as $1 \leq e^1 \leq 2 \leq e^2 \leq 4$) since H^{i+2} is a subrow of H^i for any i. Then $C_1 A_1 B_1 = H_2$, $C_1 A_1^2 B_1 = H_3, \ldots$ so that $H(\mathbf{x}) = \mathbf{Z}(A_1, B_1, C_1)\mathbf{Y}(A_1, B_1, C_1)$. It follows that $(A_1, B_1, C_1) \in S_{2,2}^2$ and that $\mathbf{Z}(A_1, B_1, C_1)\mathbf{Y}(A_1, B_1, C_1) = \mathbf{ZY}$. So there is a $g \in G$ with $\mathbf{Z}g^{-1} = \mathbf{Z}(A_1, B_1, C_1)$, $g\mathbf{Y} = \mathbf{Y}(A_1, B_1, C_1)$, $\xi_0 = \det[C_1 B_1]$, $\xi_1 = -\chi_2(A_1)$, $\xi_2 = -\chi_1(A_1)$. Let $\mathbf{C}^1 = \mathbf{Z}(A_1, B_1, C_1)^1 g$, $\mathbf{C}^2 = \mathbf{Z}(A_1, B_1, C_1)^4 g$, $\mathbf{B}_1 = g^{-1}\mathbf{Y}(A_1, B_1, C_1)_1$, $\mathbf{B}_2 = g^{-1}\mathbf{Y}(A_1, B_1, C_1)_4$, $\mathbf{A} = g^{-1} A_1 g$. Then $\psi(\mathbf{A}, \mathbf{B}, \mathbf{C}) = \mathbf{x}$. We *claim* that the bijective morphism ψ of $S_{2,2}^2$ onto $\mathbf{V}$ is an isomorphism. Consider the map $\varphi \colon \mathbf{V} \to S_{2,2}^2$ given by $\varphi(\mathbf{x}) = (\mathbf{A}(\mathbf{x}), \mathbf{B}(\mathbf{x}), \mathbf{C}(\mathbf{x}))$ where $\mathbf{A}$, $\mathbf{B}$, $\mathbf{C}$ are constructed as in the previous argument. Clearly B_1 is a regular function of $\mathbf{x}$. If H_β^α is a nonsingular 2×2 submatrix of $H^{\mathbf{e}}$, then

$$C_1 = (H_\beta^\alpha)^{-1} \begin{bmatrix} H^1 \\ H^2 \end{bmatrix}_\beta, \quad A_1 = (H_\beta^\alpha)^{-1} \begin{bmatrix} H^{e^1+2} \\ H^{e^2+2} \end{bmatrix}_\beta$$

are also regular. Since g is of the form $(\mathbf{Z}(A_1, B_1, C_1)^\lambda)^{-1}\mathbf{Z}^\lambda$, we have φ as a morphism and $\psi \circ \varphi = i$, $\varphi \circ \psi = i$. Thus $S_{2,2}^2$ is G-isomorphic to $\mathbf{V}$ and so, $S_{2,2}^2/G$ "=" $\mathbf{V}/G$. There is a natural morphism (abuse of notation) $\chi \colon \mathbf{V} \to$ Hank(2, 2, 2) with

$$\chi(\mathbf{Z}, \mathbf{Y}, \xi_0, \xi_1, \xi_2) = (\mathbf{ZY}, \xi_0, \xi_1, \xi_2) \tag{2.74}$$

such that (i) $\chi(g \cdot \mathbf{x}) = \chi(\mathbf{x})$ i.e., χ is invariant; (ii) χ is injective on G-orbits; (iii) χ is surjective; and (iv) $\chi^{-1}(H, \xi_0, \xi_1, \xi_2)$ is an orbit $O_G(\mathbf{x})$. Moreover,

Proposition 2.75 *If* $\mathbf{x} \in \mathbf{V}$, *then* $\dim O_G(\mathbf{x}) = 4$ $(= n^2)$.

Proof. Since $O_G(\mathbf{x}) \simeq G/S_G(\mathbf{x})$ where $S_G(\mathbf{x}) = \{g \in G \colon g \cdot \mathbf{x} = \mathbf{x}\}$, it follows that, for $g \in S_G(\mathbf{x})$, $Y(\mathbf{x}) = Y(g \cdot \mathbf{x}) = g\mathbf{Y}(\mathbf{x})$. If $\beta = (\beta_1, \beta_2)$ are the first

2 independent columns of $\mathbf{Y}(\mathbf{x})$, then $\mathbf{Y}_\beta(\mathbf{x}) = g\mathbf{Y}_\beta(\mathbf{x})$ and $g = I$. In other words, $O_G(\mathbf{x}) \simeq G$.

In view of the Proposition, $\chi^{-1}(H, \xi_0, \xi_1, \xi_2)$ is a *closed* orbit. Since $S_{2,2}^2$ is open in $\mathbb{A}_k^{12}$, $S_{2,2}^2$ is irreducible and so is $\mathbf{V}$. Thus, $\chi(\mathbf{V}) = \text{Hank}(2,2,2)$ is also irreducible. Let $\mathcal{H} = \{\mathcal{H}(\mathbf{H})_{i,j=1}^{\infty}\colon \mathcal{H}(\mathbf{H})_j^i = H_{i+j-1},\ \mathbf{H} = [H_1, \ldots],\ H_i$ are 2×2, $\rho(\mathcal{H}(\mathbf{H})) = 2\}$. Then, by the argument of Risannen (Lemma 1.98), $\mathcal{H}(\mathbf{H})$ can be identified with the elements of $M(6,6)$ lying in the open set $M_{*2}(6,6)$ of the linear variety defined by the symmetries of $\mathcal{H}(\mathbf{H})$. In other words, we see that Hank(2, 2, 2) is normal. Hence, χ is open on invariant sets as $4 = \dim O_G(\mathbf{x}) = \dim \chi^{-1}(H(\mathbf{x}), \xi_0(\mathbf{x}), \xi_1(\mathbf{x}), \xi_2(\mathbf{x})) = \dim \mathbf{V} - \dim \text{Hank}(2,2,2) = \dim S_{2,2}^2 - \dim \text{Hank}(2,2,2) = 4 + 2(2+2) - 2(2+2)$ $(= n^2 + n(m+p) - n(m+p))$. So to complete the proof that $(\text{Hank}(2,2,2), \chi)$ is a geometric quotient of $S_{2,2}^2$ modulo G, we must show that $k[\mathbf{V}]^G = k[\text{Hank}(2,2,2)]$ or, more precisely, that $k[\text{Hank}(2,2,2)] \simeq \chi^*(k[\text{Hank}(2,2,2)]) = k[\mathbf{V}]^G$. Now, we have not really defined these rings since the various varieties are obtained by piecing together local patches. Given a point $\boldsymbol{\xi}$ (in either Hank(2, 2, 2) or $\mathbf{V}$), there is an open affine variety $X_{\boldsymbol{\xi}}$ ($\subset \mathbf{V}$ or $\subset \text{Hank}(2,2,2)$) with $\boldsymbol{\xi} \in X_{\boldsymbol{\xi}}$ and we can consider the ring $k[X_{\boldsymbol{\xi}}]$. Provisionally then, $k[\text{Hank}(2,2,2)] = \{f\colon f \mid X_{\boldsymbol{\xi}} \in k[X_{\boldsymbol{\xi}}]$ all $\boldsymbol{\xi}\}$ and a map $\varphi\colon \mathbf{V} \to \text{Hank}(2,2,2)$ is a *morphism* if $\text{Hank}(2,2,2) = \bigcup W_i$, W_i open affine and there are open affine U_i with $\mathbf{V} = \bigcup U_i$, $\varphi(U_i) \subset W_i$ and $\varphi^*(k[W_i]) \subset k[U_i]$. Let $\boldsymbol{\alpha} = (\alpha^1, \alpha^2)$, $\boldsymbol{\beta} = (\beta_1, \beta_2)$ be row and column indices, respectively; then, $\mathcal{H}_{\boldsymbol{\beta}}^{\boldsymbol{\alpha}} = \{(H, \xi_0, \xi_1, \xi_2)\colon \det H_{\boldsymbol{\beta}}^{\boldsymbol{\alpha}} \neq 0\}$ form an affine open cover of Hank(2, 2, 2). If $U_{\boldsymbol{\beta}}^{\boldsymbol{\alpha}} \subset V$ with $U_{\boldsymbol{\beta}}^{\boldsymbol{\alpha}} = \{(\mathbf{Z}, \mathbf{Y}, \boldsymbol{\xi})\colon \det \mathbf{Z}^{\boldsymbol{\alpha}} \neq 0,\ \det \mathbf{Y}_{\boldsymbol{\beta}} \neq 0\}$, then the $U_{\boldsymbol{\beta}}^{\boldsymbol{\alpha}}$ are an affine open cover of $\mathbf{V}$ with $\chi(U_{\boldsymbol{\beta}}^{\boldsymbol{\alpha}}) \subset \mathcal{H}_{\boldsymbol{\beta}}^{\boldsymbol{\alpha}}$. Since $G \cdot U_{\boldsymbol{\beta}}^{\boldsymbol{\alpha}} \subset U_{\boldsymbol{\beta}}^{\boldsymbol{\alpha}}$, the $U_{\boldsymbol{\beta}}^{\boldsymbol{\alpha}}$ are G-invariant. We shall show that $k[U_{\boldsymbol{\beta}}^{\boldsymbol{\alpha}}]^G = k[\mathcal{H}_{\boldsymbol{\beta}}^{\boldsymbol{\alpha}}]$ i.e., $\chi^*(k[\mathcal{H}_{\boldsymbol{\beta}}^{\boldsymbol{\alpha}}])$. Since χ is invariant, it is clear that $k[\mathcal{H}_{\boldsymbol{\beta}}^{\boldsymbol{\alpha}}] \subset k[U_{\boldsymbol{\beta}}^{\boldsymbol{\alpha}}]^G$. We use the "same" classical argument of Theorem I.19.47 to show that $k[U_{\boldsymbol{\beta}}^{\boldsymbol{\alpha}}]^G \subset k[\mathcal{H}_{\boldsymbol{\beta}}^{\boldsymbol{\alpha}}]$. Let $W_{\boldsymbol{\beta}}^{\boldsymbol{\alpha}} \subset U_{\boldsymbol{\beta}}^{\boldsymbol{\alpha}}$ be the set of w with

$$w = \left(\begin{bmatrix} I^{\boldsymbol{\alpha}} \\ \hat{\mathbf{Z}}^{\boldsymbol{\alpha}} \end{bmatrix}, \mathbf{Y}, \boldsymbol{\xi} \right) \tag{2.76}$$

(here $I^{\boldsymbol{\alpha}}$ is in rows (α^1, α^2) and $\hat{\mathbf{Z}}^{\boldsymbol{\alpha}}$ is the rest). Consider $G \times W_{\boldsymbol{\beta}}^{\boldsymbol{\alpha}}$ and let G act on $G \times W_{\boldsymbol{\beta}}^{\boldsymbol{\alpha}}$ via $g \cdot (g_1, w) = (gg_1, w)$.

Claim 2.77 $k[W_{\boldsymbol{\beta}}^{\boldsymbol{\alpha}}] = k[G \times W_{\boldsymbol{\beta}}^{\boldsymbol{\alpha}}]^G$.

Proof. Let $f = \sum r_i \otimes s_i \in k[G] \otimes_k k[W_{\boldsymbol{\beta}}^{\boldsymbol{\alpha}}]$ so that $f(g, w) = \sum r_i(g) s_i(w)$. If f is invariant, then $f(g, w) = f(g \cdot I, w) = f(I, w) = \sum r_i(I) s_i(w)$ so that $f = \sum r_i(I) s_i \in k[W_{\boldsymbol{\beta}}^{\boldsymbol{\alpha}}]$.

Claim 2.78 *If* $\gamma\colon G \times W_{\boldsymbol{\beta}}^{\boldsymbol{\alpha}} \to U_{\boldsymbol{\beta}}^{\boldsymbol{\alpha}}$ *is given by* $\gamma(g, w) = g \cdot w$, *then* γ *is a* G*-isomorphism and hence* $k[U_{\boldsymbol{\beta}}^{\boldsymbol{\alpha}}]^G = k[G \times W_{\boldsymbol{\beta}}^{\boldsymbol{\alpha}}]^G = k[W_{\boldsymbol{\beta}}^{\boldsymbol{\alpha}}]$.

Proof. Clearly $\gamma(g_1,(g\cdot w)) = \gamma(g_1 g, w) = g_1 g\cdot w = g_1(g\cdot w) = g_1\gamma(g,w)$. Define $\gamma^{-1}\colon U_\beta^\alpha \to G\times W_\beta^\alpha$ by setting

$$\gamma^{-1}(\mathbf{Z},\mathbf{Y},\boldsymbol{\xi}) = \left((\mathbf{Z}^\alpha)^{-1}, \begin{pmatrix} I^\alpha \\ \hat{\mathbf{Z}}^\alpha (Z^\alpha)^{-1}\end{pmatrix}, (\mathbf{Z}^\alpha)\mathbf{Y}, \boldsymbol{\xi}\right) \tag{2.79}$$

(where I^α is in rows (α^1,α^2) and $\hat{\mathbf{Z}}^\alpha$ is the rest of $\mathbf{Z}$). Then $\gamma\circ\gamma^{-1} = i$ and $\gamma^{-1}\circ\gamma = i$.

Thus, if $f\in K[U_\beta^\alpha]^G$, then $f\in k[\hat{\mathbf{Z}}^\alpha(\mathbf{Z}^\alpha)^{-1}, \mathbf{Z}^\alpha\mathbf{Y}, \boldsymbol{\xi}] = k[\hat{\mathbf{Z}}^\alpha(\mathbf{Z}^\alpha)^{-1}, \mathbf{Z}^\alpha\hat{\mathbf{Y}}_\beta, \mathbf{Z}^\alpha\mathbf{Y}_\beta, \boldsymbol{\xi}]$ (where $\hat{\mathbf{Y}}_\beta$ is the rest of $\mathbf{Y}$ less $\mathbf{Y}_\beta$). It is therefore enough to show that $\hat{\mathbf{Z}}^\alpha(\mathbf{Z}^\alpha)^{-1}\in k[\mathcal{H}_\beta^\alpha]$. But $\hat{\mathbf{Z}}^\alpha(\mathbf{Z}^\alpha)^{-1} = \hat{\mathbf{Z}}^\alpha\mathbf{Y}_\beta(\mathbf{Z}^\alpha\mathbf{Y}_\beta)^{-1} = \hat{\mathbf{Z}}^\alpha\mathbf{Y}_\beta\mathrm{Adj}(\mathbf{Z}^\alpha\mathbf{Y}_\beta)/\det(\mathbf{Z}^\alpha\mathbf{Y}_\beta)$ which is in $k[\mathcal{H}_\beta^\alpha]$.

The various examples show that the scalar theory does not extend in a simple, straightforward way. The development of the theory for multi-input, multi-output systems is the subject of the sequel.

Exercises

(1) Show that the set $\{U(z)\colon U(z)$ a 2×2 matrix with $\det U(z) = a\in k^*\}$ forms a group (the *unimodular group*) and that $\mathrm{SO}(2,k[z])$ is a subgroup.

(2) Show that any $\mathbf{Q}(z)$ with $\det\mathbf{Q}(z)\not\equiv 0$ can be made column proper by applying elementary column operations. Show that if $\mathbf{Q}(z) = \Delta_c(\mathbf{Q})\,\mathrm{diag}[z^{r_1}\ z^{r_2}]$, then there is a $U\in\mathrm{SO}(2,k[z])$ with $\mathbf{Q}(z)U = \Delta_c(\mathbf{Q})U\cdot\mathrm{diag}[z^{r_2}\ z^{r_1}]$.

(3) Prove that the map $\mathbf{F}(z)\to\psi_{\mathbf{F}}$ of Example 2.24 is injective.

(4) Establish (2.38). [*Hint*: Apply $\mathbf{Q}^{-1}\mathbf{Q}$ to $\mathbf{M}_{\mathbf{F}}$ and compute.]

(5) Show that if $\varphi\in\mathrm{Rat}(2,2,2)$, then rank $\mathbf{H}_\varphi = 2$. [*Hint*: First show that $\rho(\mathbf{H}_\varphi) =$ rank of the matrix

$$\begin{bmatrix} H_1 & B_0 \\ B_0 & -a_{05}H_1 + a_{15}B_0\end{bmatrix}$$

with rows

$$\begin{aligned} v^1 &= (b_{12}\ \ -b_{11}\ b_{02}\ \ -b_{01}),\\ v^2 &= (b_{14}\ \ -b_{13}\ b_{04}\ \ -b_{03}),\\ v^3 &= (b_{02}\ \ -b_{01}\ a_{15}b_{01}-a_{05}b_{12}\ \ -a_{15}b_{01}+a_{05}b_{11}),\\ v^4 &= (b_{04}\ \ -b_{03}\ a_{15}b_{04}-a_{05}b_{14}\ \ -a_{15}b_{03}+a_{05}b_{13}).\end{aligned}$$

Treat cases. *case* (a): $v^1 = 0$ so $v^3 = 0$ but $\mathbf{F}_\varphi \not\equiv 0$ gives $v^2 \neq 0$, then get a contradiction to the relative primeness of the φ_i if $v^4 = \alpha v^2$; *case* (b): $v^1 \neq 0$ and $v^2 \in \text{span}\{V^1\}$ i.e., $v_2 = \alpha v^1$, then $v^4 \in \text{span}\{v^2, v^3\}$ and if $\rho = 1$, then all in $\text{span}\{v^1\}$ and again get a contradiction to the relative primeness of the φ_i; *case* (c): v^1, v^2 linearly independent so that $\rho \geq 2$, then show that $v^3 \in \text{span}\{v^1, v^2\}$ and similarly for v^4, by considering all 3×3 minors of the 3×4 matrix $(v^i)_{i=1,2,3}$ (for this the Grassmann equations (2.45) are crucial).]

(7) Show that if $\zeta_0 = 0$, $\eta_1 = 0$, $\eta_2 = 0$, then 7 of the 9 equations (2.64b), (2.64c), (2.64d) suffice and that $\rho(H_1) \leq 1$, $\rho(B_0) \leq 1$.

(8) Show that (2.66) does define a map of Rat(2, 3, 2) into Hank(2, 3, 2). Find ψ_L^{-1}.

(9) Show that for 2×2 matrices,

$$\det[B_2 \; AB_1] - \det[B_1 \; AB_2] = \det[B_1 \; B_2](-\text{Tr}\, A)$$

(i.e., $B = [B_1 \; B_2]$ is 2×2 with columns B_1, B_2).

(10) Consider the matrix

$$\mathbf{H} = \begin{bmatrix} H_1 & H_2 \\ H_2 & \xi_1 H_1 - \xi_2 H_2 \end{bmatrix}$$

where the H_i are 2×2. Show that if

$$\begin{aligned} \xi_0 &= |H_1|, \quad \xi_0 \xi_1 = |H_2 + \xi_2 H_1| \\ \xi_0 \xi_2 &= -|(H_1)_1 \; (H_2)_2| + |(H_1)_2 \; (H_2)_1| \end{aligned} \qquad (*)$$

then $\rho(H) \leq 2$. [*Hint*: Let $\mathbf{H}^i$, $i = 1, 2, 3, 4$ be the rows of $\mathbf{H}$. If $\mathbf{H}^1 = 0$, then $\mathbf{H}^3 = 0$ and $\rho \leq 2$. If $\mathbf{H}^1 \neq 0$ and $\mathbf{H}^2 = a\mathbf{H}^1$, the $\mathbf{H}^4 \in \text{span}[\mathbf{H}^2, \mathbf{H}^3] \subset \text{span}[\mathbf{H}^1, \mathbf{H}^3]$ and $\rho \leq 2$. So we need only show all 3×3 minors of $(\mathbf{H}^i)_{I=1,2,3}$, $(\mathbf{H}^i)_{i=1,2,4}$ are 0. Note that $(*)$ gives $\xi_0 \xi_1 = |H_2|$. Expand the requisite minors using the identity $|v \; w|x - |v \; x|w - |x \; w|v = 0$ for 2-vectors.]

3
The Transfer and Hankel Matrices

We begin our study of multi-variable linear systems with an analysis of the "transfer matrix" representation along the lines of the representation (7) of Chapter 1.

Let A be a ring and let $A[[x]]$ be the ring of formal power series over A. Since $A[[x]]$ is an A-module, we have $A \otimes_A A[[x]] = A[[x]]$. If $\mathbf{M}$ is an A (or $A[x]$ or $A[[x]]$) module, then $A[[x]] \otimes_A \mathbf{M} = \mathbf{M}[[x]]$ is the module of *formal power series with coefficients in* $\mathbf{M}$. An element $\mathbf{m}x^j$, $\mathbf{m} \neq 0$, is homogeneous of degree j and we write $\sum_{j=0}^{\infty} \mathbf{m}_j x^j$ for a typical element u of $\mathbf{M}[[x]]$. The first index r with $\mathbf{m}_r \neq 0$ is called the *order* $\nu(u)$ of u, and $\mathbf{m}_r x^r$ is the *initial form of* u. Note that $\nu(u+v) \geq \min(\nu(u), \nu(v))$. We let $\mathbf{M}_+[[x]]$ be the set of elements of positive order including 0.

Definition 3.1 Let z^{-1} be an indeterminate over A. An element $\mathbf{F}(z^{-1})$ of $\mathbf{M}_+[[z^{-1}]]$ is called a *strictly proper meromorphic function on* $\mathbf{M}$.

If $\mathbf{F}(z^{-1})$ is a strictly proper meromorphic function on $\mathbf{M}$, then $\mathbf{F}(z^{-1}) = \sum_{j=1}^{\infty} \mathbf{m}_j z^{-j}$ with $\mathbf{m}_j \in \mathbf{M}$, and, by abuse of notation, we write $\mathbf{F}(z) = \sum_{j=1}^{\infty} \mathbf{m}_j z^{-j}$ and we call $\sum_{j=1}^{\infty} \mathbf{m}_j z^{-j}$ the *Laurent series of* $\mathbf{F}(z)$.

Now let A be an integral domain and set $\hat{S} = A[[1/z]]$, $R = A[z]$, and let $\hat{S}_+ = (1/z)\hat{S}$ $(= A_+[[1/z]])$ be the ideal generated by $1/z$ in $\hat{S}$. The set $\{1, 1/z, 1/z^2, \dots\}$ is a multiplicative set in $\hat{S}$ and $\hat{S}_{1/z} = \hat{S}_{z^{-1}} = A[[1/z]]_{(1/z)}$. We write $A((1/z)) = A[[1/z]]_{(1/z)}$. If $A = k$ is a field, then $k((1/z))$ is the quotient field of $k[[1/z]]$ (Corollary 1.2).

P. Falb, *Methods of Algebraic Geometry in Control Theory: Part II*,
Modern Birkhäuser Classics, https://doi.org/10.1007/978-3-319-96574-1_3

Proposition 3.2 *$\hat{S}_{z^{-1}} = R \oplus \hat{S}_+$ as A-modules.*

Proof. Exactly the same as Proposition 1.3.

Since R is a subring of $\hat{S}_{z^{-1}}$, $\hat{S}_{z^{-1}}$ is an R-module.

Proposition 3.3 *The sequence*

$$0 \longrightarrow R \xrightarrow{i} \hat{S}_{z^{-1}} \xrightarrow{\pi} \hat{S}_{z^{-1}}/R \longrightarrow 0 \tag{3.4}$$

is an exact sequence of R-modules (where i is the natural injection and π the natural projection). If $\hat{S}_+$ is given an R-module structure by setting

$$z \cdot \sum_{j=1}^{\infty} a_j z^{-j} = \sum_{j=1}^{\infty} a_{j+1} z^{-j} \tag{3.5}$$

(left-shift), then $\hat{S}_+$ and $\hat{S}_{z^{-1}}/R$ are isomorphic R-modules.

Proof. Exercise.

Corollary 3.6 $0 \to R \to R \oplus \hat{S}_+ \to \hat{S}_+ \to 0$ *is an exact sequence of R-modules (with the left-shift structure).*

Corollary 3.7 *If* $\mathbf{M}$ *is an A-module, then*

$$0 \to \mathbf{M}[z] \to \mathbf{M}((1/z)) \to \mathbf{M}_+[[1/z]] \to 0 \tag{3.8}$$

is an exact sequence of R-modules.

Proof. Since $\mathbf{M}[z] = R \otimes_A M$, $\mathbf{M}((1/z)) = \hat{S}_{z^{-1}} \otimes_A \mathbf{M}$, and $\mathbf{M}_+[[1/z]] = \hat{S}_+ \otimes_A \mathbf{M}$, and since $(R \oplus \hat{S}_+) \otimes_A \mathbf{M} \simeq (R \otimes_A \mathbf{M}) \oplus (\hat{S}_+ \otimes_A \mathbf{M})$ (a property of the tensor product), the sequence

$$0 \to R \otimes_A \mathbf{M} \to (R \oplus \hat{S}_+) \otimes_A \mathbf{M} \to \hat{S}_+ \otimes_A \mathbf{M} \to 0$$

is exact.

We now make the following:

Definition 3.9 If $\mathbf{M}$ is an A-module, let $\Omega_c(\mathbf{M}) = \{\omega\colon \mathbb{Z} \to \mathbf{M}$ with $\omega(n) = 0$, $n \leq -N_\omega$, $N_\omega \geq 0\}$. If $\omega \in \Omega_c(M)$, then the *transform of* ω, $\hat{\omega}(z)$, is the element of $\mathbf{M}((1/z)) = \hat{S}_{z^{-1}} \otimes_A \mathbf{M}$ given by

$$\hat{\omega}(z) = \sum_{n=-\infty}^{\infty} \omega(n) z^{-n} = z^{N_\omega} \left(\sum_{j=0}^{\infty} \omega(j - N_\omega) \right) z^{-j} \tag{3.10}$$

(see [F-3] and cf. 1.60).

Just as before, the left-shift corresponds to multiplication by z.

Proposition 3.11 *Let* $\mathbf{M}, \mathbf{N}$ *be A-modules. If* $H = \{\mathbf{H}_j\} \in \Omega_c(\mathrm{Hom}_A(\mathbf{M}, \mathbf{N}))$ *and* $u = \{u_j\} \in \Omega_c(\mathbf{M})$, *then*

$$y = \{y_n\} = \{\mathbf{H}_j\} * \{u_j\}$$

where

$$y_n = \sum_{k=-\infty}^{\infty} \mathbf{H}_k u_{n-k} \quad \left(= \sum_{k=-\infty}^{\infty} \mathbf{H}_k(u_{n-k}) \right)$$

is a well-defined element of $\Omega_c(\mathbf{N})$ *and*

$$\hat{y}(z) = \hat{H}(z)\hat{u}(z)$$

where $\hat{H}(z) \in \mathrm{Hom}_A(\mathbf{M}, \mathbf{N})((1/z))$ *and* $\hat{u}(z) \in \mathbf{M}((1/z))$.

Proof. As in earlier proofs (e.g., Proposition 1.10)

$$y_n = \sum_{k=-N_H+1}^{n+N_u} \mathbf{H}_k u_{n-k}$$

is a *finite* sum and $y_n = 0$ for $n \leq -(N_H + N_u)$. If

$$\hat{H}(z) = z^{N_H} \left(\sum_{j=0}^{\infty} \mathbf{H}_{j-N_H} z^{-j} \right), \quad \hat{u}(z) = z^{N_u} \left(\sum_{j=0}^{\infty} u_{j-N_u} z^{-j} \right),$$

then $\hat{H}(z)\hat{u}(z) = z^{N_H+N_u} \left(\sum_{n=0}^{\infty} \alpha_n z^{-n} \right)$ where $\alpha_n = \sum_{i+j=n} \mathbf{H}_{i-N_H}(u_{j-N_u}) \in \mathbf{N}$ as $\mathbf{H}_j \in \mathrm{Hom}_A(\mathbf{M}, \mathbf{N})$.

Proposition 3.12 *Let* $\mathbf{M}, \mathbf{N}$ *be A-modules. If* Ψ *is an element of* $\mathrm{Hom}_A(\mathbf{M}, \mathbf{N})((1/z))$, *then the map* $\psi\colon \mathbf{M}((1/z)) \to \mathbf{N}((1/z))$ *given by*

$$\psi(u) = \Psi \cdot u \tag{3.13}$$

is an $\hat{S}_{z^{-1}}$*-homomorphism i.e., is an element of* $\mathrm{Hom}_{\hat{S}_{z^{-1}}}(\mathbf{M}((1/z)), \mathbf{N}((1/z))$.

Proof. If $\Psi = z^s(\psi_0 + \psi_1 z^{-1} + \cdots)$ with the ψ_i in $\mathrm{Hom}_A(\mathbf{M}, \mathbf{N})$ and $u = z^t(u_0 + u_1 z^{-1} + \cdots)$, then $\psi(u) = \Psi \cdot y = z^{s+t} \left(\sum_{n=0}^{\infty} \alpha_n z^{-n} \right)$ where $\alpha_n =$

$\sum_{i+j=n} \psi_i(u_j) \in \mathbf{N}((1/z))$. For ease of exposition, we write $\psi_i u_j$ in place of $\psi_i(u_j)$. If $u = z^t(u_0 + u_1 z^{-1} + \cdots)$, $v = z^r(v_0 + v_1 z^{-1} + \cdots)$ with (say) $t \geq r$, then $v = z^t(v_0 z^{-(t-r)} + v_1 z^{-(t-r)-1} + \cdots)$ and so we may assume that $r = t$. Then $u + v = z^t((u_0 + v_0) + (u_1 + v_1)z^1 + \cdots)$ and $\Psi \cdot (u + v) = z^{s+t} \left(\sum_{n=0}^{\infty} \tilde{\alpha}_n z^{-n} \right)$ where

$$\tilde{\alpha}_n = \sum_{i+j=n} \psi_i(u_j + v_j) = \sum_{i+j=n} \psi_i u_j + \sum_{i+j=n} \psi_i v_j$$

(as $\psi_i \in \mathrm{Hom}_A(\mathbf{M}, \mathbf{N})$). It follows that $\Psi \cdot (u + v) = \Psi \cdot u + \Psi \cdot v$. Similarly, if $\xi \in \hat{S}_{z^{-1}}$, then $\Psi \cdot (\xi u) = (\xi \Psi) \cdot u$ and $\psi \in \mathrm{Hom}_{\hat{S}_{z^{-1}}}(\mathbf{M}((1/z)), \mathbf{N}((1/z))$.

Proposition 3.14 *If* $\mathbf{M} = A^\nu$ *is a finitely-generated free A-module and* $\psi \in \mathrm{Hom}_{\hat{S}_{z^{-1}}}(\mathbf{M}((1/z)), \mathbf{N}((1/z))$, *then there is a* $\Psi \in \mathrm{Hom}_A(\mathbf{M}, \mathbf{N})((1/z))$ *with* $\psi(u) = \Psi \cdot u$.

Proof. Let $\mathbf{m}_1, \ldots, \mathbf{m}_\nu$ be a basis of $\mathbf{M}$. Then $\mathbf{m}_1, \ldots, \mathbf{m}_\nu$ generate $\hat{S}_{z^{-1}} \otimes_A \mathbf{M}$ over $\hat{S}_{z^{-1}}$. Let

$$\psi(\mathbf{m}_i) = z^{s_i}(\mathbf{n}_{0,i} + \mathbf{n}_{1,i} z^{-1} + \cdots) \tag{3.15}$$

with $\mathbf{n}_{j,i} \in \mathbf{N}$. If $s = \max s_i$, $i = 1, \ldots, \nu$, then $z^{s_i}(\mathbf{n}_{0,i} + \mathbf{n}_{1,i} z^{-1} + \cdots) = z^s(\mathbf{n}_{0,i} z^{-(s-s_i)} + \cdots) = z^s(\tilde{\mathbf{n}}_{0,i} + \cdots + \tilde{\mathbf{n}}_{s-s_i} i z^{-(s-s_i)} + \cdots)$ so that we may assume

$$\psi(\mathbf{m}_i) = z^s(\tilde{\mathbf{n}}_{0,i} + \tilde{\mathbf{n}}_{1,i} z^{-1} + \cdots) \tag{3.16}$$

for $i = 1, \ldots, \nu$. Let $\psi_j \colon \mathbf{M} \to \mathbf{N}$ be given by

$$\psi_j \left(\sum_{i=1}^{\nu} a_i \mathbf{m}_i \right) = \sum_{i=1}^{\nu} a_i \tilde{\mathbf{n}}_{j,i} \tag{3.17}$$

for $j = 0, 1, \ldots$. Clearly $\psi_j \in \mathrm{Hom}_A(\mathbf{M}, \mathbf{N})$ and $\Psi = z^s(\psi_0 + \psi_1 z^{-1} + \cdots) \in \mathrm{Hom}_A(\mathbf{M}, \mathbf{N})((1/z))$. Since $\Psi \cdot \mathbf{m}_i = \psi(\mathbf{m}_i)$, $\Psi \cdot u = \psi(u)$.

Corollary 3.18 *If* $\mathbf{M}$ *is a finitely generated A-module and* $\psi \in \mathrm{Hom}_{\hat{S}_{z^{-1}}}(\mathbf{M}((1/z)), \mathbf{N}((1/z)))$, *then there is a* $\Psi \in \mathrm{Hom}_A(\mathbf{M}, \mathbf{N})((1/z))$ *with* $\psi(u) = \Psi \cdot u$.

Proof. Let $\mathbf{m}_1, \ldots, \mathbf{m}_\nu$ generate $\mathbf{M}$ and let $\alpha \colon A^\nu \to \mathbf{M}$ be the A-homomorphism given by $\alpha(a_1, \ldots, a_\nu) = \sum_{i=1}^{\nu} a_i \mathbf{m}_i$. If $\mathbf{M}_1 = \operatorname{Ker} \alpha$, then $A^\nu / \mathbf{M}_1 \simeq \mathbf{M}$ and the sequence

$$\mathbf{M}_1 \xrightarrow{i} A^\nu \xrightarrow{\alpha} \mathbf{M} \longrightarrow 0$$

is exact. It follows from properties of the tensor product ([A-2]) that

$$\mathbf{M}_1 \otimes_A \hat{S}_{z^{-1}} \xrightarrow{i\otimes 1} A^\nu \otimes_A \hat{S}_{z^{-1}} \xrightarrow{\alpha\otimes 1} \mathbf{M} \otimes_A \hat{S}_{z^{-1}} \longrightarrow 0$$

is also exact. Since $A^\nu \otimes_A \hat{S}_{z^{-1}} \simeq A^\nu((1/z))$ and $\mathbf{M} \otimes_A \hat{S}_{z^{-1}} \simeq \mathbf{M}((1/z))$, the map $\tilde{\psi} = \psi \circ \tilde{\alpha}$ where $\tilde{\alpha} = \alpha \otimes 1$ is in $\mathrm{Hom}_{\hat{S}_{z^{-1}}}(A^\nu((1/z)), \mathbf{N}((1/z)))$ and so, there is a $\tilde{\Psi}$ in $\mathrm{Hom}_A(A^\nu, \mathbf{N})((1/z))$ with $\tilde{\Psi} \cdot \tilde{u} = \tilde{\psi}(\tilde{u})$. Let $\tilde{\Psi} = z^s(\tilde{\psi}_0 + \tilde{\psi}_1 z^{-1} + \cdots)$ with $\tilde{\psi}_i \in \mathrm{Hom}_A(A^\nu, \mathbf{N})$. If $\tilde{u} \in \mathbf{M}_1((1/z)) \simeq \mathbf{M}_1 \otimes_A \hat{S}_{z^{-1}}$ then $\tilde{u} = z^t(u_0 + u_1 z^{-1} + \cdots)$ with $u_j \in \mathbf{M}_1 = \mathrm{Ker}\,\alpha$. But $\tilde{\psi} \cdot \tilde{u} = z^{s+t}\left(\sum_{n=0}^{\infty} \tilde{\alpha}_n z^{-n}\right)$ where $\tilde{\alpha}_n = \sum_{i+j=n} \tilde{\psi}_i(u_j)$ and $\tilde{\Psi} \cdot \tilde{u} = (\psi_0 \tilde{\alpha})(\tilde{u}) = \psi((\alpha \otimes 1)(\tilde{u})) = \psi(z^t(\alpha(u_0) + \alpha(u_1)z^{-1} + \cdots) = 0$. In other words, $\tilde{\alpha}_n = 0$ for all n. It follows that $\mathrm{Ker}\,\tilde{\psi}_j \supset \mathbf{M}_1 = \mathrm{Ker}\,\alpha$ for *all* j. We now *define* $\psi_j \colon A^\nu / \mathrm{Ker}\,\alpha \to \mathbf{N}$ via $\psi_j(\bar{a}) = \tilde{\psi}_j(a)$ for $j = 0, \ldots$. Clearly if ψ_j is well-defined then $\psi_j \in \mathrm{Hom}_A(A^\nu / \mathrm{Ker}\,\alpha, \mathbf{N})$. But if $a - a' \in \mathrm{Ker}\,\alpha \subset \mathrm{Ker}\,\tilde{\psi}_j$, then $\psi_j(\bar{a}) = \tilde{\psi}_j(a) = \tilde{\psi}_j(a') = \psi_j(\bar{a}')$. Let $\Psi = z^s(\psi_0 + \psi_1 z^{-1} + \cdots) \in \mathrm{Hom}_A(\mathbf{M}, \mathbf{N})((1/z))$. If $v \in \mathbf{M}((1/z)) \simeq [A^\nu / \mathrm{Ker}\,\alpha]((1/z))$ then $v = z^t(\bar{v}_0 + \bar{v}_1 z^{-1} + \cdots)$ with $\bar{v}_j = \alpha(v_j)$, $v_j \in A^\nu$ and $\Psi \cdot v = z^{s+t}\left(\sum_{n=0}^{\infty} \alpha_n z^{-n}\right)$, $\alpha_n = \sum_{i+j=n} \psi_i(\bar{v}_j)$. But $\psi_i(\bar{v}_j) = \tilde{\psi}_i(v_j)$ so that $\Psi \cdot v = \tilde{\Psi} \cdot \tilde{v}$ where $\tilde{v} = z^t(v_0 + v_1 z^{-1} + \cdots)$. However, $\tilde{\Psi} \cdot \tilde{v} = \psi(\tilde{\alpha}(\tilde{v})) = \psi(v)$ since $(\alpha \otimes 1)(\tilde{v}) = z^t(\alpha(v_0) + \alpha(v_1)z^{-1} + \cdots) = z^t(\bar{v}_0 + \bar{v}_1 z^{-1} + \cdots) = v$. (The technique of this proof is probably worth remembering.)

Definition 3.19 An element ψ of $\mathrm{Hom}_{\hat{S}_{z^{-1}}}(\mathbf{M}((1/z)), \mathbf{N}((1/z))$ is *causal* if $\psi(\mathbf{M}[[1/z]]) \subset \mathbf{N}_+[[1/z]]$.

Since $R = A[z]$ is a subring of $\hat{S}_{z^{-1}}$, we can also consider causal elements of $\mathrm{Hom}_R(\mathbf{M}((1/z)), \mathbf{N}((1/z))$.

Corollary 3.20 *If ψ is causal and $\mathbf{M}$ is finitely generated, then* $\Psi \in \mathrm{Hom}_A(\mathbf{M}, \mathbf{N})_+[[1/z]]$.

Proof. Let $\mathbf{m}_1, \ldots, \mathbf{m}_\nu$ generate $\mathbf{M}$. Then

$$\psi(\mathbf{m}_i) = \mathbf{n}_{1,i} z^{-1} + \cdots = \Psi \cdot \mathbf{m}_i = z^s \left(\sum_{n=0}^{\infty} \alpha_n z^{-n}\right) \tag{3.21}$$

where $\alpha_n = \sum_{i+j=n} \psi_i(v_j) = \psi_n(\mathbf{m}_i)$. It follows that $\psi_0(\mathbf{m}_i) = 0, \ldots, \psi_s(\mathbf{m}_i) = 0$ and hence that $\psi_0(\mathbf{m}) = 0, \ldots, \psi_s(\mathbf{m}) = 0$ for all $\mathbf{m} \in \mathbf{M}$. Thus

$$\Psi = \psi_{s+1} z^{-1} + \cdots \in \mathrm{Hom}_A(\mathbf{M}, \mathbf{N})_+[[1/z]].$$

Corollary 3.22 *If $\Psi \in \mathrm{Hom}_A(\mathbf{M},\mathbf{N})_+[[1/z]]$, then ψ is causal.*

Proof. If $\Psi = \psi_1 z^{-1} + \cdots$ and $u = u_0 + u_1 z^{-1} + \cdots$, then $\Psi \cdot u = \sum_{n=0}^{\infty} \alpha_n z^{-n}$, $\alpha_n = \sum_{i+j=n} \psi_i(u_j)$. But $\alpha_0 = \psi_0(u_0) = 0$ so that $\Psi \cdot u \in \mathbf{N}_+[[1/z]]$.

If ψ is a causal element of $\mathrm{Hom}_R(\mathbf{M}((1/z)), \mathbf{N}((1/z))$ and $i_M \colon \mathbf{M}[z] \to \mathbf{M}((1/z))$ is the natural injection and $\pi_N \colon \mathbf{N}((1/z)) \to \mathbf{N}((1/z))/\mathbf{N}[z] \simeq \mathbf{N}_+[[1/z]]$ is the natural projection, then $\boldsymbol{\psi} = \pi_N \circ \psi \circ i_M \colon \mathbf{M}[z] \to \mathbf{N}_+[[1/z]]$ is an R-homomorphism. Conversely, we have:

Proposition 3.23 *If $\boldsymbol{\psi} \in \mathrm{Hom}_R(\mathbf{M}[z], \mathbf{N}_+[[1/z]])$ then there is a (unique) causal $\psi \in \mathrm{Hom}_R(\mathbf{M}((1/z)), \mathbf{N}((1/z)))$ with $\boldsymbol{\psi} = \pi_N \circ \psi \circ i_M$.*

Proof. Let $w = z^t(\mathbf{m}_0 + \mathbf{m}_1 z^{-1} + \cdots)$ be an element of $\mathbf{M}((1/z))$ and suppose that $\boldsymbol{\psi}(\mathbf{m}_j) = \sum_{i=1}^{\infty} \mathbf{n}_{i,j} z^{-i}$. Let $\alpha_r = \sum_{i+j=r} \mathbf{n}_{i,j} \in \mathbf{N}$ and *define* $\psi(w)$ by

$$\psi(w) = z^t \left(\sum_{r=1}^{\infty} \alpha_r z^{-r} \right) = \sum_{-n \leq j} \boldsymbol{\psi}(\mathbf{m}_j) z^{-j}. \tag{3.24}$$

Then ψ is a causal R-homomorphism.

Proposition 3.25 *If* $\mathbf{M}$ *is finitely generated and $\psi, \tilde{\psi}$ are elements of*

$$\mathop{\mathrm{Hom}}_{A}(\mathbf{M}((1/z)), \mathbf{N}((1/z)),$$

then (i) $\psi = \tilde{\psi}$ *on* $\mathbf{M}[z]$ *implies that* $\psi = \tilde{\psi}$*; and,* (ii) $\psi = \tilde{\psi}$ *on* $\mathbf{M}_+[[1/z]]$ *implies that* $\psi = \tilde{\psi}$.

Proof. Let $\mathbf{m}_1, \ldots, \mathbf{m}_\nu$ generate $\mathbf{M}$. (i) Then $\psi(\mathbf{m}_i) = \tilde{\psi}(\mathbf{m}_i)$, $i = 1, \ldots, \nu$ and so, $\psi = \tilde{\psi}$. (ii) Since $\psi(\mathbf{m}/z) = \tilde{\psi}(\mathbf{m}/z)$ for all $\mathbf{m}$, we have $\psi(\mathbf{m}_i) = \psi(z \cdot \mathbf{m}_i/z) = z\psi(\mathbf{m}_i/z) = z\tilde{\psi}(\mathbf{m}_i/z) = \tilde{\psi}(z \cdot \mathbf{m}_i/z) = \tilde{\psi}(\mathbf{m}_i)$ for $i = 1, \ldots, \nu$.

So we can view a causal (linear) system as either an R-homomorphism of $\mathbf{M}[z]$ into $\mathbf{N}_+[[1/z]]$ ([F-3]) or as a formal Laurent series with coefficients in $\mathrm{Hom}_A(\mathbf{M},\mathbf{N})$ or as a strictly proper meromorphic function on $\mathrm{Hom}_A(\mathbf{M},\mathbf{N})$ (for $\mathbf{M}$ finitely generated). If $\hat{H}(z) = \sum_{j=1}^{\infty} \mathbf{H}_j z^{-j}$ is an element of $\mathrm{Hom}_A(\mathbf{M},\mathbf{N})_+[[1/z]]$, then $\{\mathbf{H}_j\} \in \Omega_c(\mathrm{Hom}_A(\mathbf{M},\mathbf{N}))$ as $\omega_H(j) = 0$ for $j \leq 0$ and $\omega_H(j) = \mathbf{H}_j$, $j \geq 1$.

Definition 3.26 If $\mathbf{F}$ is causal, then the corresponding strictly proper meromorphic function $\mathbf{F}(z)$ on $\mathrm{Hom}_A(\mathbf{M},\mathbf{N})$ is the *transfer matrix* (abuse of language) *of* $\mathbf{F}$.

We shall, for simplicity, suppose from now on that $A = k$, an algebraically closed field, that $\mathbf{M} = k^m = \mathbb{A}_k^m$ and that $\mathbf{N} = k^p = \mathbb{A}_k^p$ so that $\mathrm{Hom}_A(\mathbf{M}, \mathbf{N})$ can be identified with $M(p, m; k)$ ($p \times m$ matrices with entries in k). Then $\mathbf{F}(z) = \sum_{j=1}^{\infty} \mathbf{H}_j z^{-j}$ with $\mathbf{H}_j \in M(p, m)$ $(= M(p, m; k))$.

Definition 3.27 Let $\{\mathbf{H}_j\}_1^{\infty}$ be a sequence of elements of $M(p, m)$. The sequence is *recurrent of order* n if there are $a_0, \ldots, a_n \in k$, not all 0, such that

$$a_0 \mathbf{H}_j + a_1 \mathbf{H}_{j+1} + \cdots + a_n \mathbf{H}_{j+n} = 0 \tag{3.28}$$

for $j = 1, 2, \ldots$. A polynomial $r(z) = r_0 + r_1 z + \cdots + r_\nu z^\nu \in k[z]$ *annihilates* the sequence if

$$r_0 \mathbf{H}_j + r_j \mathbf{H}_{j+1} + \cdots + r_\nu \mathbf{H}_{j+\nu} = 0 \tag{3.29}$$

for $j = 1, 2, \ldots$. Let $\mathrm{Ann}\{\mathbf{H}_j\} = \{r(z)\colon r$ annihilates the sequence $\{\mathbf{H}_j\}\}$. $\mathrm{Ann}\{\mathbf{H}_j\}$ is called the *annihilator of* $\{\mathbf{H}_j\}$.

Proposition 3.30 (i) $\mathrm{Ann}\{\mathbf{H}_j\}$ *is an ideal in* $k[z]$; (ii) $\mathrm{Ann}\{\mathbf{H}_j\} \neq (0)$ *if and only if* $\{\mathbf{H}_j\}$ *is recurrent; and,* (iii) *if* $\mathrm{Ann}\{\mathbf{H}_j\} \neq (0)$, *then there is a monic* $q(z)$ *of lowest degree such that* $\mathrm{Ann}\{\mathbf{H}_j\} = (q(z))$.

Proof. Obvious.

Proposition 3.31 *If* $\mathbf{F}(z) = \sum_{j=1}^{\infty} \mathbf{H}_j z^{-j}$ *and* $\{\mathbf{H}_j\}$ *is recurrent, then there is a* $q(z) \in k[z]$ *and a* $\mathbf{P}(z) \in M(p, m)[z]$ *such that* $q(z)\mathbf{F}(z) = \mathbf{P}(z)$. *Conversely, if there are such a* $q(z)$ *and* $\mathbf{P}(z)$, *then* $\{\mathbf{H}_j\}$ *is recurrent. [Note that such an* $\mathbf{F}(z)$ *is an element of* $M(p, m; k(z))$ *i.e., is a* $p \times m$ *matrix with entries in* $k(z)$.*]*

Proof. Suppose $\{\mathbf{H}_j\}$ is recurrent and let $q(z) = a_0 + a_1 z + \cdots + z^n$ be the monic generator of $\mathrm{Ann}\{\mathbf{H}_j\}$ so that

$$a_0 \mathbf{H}_j + a_1 \mathbf{H}_{j+1} + \cdots + \mathbf{H}_{j+n} = 0 \tag{3.32}$$

for $j = 1, 2, \ldots$. Then, setting $a_n = 1$, and

$$\mathbf{P}_t = \sum_{s=1}^{n-t} a_{s+t} \mathbf{H}_s \tag{3.33}$$

for $t = 0, 1, \ldots, n-1$, we have

$$
\begin{aligned}
q(z)\mathbf{F}(z) &= \sum_{j=1}^{\infty} a_0 \mathbf{H}_j z^{-j} + \sum_{j=1}^{\infty} a_1 \mathbf{H}_j z^{1-j} + \cdots + \sum_{j=1}^{\infty} \mathbf{H}_j z^{n-j} \\
&= \sum_{j=1}^{\infty} a_0 \mathbf{H}_j z^{-j} + \sum_{j=1}^{\infty} a_1 \mathbf{H}_{j+1} z^{-j} + \cdots + \sum_{j=1}^{\infty} \mathbf{H}_{j+n} z^{-j} \\
&\quad + \sum_{t=0}^{n-1} \left(\sum_{s=1}^{n-t} a_{s+t} \mathbf{H}_s \right) z^t \\
&= \sum_{j=1}^{\infty} (a_0 \mathbf{H}_j + \cdots + \mathbf{H}_{j+n}) z^{-j} + \sum_{t=0}^{n-1} \mathbf{P}_t z^t \\
&= \sum_{t=0}^{n-1} \mathbf{P}_t z^t
\end{aligned}
\tag{3.34}
$$

in view of (3.32). Conversely, since $q(z) \neq 0$, we may suppose that $q(z) = a_0 + a_1 z + \cdots + z^n$. Then

$$
\begin{aligned}
q(z)\mathbf{F}(z) &= \sum_{j=1}^{\infty} (a_0 \mathbf{H}_j + \cdots + \mathbf{H}_{j+n}) z^{-j} + \sum_{t=0}^{n-1} \left(\sum_{s=1}^{n-1} a_{s+t} \mathbf{H}_s \right) z^t \\
&= \sum_{r=0}^{\nu} \mathbf{P}_r z^r
\end{aligned}
\tag{3.35}
$$

which implies that $\sum_{j=1}^{\infty} (a_0 \mathbf{H}_j + \cdots + \mathbf{H}_{j+n}) z^{-j} \in M(p,m)[z]$. It follows that $a_0 \mathbf{H}_j + \cdots + \mathbf{H}_{j+n} = 0$ for $j = 1, 2, \ldots$.

Definition 3.36 An $\mathbf{F}(z)$ satisfying the conditions of Proposition 3.31 is called a *strictly proper rational transfer matrix.*

A strictly proper rational transfer matrix $\mathbf{F}(z) = (f_{ij}(z))_{i=1,\ldots,p, j=1,\ldots,m}$ has entries $f_{ij}(z)$ which are transfer functions. If $\mathbf{G}(z) = (g_{ij}(z))_{\substack{i=1,\ldots,p \\ j=1,\ldots,m}}$ is an element of $M(p, m; k(z))$ where each $g_{ij}(z)$ is a transfer function (i.e., a strictly proper rational meromorphic function in $k(z)$), then $g_{ij}(z) = \sum_{\ell=1}^{\infty} h_{ij}^{\ell} z^{-\ell}$ and $\mathbf{G}(z) = \sum_{s=1}^{\infty} \mathbf{H}_s z^{-s}$ where $\mathbf{H}_s = (h_{ij}^{s})_{i=1,\ldots,p, j=1,\ldots,m}$ so that $\mathbf{G}(z)$ is a strictly proper rational transfer matrix. We shall use either interpretation and shall speak simply of "transfer matrices."

Let $\{\mathbf{H}_j\}_{j=1}^{\infty}$ be a sequence of elements of $M(p, m)$ and let

$$
\mathcal{H}_{\infty}^{\infty}(\{\mathbf{H}_j\}) = (\mathbf{H}_{i+j-1})_{i,j=1}^{\infty}
\tag{3.37}
$$

be an $\infty \times \infty$ block Hankel matrix generated by the sequence $\{\mathbf{H}_j\}$. If $\mathbf{F}(z) = \sum_{j=1}^{\infty} \mathbf{H}_j z^{-j}$ is a transfer matrix, we often write $\mathcal{H}_F$ for $\mathcal{H}_\infty^\infty(\{\mathbf{H}_j\})$.

Theorem 3.38 (Hankel) *Let $\mathbf{F}(z)$ be a strictly proper meromorphic function and let $\mathcal{H}_F = (\mathbf{H}_{i+j-1})$. Then $\mathbf{F}(z)$ is rational if and only if $\mathcal{H}_F$ has finite rank.*

Proof 1. If $\mathbf{F}(z)$ is rational, then $\{\mathbf{H}_j\}$ is recurrent and all columns of $\mathcal{H}_F$ are in the span of the first nm columns (where n is the order of a recurrence). Conversely, if $\mathcal{H}_F$ has finite rank, then $\{\mathbf{H}_j\}$ is recurrent and $\mathbf{F}(z)$ is rational.
Proof 2. Let $\mathbf{H}_j = (h_{\alpha\beta}^j)$, $\alpha = 1, \dots, p$, $\beta = 1, \dots, m$. Then $\mathbf{F}(z) = f_{\alpha\beta}(z))$ where $f_{\alpha\beta}(z) = \sum_{j=1}^{\infty} h_{\alpha\beta}^i z^{-j}$ and $\mathbf{F}$ is rational if and only if all $f_{\alpha\beta}$ are rational. If $\mathbf{F}$ is rational, then $f_{\alpha\beta}(z) = p_{\alpha\beta}(z)/q_{\alpha\beta}(z)$ with $q_{\alpha\beta}(z)$ monic. Let $q(z)$ be the least common multiple of the $q_{\alpha\beta}(z)$; then $q(z)$ is monic, $(q(z)) = \bigcap(q_{\alpha\beta}(z))$ and $q(z) = a_0 + a_1 z + \cdots + z^r$. But $\mathrm{Ann}\{\mathbf{H}_j\} = (q(z))$ since $a_0\mathbf{H}_j + \cdots + \mathbf{H}_{j+n} = 0$ if and only if $a_0 h_{\alpha\beta}^j + \cdots + h_{\alpha\beta}^{j+n} = 0$ for all α, β i.e., if and only if $q(z)$ annihilates all $\{h_{\alpha\beta}^j\}$ or, equivalently, if and only if $q(z) \in \bigcap\, (q_{\alpha\beta}(z))$. It follows that all columns of $\mathcal{H}_F$ are in the span of the first nm columns. If $\mathcal{H}_F$ has finite rank, then each $\mathcal{H}_{f_{\alpha\beta}}$ has finite rank and all $f_{\alpha\beta}(z)$ are rational.

Definition 3.39 $\mathcal{H}_F$ is the *Hankel matrix of* $\mathbf{F}$. An *∞-block Hankel matrix* $\mathcal{H}$ (with $p \times m$ blocks) is a matrix $(\mathbf{H}_{i+j-1})_{i,j=1}^{\infty}$ of the form

$$\begin{bmatrix} \mathbf{H}_1 & \mathbf{H}_2 & \mathbf{H}_3 & \cdots \\ \mathbf{H}_2 & \mathbf{H}_3 & \mathbf{H}_4 & \cdots \\ \mathbf{H}_3 & \mathbf{H}_4 & \mathbf{H}_5 & \cdots \\ \vdots & \vdots & \vdots & \cdots \end{bmatrix} \tag{3.40}$$

where $\mathbf{H}_i \in M(p, m)$.

We write h_j^i for the entries of $\mathcal{H}$, $\mathcal{H}_\infty^i = \mathbf{v}^i = [H_i \; H_{i+1} \; \cdots]$ for the ∞-$p \times m$ block matrix with blocks $\mathbf{H}_i, \mathbf{H}_{i+1}, \dots$; $\mathcal{H}_\mu^\lambda$ for the $\lambda p \times \mu m$ block-Hankel matrix $(\mathbf{H}_{i+j-1})_{i=1,\dots,\lambda; j=1,\dots,\mu}$ and we write $H^r = (h_j^r)_{j=1,\dots}$ for the rth row of $\mathcal{H}$, $(\mathcal{H}_\infty^i)^r$ for the rth row of $(\mathcal{H}_\infty^i)$, etc. We note that

$$h_{j+m}^r = h_j^{r+p} \tag{3.41}$$

for all $r = 1, \dots$ and $j = 1, \dots$ (by the form of $\mathcal{H}$) and hence, that H^{r+p} is a subrow of H^r.

Lemma 3.42 (Risannen) *Let $\mathcal{H}$ be a block Hankel matrix.* (a) *If $H^r \in \text{span}\{H^s\colon s = 1, \dots, r-1\}$, then $(\mathcal{H}_\mu^\lambda)^r \in \text{span}\{(\mathcal{H}_\mu^\lambda)^s\colon s = 1, \dots, r-1\}$;* (b) *If $\rho(\mathcal{H}) = n < \infty$ and $\boldsymbol{\alpha} = (\alpha_1, \dots, \alpha_n)$ indexes the first n independent rows of $\mathcal{H}$, then $r \notin \boldsymbol{\alpha}$ (i.e., $r \neq \alpha_i$, $i = 1, \dots, n$) implies $H^r \in \text{span}\{H^s\colon s = 1, \dots, r-1\}$ and $r + p \in \boldsymbol{\alpha}$ implies $r \in \boldsymbol{\alpha}$; and,* (c) *If λ is the smallest integer such that $\rho(\mathcal{H}_\lambda^\lambda) = n$, then $\lambda \leq n$.*

Proof. (a) If $H^r \in \text{span}\{H^s\colon s = 1, \dots, r-1\}$, then $H^r = \sum_{s=1}^{r-1} a_{rs} H^s$ so that $h_j^r = \sum_{s=1}^{r-1} a_{rs} h_j^s$ for all j and hence, $(\mathcal{H}_\mu^\lambda)^r \in \text{span}\{(\mathcal{H}_\mu^\lambda)^s\colon s = 1, \dots, r-1\}$. (b) Observe that $1 \leq \alpha_1 < \cdots < \alpha_n$. If $r > \alpha_n$, then clearly $H^r \in \text{span}\{H^s\colon s = 1, \dots, r-1\}$. If (say) $\alpha_i < r < \alpha_{i+1}$, then $H^r \in \text{span}\{H^{\alpha_j}\colon j = 1, \dots, i\} = \text{span}\{H^s\colon s = 1, \dots, r-1\}$. If $r + p \in \boldsymbol{\alpha}$ but $r \notin \boldsymbol{\alpha}$, then, by what was just shown, $H^r \in \text{span}\{H^s\colon s = 1, \dots, r-1\}$ so that $h_j^{r+p} = h_{j+m}^r$ (3.41) $= \sum_{s=1}^{r-1} a_{rs} h_{j+m}^s = \sum_{s=1}^{r-1} a_{rs} h_j^{s+p}$ (3.41) and so,

$$H^{r+p} \in \text{span}\{H^{1+p}, \dots, H^{r-1+p}\} \subset \text{span}\{H^s\colon s = 1, \dots, r+p-1\}$$

which contradicts $r + p \in \boldsymbol{\alpha}$. (Note the critical role played by the form of $\mathcal{H}$ i.e., (3.41)). (c) Let $\boldsymbol{\alpha}_\lambda = \{\alpha_t \in \boldsymbol{\alpha}\colon \alpha_t \leq \lambda p\}$. If $r \notin \boldsymbol{\alpha}_\lambda$ and $r \leq \lambda p$, then $(\mathcal{H}_\lambda^\lambda)^r \in \text{span}\{(\mathcal{H}_\lambda^\lambda)^s\colon s = 1, \dots, r-1\}$. Thus, $\rho(\mathcal{H}_\lambda^\lambda) \leq \#\boldsymbol{\alpha}_\lambda \leq n$. Since $\rho(\mathcal{H}_\lambda^\lambda) = n$, we have $\boldsymbol{\alpha}_\lambda = \boldsymbol{\alpha}$ and so $1 \leq \alpha_1 < \alpha_2 < \cdots < \alpha_n \leq \lambda p$ are the first n independent rows of $\mathcal{H}_\lambda^\lambda$. Since λ is minimal, $(\mathcal{H}_\lambda^\lambda)^{\alpha_n} \in [H_\lambda\ H_{\lambda+1} \cdots H_{2\lambda-1}]$ for otherwise $\rho(\mathcal{H}_{\lambda-1}^{\lambda-1}) = n$. Therefore, $\rho(\mathcal{H}_{\lambda-1}^{\lambda-1}) \leq n-1$ and $\alpha_n = (\lambda-1)p + j$ for some j in $(1, \dots, p)$. If $\lambda = 1$, then $\lambda \leq n$. So, if $\lambda > 1$, then $\alpha_n = (\lambda-2)p+p+j_0$ for some j_0 in $\mathbf{p} = (1, \dots, p)$. In view of (b), we have for $t = 0, 1, \dots, \lambda-1$, $\beta_t = (\lambda - 1 - t)p + j_0 \in \boldsymbol{\alpha}$ and so $\lambda \leq n$.

Corollary 3.43 *If $\rho(\mathcal{H}_{\mathbf{F}}) = n$, then $\rho(\mathcal{H}_{\mathbf{F}n}^n) = n$.*

Let $\mathcal{H}$ be an ∞-block-Hankel matrix and recall that $\mathcal{H}_\infty^i = \mathbf{v}^i = [H_i\ H_{i+1} \cdots]$ so that $\text{span}\{\mathbf{v}^i\colon i = 1, 2, \dots\}$ is the row space $\mathfrak{r} = \mathfrak{r}(\mathcal{H})$ of $\mathcal{H}$. If $R = k[z]$, then $\mathfrak{r}$ becomes an R-module via $z\mathbf{v}^i = \mathbf{v}^{i+1}$. Let $\text{Ann}_R(\mathfrak{r})$ be the annihilator of $\mathfrak{r}$ as an R-module. If $\{\mathbf{H}_j\}$ is recurrent of order n (with $a_n = 1$), then

$$\mathbf{H}_{n+j} = -a_0 \mathbf{H}_j - \cdots - a_{n-1} \mathbf{H}_{j+n-1} \tag{3.44}$$

for $j = 1, 2, \dots$, or, equivalently,

$$\mathbf{v}^{n+j} = -a_0 \mathbf{v}^j - \cdots - a_{n-1} \mathbf{v}^{j+n-1} \tag{3.45}$$

or, equivalently,

$$z^n \mathbf{v}^j = (-a_0 - a_1 z - \cdots - a_{n-1} z^{n-1})\mathbf{v}^j \tag{3.46}$$

for $j = 1, \ldots$. In other words, $z^n + a_{n-1}z^{n-1} + \cdots + a_0 \in \mathrm{Ann}_R(\mathfrak{r})$ which is a proper ideal. Conversely, if $\mathrm{Ann}_R(\mathfrak{r})$ is a proper ideal, then $\{\mathbf{H}_j\}$ is recurrent. We note also that if $\{\mathbf{H}_j\}$ is recurrent of order n, then $\mathbf{v}^{n+j} \in \mathrm{span}\{\mathbf{v}^1, \ldots, \mathbf{v}^n\}$ so that $\rho(\mathcal{H}_\infty^\infty) < \infty$ as $\dim_k \mathfrak{r} \leq np$.

Suppose $\rho(\mathcal{H}_\infty^\infty) = n$ and let λ be minimal with $\rho(\mathcal{H}_\lambda^\lambda) = n$. Let $\boldsymbol{\alpha} = (\alpha_1, \ldots, \alpha_n)$ be the first n independent rows. Since $\lambda \leq n$, $\boldsymbol{\alpha} \subset (1, \ldots, \lambda p)$. Let $(\alpha_1, \ldots, \alpha_{n_1}) \in (1, \ldots, p)$,

$$(\alpha_{n_1+1}, \ldots, \alpha_{n_1+n_2}) \in (p+1, \ldots, 2p), \ldots, (\alpha_{n_1+n_2+\cdots+n_{\lambda_1}+1}, \ldots, \alpha_{n_1+n_2+\cdots+n_\lambda}) \in ((\lambda-1)p+1, \ldots, \lambda p)$$

so that

$$n_1 + n_2 + \cdots + n_\lambda = n. \tag{3.47}$$

By the lemma of Risannen, $\alpha_{n_1+1} - p, \ldots, \alpha_{n_1+n_2} - p$ are all in $(\alpha_1, \ldots, \alpha_{n_1})$ and $\alpha_{n_1+n_2+1} - p, \ldots, \alpha_{n_1+n_2+n_3} - p$ are all in $(\alpha_{n_1+1}, \ldots, \alpha_{n_1+n_2})$, etc. It follows that $n_1 \geq n_2 \geq \cdots \geq n_\lambda$. Since $\mathbf{v}^{\alpha_1}, \ldots, \mathbf{v}^{\alpha_n}$ span $\mathfrak{r}$, there is a (unique) $n \times n$ matrix A such that

$$\begin{bmatrix} H^{\alpha_1+p} \\ \vdots \\ H^{\alpha_n+p} \end{bmatrix} = A \begin{bmatrix} H^{\alpha_1} \\ \vdots \\ H^{\alpha_n} \end{bmatrix}. \tag{3.48}$$

But $H^{\alpha_j+p} = zH^{\alpha_j}$ and so

$$(zI - A)\begin{bmatrix} H^{\alpha_1} \\ \vdots \\ H^{\alpha_n} \end{bmatrix} = 0 \tag{3.49}$$

which means that $\psi(z) = \det(zI - A) \in \mathrm{Ann}_R(\mathfrak{r})$ and that $\{\mathbf{H}_j\}$ is recurrent. This, of course, provides an alternate proof of Hankel's Theorem (3.38).

Example 3.50 Let $n = 4$, $p = 2$, $\lambda = 3$, $m = m$ and $\boldsymbol{\alpha} = (\alpha_1, \alpha_2, \alpha_3, \alpha_4) = (1, 2, 3, 5)$. Then $n_1 = 2$, $n_2 = 1$, $n_3 = 1$. We have

$$\begin{bmatrix} z\mathbf{H}^1 \\ z\mathbf{H}^2 \\ z\mathbf{H}^3 \\ z\mathbf{H}^4 \end{bmatrix} = \begin{bmatrix} \mathbf{H}^{1+2} \\ \mathbf{H}^{2+2} \\ \mathbf{H}^{3+2} \\ \mathbf{H}^{5+2} \end{bmatrix} = \begin{bmatrix} 0 & 0 & 1 & 0 \\ a & b & c & 0 \\ 0 & 0 & 0 & 1 \\ d & e & f & g \end{bmatrix} \begin{bmatrix} \mathbf{H}^1 \\ \mathbf{H}^2 \\ \mathbf{H}^3 \\ \mathbf{H}^4 \end{bmatrix}$$

(eq. (3.49)) and

$$\begin{aligned}\psi(z) &= \det \begin{bmatrix} z & 0 & -1 & 0 \\ -a & z-b & -c & 0 \\ 0 & 0 & z & -1 \\ -d & -e & -f & z-g \end{bmatrix} \\ &= (z^3 - gz^2 - fz - d)(z-b) - e(cz+a).\end{aligned}$$

Then $\psi(z) \in \mathrm{Ann}_R(\mathfrak{r})$. For instance, $\psi(z)\mathbf{H}^1 = (z-b)[\mathbf{H}^7 - g\mathbf{H}^5 - f\mathbf{H}^3 - d\mathbf{H}^1] = e(c\mathbf{H}^3 + a\mathbf{H}^1) = (z-b)e\mathbf{H}^2 - e(c\mathbf{H}^3 + a\mathbf{H}^1) = e(\mathbf{H}^4 - a\mathbf{H}^1 - b\mathbf{H}^2 - c\mathbf{H}^3) = 0$.

Definition 3.51 If $\mathbf{F}(z)$ is a rational $p \times m$ transfer matrix, then the *degree of* $\mathbf{F}$ is the rank $\rho(\mathcal{H}_{\mathbf{F}})$ of $\mathcal{H}_{\mathbf{F}}$.

If $q_F(z)$ is the minimal annihilator of $\mathbf{F}$, then the degree n_q of $q_F(z)$ is the order of the shortest recurrence for $\{\mathbf{H}_j\}$ (where $\mathbf{F}(z) = \sum_{j=1}^{\infty} \mathbf{H}_j z^{-j}$). We have noted earlier (2.2) that n_q can be less than the degree of $\mathbf{F}$.

Definition 3.52 Let

$$\begin{aligned}\mathbf{Rat}(n;p,m) &= \{\mathbf{F}(z) \colon \mathbf{F}(z) \text{ rational } p \times m \text{ of degree } n\} \\ \mathbf{Hank}(n;p,m) &= \{\mathcal{H}_\infty^\infty \colon \mathcal{H}_\infty^\infty \text{ block } p \times m \text{ Hankel of rank } n\}\end{aligned}$$

be the set of *rational* $p \times m$ *transfer matrices of degree* n and the *set of (block) Hankel matrices of degree* n. The map $\mathbf{L} \colon \mathbf{Rat}(n;p,m) \to \mathbf{Hank}(n;p,m)$ given by

$$\mathbf{L}(\mathbf{F}) = \mathcal{H}_{\mathbf{F}} \tag{3.53}$$

is called the *Laurent map.*

Our aim will be to give both $\mathbf{Rat}(n;p,m)$ and $\mathbf{Hank}(n;p,m)$ an appropriate algebraic structure and to show that $\mathbf{L}$ is an isomorphism. Although both **Rat** and **Hank** appear to involve infinite objects, there are actually only a finite number of parameters involved because of the degree n condition. We shall ultimately show that both $\mathbf{Rat}(n;p,m)$ and $\mathbf{Hank}(n;p,m)$ are of dimension $n(m+p)$. Here in this section we shall simply show that both can be viewed as (naturally) imbedded in $\mathbb{A}_k^{n(pm+1)}$.

Definition 3.54 Let $\mathcal{H}_\infty^\infty = \mathcal{H}(\mathbf{H}_j)$ be a $p \times m$ block Hankel matrix. A triple $x = (A, B, C)$ with $A \in M(\nu,\nu)$, $B \in M(\nu, m)$, and $C \in M(p,\nu)$ is a *realization of* $\mathcal{H}_\infty^\infty$ if $\mathbf{H}_j = CA^{j-1}B$ for $j = 1, \ldots$. The *order of* x is ν.

Lemma 3.55 *If* $\mathcal{H}_\infty^\infty$ *has a realization* x *of order* ν, *then* $\rho(\mathcal{H}_\infty^\infty) \leq \nu$.

Proof. Let $Y_\nu^\infty(x) = (CA^{j-1})^{j=1,\dots,\infty}$, $Z_\infty^\nu(x) = (A^{j-1}B)_{j=1,\dots,\infty}$ so that $\mathcal{H}_\infty^\infty = Y_\nu^\infty(x)Z_\infty^\nu(x)$ and $\rho(\mathcal{H}_\infty^\infty) \le \min(\rho(Y_\nu^\infty(x)), \rho(Z_\infty^\nu(x))) \le \nu$.

Lemma 3.56 ([K-2]) *If $\rho(\mathcal{H}_\infty^\infty) = n$, then there is a realization x of order n.*

Proof. Let $\boldsymbol{\alpha} = (\alpha_1, \dots, \alpha_n)$ index the first n independent rows of $\mathcal{H}_\infty^\infty$. By the lemma of Risannen, $1 \le \alpha_1 < \cdots < \alpha_n \le np$ and $r + p \in \boldsymbol{\alpha}$ implies $r \in \boldsymbol{\alpha}$. By the Hankel form,

$$h_{j+m}^i = h_j^{i+p} \tag{3.57}$$

for all i, j. Hence, there is a (unique) $p \times n$ matrix C with $\mathcal{H}_\infty^{1,\dots,p} = C\mathcal{H}_\infty^{\alpha_1,\dots,\alpha_n}$ so that

$$\mathbf{H}_j = C(h_{m(j-1)+s}^{\alpha_i}) \tag{3.58}$$

for $i = 1, \dots, n$, $s = 1, \dots, m$ and $j = 1, 2, \dots$. Let B be the $n \times m$ matrix $(h_s^{\alpha_i})$, $i = 1, \dots, n$, $s = 1, \dots, m$. Then $\mathbf{H}_1 = CB$ by (3.58). Since $H^{\alpha_1}, \dots, H^{\alpha_n}$ are a basis of the row space of $\mathcal{H}_\infty^\infty$, there is a (unique) $n \times n$ matrix A such that

$$\mathcal{H}_\infty^{\alpha_1+p,\dots,\alpha_n+p} = A\mathcal{H}_\infty^{\alpha_1,\dots,\alpha_n} \tag{3.59}$$

or, equivalently,

$$(h_{m(j-1)+s}^{\alpha_i+p}) = A(h_{m(j-1)+s}^{\alpha_i}) \tag{3.60}$$

for $i = 1, \dots, n$, $s = 1, \dots, m$ and $j = 1, 2, \dots$. It follows from (3.57) that

$$(h_{mj+s}^{\alpha_i}) = (h_{m(j-1)+s}^{\alpha_i+p}) = A(h_{m(j-1)+s}^{\alpha_i}) \tag{3.61}$$

and from (3.58) that

$$\mathbf{H}_2 = C(h_{m+s}^{\alpha_i}) = CA(h_s^{\alpha_i}) = CAB \tag{3.62}$$

and

$$\mathbf{H}_{t+1} = C(h_{mt+s}^{\alpha_i}) = CA(h_{m(t-1)+s}^{\alpha_i}) = CA^tB \tag{3.63}$$

by induction for $t > 1$. Thus, $x = (A, B, C)$ is a realization of order n.

Lemma 3.64 *Suppose $\rho(\mathcal{H}) = n$ and let $\mathbf{F}_\mathcal{H}(z) = \sum_{j=1}^\infty \mathbf{H}_j z^{-j}$. If $x = (A, B, C)$ is a realization of $\mathcal{H}$, then*

$$\begin{aligned}\mathbf{F}_\mathcal{H}(z) &= \sum_{j=1}^\infty (CA^{j-1}B)z^{-j} = C(zI - A)^{-1}B \\ &= C\,\mathrm{Adj}(zI - A)B/\det(zI - A).\end{aligned}$$

Proof. Simply note that $I + A/z + A^2/z^2 + = (I - A/z)^{-1}$ so that

$$\begin{aligned}\mathbf{F}_{\mathcal{H}}(z) &= \sum_{j=1}^{\infty}(CA^{j-1}B)z^{-j} = C(I - A/z)^{-1}B/z \\ &= C[(zI)(I - A/z)]^{-1}B = C(zI - A)^{-1}B.\end{aligned}$$

Corollary 3.65 *If $\mathbf{F}$ is rational of degree n, then there is an $x = (A, B, C)$ of order n with $\mathbf{F}(z) = \mathbf{F}_x(z) = C(zI - A)^{-1}B$.*

Proof. $\mathbf{F}$ rational of degree n means $\rho(\mathcal{H}_{\mathbf{F}}) = n$ and apply Lemma 3.64.

Theorem 3.66 *The Laurent map $\mathbf{L}$ is bijective.*

Proof. If $\mathbf{L}(\mathbf{F}) = \mathbf{L}(\tilde{\mathbf{F}})$, then $\mathbf{F}(z) = \sum_{j=1}^{\infty} \mathbf{H}_j z^{-j} = \sum_{j=1}^{\infty} (\tilde{\mathbf{H}})_j z^{-j} = \tilde{\mathbf{F}}(z)$ and $\mathbf{L}$ is injective. If $\mathcal{H} \in \mathbf{Hank}(n; p, m)$, then there is a realization x of order n and $\mathbf{F}_x(z) = \sum \mathbf{H}_j z^{-j}$ is rational. If $\deg \mathbf{F}_x < n$, then $\rho(\mathcal{H}_{\mathbf{F}_x}) < n$. But $\mathcal{H} = \mathcal{H}_{\mathbf{F}_x}$ so that $\mathbf{F}_x \in \mathbf{Rat}(n; p, m)$.

Example 3.67 Let $p = m = 2$ and let $\mathbf{H}_1 = I_2$, $\mathbf{H}_j = 0$ for $j = 2, 3, \ldots$. Then $\mathcal{H}$ is of degree 2 and $x = (O_2, I_2, I_2)$ is a realization of order 2. $\mathbf{F}_{\mathcal{H}}(z) = I_2/z = \mathbf{F}_x(z)$ and $\det(zI_2 - O_2) = 0$. Here $\chi_1(A) = 0$, $\chi_2(A) = 0$ and $\mathbf{H}_3 = \chi_2(A)\mathbf{H}_1 + \chi_1(A)\mathbf{H}_2 = 0$, etc. Thus, $(I_2, O_2, 0, 0) \in \mathbb{A}_k^{10}$ represents the element $\mathcal{H}$ of $\mathbf{Hank}(2; 2, 2)$. If $\tilde{\mathbf{H}}_1 = I_2$, $\tilde{\mathbf{H}}_2 = O_2$, $\xi_1 = 0$, $\xi_2 = 1$, then $\tilde{\mathbf{H}}_{2+j} = -\xi_1\tilde{\mathbf{H}}_j - \xi_2\tilde{\mathbf{H}}_{j+1}$ and $\mathcal{H}(\tilde{\mathbf{H}}_j) = \mathcal{H}$ with $\rho(\mathcal{H}) = 2$. Moreover, there is a recurrence of order 2 generated by ξ_1, ξ_2. However, $(I_2, O_2, 0, 1)$ does *not* represent an element of $\mathbf{Hank}(2; 2, 2)$. We recall (2.57) that the Grassmann equations

$$\begin{aligned}|\tilde{\mathbf{H}}_1|\xi_1 &= |\tilde{\mathbf{H}}_2 + \xi_2\mathbf{H}_1| \\ |\tilde{\mathbf{H}}_1|\xi_2 &= -|\tilde{\mathbf{H}}_{1,1}\ \tilde{\mathbf{H}}_{2,2}| + |\tilde{\mathbf{H}}_{1,2}\ \tilde{\mathbf{H}}_{2,1}|\end{aligned}$$

must be satisfied if $(\tilde{\mathbf{H}}_1, \tilde{\mathbf{H}}_2, \xi_1, \xi_2)$ is to represent an element of $\mathbf{Hank}(2; 2, 2)$. Clearly, $(I_2, O_2, 0, 1)$ does not satisfy these equations.

If $\mathbf{F} \in \mathbf{Rat}(n; p, m)$, then $\mathbf{F}(z) = \mathbf{F}_x(z) = C\,\mathrm{Adj}(zI - A)B/\det(zI - A)$ (by Lemma 3.64 and Corollary 3.65) so that each entry $f_{ij}(z)$ of $\mathbf{F}$ may be written in the form $f_{ij}(z) = p_{ij}(z)/q(z)$ where $q(z)$ is monic of degree n and $p_{ij}(z)$ is of degree $\leq n - 1$. Thus,

$$\mathbf{F}(z) = \mathbf{P}_0 + \mathbf{P}_1 z + \cdots + \mathbf{P}_{n-1}z^{n-1}/a_0 + a_1 z + \cdots + a_{n-1}z^{n-1} + z^n \tag{3.68}$$

where the $\mathbf{P}_i$ are in $M(p, m)$.

Definition 3.69 Let $\text{Hank}(n;p,m) \subset \mathbb{A}_k^{n(pm+1)}$ be the set of $(\mathbf{H}_1,\ldots,\mathbf{H}_n, \xi_1,\ldots,\xi_n)$ such that (i) if $\mathbf{H}_{n+j} = -\xi_1\mathbf{H}_j - \cdots - \xi_n\mathbf{H}_{n+j-1}$, $j = 1,2,\ldots,$ then $\rho(\mathcal{H}(\mathbf{H}_j)) = n$; and, (ii) there is a realization x of $\mathcal{H}(\mathbf{H}_j)$ of order n with $\chi_1(x) = -\xi_n$, $\chi_2(x) = -\xi_{n-1},\ldots,\chi_n(x) = -\xi_1$. Let $\text{Rat}(n;p,m) \subset \mathbb{A}_k^{n(pm+1)}$ be the set of $(\mathbf{P}_0,\ldots,\mathbf{P}_{n-1}, a_0,\ldots,a_{n-1})$ such that (i) if

$$\begin{aligned} \mathbf{H}_1(\mathbf{P}_0,\ldots,\mathbf{P}_{n-1}) &= \mathbf{P}_{n-1} \\ \mathbf{H}_r(\mathbf{P}_0,\ldots,\mathbf{P}_{n-1}) &= \mathbf{P}_{n-r} - \sum_{j=1}^{r-1} \mathbf{H}_j(\mathbf{P}_0,\ldots,\mathbf{P}_{n-1})a_{r-j-1} \end{aligned} \tag{3.70}$$

for $r = 2,\ldots,n$ and $\mathbf{H}_{n+j}(\mathbf{P}_0,\ldots,\mathbf{P}_{n-1}) = -a_0\mathbf{H}_j(\mathbf{P}_0,\ldots,\mathbf{P}_{n-1}) - \cdots - a_{n-1}\mathbf{H}_{n+j-1}(\mathbf{P}_0,\ldots,\mathbf{P}_{n-1})$, then $\rho(\mathcal{H}(\mathbf{H}_j)) = n$; (ii) there is a realization x of $\mathcal{H}(\mathbf{H}_j(\mathbf{P}_0,\ldots,\mathbf{P}_{n-1}))$ of order n with $\chi_1(x) = -a_{n-1},\ldots,\chi_n(x) = -a_0$.

We shall now show that $\mathbf{Hank}(n;p,m)$ can be identified with $\text{Hank}(n;p,m)$ and that, likewise, $\mathbf{Rat}(n;p,m)$ can be identified with $\text{Rat}(n;p,m)$. Define a map ψ: $\mathbf{Hank}(n;p,m) \to \mathbb{A}_k^{n(mp+1)}$ by

$$\psi(\mathcal{H}) = (\mathbf{H}_1,\ldots,\mathbf{H}_n, -\chi_n(x),\ldots,-\chi_1(x)) \tag{3.71}$$

where x is a realization of $\mathcal{H}$ of order n (Lemma 3.56). Since $\mathbf{H}_{n+j} = CA^{n+j-1}B = \chi_1(x)CA^{n+j-2}B+\cdots+\chi_n(x)CA^{j-1}B = -\xi_1\mathbf{H}_j-\cdots-\xi_n\mathbf{H}_{n+j-1}$ (where $x = (A,B,C)$ and $\mathbf{H}_j$, ξ_i are coordinates on $\mathbb{A}_k^{n(pm+1)}$), we have $\psi(\mathcal{H}) \in \text{Hank}(n;p,m)$ if ψ is well-defined. Assuming this for the moment, we have:

Claim 3.72 *ψ is injective.*

Proof. If $\psi(\mathcal{H}) = \psi(\tilde{\mathcal{H}})$, then $\mathbf{H}_1 = \tilde{\mathbf{H}}_1,\ldots,\mathbf{H}_n = \tilde{\mathbf{H}}_n$ and $\xi_1 = \tilde{\xi}_1,\ldots,\xi_n = \tilde{\xi}_n$ so that $\mathbf{H}_{n+j} = -\xi_1\mathbf{H}_j - \cdots - \xi_n\mathbf{H}_{n+j-1} = -\tilde{\xi}_1\tilde{\mathbf{H}}_1 - \cdots - \tilde{\xi}_n\tilde{\mathbf{H}}_{n+j-1} = \tilde{\mathbf{H}}_{n+j}$ for $j = 1,\ldots$ (by induction on j). Hence $\mathcal{H} = \tilde{\mathcal{H}}$.

Claim 3.73 *ψ is surjective.*

Proof. If $(\mathbf{H}_1,\ldots,\mathbf{H}_n,\xi_1,\ldots,\xi_n) \in \text{Hank}(n;p,m)$, then $\mathcal{H} = \mathcal{H}(\mathbf{H}_j) \in \mathbf{Hank}(n;p,m)$ and $\psi(\mathcal{H}) = (\mathbf{H}_1,\ldots,\mathbf{H}_n, \xi_1,\ldots,\xi_n)$ since ψ is assumed well-defined. So, to complete the identification, we need only prove the following:

Theorem 3.74 *Let $\mathcal{H}$ be an element of $\mathbf{Hank}(n;p,m)$ and let $x = (A,B,C)$, $x_1 = (A_1,B_1,C_1)$ be order n realizations of $\mathcal{H}$. Then $\chi_i(x) = \chi_i(x_1)$, $i = 1,\ldots,n$.*

Proof. Since x, x_1 are realizations of $\mathcal{H}$,

$$CA^{j-1}B = C_1A_1^{j-1}B_1 \tag{3.75}$$

for all $j = 1, \ldots$. Let $Y(x) = (CA^{j-1})^{j=1,\ldots,n}$, $Z(x) = (A^{j-1}B)_{j=1,\ldots,n}$, $Y(x_1) = (C_1 A_1^{j-1})^{j=1,\ldots,n}$, and $Z(x_1) = (A_1^{j-1}B_1)_{j=1,\ldots,n}$. Then the $np \times nm$ matrix $\mathcal{H}_n^n$ is given by

$$\mathcal{H}_n^n = Y(x)Z(x) = Y(x_1)Z(x_1) \tag{3.76}$$

and $\rho(\mathcal{H}_n^n) = n$ by the lemma of Risannen. It follows that $\rho(Y(x)) = n = \rho(Y(x_1))$ and $\rho(Z(x)) = n = \rho(Z(x_1))$. Let $\mathbf{e} = (e_1, \ldots, e_n)$ be (say) the first n independent rows of $Y(x)$ so that $Y^{\mathbf{e}}(x)$ is nonsingular and let $\mathbf{f} = (f_1, \ldots, f_n)$ be the first n independent columns of $Z(x)$ so that $Z_{\mathbf{f}}(x)$ is nonsingular. Then $Y^{\mathbf{e}}(x)Z_{\mathbf{f}}(x) = Y^{\mathbf{e}}(x_1)Z_{\mathbf{f}}(x_1)$ so that $Y^{\mathbf{e}}(x_1)$, $Z_{\mathbf{f}}(x_1)$ are also nonsingular. Let $g \in G = \mathrm{GL}(n, k)$ be given by

$$g = Y^{\mathbf{e}}(x_1)^{-1}Y^{\mathbf{e}}(x) = Z_{\mathbf{f}}(x_1)Z_{\mathbf{f}}(x)^{-1}. \tag{3.77}$$

Then it follows from (3.75) that

$$\begin{aligned} gB &= Y^{\mathbf{e}}(x_1)^{-1}Y^{\mathbf{e}}(x)B = Y^{\mathbf{e}}(x_1)^{-1}Y^{\mathbf{e}}(x_1)B_1 = B_1 \\ Cg^{-1} &= CZ_{\mathbf{f}}(x)Z_{\mathbf{f}}(x_1)^{-1} = C_1 Z_{\mathbf{f}}(x_1)Z_{\mathbf{f}}(x_1)^{-1} = C_1 \\ Y^{\mathbf{e}}(x)AZ_{\mathbf{f}}(x) &= Y^{\mathbf{e}}(x_1)AZ_{\mathbf{f}}(x_1) \end{aligned} \tag{3.78}$$

and hence that $gAg^{-1} = A_1$. Thus, $\chi_i(A) = \chi_i(A_1)$ for $i = 1, \ldots, n$.

Now define a map φ: $\mathbf{Rat}(n; p, m) \to \mathbb{A}_k^{n(pm+1)}$ by

$$\varphi(\mathbf{F}) = (\mathbf{P}_0(x), \ldots, \mathbf{P}_{n-1}(x), a_0(x), \ldots, a_{n-1}(x)) \tag{3.79}$$

where $x = (A, B, C)$ is a realization of $\mathcal{H}_{\mathbf{F}}$ of order n and

$$\begin{aligned} \mathbf{P}_{n-1}(x) &= \mathbf{H}_1(x) = CB \\ \mathbf{P}_r(x) &= \mathbf{H}_{n-r}(x) + \sum_{j=1}^{n-r-1} \mathbf{H}_j(x)a_{n-j-r-1}(x) \\ a_t(x) &= -\chi_{n-t}(x) \end{aligned} \tag{3.80}$$

for $r = n-2, n-3, \ldots, 0$, $t = 0, 1, \ldots, n-1$ and $\mathbf{H}_j(x) = CA^{j-1}B$. If we let

$$\begin{aligned} \tilde{\mathbf{H}}_1 &= \mathbf{P}_{n-1}(x) = \mathbf{H}_1 \\ \tilde{\mathbf{H}}_s &= \mathbf{P}_{n-s}(x) - \sum_{j=1}^{s-1} \tilde{\mathbf{H}}_j a_{s-j-1} \end{aligned} \tag{3.81}$$

for $s = 2, \ldots, n$ and $\tilde{\mathbf{H}}_{n+j} = -a_0\tilde{\mathbf{H}}_j - \cdots - a_{n-1}\tilde{\mathbf{H}}_{n+j-1}$. Then, substituting (3.80), we have

$$\tilde{\mathbf{H}}_s = \mathbf{H}_s + \sum_{j=1}^{s-1} \mathbf{H}_j a_{s-j-1} - \sum_{j=1}^{s-1} \tilde{\mathbf{H}}_j a_{s-j-1} \tag{3.82}$$

for $s = 2, \dots, n$. By induction on s, $\tilde{\mathbf{H}}_s = \mathbf{H}_s$ for all s. In other words, $\mathcal{H}(\tilde{\mathbf{H}}_j) = \mathcal{H}_{\mathbf{F}}$ and so, $\varphi(\mathbf{F}) \in \mathrm{Rat}(n; p, m)$ if φ is well-defined. Again, assuming this for the moment, we have:

Claim 3.83 *φ is injective.*

Proof. If $\varphi(\mathbf{F}) = \varphi(\tilde{\mathbf{F}})$, then $\mathcal{H}_{\mathbf{F}} = \mathcal{H}_{\tilde{\mathbf{F}}}$ and $\mathbf{F} = \tilde{\mathbf{F}}$.

Claim 3.84 *φ is surjective.*

Proof. If $(\mathbf{P}_0, \dots, \mathbf{P}_{n-1}, a_0, \dots, a_{n-1}) \in \mathrm{Rat}(n; m, p)$, then $\mathcal{H} = \mathcal{H}(\mathbf{H}_j(\mathbf{P}_0, \dots, \mathbf{P}_{n-1}))$ has rank n and $\mathbf{F}_{\mathcal{H}}(z) \in \mathbf{Rat}(n; p, m)$. Then $\varphi(\mathbf{F}_{\mathcal{H}}) = (\mathbf{P}_0(x), \dots, \mathbf{P}_{n-1}(x), a_0(x), \dots, a_{n-1}(x))$ where x is a realization of $\mathcal{H}$ defining $(\mathbf{P}_0, \dots, \mathbf{P}_{n-1}, a_0, \dots, a_{n-1})$ i.e., $\varphi(\mathbf{F}_{\mathcal{H}}) = (\mathbf{P}_0(x), \dots, a_{n-1}(x)) = (\mathbf{P}_0, \dots, a_{n-1})$.

So to complete the identification, we must show that if x, x_1 are realizations of $\mathcal{H}_F$ of order n, then $a_j(x) = a_j(x_1)$, $j = 0, \dots, n-1$ and $\mathbf{P}_s(x) = \mathbf{P}_s(x_1)$, $s = 0, \dots, n-1$ where the $\mathbf{P}_s$ are given by (3.80). Clearly then we need only show that $\chi_t(x) = \chi_t(x_1)$ for $t = 1, \dots, n$ which is Theorem 3.74.

Consider the diagram

$$\begin{array}{ccc} \mathbf{Rat}(n; p, m) & \xrightarrow{\varphi} & \mathrm{Rat}(n; p, m) \\ \mathbf{L} \downarrow & & L \downarrow \\ \mathbf{Hank}(n; p, m) & \xrightarrow{\psi} & \mathrm{Hank}(n; p, m) \end{array} \tag{3.85}$$

where the map L, which is to be determined, makes the diagram commute i.e., $L \circ \varphi = \psi \circ \mathbf{L}$. Let $\psi_L : \mathbb{A}_k^{n(mp+1)} \to \mathbb{A}_k^{n(mp+1)}$ be the morphism given by $\psi_L(\mathbf{X}_1, \dots, \mathbf{X}_n, x_{n+1}, \dots, x_{2n}) = (\mathbf{Y}_1, \dots, \mathbf{Y}_n, y_{n+1}, \dots, y_{2n})$ where

$$\begin{aligned} y_{n+j} &= x_{n+j}, \quad j = 1, 2, \dots, n \\ \mathbf{Y}_1 &= \mathbf{X}_n \\ \mathbf{Y}_s &= \mathbf{X}_{n+1-s} - \sum_{j=1}^{s-1} \mathbf{Y}_j x_{n+s-j} \end{aligned} \tag{3.86}$$

for $s = 2, \dots, n$. The map ψ_L is an isomorphism with inverse $\psi_L^{-1} : \mathbb{A}_k^{n(mp+1)} \to \mathbb{A}_k^{n(mp+1)}$ given by

$$\psi_L^{-1}(\mathbf{Y}_1, \dots, \mathbf{Y}_n, y_{n+1}, \dots, y_n) = (\mathbf{X}_1, \dots, \mathbf{X}_n, x_{n+1}, \dots, x_{2n})$$

where

$$\begin{aligned} x_{n+j} &= y_{n+j}, \quad j = 1, 2, \dots, n \\ \mathbf{X}_n &= \mathbf{Y}_1 \\ \mathbf{X}_t &= \mathbf{Y}_{n+1-t} + \sum_{j=1}^{n-t} \mathbf{Y}_j x_{2n+1-j-t} \end{aligned} \tag{3.87}$$

for $t = n-1, \dots, 1$. Let L be the restriction of ψ_L to $\mathrm{Rat}(n;p,m)$.

Proposition 3.88 $L \circ \varphi = \psi \circ \mathbf{L}$.

Proof. Let $\mathbf{F} \in \mathbf{Rat}(n;p,m)$ and $\mathcal{H}_{\mathbf{F}} = \mathcal{H}(\mathbf{H}_j)$. Then $\mathbf{L}(\mathbf{F}) = \mathcal{H}_{\mathbf{F}}$ and $(\psi \circ \mathbf{L})(\mathbf{F}) = (\mathbf{H}_1, \dots, \mathbf{H}_n, -\chi_n(x), \dots, -\chi_1(x))$ (where x is an order n realization of $\mathcal{H}_{\mathbf{F}}$). Now, $\varphi(\mathbf{F}) = (\mathbf{P}_0, \dots, \mathbf{P}_{n-1}, -\chi_n(x_1), \dots, -\chi_1(x_1))$ where x_1 is an order n realization of $\mathcal{H}_{\mathbf{F}}$ and

$$\mathbf{P}_{n-1} = \mathbf{H}_1$$

$$\mathbf{P}_r = \mathbf{H}_{n-r} + \sum_{j=1}^{n-r-1} \mathbf{H}_j(-\chi_{j+r+1}(x_1))$$

for $r = n-2, \dots, 0$. But then $(\psi_L \circ \varphi)(\mathbf{F}) = (\mathbf{H}_1, \dots, \mathbf{H}_n, -\chi_n(x_1), \dots, -\chi_1(x_1)) = (\psi \circ \mathbf{L})(\mathbf{F})$ in view of Theorem 3.74:

Example 3.89 Let $n = 3$. Then $\psi_L(\mathbf{X}_1, \mathbf{X}_2, \mathbf{X}_3, x_4, x_5, x_6) = (\mathbf{X}_3, \mathbf{X}_2 - \mathbf{X}_3 x_4, \mathbf{X}_1 - \mathbf{X}_3 x_5 - \mathbf{X}_2 x_4 + \mathbf{X}_3 x_4^2, x_4, x_5, x_6) = \mathbf{Y}$ and $\psi_L^{-1}(\mathbf{Y}) = (\tilde{\mathbf{X}}_1, \tilde{\mathbf{X}}_2, \tilde{\mathbf{X}}_3, x_4, x_5, x_6)$ where

$$\begin{aligned}
\tilde{\mathbf{X}}_3 &= \mathbf{X}_3 \\
\tilde{\mathbf{X}}_2 &= \mathbf{X}_2 - \mathbf{X}_3 x_4 + x_4 \mathbf{X}_3 = \mathbf{X}_2 \\
\tilde{\mathbf{X}}_1 &= \mathbf{X}_1 - \mathbf{X}_3 x_5 - \mathbf{X}_2 x_4 + \mathbf{X}_3 x_4^2 + \mathbf{X}_3 x_5 + (\mathbf{X}_2 - \mathbf{X}_3 x_4) x_4 \\
&= \mathbf{X}_1.
\end{aligned}$$

Hence, $\psi_L^{-1} \circ \psi_L = I$. Also, $(\psi \circ \mathbf{L})(\mathbf{F}) = (\mathbf{H}_1, \mathbf{H}_2, \mathbf{H}_3, -\chi_3, -\chi_2, -\chi_1)$ so that $\psi_L^{-1} \circ (\psi \circ \mathbf{L}) = (\mathbf{H}_3 - \chi_2 \mathbf{H}_1 - \chi_3 \mathbf{H}_2, \mathbf{H}_2 - \chi_3 \mathbf{H}_1, \mathbf{H}_1, -\chi_3, -\chi_2, -\chi_1)$. But, by (3.80), $\varphi(\mathbf{F}) = (\mathbf{P}_0, \mathbf{P}_1, \mathbf{P}_2, -\chi_2, -\chi_2, -\chi_1)$ where $\mathbf{P}_0 = \mathbf{H}_3 - \mathbf{H}_1 \chi_2 - \mathbf{H}_2 \chi_3$, $\mathbf{P}_1 = \mathbf{H}_2 - \mathbf{H}_1 \chi_3$, $\mathbf{P}_2 = \mathbf{H}_1$ and so, $\varphi(\mathbf{F}) = \psi_L^{-1} \circ \psi \circ \mathbf{L}$. In other words, $L \circ \varphi = \psi \circ \mathbf{L}$.

Since L is the restriction of the isomorphism ψ_L, L is bijective and we also call $L\colon \mathrm{Rat}(n;p,m) \to \mathrm{Hank}(n;p,m)$, the *Laurent map*. We need to give $\mathrm{Rat}(n;p,m)$ and $\mathrm{Hank}(n;p,m)$ an algebraic structure which makes L an isomorphism. Note that it is by no means clear from Definition 3.69 that $\mathrm{Rat}(n;p,m)$ and $\mathrm{Hank}(n;p,m)$ have a natural (and obvious) algebraic structure. As we observed in Chapter 2, this will require projective algebraic geometry.

Exercises

(1) Prove Proposition 3.3 and Corollary 3.6.

(2) Prove Proposition 3.30 and attempt to extend it to the case of A-modules $\mathbf{M}, \mathbf{N}$ (i.e., $M(p, m)$ is replaced by $\mathrm{Hom}_A(\mathbf{M}, \mathbf{N})$). *Hint*: recall (I.16.18) that a module is faithful if $\mathrm{Ann}\,\mathbf{M} = \{x \colon x\mathbf{M} = 0\} = (0)$.

(3) Let A be a ring and let $\mathbf{H}_j \in \mathrm{Hom}_A(\mathbf{M}, \mathbf{N})$ for A-modules $\mathbf{M}, \mathbf{N}$. Consider the Hankel operator $\mathcal{H}(\mathbf{H}_j) = (\mathbf{H}_{i+j-1})$ and attempt to develop the theory in this setting ([F-3], [S-3]). You may assume finite generation (even free as a start).

(4) Show that ψ_L of (3.86) is indeed an isomorphism with inverse given by (3.87).

4
Polynomial Matrices

Let $\mathbf{F}(z)$ be an element of $\mathbf{Rat}(n;p,m)$. We shall develop the representation $\mathbf{F}(z) = \mathbf{P}(z)\mathbf{Q}^{-1}(z)$ where $\mathbf{P}(z)$ is a $p \times m$ matrix with entries in $R = k[z]$ and $\mathbf{Q}(z)$ is an $m \times m$ matrix with entries in R for which $\det \mathbf{Q}(z) \not\equiv 0$. We let $M_{r,s} = M(r,s;R) = M(r,s;k[z])$ be the set of $r \times s$ matrices with polynomial entries. Elements of $M_{r,s}$ are called *polynomial matrices.* We observe that if $\mathbf{Q}(z) \in M_{m,m}$, then $\mathbf{Q}(z)$ has an inverse in $M(m,m;k(z))$ if and only if $\det \mathbf{Q}(z) \neq 0$ as an element of R. Furthermore, $\mathbf{Q}^{-1}(z) = \text{Adj } \mathbf{Q}(z)/\det \mathbf{Q}(z)$ where Adj $\mathbf{Q}(z)$ is the adjoint of $\mathbf{Q}(z)$ and is a polynomial matrix.

Definition 4.1 $\mathbf{U}(z) \in M_{r,r}$ is *unimodular* if $\mathbf{U}(z)$ is a unit in the ring $M_{r,r}$ i.e., if there is a $\tilde{\mathbf{U}}(z)$ in $M_{r,r}$ such that $\mathbf{U}\tilde{\mathbf{U}} = I_r = \tilde{\mathbf{U}}\mathbf{U}$.

Proposition 4.2 *$\mathbf{U}(z)$ is unimodular if and only if* $\det \mathbf{U}(z)$ *is a unit in R i.e., is in k^*.*

Proof. If $\mathbf{U}(z)\tilde{\mathbf{U}}(z) = I_r$, then $\det \mathbf{U}(z) \det \tilde{\mathbf{U}}(z) = 1$ and $\det \mathbf{U}(z)$, $\det \tilde{\mathbf{U}}(z) \in R$ together give $\det \mathbf{U}(z) \in k^*$. Conversely, if $\det \mathbf{U}(z) \in k^*$, then $\det \mathbf{U}(z) \neq 0$ and $\mathbf{U}^{-1}(z) = \text{Adj } \mathbf{U}(z)/\det \mathbf{U}(z)$ has entries in R.

Let $G(r;R) = \text{GL}(r;R) = \text{GL}(r;k[z])$ be the set of $r \times r$ unimodular matrices. $G(r;R)$ is a group, called the *unimodular group*, under multiplication. We have:

Lemma 4.3 *Let $a^1(z), \ldots, a^r(z)$ be elements of R and let $d(z)$ be the greatest common divisor of the $a^i(z)$. Then there is an $\mathbf{A}(z)$ in $M_{r,r}$ with* $\det \mathbf{A}(z) = d(z)$ *and $\mathbf{A}(z)_1 = [a^i(z)]$ (i.e., the first column of $\mathbf{A}(z)$ has entries $a^i(z)$).*

P. Falb, *Methods of Algebraic Geometry in Control Theory: Part II*,
Modern Birkhäuser Classics, https://doi.org/10.1007/978-3-319-96574-1_4

Proof. We use induction on r. The result is trivial for $r = 1$. Let $d_1(z) = \gcd(a^1(z), \ldots, a^{r-1}(z))$ and, by the induction hypothesis, let $\tilde{\mathbf{A}}(z) \in M_{r-1,r-1}$ with $\det \tilde{\mathbf{A}}(z) = d_1(z)$ and $\tilde{\mathbf{A}}(z)_1 = [a^i(z)]_1^{r-1}$. Since $d(z) = \gcd(d_1(z), a^r(z))$, there are $a(z)$, $b(z)$ in R with $a(z)d_1(z) - b(z)a^r(z) = d(z)$. Set

$$\mathbf{A}(z) = \begin{bmatrix} & & \vdots & a^1(z)b(z)/d_1(z) \\ & \tilde{\mathbf{A}}(z) & \vdots & \\ & & \vdots & a^{r-1}(z)b(z)/d_1(z) \\ \cdots & \cdots & & \\ a^r(z) \; 0 & \cdots & \vdots & a(z) \end{bmatrix}. \tag{4.4}$$

Since $d_1(z)$ divides $a^i(z)$, $i = 1, \ldots, r-1$, $\mathbf{A}(z) \in M_{r,r}$ and $\mathbf{A}(z)_1 = [a^i(z)]_1^r$. Expanding $\det \mathbf{A}(z)$ via the last row, we have

$$\det \mathbf{A}(z) = a(z) \det \tilde{\mathbf{A}}(z) + (-1)^{r+1} a^r(z) \det \tilde{\mathbf{B}}(z) \tag{4.5}$$

where $\tilde{\mathbf{B}}(z)$ is obtained from $\mathbf{A}(z)$ by deleting the first column and last row. In other words,

$$\mathbf{A}(z) = \begin{bmatrix} a^1(z) & \vdots & \\ \vdots & \vdots & \tilde{\mathbf{B}}(z) \\ \vdots & \vdots & \cdots \\ a_r(z) & \vdots & 0 \cdots a(z) \end{bmatrix}. \tag{4.6}$$

Expanding $\det \tilde{\mathbf{B}}(z)$ in terms of its last column and expanding $\det \tilde{\mathbf{A}}(z)$ in terms of its first column, we get

$$\det \tilde{\mathbf{B}}(z) = (-1)b(z)d_1(z)/d_1(z) = (-1)^r b(z) \tag{4.7}$$

and so,

$$\det \mathbf{A}(z) = a(z)d_1(z) - a^r(z)b(z) = d(z) \tag{4.8}$$

which completes the proof.

Corollary 4.9 *If $a^1(z), \ldots, a^r(z)$ are relatively prime (i.e., $d(z) = 1$), then there is a $\mathbf{U}(z) \in G(r; R)$ with $\det \mathbf{U}(z) = 1$ and $(a^1(z), \ldots, a^r(z))$ as any row or column.*

Definition 4.10 An *elementary column operation* is one of the following: (i) an interchange of columns; (ii) multiplication of a column by a unit in R; and, (iii) addition of $a(z)$ times a column to another column.

Example 4.11 Let $r = 3$ and let

$$\mathbf{U}_1(z) = \begin{bmatrix} 0 & 1 & 0 \\ 1 & 0 & 0 \\ 0 & 0 & 1 \end{bmatrix}, \quad \mathbf{U}_2(z) = \begin{bmatrix} 1 & 0 & 0 \\ 0 & u & 0 \\ 0 & 0 & 1 \end{bmatrix},$$

$$\mathbf{U}_3(z) = \begin{bmatrix} 1 & a(z) & 0 \\ 0 & 1 & 0 \\ 0 & 0 & 1 \end{bmatrix}$$

then $\mathbf{A}(z)\mathbf{U}_1(z)$ is $\mathbf{A}(z)$ with columns 1 and 2 interchanged, $\mathbf{A}(z)\mathbf{U}_2(z)$ is $\mathbf{A}(z)$ with column 2 multiplied by u, and $\mathbf{A}(z)\mathbf{U}_3(z)$ has column 2 equal to the sum of $a(z)$ times column 1 of $\mathbf{A}(z)$ and column 2 of $\mathbf{A}(z)$.

Proposition 4.12 *All elementary column operations can be achieved via right multiplication by a $\mathbf{U}(z)$ in $G(r; R)$ (called an elementary matrix) and conversely, every $U(z)$ is a product of elementary matrices.*

Proof. Exercise or [B-1], [W-4].

Of course, similar definitions and results apply to row operations.

Definition 4.13 Let $p(z)$, $p_1(z)$ be polynomials. $p_1(z)$ is *reduced modulo $p(z)$* if: (a) $\deg p_1(z) < \deg p(z)$ when $\deg p(z) \neq 0$; and, (b) $p_1(z) = \epsilon p(z)$, ϵ a unit, when $\deg p(z) = 0$. If $\mathbf{v}(z) = [\psi(z)\ a_1(z) \cdots a_{r-1}(z)]$ is a polynomial row vector, then $\mathbf{v}(z)$ is *row reduced modulo $\psi(z)$* (or, simply row reduced) if all $a_i(z)$ are reduced modulo $\psi(z)$.

Proposition 4.14 *If $\mathbf{A}(z)$ is an upper triangular element of $M_{r,r}$, then there is a $\mathbf{U}(z)$ in $G(r; R)$ such that $\mathbf{A}(z)\mathbf{U}(z)$ is row reduced modulo the diagonal. In other words, there is a $\mathbf{U}(z)$ in $G(r; R)$ with*

$$\mathbf{A}(z)\mathbf{U}(z) = \begin{bmatrix} \psi_1(z) & b_{12}(z) & \cdots & b_{1r}(z) \\ 0 & \psi_2(z) & \cdots & b_{2r}(z) \\ \vdots & \vdots & & \\ 0 & 0 & & \psi_r(z) \end{bmatrix} \tag{4.15}$$

where $b_{ij}(z)$, $i < j$ is reduced modulo $\psi_i(z)$.

Proof. Let $r = 2$ so that

$$\mathbf{A}(z) = \begin{bmatrix} \psi_1(z) & a_{12}(z) \\ 0 & \psi_2(z) \end{bmatrix}.$$

Then $a_{12}(z) = a(z)\psi_1(z) + r(z)$ with $\deg r(z) < \deg \psi(z)$. If

$$\mathbf{U}(z) = \begin{bmatrix} 1 & -a(z) \\ 0 & 1 \end{bmatrix}$$

then $\mathbf{U}(z) \in G(2; R)$ and

$$\mathbf{A}(z)\mathbf{U}(z) = \begin{bmatrix} \psi_1(z) & -a(z)\psi_1(z) + a_{12}(z) \\ 0 & \psi_2(z) \end{bmatrix} = \begin{bmatrix} \psi_1(z) & r(z) \\ 0 & \psi_2(z) \end{bmatrix}$$

is of the required form. Now use induction on r. Let

$$\mathbf{A}(z) = \begin{bmatrix} & \vdots & a_{1r}(z) \\ \mathbf{A}_1(z) & \vdots & \\ \cdots & \vdots & a_{r-1r}(z) \\ 0 & \vdots & \\ & & \psi_r(z) \end{bmatrix}$$

and let $\tilde{\mathbf{U}}_1(z)$ be $r-1 \times r-1$ in $G(r-1; R)$ which reduces $\mathbf{A}_1(z)$ and let

$$\mathbf{U}_1(z) = \begin{bmatrix} \tilde{\mathbf{U}}_1(z) & \vdots & 0 \\ & \vdots & 0 \\ \cdots\cdots & & \\ 0 & \vdots & 1 \end{bmatrix}.$$

Then $\mathbf{U}_1(z) \in G(r; R)$ and

$$\mathbf{A}(z)\mathbf{U}_1(z) = \begin{bmatrix} & \vdots & a_{1r}(z) \\ \mathbf{A}_1(z)\tilde{\mathbf{U}}_1(z) & \vdots & \\ \cdots\cdots & \vdots & a_{r-1r}(z) \\ 0 & \vdots & \psi_r(z) \end{bmatrix}.$$

Now $a_{jr}(z) = a_j(z)\psi_j(z) + r_j(z)$ with $\deg r_j(z) < \deg \psi_j(z)$. Let $\mathbf{U}_2(z)$ be the element of $G(r; R)$ with

$$\mathbf{U}_2(z) = \left[\begin{array}{c:c} & -a_1(z) \\ I_{r-1} & \vdots \\ \cdots\cdots & -a_{r-1}(z) \\ 0 & 1 \end{array}\right].$$

Then $\mathbf{A}(z)\mathbf{U}_1(z)\mathbf{U}_2(z)$ is of the required form.

Theorem 4.15 *Let $\mathbf{A}(z) \in M_{r,r}$. Then there is a $\mathbf{U}(z)$ in $G(r; R)$ such that $\mathbf{A}(z)\mathbf{U}(z)$ is upper triangular and row reduced modulo the diagonal.*

Proof. In view of the previous Proposition, it is enough to show that there is a unimodular $\mathbf{U}(z)$ with $\mathbf{A}(z)\mathbf{U}(z)$ upper triangular. Now either the last row of $\mathbf{A}(z)$ is 0 (in which case the result follows by induction) or there is an element of lowest degree which we may assume (by column interchange) is $a_{rr}(z)$. Let $a_{rj}(z) = a_j(z)a_{rr}(z) + r_j(z)$ with $\deg r_j(z) < \deg a_{rr}(z)$. If $r_j(z) \neq 0$, then, by elementary column operations, we eventually have a unimodular $\mathbf{U}_1(z)$ (as degrees decrease) such that

$$\mathbf{A}(z)\mathbf{U}_1(z) = \mathbf{A}_1(z) = (a_{ij}^1(z))$$

with

$$a_{rj}^1(z) = \alpha_j^1(z)a_{rr}^1(z)$$

for $j = 1, \ldots, r-1$ (i.e., $a_{rr}^1(z)$ is the greatest common divisor of the last row). Hence there is a unimodular $\mathbf{U}_2(z)$ with

$$\mathbf{A}_1(z)\mathbf{U}_2(z) = \left[\begin{array}{c:c} \mathbf{B}_{r-1,r-1}(z) & * \\ & \\ 0 & a_{rr}^1(z) \end{array}\right].$$

By induction, there is a $\tilde{\mathbf{U}}(z)$ in $G(r-1; R)$ with $\mathbf{B}\tilde{\mathbf{U}}$ upper triangular. If

$$\mathbf{U}_3(z) = \left[\begin{array}{c:c} \tilde{\mathbf{U}}(z) & 0 \\ \cdots\cdots & \\ 0 & 1 \end{array}\right]$$

then $\mathbf{U}_3(z) \in G(r; R)$ and

$$\mathbf{A}(z)\mathbf{U}_1(z)\mathbf{U}_2(z)\mathbf{U}_3(z) = \begin{bmatrix} \mathbf{B}(z)\tilde{\mathbf{U}}(z) & * \\ 0 & \psi_r(z) \end{bmatrix}$$

is upper triangular.

Corollary 4.16 *If $\mathbf{A}(z) \in M_{q,r}$ with $q \geq r$, then there is a $\mathbf{U}(z)$ in $G(r; R)$ such that*

$$\mathbf{A}(z)\mathbf{U}(z) = \begin{bmatrix} \mathbf{A}_t(z) \\ 0 \end{bmatrix}$$

where $\mathbf{A}_t(z)$ is an upper triangular element of $M_{r,r}$ which is row reduced modulo the diagonal.

Analogous results hold for row operations.

Definition 4.17 If $\mathbf{A}(z) \in M_{r,s}$, then $\mathbf{D}(z) \in M_{s,s}$ is a *right divisor* of $\mathbf{A}(z)$ if $\mathbf{A}(z) = \mathbf{A}_1(z)\mathbf{D}(z)$ for some $\mathbf{A}_1(z) \in M_{r,s}$. If $\mathbf{A}(z), \mathbf{B}(z) \in M_{r,s}$, then $\mathbf{D}(z)$ is a *common right divisor* if $\mathbf{D}(z)$ is a right divisor of both $\mathbf{A}(z)$ and $\mathbf{B}(z)$. $\mathbf{D}(z)$ is a *greatest common right divisor* (gcrd) of $\mathbf{A}(z)$ and $\mathbf{B}(z)$ if $\mathbf{D}(z)$ is a common right divisor and is a left multiple of every common right divisor.

Theorem 4.18 *If $\mathbf{P}(z) \in M_{p,m}$ and $\mathbf{Q}(z) \in M_{m,m}$, then there is a gcrd $\mathbf{D}(z)$ of $\mathbf{P}(z)$ and $\mathbf{Q}(z)$ with $\mathbf{D}(z) = \mathbf{A}(z)\mathbf{P}(z) + \mathbf{B}(z)\mathbf{Q}(z)$, $\mathbf{A}(z) \in M_{m,p}$ and $\mathbf{B}(z) \in M_{m,m}$.*

Proof. Consider the element $\mathbf{L}(z)$ of $M_{p+m,2m}$ given by

$$\mathbf{L}(z) = \begin{bmatrix} \mathbf{P}(z) & \mathbf{O}_{p,m} \\ \mathbf{Q}(z) & \mathbf{O}_{m,m} \end{bmatrix}.$$

Then, by virtue of the analog of Corollary 4.16 for row operations, there is a $\mathbf{U}(z) \in G(p+m; R)$ such that

$$\mathbf{U}(z)\mathbf{L}(z) = \begin{bmatrix} \mathbf{D}(z) & 0 \\ 0 & 0 \end{bmatrix}.$$

If

$$\mathbf{U}(z) = \begin{bmatrix} \mathbf{U}_{11}(z) & \mathbf{U}_{12}(z) \\ \mathbf{U}_{21}(z) & \mathbf{U}_{22}(z) \end{bmatrix}$$

then $\mathbf{D}(z) = \mathbf{U}_{11}(z)\mathbf{P}(z) + \mathbf{U}_{12}(z)\mathbf{Q}(z)$. We claim $\mathbf{D}(z)$ is a gcrd. If $\mathbf{U}^{-1}(z) = \mathbf{V}(z)$ with

$$\mathbf{V}(z) = \begin{bmatrix} \mathbf{V}_{11}(z) & \mathbf{V}_{12}(z) \\ \mathbf{V}_{21}(z) & \mathbf{V}_{22}(z) \end{bmatrix}$$

then

$$\mathbf{L}(z) = \begin{bmatrix} \mathbf{P}(z) & 0 \\ \mathbf{Q}(z) & 0 \end{bmatrix} = \mathbf{V}(z)(\mathbf{U}(z)\mathbf{L}(z)) = \begin{bmatrix} \mathbf{V}_{11}(z)\mathbf{D}(z) & 0 \\ \mathbf{V}_{21}(z)\mathbf{D}(z) & 0 \end{bmatrix}$$

so that $\mathbf{P}(z) = \mathbf{V}_{11}(z)\mathbf{D}(z)$, $\mathbf{Q}(z) = \mathbf{V}_{21}(z)\mathbf{D}(z)$ and $\mathbf{D}(z)$ is a common right divisor. If $\mathbf{D}_1(z)$ is any common right divisor, then $\mathbf{P}(z) = \mathbf{P}_1(z)\mathbf{D}_1(z)$, $\mathbf{Q}(z) = \mathbf{Q}_1(z)\mathbf{D}_1(z)$ and $\mathbf{D}(z) = (\mathbf{U}_{11}(z)\mathbf{P}_1(z) + \mathbf{U}_{12}(z)\mathbf{Q}_1(z)).\mathbf{D}_1(z)$ is a left-multiple of $\mathbf{D}_1(z)$.

Definition 4.19 If $\mathbf{P}(z) \in M_{p,m}$ and $\mathbf{Q}(z) \in M_{m,m}$, then $\mathbf{P}(z)$ and $\mathbf{Q}(z)$ are *(right) coprime* if the gcrd $\mathbf{D}(z)$ of $\mathbf{P}(z)$ and $\mathbf{Q}(z)$ is unimodular.

Corollary 4.20 *$\mathbf{P}(z)$ and $\mathbf{Q}(z)$ are coprime if and only if I is a gcrd. In that case, $I = \mathbf{A}(z)\mathbf{P}(z) + \mathbf{B}(z)\mathbf{Q}(z)$ for suitable $\mathbf{A}(z)$, $\mathbf{B}(z)$.*

Proof. If I is a gcrd, then $\mathbf{P}(z)$, $\mathbf{Q}(z)$ are coprime since I is unimodular. Conversely, if $\mathbf{P}(z)$, $\mathbf{Q}(z)$ are coprime, then $\mathbf{D}(z) = \mathbf{A}(z)\mathbf{P}(z) + \mathbf{B}(z)\mathbf{Q}(z)$ is unimodular and $I = (\mathbf{D}^{-1}(z)\mathbf{A}(z))\mathbf{P}(z) + (\mathbf{D}^{-1}(z)\mathbf{B}(z))\mathbf{Q}(z)$. Clearly I is a common right divisor. If $\mathbf{D}_1(z)$ is a common right divisor, then $\mathbf{D}(z) = \mathbf{E}(z)\mathbf{D}_1(z)$ and $I = (\mathbf{D}^{-1}(z)\mathbf{E}(z))\mathbf{D}_1(z)$ is a left multiple of $\mathbf{D}_1(z)$.

Corollary 4.21 *If $I = \mathbf{A}(z)\mathbf{P}(z) + \mathbf{B}(z)\mathbf{Q}(z)$, then $\mathbf{P}(z)$ and $\mathbf{Q}(z)$ are coprime.*

Proof. If $I = \mathbf{A}(z)\mathbf{P}(z) + \mathbf{B}(z)\mathbf{Q}(z)$ and $\mathbf{D}(z)$ is a gcrd, then $I = (\mathbf{A}(z)\mathbf{P}_1(z) + \mathbf{B}(z)\mathbf{Q}_1(z))\mathbf{D}(z)$ where $\mathbf{P}(z) = \mathbf{P}_1(z)\mathbf{D}(z)$, $\mathbf{Q}(z) = \mathbf{Q}_1(z)\mathbf{D}(z)$ and $\mathbf{D}(z)$ is unimodular.

Definition 4.22 If $\mathbf{F}(z)$ is a $p \times m$ transfer matrix, then $(\mathbf{P}(z), \mathbf{Q}(z))$ with $\mathbf{P}(z) \in M_{p,m}$ and $\mathbf{Q}(z) \in M_{m,m}$ is a *(right) realization of* $\mathbf{F}(z)$ if $\mathbf{F}(z) = \mathbf{P}(z)\mathbf{Q}^{-1}(z)$.

We have:

Proposition 4.23 *If $\mathbf{F}(z)$ is rational, then there is a realization $(\mathbf{P}(z), \mathbf{Q}(z))$.*

Proof. Let $q(z)$ be the (monic) least common multiple of the denominators of the $f_{ij}(z)$. Then $q(z)\mathbf{F}(z) = \mathbf{P}_0 + \mathbf{P}_1 z + \cdots + \mathbf{P}_{n-1}z^{n-1}$. Let $\mathbf{P}(z) = \mathbf{P}_0 + \cdots + \mathbf{P}_{n-1}z^{n-1}$ and $\mathbf{Q}(z) = q(z)I_m$. Then $\mathbf{Q}^{-1}(z) = 1/q(z) \cdot I_m$ and $\mathbf{P}(z)\mathbf{Q}^{-1}(z) = \mathbf{F}(z)$.

Corollary 4.24 *If $\mathbf{F}(z)$ is rational, then there is a realization $(\mathbf{P}(z), \mathbf{Q}(z))$ with $\mathbf{P}(z)$ and $\mathbf{Q}(z)$ coprime. If $(\mathbf{P}(z), \mathbf{Q}(z))$ and $(\mathbf{P}_1(z), \mathbf{Q}_1(z))$ are coprime realizations of $\mathbf{F}(z)$, then there is a $\mathbf{U}(z) \in G(m; R)$ with $\mathbf{P}(z) = \mathbf{P}_1(z)\mathbf{U}(z)$ and $\mathbf{Q}(z) = \mathbf{Q}_1(z)\mathbf{U}(z)$ (i.e., $\mathbf{P}(z) \equiv_{G(m;R)} \mathbf{P}_1(z)$ and $\mathbf{Q}(z) \equiv_{G(m;R)} \mathbf{Q}_1(z)$, equivalent under the action of $G(m; R)$). Moreover, $\deg \det \mathbf{Q}(z) = \deg \det \mathbf{Q}_1(z)$.*

Proof. Let $(\tilde{\mathbf{P}}(z), \tilde{\mathbf{Q}}(z))$ be a realization with $\tilde{\mathbf{D}}(z)$ a gcrd of $\tilde{\mathbf{P}}(z)$, $\tilde{\mathbf{Q}}(z)$. Then $\mathbf{P}(z)\tilde{\mathbf{D}}(z) = \tilde{\mathbf{P}}(z)$ and $\mathbf{Q}(z)\tilde{\mathbf{D}}(z) = \tilde{\mathbf{Q}}(z)$ for some $\mathbf{P}(z)$, $\mathbf{Q}(z)$. Since $\tilde{\mathbf{Q}}^{-1}(z)$ exists, $\mathbf{Q}(z)$ and $\tilde{\mathbf{D}}(z)$ are invertible. It follows that

$$\begin{aligned}
\mathbf{P}(z)\mathbf{Q}^{-1}(z) &= \mathbf{P}(z)(\tilde{\mathbf{D}}(z)\tilde{\mathbf{D}}^{-1}(z))\mathbf{Q}^{-1}(z) \\
&= (\mathbf{P}(z)\tilde{\mathbf{D}}(z))(\tilde{\mathbf{D}}^{-1}(z)\mathbf{Q}^{-1}(z)) \\
&= \tilde{\mathbf{P}}(z)\tilde{\mathbf{Q}}^{-1}(z) = \mathbf{F}(z).
\end{aligned}$$

Hence, $(\mathbf{P}(z), \mathbf{Q}(z))$ is a realization. If $\mathbf{D}(z)$ is a gcrd of $\mathbf{P}(z)$, $\mathbf{Q}(z)$, then $\mathbf{D}(z)\tilde{\mathbf{D}}(z)$ is a common right divisor of $\tilde{\mathbf{P}}(z)$, $\tilde{\mathbf{Q}}(z)$ and so $\tilde{\mathbf{D}}(z)$ is a left multiple of $\mathbf{D}(z)\tilde{\mathbf{D}}(z)$ i.e., $\tilde{\mathbf{D}}(z) = \mathbf{V}(z)\mathbf{D}(z)\tilde{\mathbf{D}}(z)$ for some $\mathbf{V}(z) \in M_{m,m}$. But then $\mathbf{V}(z)\mathbf{D}(z) = I$ and the polynomial matrix $\mathbf{D}(z)$ has a polynomial (not a rational) inverse. In other words, $\mathbf{D}(z)$ is unimodular. Now, for the second part, let $\mathbf{U}(z) = \mathbf{Q}_1^{-1}(z)\mathbf{Q}(z)$. Suppose we show that $\mathbf{U}(z)$ is a polynomial matrix. Then $\mathbf{U}(z)$ would a common right divisor of $\mathbf{P}(z)$, $\mathbf{Q}(z)$ (since $\mathbf{Q}(z) = \mathbf{Q}_1(z)\mathbf{U}(z)$ and $\mathbf{P}(z) = \mathbf{P}(z)\mathbf{Q}^{-1}(z)\mathbf{Q}(z) = \mathbf{P}_1(z)\mathbf{Q}_1^{-1}(z)\mathbf{Q}(z) = \mathbf{P}_1(z)\mathbf{U}(z)$), and then would be unimodular. Now $\mathbf{Q}_1^{-1}(z) = (\text{Adj } \mathbf{Q}_1(z))/\det \mathbf{Q}_1(z)$, $\mathbf{P}_1(z)\mathbf{Q}_1^{-1}(z) = \mathbf{P}_1(z)\text{Adj } \mathbf{Q}_1(z)/\det \mathbf{Q}_1(z)$, $\mathbf{Q}_1(z)\text{Adj } \mathbf{Q}_1(z) = \det \mathbf{Q}_1(z) I$ and $\mathbf{P}_1(z)\mathbf{Q}_1^{-1}(z) = \mathbf{P}(z)\mathbf{Q}^{-1}(z)$. Since $\mathbf{P}_1(z)$, $\mathbf{Q}_1(z)$ are coprime, there are polynomial matrices $\mathbf{A}_1(z)$, $\mathbf{B}_1(z)$ with $I = \mathbf{A}_1(z)\mathbf{P}_1(z) + \mathbf{B}_1(z)\mathbf{Q}_1(z)$. It follows that

$$\begin{aligned}
(\text{Adj } \mathbf{Q}_1(z))\mathbf{Q}(z) &= \mathbf{A}_1(z)\mathbf{P}_1(z)\text{Adj } \mathbf{Q}_1(z)\mathbf{Q}(z) \\
&\quad + \mathbf{B}_1(z)\mathbf{Q}_1(z)(\text{Adj } \mathbf{Q}_1(z))\mathbf{Q}(z) \\
&= \mathbf{A}_1(z)\det \mathbf{Q}_1(z)(\mathbf{P}_1(z)\mathbf{Q}_1^{-1}(z))\mathbf{Q}(z) + \mathbf{B}_1(z)\det \mathbf{Q}_1(z)\mathbf{Q}(z) \\
&= \mathbf{A}_1(z)\det \mathbf{Q}_1(z)\mathbf{P}(z)\mathbf{Q}_1^{-1}(z)\mathbf{Q}(z) + \mathbf{B}_1(z)(\det \mathbf{Q}_1(z))\mathbf{Q}(z) \\
&= (\det \mathbf{Q}_1(z))(\mathbf{A}_1(z)\mathbf{P}(z) + \mathbf{B}_1(z)\mathbf{Q}(z).
\end{aligned}$$

and hence, that

$$\mathbf{U}(z) = \frac{(\text{Adj } \mathbf{Q}_1(z))\mathbf{Q}(z)}{\det \mathbf{Q}_1(z)} = \mathbf{A}_1(z)\mathbf{P}(z) + \mathbf{B}_1(z)\mathbf{Q}(z)$$

is polynomial. But $\det \mathbf{Q}(z)\mathbf{U}(z) = \det \mathbf{Q}(z) \cdot \det \mathbf{U}(z) = \det \mathbf{Q}_1(z)$ gives the last assertion.

We let $I_m^n = \{(i_1, \dots, i_m) \colon 1 \le i_1 < \cdots < i_m \le n\}$. If $\boldsymbol{\alpha} \in I_p^r$, $\boldsymbol{\beta} \in I_p^s$, $\boldsymbol{\alpha} = (i_1, \dots, i_p)$, $\boldsymbol{\beta} = (j_1, \dots, j_p)$ and $\mathbf{X} \in M_{r,s}$, then we let $|\mathbf{X}|_{\boldsymbol{\beta}}^{\boldsymbol{\alpha}}$ denote the determinant of the $p \times p$ submatrix of $\mathbf{X}$ indexed by rows $(i_1, \dots, i_p)$ and columns $(j_1, \dots, j_p)$.

Theorem 4.25 (Binet-Cauchy Formula [R-2], [W-4]) *If $\mathbf{A} \in M_{r,m}$ and $\mathbf{B} \in M_{m,s}$ and if $\boldsymbol{\alpha} \in I_p^r$, $\boldsymbol{\beta} \in I_p^s$, then*

$$|\mathbf{AB}|_{\boldsymbol{\beta}}^{\boldsymbol{\alpha}} = \sum_{\boldsymbol{\gamma} \in I_p^m} |\mathbf{A}|_{\boldsymbol{\gamma}}^{\boldsymbol{\alpha}} |\mathbf{B}|_{\boldsymbol{\beta}}^{\boldsymbol{\gamma}}. \tag{4.26}$$

[Note if $r = m = s$ and $p = m$, then this is the familiar $\det(\mathbf{AB}) = (\det \mathbf{A}) \cdot (\det \mathbf{B})$].

The proof is left as an exercise.

Theorem 4.27 (The Smith Form) *Suppose that* $\mathbf{D}(z) \in M_{r,s}$ *has rank* ρ. *Then there are* $\mathbf{U}(z) \in G(r; R)$, $\mathbf{V}(z) \in G(s; R)$ *such that*

$$\mathbf{U}(z)\mathbf{D}(z)\mathbf{V}(z) = \begin{bmatrix} d_1(z) & 0 & & & 0 \\ 0 & d_2(z) & & & \\ & & \ddots & & \\ \vdots & & & d_\rho(z) & \\ 0 & & & & 0 \end{bmatrix} \tag{4.28}$$

where $d_i(z)$ *divides* $d_{i+1}(z)$ *for* $i = 1, \ldots, \rho - 1$ *and* $\Delta_\tau(z) = d_1(z) \cdots d_\tau(z)$ *is a greatest common divisor of all* $\tau \times \tau$ *minors of* $\mathbf{D}(z)$. *The* $d_i(z)$ *are unique to within unit factors and are called the invariant factors of* $\mathbf{D}(z)$.

Proof. Suppose there are $\tilde{\mathbf{U}}(z)$, $\tilde{\mathbf{V}}(z)$ such that $\tilde{\mathbf{U}}(z)\mathbf{D}(z)\tilde{\mathbf{V}}(z)$ is diagonal i.e.,

$$\tilde{\mathbf{U}}(z)\mathbf{D}(z)\tilde{\mathbf{V}}(z) = \operatorname{diag}[\delta_1(z) \cdots \delta_\rho(z)\ 0 \cdots 0]$$

with $\delta_i(z) \neq 0$. We claim then that $\operatorname{diag}[\delta_1(z) \cdots \delta_\rho(z)\ 0 \cdots 0]$ is "equivalent" to $\operatorname{diag}[d_1(z) \cdots d_\rho(z)\ 0 \cdots 0]$. Let $d_1(z) = \gcd(\delta_1(z), \ldots, \delta_\rho(z))$. By elementary column operations,

$$\operatorname{diag}[\delta_1(z) \cdots \delta_\rho(z)\ 0 \cdots 0]\mathbf{V}_1(z) = \begin{bmatrix} \delta_1(z) & 0 & & 0 & 0 \\ \delta_2(z) & \delta_2(z) & & & \cdot \\ \vdots & 0 & \ddots & & \cdot \\ \delta_p(z) & 0 & & \delta_p(z) & \cdot \\ 0 & \cdots & & 0 & 0 \end{bmatrix}$$

and there is a unimodular $\mathbf{U}_1(z)$ such that

$$\mathbf{U}_1(z) \operatorname{diag}[\delta_1(z) \cdots \delta_p(1)\ 0 \cdots 0]\mathbf{V}_1(z) = \begin{bmatrix} d_1(z) & * \\ 0 & \mathbf{D}_1(z) \end{bmatrix}$$

where all entries in $*$ and $\mathbf{D}_1(z)$ are in the ideal $(\delta_2(z), \ldots, \delta_\rho(z))$ and are, therefore, multiples of $d_1(z)$. It follows that there are unimodular $\mathbf{U}_2(z)$, $\mathbf{V}_2(z)$ with

$$\mathbf{U}_2(z) \begin{bmatrix} d_1(z) & * \\ 0 & \mathbf{D}_1(z) \end{bmatrix} \mathbf{V}_2(z) = \begin{bmatrix} d_1(z) & 0 \\ 0 & \operatorname{diag}[\delta_1'(z) \cdots \delta_{\rho-1}'(z)\ 0 \cdots 0] \end{bmatrix}$$

and the result follows by induction. So it remains to show that $\mathbf{D}(z)$ is "equivalent" to a diagonal matrix. But, by elementary row and column operations, we can reduce $\mathbf{D}(z)$ to the form

$$\begin{bmatrix} \delta_1(z) & 0 \\ 0 & \mathbf{D}'(z) \end{bmatrix}$$

and then apply induction. The assertion about $\Delta_\tau(z)$ follows from the Binet-Cauchy formula.

Corollary 4.29 *$\mathbf{P}(z) \in M_{p,m}$ and $\mathbf{Q}(z) \in M_{m,m}$ are (right) coprime if and only if* 1 *is a* gcd *of all $m \times m$ minors of*

$$\mathbf{L}(z) = \begin{bmatrix} \mathbf{P}(z) \\ \mathbf{Q}(z) \end{bmatrix}$$

i.e., $\{|\mathbf{L}(z)|^{\boldsymbol{\alpha}}_{1\cdots m} : \boldsymbol{\alpha} \in I^{p+m}_m\}$ *are relatively prime.*

Proof. If 1 is a gcd, then all invariant factors of $\mathbf{L}(z)$ are units and there are $\mathbf{U}(z) \in G(p+m; R)$, $\mathbf{V}(z) \in G(m; R)$ such that

$$\mathbf{U}(z) \begin{bmatrix} \mathbf{P}(z) \\ \mathbf{Q}(z) \end{bmatrix} \mathbf{V}(z) = \begin{bmatrix} I_m \\ 0 \end{bmatrix}.$$

If

$$\mathbf{U}(z) = \begin{bmatrix} \mathbf{U}_{11}(z) & \mathbf{U}_{12}(z) \\ \mathbf{U}_{21}(z) & \mathbf{U}_{22}(z) \end{bmatrix}$$

where $\mathbf{U}_{11}(z) \in M_{m,p}$, $\mathbf{U}_{12}(z) \in M_{m,m}$, $\mathbf{U}_{21}(z) \in M_{p,p}$ and $\mathbf{U}_{22}(z) \in M_{p,m}$, then $(\mathbf{U}_{11}(z)\mathbf{P}(z) + \mathbf{U}_{12}(z)\mathbf{Q}(z)) \cdot \mathbf{V}(z) = I$. Then

$$\mathbf{V}(z)(\mathbf{U}_{11}(z)\mathbf{P}(z) + \mathbf{U}_{12}(z)\mathbf{Q}(z))\mathbf{V}(z)\mathbf{V}^{-1}(z) = \mathbf{V}(z)I\mathbf{V}^{-1}(z) = I$$

and so, $(\mathbf{V}(z)\mathbf{U}_{11}(z))\mathbf{P}(z) + (\mathbf{V}(z)\mathbf{U}_{12}(z)) \cdot \mathbf{Q}(z) = I$ i.e., $\mathbf{P}(z)$, $\mathbf{Q}(z)$ are coprime by Corollary 4.21. On the other hand, if $\mathbf{P}(z)$, $\mathbf{Q}(z)$ are coprime, then, by Corollary 4.20, there is a unimodular $\mathbf{U}(z)$ such that

$$\mathbf{U}(z) \begin{bmatrix} \mathbf{P}(z) \\ \mathbf{Q}(z) \end{bmatrix} = \begin{bmatrix} \mathbf{D}(z) \\ 0 \end{bmatrix}$$

with $\mathbf{D}(z)$ unimodular. Let $\mathbf{V}(z) = \mathbf{D}^{-1}(z)$. Then

$$\mathbf{U}(z)\mathbf{L}(z)\mathbf{V}(z) = \begin{bmatrix} I_m \\ 0 \end{bmatrix}$$

and, by the Smith Form theorem, 1 is a gcd of the $m \times m$ minors of $\mathbf{L}(z)$.

The condition on the $m \times m$ minors is called *zero coprimeness* (over general integral domains this is not equivalent to being coprime).

Definition 4.30 A realization $(\mathbf{P}(z), \mathbf{Q}(z))$ of $\mathbf{F}(z)$ is *minimal* if $\mathbf{P}(z)$ and $\mathbf{Q}(z)$ are coprime.

We have shown that if $\mathbf{F}(z) \in \mathbf{Rat}(n; p, m)$, then there are minimal realizations of $\mathbf{F}(z)$ and any two minimal realizations are equivalent under the action (on the right) of the unimodular group $G(m; R)$.

Corollary 4.31 *If $\mathbf{F}(z)$ is a $p \times m$ transfer matrix, then there is a $\mathbf{U}(z)$ in $G(p; R)$ and a $\mathbf{V}(z)$ in $G(m; R)$ such that*

$$\mathbf{U}(z)\mathbf{F}(z)\mathbf{V}(z) = \operatorname{diag}\left[\epsilon_1(z)/\psi_1(z), \ldots, \epsilon_t(z)/\psi_t(z)\ 0 \cdots 0\right] \tag{4.32}$$

where $\psi_{i+1}(z)$ divides $\psi_i(z)$, $\epsilon_i(z)$ divides $\epsilon_{i+1}(z)$, and $\psi_1(z) = q_F(z)$ is the least common multiple of the denominators of the $(f_{ij}(z))$ i.e., the minimal annihilator of $\mathcal{H}_{\mathbf{F}}$.

Proof. In view of Proof 2 of Theorem 3.38, $q_F(z)\mathbf{F}(z) = \mathbf{N}(z)$ is a polynomial matrix and there are unimodular $\mathbf{U}(z)$, $\mathbf{V}(z)$ with

$$\mathbf{U}(z)\mathbf{N}(z)\mathbf{V}(z) = \operatorname{diag}\left[\lambda_1(z) \cdots \lambda_i(z)\ 0 \cdots\right] \tag{4.33}$$

where $\lambda_i(z)$ divides $\lambda_{i+1}(z)$. If $\Delta_i(z) =$ gcd of the $i \times i$-minors of $\mathbf{N}(z)$, then $\lambda_i(z) = \Delta_i(z)/\Delta_{i-1}(z)$ (with $\Delta_0(z) \equiv 1$). Let $\lambda_i(z)/q_F(z) = \epsilon_i(z)/\psi_i(z)$ with $\epsilon_i(z)$, $\psi_i(z)$ relatively prime. Then $\lambda_1/q_F = \epsilon_1/\psi_1$. If $\psi_1 \neq q_F$, then λ_1, q_F are not relatively prime and hence, Δ_1, q_F are not relatively prime which contradicts the fact that $q_F(z)$ is the least common multiple of the denominators of the $f_{ij}(z)$. Hence, $\lambda_1 = q_F\epsilon_1/\psi_1 = \epsilon_1$. Now $\lambda_1 = \epsilon_1\psi_1/\psi_2$ so that $\lambda_1 \mid \lambda_2$ implies $\epsilon_1 \mid \epsilon_2\psi_1/\psi_2$ or, equivalently, $\epsilon_1\psi_2 \mid \epsilon_2\psi_1$. Since ϵ_1, ψ_1 are relatively prime and ϵ_2, ψ_2 are relatively prime, $\psi_2 \mid \psi_1$ and $\epsilon_1 \mid \epsilon_2$. The rest follows similarly.

Corollary 4.34 *Let*

$$\mathbf{E}(z) = \operatorname{diag}[\epsilon_1(z) \cdots \epsilon_t(z)\ 0 \cdots 0]$$

and

$$\mathbf{D}(z) = \operatorname{diag}[\psi_1(z) \cdots \psi_t(z)\ 1 \cdots 1].$$

Then $\mathbf{E}(z)$, $\mathbf{D}(z)$ are (right) coprime, $\det \mathbf{D}(z) \neq 0$, and $\mathbf{F}(z) = \mathbf{P}(z)\mathbf{Q}^{-1}(z)$ where $\mathbf{P}(z) = \mathbf{U}^{-1}(z)\mathbf{E}(z)$, $\mathbf{Q}(z) = \mathbf{V}(z)\mathbf{D}(z)$ ($\mathbf{U}(z)$, $\mathbf{V}(z)$ unimodular as in (4.33)) is a coprime (minimal) realization of $\mathbf{F}(z)$ with $\deg \det \mathbf{Q}(z) = \Sigma \deg \psi_i(z)$.

We recall that the degree of $\mathbf{F}(z)$ is the rank of $\mathcal{H}_{\mathbf{F}}$. We shall eventually show that $\rho(\mathcal{H}_{\mathbf{F}}) = \deg\det \mathbf{Q}(z)$ where $(\mathbf{P}(z), \mathbf{Q}(z))$ is a minimal realization of $\mathbf{F}(z)$. We begin with:

Definition 4.35 The *degree of a polynomial vector* is the maximum degree of its entries. If $\mathbf{A}(z) \in M_{r,s}$, then $\partial(\mathbf{A}(z)_i) = \partial_i$, $i = 1, \ldots, s$ are the *column degrees of* $\mathbf{A}(z)$ (and $\partial(\mathbf{A}(z)^i) = \partial^i$, $i = 1, \ldots, r$ are the *row degrees*). If $\mathbf{Q}(z) \in M_{m,m}$ then

$$\mathbf{Q}(z) = \Delta_c(\mathbf{Q})\,\mathrm{diag}\big[z^{\partial_1}\cdots z^{\partial_m}\big] + \hat{\mathbf{Q}}(z) \tag{4.36}$$

where $\partial_i(\hat{\mathbf{Q}}(z)) < \partial_i(\mathbf{Q}(z)) = \partial_i$ and $\Delta_c(\mathbf{Q})$ is an $m \times m$ matrix with entries in k called the *column coefficient of* $\mathbf{Q}(z)$. $\mathbf{Q}(z)$ is *column proper* ([W-4]) if $\deg\det \mathbf{Q}(z) = \partial_1 + \cdots + \partial_m$ (and similarly for row coefficient and row proper).

Proposition 4.37 *If* $\mathbf{F}(z)$ *is strictly proper and* $\mathbf{F}(z) = \mathbf{P}(z)\mathbf{Q}^{-1}(z)$, *then* $\partial_j(\mathbf{P}(z)) < \partial_j(\mathbf{Q}(z))$, $j = 1, \ldots, m$.

Proof. Since $\mathbf{FQ} = \mathbf{P}$, $\partial_j(\mathbf{P}) = \partial_j(\mathbf{FQ})$. But $(\mathbf{FQ})_j = \mathbf{F}(\mathbf{Q}_j)$. Let $\mathbf{Q}_j = (q_j^\alpha)$ so that $\mathbf{F}^i\mathbf{Q}_j = \sum_{\alpha=1}^{m} F_\alpha^i q_j^\alpha$. If $F_\alpha^i = f_{i\alpha} = n_\alpha^i/d_\alpha^i$ with n_α^i, d_α^i coprime, then $\deg f_{i\alpha}q_j^\alpha = \deg n_\alpha^i - \deg d_\alpha^i + \deg q_j^\alpha < \deg q_j^\alpha$ (since $\deg n_\alpha^i - \deg d_\alpha^i < 0$). Hence, $\deg(\mathbf{F}^i\mathbf{Q}_j) \le \max(\deg f_{i\alpha}q_j^\alpha) < \max_\alpha(\deg q_j^\alpha) = \partial_j(\mathbf{Q}(z))$.

Proposition 4.38 *If* $\mathbf{Q}(z)$ *is column proper and* $\partial_j(\mathbf{P}(z)) < \partial_j(\mathbf{Q}(z))$, $j = 1, \ldots, m$, *then* $\mathbf{F}(z) = \mathbf{P}(z)\mathbf{Q}^{-1}(z)$ *is strictly proper.*

Proof. Let $\deg\det \mathbf{Q} = n = \sum_{j=1}^{m} \partial_j(\mathbf{Q})$. Since $f_{ij}(z) = \sum_{\alpha=1}^{m} \mathbf{P}_\alpha^i(\mathrm{Adj}\;\mathbf{Q})_j^\alpha/\det \mathbf{Q}$, it suffices to show that $\deg \sum_{\alpha=1}^{m} \mathbf{P}_\alpha^i(\mathrm{Adj}\;\mathbf{Q})_j^\alpha < n$. But we have $\deg \mathbf{P}_\alpha^i(\mathrm{Adj}\;\mathbf{Q})_j^\alpha \le \partial_\alpha(\mathbf{P}) + n - \partial_\alpha(\mathbf{Q}) < n$ for all α.

Theorem 4.39 ([W-4]) *If* $\mathbf{Q}(z) \in M_{m,m}$ *is nonsingular then there is a* $\mathbf{U}(z)$ *in* $G(m; R)$ *with* $\mathbf{Q}(z)\mathbf{U}(z)$ *column proper.*

Proof. If $\Delta_c(\mathbf{Q})$ is nonsingular, then $\deg\det \mathbf{Q}(z) = \partial_1 + \cdots + \partial_m$. So suppose that $\Delta_c(\mathbf{Q})$ is singular. Then $\Delta_c(\mathbf{Q})\,\mathrm{diag}[z^{\partial_1}\cdots z^{\partial_m}]$ is also singular and the columns of $\Delta_c(\mathbf{Q})\,\mathrm{diag}[z^{\partial_1}\cdots z^{\partial_m}]$ are dependent over $R = k[z]$. If $\mathbf{Q}_1^c(z), \ldots, \mathbf{Q}_m^c(z)$ are these columns, then, by shifting columns, we may suppose

$$\mathbf{Q}_1^c(z) = \sum_{j=2}^{m} a_j(z)\mathbf{Q}_j^c(z) \tag{4.40}$$

where the $a_j(z)$ are not all 0. Set

$$\mathbf{U}_1(z) = \begin{bmatrix} 1 & 0 & & \\ -a_2(z) & 1 & & \\ & 0 & & \\ \vdots & \vdots & & \\ -a_m(z) & 0 & \dots & 1 \end{bmatrix}$$

so that $\mathbf{U}_1(z)$ is unimodular. Then $\mathbf{Q}(z)\mathbf{U}_1(z) = \mathbf{Q}_1(z)$ has $\partial_1(\mathbf{Q}) < \partial_1(\mathbf{Q}_1)$ and $\partial_j(\mathbf{Q}) = \partial_j(\mathbf{Q}_1)$ for $j = 2, \dots, m$. The result follows by induction after at most $\partial_1 + \cdots + \partial_m$ steps.

Corollary 4.41 *If* $\mathbf{F}(z)$ *is a* $p \times m$ *transfer matrix, then* $\mathbf{F}(z)$ *has a minimal realization* $(\mathbf{P}(z), \mathbf{Q}(z))$ *with* $\mathbf{Q}(z)$ *column proper.*

Example 4.42 Let

$$\mathbf{Q}(z) = \begin{bmatrix} z^2+3z+1 & z(z^2+3z+1) \\ 2z+3 & 3z^2+3z+6 \end{bmatrix}.$$

Then

$$\Delta_c(\mathbf{Q}) = \begin{bmatrix} 1 & 1 \\ 0 & 0 \end{bmatrix}$$

is singular. Let

$$\mathbf{U}_1(z) = \begin{bmatrix} 1 & -z \\ 0 & 1 \end{bmatrix}$$

so that

$$\mathbf{Q}(z)\mathbf{U}(z) = \begin{bmatrix} z^2+3z+1 & 0 \\ 2z+3 & z^2+6 \end{bmatrix}$$

and $\Delta_c(\mathbf{Q}(z)\mathbf{U}(z)) = I$ is nonsingular.

Example 4.43 Consider $\operatorname{diag}[z^{\partial_1}\ z^{\partial_2}]$ and let

$$\mathbf{U} = \begin{bmatrix} 0 & 1 \\ -1 & 0 \end{bmatrix}, \quad \mathbf{U}^{-1} = \begin{bmatrix} 0 & -1 \\ 1 & 0 \end{bmatrix}$$

so that $\det \mathbf{U} = \det \mathbf{U}^{-1} = 1$. Moreover, $\mathbf{U}^{-1} \operatorname{diag}[z^{\partial_1}\ z^{\partial_2}]\mathbf{U} = \operatorname{diag}[z^{\partial_2}\ z^{\partial_1}]$. If $\mathbf{Q}$ is any 2×2 matrix, then $\mathbf{QU} = [-\mathbf{Q}_2\ \mathbf{Q}_1]$ where the $\mathbf{Q}_i$ are the columns of $\mathbf{Q}$. Thus, if $\tilde{\mathbf{Q}}(z) = \Delta_c(\tilde{\mathbf{Q}}) \operatorname{diag}[z^{\partial_1}\ z^{\partial_2}] + \tilde{\mathbf{Q}}_1(z)$ with $\partial_i(\tilde{\mathbf{Q}}_1) < \partial_i$, then

$$\begin{aligned} \tilde{\mathbf{Q}}(z)\mathbf{U} &= (\Delta_c(\tilde{\mathbf{Q}})\mathbf{U})(\mathbf{U}^{-1} \operatorname{diag}[z^{\partial_1}\ z^{\partial_2}]\mathbf{U}) + \tilde{\mathbf{Q}}_1(z)\mathbf{U} \\ &= (\Delta_c(\tilde{\mathbf{Q}})\mathbf{U}) \operatorname{diag}[z^{\partial_2}\ z^{\partial_1}] + \tilde{\mathbf{Q}}_2(z) \end{aligned}$$

with $\partial_1(\tilde{\mathbf{Q}}_2) < \partial_2$ and $\partial_2(\tilde{\mathbf{Q}}_2) < \partial_1$.

Proposition 4.44 *If* $\mathbf{Q}(z) = \Delta_c(\mathbf{Q})\,\mathrm{diag}[z^{\partial_1}\cdots z^{\partial_m}] + \mathbf{Q}_1(z)$ *with* $\partial_i(\mathbf{Q}_1) < \partial_i(\mathbf{Q})$*, then there is a* $\mathbf{U}$ *with entries in* k *and* $\det \mathbf{U} = 1$ *such that* $\mathbf{Q}(z)\mathbf{U} = (\Delta_c(\mathbf{Q})\mathbf{U})\ (\mathbf{U}^{-1}\,\mathrm{diag}[z^{\partial_1}\cdots z^{\partial_m}]\mathbf{U}) + \mathbf{Q}_1(z)\mathbf{U} = (\Delta_c(\mathbf{Q})\mathbf{U})\,\mathrm{diag}[z^{\partial_1'}\cdots z^{\partial_m'}] + \mathbf{Q}_2(z)$ *where* $\partial_1' \geq \partial_2' \geq \cdots \geq \partial_m'$, $\partial_i(\mathbf{Q}_2) < \partial_i'$ *and the sets* $\{\partial_1', \ldots, \partial_m'\}$ *and* $\{\partial_1, \ldots, \partial_m\}$ *are the same.*

Corollary 4.45 *If* $\mathbf{F}(z)$ *is a* $p \times m$ *transfer matrix, then* $\mathbf{F}(z)$ *has a minimal realization* $(\mathbf{P}(z), \mathbf{Q}(z))$ *with* $\mathbf{Q}(z)$ *column proper,* $\det \mathbf{Q}(z)$ *monic, and* $\partial_1(\mathbf{Q}) \geq \partial_2(\mathbf{Q}) \geq \cdots \geq \partial_m(\mathbf{Q})$.

We are naturally led to the following:

Definition 4.46 If $\mathbf{F}(z)$ is a $p \times m$ transfer matrix and $(\mathbf{P}(z), \mathbf{Q}(z))$ is a minimal realization with $\mathbf{Q}(z)$ column proper, then $\{\partial_1(\mathbf{Q}), \ldots, \partial_m(\mathbf{Q})\}$ is the *(column) Kronecker set of* $\mathbf{F}(z)$. (A similar treatment gives the *(row) Kronecker set of* $\mathbf{F}(z)$.)

Proposition 4.47 *If* $\mathbf{Q}(z) = \mathbf{Q}_1(z)\mathbf{U}(z)$ *with* $\mathbf{Q}(z)$, $\mathbf{Q}_1(z)$ *column proper and* $\mathbf{U}(z) \in G(m; R)$ *then* $\{\partial_1(\mathbf{Q}), \ldots, \partial_m(\mathbf{Q})\} = \{\partial_1(\mathbf{Q}_1), \ldots, \partial_m(\mathbf{Q}_1)\}$ *as sets. Proof.* Since $\mathbf{Q} = \mathbf{Q}_1\mathbf{U}_1$, we have

$$\mathbf{Q}_j^i = \sum_\alpha \mathbf{u}_\alpha^i (\mathbf{Q}_1)_j^\alpha.$$

But $\mathbf{U} = \mathbf{Q}_1^{-1}\mathbf{Q} = (\mathrm{Adj}\ \mathbf{Q}_1)\mathbf{Q}/\det \mathbf{Q}_1$ and $\deg(\mathrm{Adj}\ \mathbf{Q}_1)_\beta \leq n - \partial_i(\mathbf{Q}_1)$. It follows that $\mathbf{u}_\alpha^i = \sum_\beta (\mathrm{Adj}\ \mathbf{Q}_1)_\beta^i \mathbf{Q}_\alpha^\beta / \det \mathbf{Q}_1$ has degree $\leq \partial_\alpha(\mathbf{Q}) - \partial_i(\mathbf{Q}_1)$. For a fixed α, not all $\mathbf{u}_\alpha^i$ are zero (otherwise $\mathbf{U}$ has a zero column) and so, $\partial_\alpha(\mathbf{Q}) - \partial_{i_\alpha}(\mathbf{Q}_1) \geq 0$ for some i_α. In other words, for each $\partial_i(\mathbf{Q})$, there is a $\partial_{j(i)}(\mathbf{Q}_1)$ such that $\partial_i(\mathbf{Q}) \geq \partial_{j(i)}(\mathbf{Q}_1)$ and vice versa. Thus the sets $\{\partial_i(\mathbf{Q})\}$ and $\{\partial_i(\mathbf{Q}_1)\}$ coincide (as all $\partial_i(\mathbf{Q}) \geq 0$ and all $\partial_i(\mathbf{Q}_1) \geq 0$ and $\sum \partial_i(\mathbf{Q}) = \sum \partial_i(\mathbf{Q}_1)$).

Corollary 4.48 *The Kronecker set of* $\mathbf{F}(z)$ *is well-defined.*

Let $\mathbf{F}(z)$ be a $p \times m$ transfer matrix and let $x = (A, B, C)$ be a realization of $\mathcal{H}_\mathbf{F}$ (3.54). Then, by Lemma 3.64, $\mathbf{F}(z) = C(zI - A)^{-1}B$ and the order ν of x satisfies the inequality

$$\text{degree}\ \mathbf{F}(z) = \rho(\mathcal{H}_\mathbf{F}) \leq \nu. \tag{4.49}$$

We shall also call x a *realization of* $\mathbf{F}(z)$.

Lemma 4.50 (Transfer Lemma) *$(\mathbf{P}(z), \mathbf{Q}(z))$ and $x = (A, B, C)$ are realizations of $\mathbf{F}(z)$ if and only if there is a $\nu \times m$ polynomial matrix $\mathbf{X}(z)$ such that*

$$(zI - A)\mathbf{X}(z) = B\mathbf{Q}(z), \quad C\mathbf{X}(z) = \mathbf{P}(z). \tag{4.51}$$

Proof. If there is such an $\mathbf{X}(z)$, then $\mathbf{X}(z) = (zI - A)^{-1}B\mathbf{Q}(z)$ and $\mathbf{P}(z) = C\mathbf{X}(z) = C(zI - A)^{-1}B\mathbf{Q}(z)$ so that $\mathbf{F}(z) = \mathbf{P}(z)\mathbf{Q}^{-1}(z) = C(zI - A)^{-1}B$. Conversely, if $C(zI - A)^{-1}B = \mathbf{P}(z)\mathbf{Q}^{-1}(z)$, then $\mathbf{X}(z) = (zI - A)^{-1}B\mathbf{Q}(z)$ satisfies (4.51).

The Transfer Lemma is a key to showing that the degree of $\mathbf{F}(z)$ is the same as the degree $\det \mathbf{Q}(z)$ of a minimal realization $(\mathbf{P}(z), \mathbf{Q}(z))$ of $\mathbf{F}(z)$. In fact, what we will do is first show that given a minimal realization $(\mathbf{P}(z), \mathbf{Q}(z))$, we can find $x = (A, B, C)$ with order degree $\det \mathbf{Q}(z)$ (thus $\rho(\mathcal{H}_F) \leq \deg \det \mathbf{Q}(z)$). Then we will show that there is a minimal realization x of $\mathcal{H}_\mathbf{F}$ and a solution $(\mathbf{P}_x(z), \mathbf{Q}_x(z))$ of (4.51) such that $\deg \det \mathbf{Q}(z) =$ order of $x = \rho(\mathcal{H}_\mathbf{F}) = \deg \mathbf{F}(z)$.

Proposition 4.52 *If $\rho(\mathcal{H}_\mathbf{F}) = n$ and $x = (A, B, C)$ is a minimal realization of $\mathcal{H}_\mathbf{F}$, then $g \cdot x = (gAg^{-1},\ gB,\ Cg^{-1})$ is also a minimal realization of $\mathcal{H}_\mathbf{F}$ for all $g \in \mathrm{GL}(n; k)$.*

Proof. $(Cg^{-1})(gAg^{-1})^{j-1}(gB) = C(g^{-1})(gA^{j-1})g^{-1}gB = CA^{j-1}B$ for all j so that $g \cdot x$ is also an order n realization of $\mathcal{H}_\mathbf{F}$.

We recall also (Corollary 4.24) that if $(\mathbf{P}(z), \mathbf{Q}(z))$ and $(\mathbf{P}_1(z), \mathbf{Q}_1(z))$ are minimal realizations of $\mathbf{F}(z)$ then $\deg \det \mathbf{Q}(z) = \deg \det \mathbf{Q}_1(z)$.

Let $\partial_1 \geq \partial_2 \geq \cdots \geq \partial_t > 0$ with $t \leq m$ be integers. Suppose $\partial_1 + \cdots + \partial_t = n$ and set $\partial_{t+1} = 0, \ldots, \partial_m = 0$. We let

$$d_0 = 0, \quad d_j = \sum_{i=1}^{j} \partial_i \tag{4.53}$$

for $j = 1, \ldots, t$. Then

$$d_{j+1} - d_j = \partial_{j+1} \tag{4.54}$$

for $j = 0, \ldots, t-1$ and $d_t = \sum\limits_{j=1}^{t} \partial_j = n$. We let $\mathbf{X}_\partial(z)$ be the $n \times m$ matrix

$$
\mathbf{X}_{\partial}(z) = \begin{bmatrix}
1 & 0 & \cdots & 0 & \\
z & \cdot & & & \\
\vdots & \cdot & & & \\
z^{\partial_1 - 1} & 0 & & 0 & \\
0 & 1 & & & O_{n,m-t} \\
\cdot & z & & & \\
\cdot & \vdots & & & \\
\cdot & z^{\partial_2 - 1} & & 0 & \\
\cdot & \cdot & & 1 & \\
\cdot & \cdot & & 2 & \\
\cdot & \cdot & & \vdots & \\
0 & 0 & & z^{\partial_1 - 1} &
\end{bmatrix} \tag{4.55}
$$

and we let $\mathbf{Y}_{\partial}(z)$ be the $n \times m$ matrix

$$
\mathbf{Y}_{\partial}(z) = \begin{bmatrix}
O_{\partial_1 - 1, m} \\
\epsilon^1 \\
O_{\partial_2 - 1, m} \\
\epsilon^2 \\
\vdots \\
\epsilon^t
\end{bmatrix} \tag{4.56}
$$

where ϵ^i is a unit $1 \times m$ vector with 1 in the ith column. We observe that

$$
(zI)\mathbf{X}_{\partial}(z) - \mathbf{Y}_{\partial}(z)\,\mathrm{diag}[z^{1} \cdots z^{\partial_t}\ 1 \cdots 1]
$$
$$
= \left[\begin{array}{cccc}
z & 0 \ \cdots & 0 & \\
\vdots & . & . & \\
z^{\partial_1 - 1} & \cdots & . & \\
0 & 0 & & \\
. & z & & \\
. & \vdots & & O_{n,m-t} \\
. & z^{\partial_2 - 1} & & \\
. & 0 & & \\
. & . & z^{\partial_t - 1} & \\
0 & 0 & 0 &
\end{array}\right]. \tag{4.57}
$$

An $n \times n$ matrix is in *block companion form* if $A = [A_{ij}]$ where the A_{ij}, $i, j = 1, \ldots, t$ are $\partial_i \times \partial_j$ matrices with

$$
A_{ii} = \left[\begin{array}{ccccc}
0 & 1 & 0 & \cdots & 0 \\
0 & 0 & & & \\
\vdots & \vdots & & & \vdots \\
0 & & & 1 & \\
a_{d_i, d_{i-1}+1} & & & & a_{d_i, d_i}
\end{array}\right], \quad
A_{ij} = \left[\begin{array}{c}
O_{\partial_i - 1, \partial_j} \\
a_{d_i, d_{j-1}+1} \ \cdots \ a_{d_i, d_j}
\end{array}\right] \tag{4.58}
$$

for $i \neq j$. Then

$$
A\mathbf{X}_{\partial}(z) = \left[\begin{array}{cccc}
z & 0 & & \\
\vdots & & . & \\
z^{\partial_1 - 1} & 0 & . & \\
[A\mathbf{X}_{\partial}(z)]^{d_1} & & & \\
0 & z & \cdots & O_{n,m-t} \\
. & \vdots & & \\
. & z^{\partial_2 - 1} & & \\
[A\mathbf{X}_{\partial}(z)]^{d_2} & & & \\
\vdots & \vdots & &
\end{array}\right] \tag{4.59}
$$

where the d_ith row of $A\mathbf{X}_\partial(z)$ is given by

$$[A\mathbf{X}_\partial(z)]^{d_i} = \big(a_{d_i,1} + \cdots + a_{d_i,\partial_1} z^{\partial_1 - 1}, a_{d_i,d_i+1} \\ + \cdots + a_{d_1,d_2} z^{\partial_2 - 1}, \ldots + a_{d_1,n} z^{\partial_t - 1}, \ldots 0\big). \tag{4.60}$$

It follows that, for any such A,

$$(zI - A)\mathbf{X}_\partial(z) = \mathbf{Y}_\partial(z) \operatorname{diag}[z^{\partial_1} \cdots z^{\partial_t} \; 1 \cdots 1] \\ - \begin{bmatrix} O_{\partial_1 - 1,m} \\ [A\mathbf{X}_\partial(z)]^{d_1} \\ O_{\partial_2 - 1,m} \\ [A\mathbf{X}_\partial(z)]^{d_2} \\ \vdots \\ [A\mathbf{X}_\partial(z)]^{d_1} \end{bmatrix}. \tag{4.61}$$

If Q_1 is any $m \times n$ matrix with entries in k, then

$$\mathbf{Y}_\partial(z) Q_1 \mathbf{X}_\partial(z) = \begin{bmatrix} O_{\partial_1 - 1,m} \\ \epsilon^1 Q_1 \mathbf{X}_\partial(z) \\ O_{\partial_2 - 1,m} \\ \epsilon^2 Q_1 \mathbf{X}_\partial(z) \\ \vdots \\ \epsilon^t Q_1 \mathbf{X}_\partial(z) \end{bmatrix}. \tag{4.62}$$

We note that $\epsilon^i Q_1 = Q_1^i$ (the ith row of Q_1) for $i = 1, \ldots, t$. Thus, for example, $\epsilon^1 Q_1 \mathbf{X}_\partial(z) = Q_1^1 \mathbf{X}_\partial(z) = (q_{11}^1 + \cdots + q_{1\partial_1}^1 z^{\partial_1 - 1}, \ldots, \cdots + q_{1d_t}^1 z^{\partial_t - 1}, \cdots 0)$ is the d_1th row of $\mathbf{Y}_\partial(z) Q_1 \mathbf{X}_\partial(z)$.

Now let $(\mathbf{P}(z), \mathbf{Q}(z))$ be a column proper minimal realization of $\mathbf{F}(z)$ with $\partial_1(\mathbf{Q}) \geq \partial_2(\mathbf{Q}) \geq \cdots \geq \partial_t(\mathbf{Q}) > 0$ and $\partial_{t+1}(\mathbf{Q}) = 0, \ldots, \partial_m(\mathbf{Q}) = 0$. Set $\partial_i = \partial_i(\mathbf{Q})$. Then

$$\mathbf{Q}(z) = \Delta_c(\mathbf{Q}) \operatorname{diag}\big[z^{\partial_1} \cdots z^{\partial_2} \; 1 \cdots 1\big] + Q_2 \mathbf{X}_\partial(x) \tag{4.63}$$

and, since $\partial_i(\mathbf{P}) < \partial_i(\mathbf{Q}) = \partial_i$ for $i = 1, \ldots, m$, there is a $p \times n$, C with $C\mathbf{X}_\partial(z) = \mathbf{P}(z)$. Let $B_m = \Delta_c(\mathbf{Q})^{-1}$ and let B_m^i be its ith row. We let B be the $n \times m$ matrix given by

$$B = \begin{bmatrix} \mathbf{O}_{\partial_1 - 1, m} \\ B_m^1 \\ \mathbf{O}_{\partial_2 - 1, m} \\ B_m^2 \\ \vdots \\ B_m^t \end{bmatrix} \tag{4.64}$$

so that $B\Delta_c(\mathbf{Q}) = \mathbf{Y}_\partial(z)$. If $Q_1 = \Delta_c(\mathbf{Q})^{-1}\mathbf{Q}_2$ then $\Delta_c(\mathbf{Q})Q_1 = Q_2$ and $BQ_2 = B\Delta_c(\mathbf{Q})Q_1 = \mathbf{Y}_\partial(z)Q_1$. Now let A be the $n \times n$ block companion matrix with d_ith row $-\epsilon^i Q_1 = -Q_1^i$ for $i = 1, \dots, t$. It follows from (4.61) and (4.62) that $(zI - A)\mathbf{X}_\partial(z) = B\mathbf{Q}(z)$. By the Transfer Lemma 4.50, $x = (A, B, C)$ is a realization of order $n = \deg\det \mathbf{Q}(z)$ and so $\rho(\mathcal{H}_\mathbf{F}) \leq \deg\det \mathbf{Q}(z)$.

Now let us suppose there is a minimal realization $x = (A, B, C)$ of $\mathbf{F}(z)$ with A in block companion form and B in the form (4.64) i.e.,

$$B = \begin{bmatrix} O_{\partial_1 - 1, m} \\ B^1 \\ \vdots \\ O_{\partial_t - 1, m} \\ B^t \end{bmatrix} \tag{4.65}$$

where the $t \times m$ matrix $[B^i]$ has rank t and the $m \times m$ matrix

$$B_m = \begin{bmatrix} B^1 \\ \vdots \\ B^t \\ \epsilon^{t+1} \\ \epsilon^m \end{bmatrix}$$

has rank m (i.e., is nonsingular). Let $\Delta_c = B_m^{-1}$ and *set* $\mathbf{P}(z) = C\mathbf{X}_\partial(z)$ and $\mathbf{Q}(z) = \Delta_c \operatorname{diag}[z^{\partial_1} \cdots z^{\partial_t}\ 1 \cdots 1] + \Delta_c Q_1 \mathbf{X}_\partial(z)$ where $-\epsilon^i Q_1 = -Q_1^i = d_i$th row of A. Then, by (4.61) and (4.62), $(zI - A)\mathbf{X}_\partial(z) = B\mathbf{Q}(z)$ and so, by the Transfer Lemma 4.50, $(\mathbf{P}(z), \mathbf{Q}(z))$ is a (column proper) realization of $\mathbf{F}(z)$ with $\deg\det \mathbf{Q}(z) = n = \text{order } x = \rho(\mathcal{H}_\mathbf{F})$. Hence if $(\mathbf{P}_1(z), \mathbf{Q}_1(z))$ is a minimal realization (actually $(\mathbf{P}(z), \mathbf{Q}(z))$ is), then $\deg\det \mathbf{Q}_1(z) \leq \deg\det \mathbf{Q}(z) = \rho(\mathcal{H}_\mathbf{F})$. We note also that if A is in the block companion form and B in the form (4.65), then the Kronecker set of $C(zI - A)^{-1}B$ is precisely $\{\partial_1, \dots, \partial_t, 0, \dots, 0\}$ and moreover, $B_1, \ldots, A^{\partial_1-1}B_1, \ldots, B_2, \ldots, A^{\partial_2-1}B_2, \ldots, A^{\partial_t-1}B_t$ are a basis of k^n. Thus, to complete things, we must show that there is such a minimal realization in what we shall call the *controllable companion form.* (Cf. Chapter 13).

Suppose that $x_1 = (A_1, B_1, C_1)$ is a minimal realization of $\mathcal{H}_{\mathbf{F}}$, then $\mathbf{Y}_{\infty}^n(A,B) = [B \; AB \cdots]$ has rank $n = \rho(\mathcal{H}_{\mathbf{F}})$. Hence, the $n \times nm$ matrix $\mathbf{Y}(A,B) = [B \; AB \cdots A^{n-1}B]$ has rank n. Let us take a basis of k^n consisting of the first n linearly independent columns of $Y(A,B)$ (we may and do assume $B_1, \ldots, B_t$ are independent with B of rank t). We let

$$L = \left[B_1 \; AB_1 \cdots A^{\tilde{\partial}_1-1}B_1 \cdots A^{\tilde{\partial}_t-1}B_t\right]$$

be the nonsingular $n \times n$ matrix of the elements of this basis. Let $\tilde{d}_0 = 0$, $\tilde{d}_j = \sum_{i=1}^{j} \tilde{\partial}_j$ and let M^j be the $\tilde{d}_j$th row of L^{-1}. If

$$g = \begin{bmatrix} M^1 \\ M^1 A \\ \vdots \\ M^1 A^{\tilde{\partial}_j-1} \\ \vdots \\ M^t A^{\tilde{\partial}_t-1} \end{bmatrix} \tag{4.66}$$

then g is nonsingular since (for example) we can write

$$g = \begin{array}{c} \\ \partial_1 \\ \partial_2 \\ \\ \partial_t \\ \\ \end{array} \begin{bmatrix} M^1 & 0 & & & \\ 0 & M^1 & & & \\ & & M^2 & & \\ & & & \ddots & \\ & & & & M^t \end{bmatrix} \begin{bmatrix} I_n \\ A \\ A^{\tilde{\partial}_1-1} \\ I_n \\ \vdots \\ A^{\tilde{\partial}_t-1} \end{bmatrix} \tag{4.67}$$
$$\qquad\qquad n\partial_1 \qquad n\partial_2 \quad \cdots \quad n\partial_t \qquad n\left(\textstyle\sum \partial_i\right) \times n$$

a product of matrices of rank n. The fact that $g \cdot x_1$ is in controllable companion form is a straightforward computation which is left to the reader. We note also that, in view of our earlier discussion, $\{\partial_1, \ldots, \partial_t, 0, \ldots 0\}$ is the (column) Kronecker set of $C(zI - A)^{-1}B$.

Now let us sketch an alternative approach based on the Smith form. Let $\mathbf{F}(z)$ be an element of $\mathbf{Rat}(n; p, m)$. Then, by Corollary 4.34, $\mathbf{F}(z) = \mathbf{P}(z)\mathbf{Q}^{-1}(z)$ where $\mathbf{P}(z) = \mathbf{U}^{-1}(z)\mathbf{E}(z)$, $\mathbf{Q}(z) = \mathbf{V}(z)\mathbf{P}(z)$ is a coprime realization with

$\mathbf{U}(z)\mathbf{F}(z)\mathbf{V}(z) = \mathbf{E}(z)\mathbf{D}^{-1}(z) = \text{diag}[\epsilon_1(z)/\psi_1(z) \cdots \epsilon_t(z)/\psi_t(z)\ 0 \cdots 0] = \mathbf{G}(z)$ and $\deg\det \mathbf{Q}(z) = \Sigma \deg \psi_i$. Suppose that it could be shown that $\rho(\mathcal{H}_{\mathbf{F}}) = \rho(\mathcal{H}_{\mathbf{G}})$. Then it would be sufficient to show that $\rho(\mathcal{H}_{\mathbf{G}}) = \Sigma \deg \psi_i$. Since $\epsilon_i(z)/\psi_i(z)$ is a scalar transfer function of degree $n_i = \deg \psi_i$ i.e., $\epsilon_i/\psi_i \in \mathbf{Rat}(n; 1, 1)$, there is a minimal realization $x_i = (A_i, b_i, c^i)$ of ϵ_i/ψ_i (part I). Let $x = (A, B, C)$ where

$$A = \begin{bmatrix} A_1 & 0 & \cdots & 0 \\ 0 & A_2 & & \vdots \\ \vdots & & & \\ 0 & 0 & \cdots & A_t \end{bmatrix}, \quad B = \begin{bmatrix} b_1 & O_{n_1,1} & \\ \mathbf{O}_{n-n_1,1} & b_2 & \\ \vdots & \mathbf{O}_{n-n_1-n_2,1} & \cdots \end{bmatrix},$$

$$C = \begin{bmatrix} c^1 & \mathbf{O}_{1,n_2} & \\ & c^2 & \cdots \\ \mathbf{O}_{p-1,n_1} & \mathbf{O}_{p-2,n_2} & \end{bmatrix} \tag{4.68}$$

$x = \oplus \sum_{i=1}^{t} x_i$ is called the *direct sum of the* x_i. It is easy to check that x is a (minimal) realization of $\mathbf{G}(z)$ and that $\mathcal{H}_{\mathbf{G}} = \oplus\Sigma\mathcal{H}_{\epsilon_i/\psi_i}$. It follows that $\rho(\mathcal{H}_{\mathbf{G}}) = \Sigma\rho(\mathcal{H}_{\epsilon_i/\psi_i}) = \Sigma \deg \psi_i$.

Example 4.69 Let

$$\mathbf{G}(z) = \begin{bmatrix} \epsilon_1(z)/\psi_1(z) & 0 \\ 0 & \epsilon_2(z)/\psi_2(z) \end{bmatrix}.$$

Let $x_1 = (A_1, b_1, c^1)$ be a minimal realization of ϵ_1/ψ_1 so that $c^1(zI - A_1)^{-1}b_1 = \epsilon_1/\psi_1$ and $n_1 = \deg \psi_1 = \text{order } A_1$; and, let $x_2 = (A_2, b_2, c^2)$ be a minimal realization of ϵ_2/ψ_2 so that $c^2(zI - A_2)^{-1}b_2 = \epsilon_2/\psi_2$ and $n_2 = \deg \psi_2 = \text{order } A_2$. Set

$$A = \begin{bmatrix} A_1 & O_{n_1,n_2} \\ O_{n_2,n_1} & A_2 \end{bmatrix}, \quad B = \begin{bmatrix} b_1 & O \\ O_{n-n_1,1} & b_2 \end{bmatrix},$$

$$C = \begin{bmatrix} c^1 & O_{1,n-n_1} \\ O_{1,n_1} & c^2 \end{bmatrix}.$$

Then $C(zI - A)^{-1}B = \mathbf{G}(z)$ and

$$CB = \begin{bmatrix} c^1 b_1 & 0 \\ 0 & c^2 b_2 \end{bmatrix}, \quad CAB = \begin{bmatrix} c^1 A_1 b_1 & 0 \\ 0 & c^2 A_2 b_2 \end{bmatrix}$$

so that

$$\mathcal{H}_{\mathbf{G}} = \begin{bmatrix} CB & CAB \\ CAB & CA^2B \\ \vdots & \vdots \end{bmatrix}$$

and $\rho(\mathcal{H}_{\mathbf{G}}) = \rho(H_{\mathbf{G}})$ where $H_{\mathbf{G}} = (\mathcal{H}_{\mathbf{G}})_n^n$. But $\rho(H_{\mathbf{G}}) = \rho(\tilde{H}_{\mathbf{G}})$ where

$$\tilde{H}_{\mathbf{G}} = \begin{bmatrix} c^1b_1 & \cdots & c^1A_1^{n-1}b_1 & & & \\ \vdots & & & & O & \\ c^1A_1^{n-1}b_1 & & c^1A_1^{2n-1}b_1 & & & \\ & & & c^2b_2 & \cdots & c^2A_2^{n-1}b_2 \\ & & & \vdots & & \vdots \\ & & & c^2A_2^{n-1}b_2 & & c_2A_2^{2n-1}b_2 \end{bmatrix}.$$

Thus $\rho(\mathcal{H}_{\mathbf{G}}) = \rho(\tilde{H}_{\mathbf{G}}) = \rho(\mathcal{H}_{\epsilon_1/\psi_1}) + \rho(\mathcal{H}_{\epsilon_2/\psi_2})$.

Now we give a "quasi-proof" that $\rho(\mathcal{H}_{\mathbf{F}}) = \rho(\mathcal{H}_{\mathbf{G}})$ (a rigorous proof is left to the reader as an exercise). Since

$$\mathbf{F}(z) = \sum H_i z^{-i},$$

$$\mathbf{U}(z)\mathbf{F}(z)\mathbf{V}(z) = \sum_{i=1}^{\infty} (\mathbf{U}(z)H_i\mathbf{V}(z))z^{-i} = \sum \mathbf{G}_i z^{-i} = \mathbf{G}(z),$$

we have (sic!) $\mathcal{H}_{\mathbf{G}} = [\mathbf{U}(z)H_{i+j-1}\mathbf{V}(z)]_{i,j=1}^{\infty}$ so that

$$H_{\mathbf{G}} = \text{block diag}[\mathbf{U}\cdots\mathbf{U}]H_{\mathbf{F}}\text{block diag}[\mathbf{V}\cdots\mathbf{V}]$$

and $\rho(H_{\mathbf{G}}) = \rho(H_{\mathbf{F}})$ as block diag$[\mathbf{U}\cdots\mathbf{U}]$, block diag$[\mathbf{V}\cdots\mathbf{V}]$ are unimodular. (Of course since $\mathbf{U}(z)$, $\mathbf{V}(z)$ depend on z, this is *not* a real proof).

In any case, we have, by our earlier argument, established that if $\mathbf{F}(z) \in \mathbf{Rat}(n;p,m)$ then there is a minimal realization $(\mathbf{P}(z), \mathbf{Q}(z))$ of $\mathbf{F}(z)$ with (i) $\mathbf{Q}(z)$ column proper; (ii) $\partial_1(\mathbf{Q}) \geq \cdots \geq \partial_m(\mathbf{Q}) \geq 0$; and, (iii) $\det \mathbf{Q}(z)$ monic of degree n. Conversely, for any such coprime $(\mathbf{P}(z), \mathbf{Q}(z))$, $\mathbf{F}(z) = \mathbf{P}(z)\mathbf{Q}^{-1}(z)$ is an element of $\mathbf{Rat}(n;p,m)$. Similarly, if $\mathbf{F}(z) \in \mathbf{Rat}(n;p,m)$ then there is a (row) minimal realization $(\mathbf{Q}_L(z), \mathbf{R}(z))$ of $\mathbf{F}(z)$ with (i) $\mathbf{Q}_L(z)$ row proper; (ii) $\partial^1(\mathbf{Q}_L) \geq \cdots \geq \partial^p(\mathbf{Q}_L) \geq 0$ (the $\partial^i(\mathbf{Q}_L)$ are the row degrees of $\mathbf{Q}_L(z)$ which is $p \times p$); and, (iii) $\det \mathbf{Q}_L(z)$ is monic of degree n. Conversely, for any such (left) coprime $(\mathbf{Q}_L(z), \mathbf{R}(z))$, $\mathbf{F}(z) = \mathbf{Q}_L^{-1}(z)\mathbf{R}(z) \in \mathbf{Rat}(n;p,m)$.

Suppose that $\mathbf{F}(z)$ is in $\mathbf{Rat}(n;p,m)$ and that $\mathbf{F}(z)$ is of rank r. Then, by Corollary 4.31 (say), there are unimodular $\mathbf{U}(z)$, $\mathbf{V}(z)$ with

$$\mathbf{F}(z) = \mathbf{U}^{-1}(z)\mathbf{E}(z)\mathbf{D}(z)\mathbf{V}^{-1}(z)$$

where

$$\mathbf{E}(z) = \begin{bmatrix} \epsilon_1(z) & & & & & & \\ & \epsilon_2(z) & & & & & \\ & & \ddots & & & & \\ & & & \epsilon_r(z) & & & \\ & & & & 0 & & \\ & & & & & \ddots & \\ & & & & & & 0 \end{bmatrix},$$

$$\mathbf{D}(z) = \begin{bmatrix} \psi_1(z) & & & \\ & \ddots & & \\ & & \psi_r(z) & \\ & & & I_{m-r} \end{bmatrix}$$

and $\sum \deg \psi_i = n$. Let $\mathbf{E}_r(z) = \operatorname{diag}[\epsilon_1(z) \cdots \epsilon_r(z)]$ and

$$\mathbf{D}_r(z) = \operatorname{diag}[\psi_1(z) \cdots \psi_r(z)].$$

Then

$$\mathbf{F}(z) = \left(\mathbf{U}^{-1}(z) \begin{bmatrix} \mathbf{E}_r(z) \\ \mathbf{O}_{p-r,r} \end{bmatrix}\right) \mathbf{D}_r^{-1}(z) \left([I_r \ \mathbf{O}_{r,m-r}]\mathbf{V}^{-1}(z)\right)$$

is of the form $\mathbf{P}(z)\mathbf{Q}^{-1}(z)\mathbf{R}(z)$ where $\mathbf{P}(z) \in M_{p,r}$, $\mathbf{Q}(z) \in M_{r,r}$ and $\mathbf{R}(z) \in M_{r,m}$. We observe that since $\mathbf{V}^{-1}(z)\mathbf{V}(z) = I_m$, there is an $m \times r$ $\mathbf{V}_1(z)$ with $\mathbf{V}^{-1}(z)\mathbf{V}_1(z) = \begin{bmatrix} I_r \\ O_{m-r,r} \end{bmatrix}$ and hence, $\mathbf{D}_r(z) \cdot O + ([\mathbf{I}_r \ O_{m-r,r}]\mathbf{V}^{-1}(z))\mathbf{V}_1(z) = I_r$. In other words, $(\mathbf{Q}(z), \mathbf{R}(z))$ are left coprime. Since $\epsilon_i(z)$, $\psi_i(z)$ are relatively prime for each i, there are $r \times r$ (diagonal) matrices A_r, B_r with $A_r\mathbf{E}_r + B_r\mathbf{D}_r = I_r$. Let X be the $r \times p$ matrix $X = [A_r \ O_{r,p-r}]\mathbf{U}$ and let Y be the $r \times r$ matrix B_r. Then

$$X\left(\mathbf{U}^{-1} \begin{bmatrix} \mathbf{E}_r \\ \mathbf{O}_{p-r,r} \end{bmatrix}\right) + Y\mathbf{D}_r = I_r$$

and so $(\mathbf{R}(z), \mathbf{Q}(z))$ are right coprime.

Definition 4.70 $(\mathbf{P}(z), \mathbf{Q}(z), \mathbf{R}(z))$ is a (two-sided) *realization* of $\mathbf{F}(z) \in \mathbf{Rat}(n; p, m)$ if $\mathbf{F}(z) = \mathbf{P}(z)\mathbf{Q}^{-1}(z)\mathbf{R}(z)$. The realization is *prime* if $(\mathbf{P}(z), \mathbf{Q}(z))$ are right prime and $(\mathbf{Q}(z), \mathbf{R}(z))$ are left prime. A prime realization is *minimal* if $\mathbf{Q}(z)$ is $r \times r$ where $r = \operatorname{rank} \mathbf{F}(z)$ (in all cases, it is assumed that $\deg \det \mathbf{Q}(z) = n$).

We note that there are minimal realizations by the previous discussion. Moreover, we have:

Proposition 4.71 *Let* $(\mathbf{P}(z), \mathbf{Q}(z), \mathbf{R}(z))$ *be a prime realization of* $\mathbf{F}(z)$. *Then:* A) *there are* $\mathbf{X}(z)$, $\mathbf{Y}(z)$, $\mathbf{W}(z)$, $\mathbf{Z}(z)$ *with* $\mathbf{Z}(z)$ *nonsingular,* $\deg\det \mathbf{Z}(z) = n = \deg\det \mathbf{Q}(z)$ *and* $\mathbf{F}(z) = \mathbf{Z}^{-1}(z)(\mathbf{W}(z)\mathbf{R}(z))$ *with* $(\mathbf{Z}(z), \mathbf{W}(z)\mathbf{R}(z))$ *left coprime and* $\mathbf{X}(z)\mathbf{Q}(z) + \mathbf{Y}(z)\mathbf{P}(z) = I$;
B) *there are* $\tilde{\mathbf{X}}(z)$, $\tilde{\mathbf{Y}}(z)$, $\tilde{\mathbf{W}}(z)$, $\tilde{\mathbf{Z}}(z)$ *with* $\tilde{\mathbf{Z}}(z)$ *nonsingular,* $\deg\det \tilde{\mathbf{Z}}(z) = \deg\det \mathbf{Q}(z)$ *and* $\mathbf{F}(z) = (\mathbf{P}(z)\tilde{\mathbf{W}}(z))\tilde{\mathbf{Z}}^{-1}(z)$ *with* $(\mathbf{P}(z)\tilde{\mathbf{W}}(z), \tilde{\mathbf{Z}}(z))$ *right coprime and* $\mathbf{Q}(z)\tilde{\mathbf{X}}(z) + \mathbf{R}(z)\tilde{\mathbf{Y}}(z) = I$.

Proof. Since $\mathbf{P}(z)$, $\mathbf{Q}(z)$ are right coprime, there are $\mathbf{X}(z)$, $\mathbf{Y}(z)$ with $\mathbf{X}(z)\mathbf{Q}(z) + \mathbf{Y}(z)\mathbf{P}(z) = I$. Since $\mathbf{P}(z)\mathbf{Q}^{-1}(z)$ is a transfer matrix, there are $(\mathbf{W}(z), \mathbf{Z}(z))$ left coprime with $\mathbf{Z}^{-1}(z)\mathbf{W}(z) = \mathbf{P}(z)\mathbf{Q}^{-1}(z)$ and $\deg\det \mathbf{Q}(z) = \deg\det \mathbf{Z}(z)$. Clearly, $\mathbf{F}(z) = \mathbf{Z}^{-1}(z)(\mathbf{W}(z)\mathbf{R}(z))$. Since $(\mathbf{Q}(z), \mathbf{R}(z))$ and $(\mathbf{W}(z), \mathbf{Z}(z))$ are left coprime, there are $\mathbf{X}_1(z)$, $\mathbf{Y}_1(z)$, $\mathbf{X}_2(z)$, $\mathbf{Y}_2(z)$ such that $\mathbf{Q}\mathbf{X}_1 + \mathbf{R}\mathbf{Y}_1 = I$ and $\mathbf{Z}\mathbf{X}_2 + \mathbf{W}\mathbf{Y}_2 = I$. Then $\mathbf{Z}(\mathbf{P}\mathbf{X}_1\mathbf{Y}_2 + \mathbf{X}_2) + (\mathbf{W}\mathbf{R})\mathbf{Y}_1\mathbf{Y}_2 = (\mathbf{Z}\mathbf{P}\mathbf{X}_1 + \mathbf{W}\mathbf{R}\mathbf{Y}_1)\cdot \mathbf{Y}_2 + \mathbf{Z}\mathbf{X}_2 = \mathbf{W}(\mathbf{Q}\mathbf{X}_1 + \mathbf{R}\mathbf{Y}_1)\mathbf{Y}_2 + \mathbf{Z}\mathbf{X}_2 = \mathbf{W}\mathbf{Y}_2 + \mathbf{Z}\mathbf{X}_2 = I$ i.e., $(\mathbf{Z}(z), \mathbf{W}(z)\mathbf{R}(z))$ are left prime. Note also that

$$\begin{bmatrix} \mathbf{X}(z) & \mathbf{Y}(z) \\ \mathbf{W}(z) & -\mathbf{Z}(z) \end{bmatrix} \begin{bmatrix} \mathbf{Q}(z) \\ \mathbf{P}(z) \end{bmatrix} = \begin{bmatrix} I \\ 0 \end{bmatrix}. \tag{4.72}$$

This establishes A). As for B), the argument is entirely similar and we note that

$$[\mathbf{Q}(z)\ \mathbf{R}(z)] \begin{bmatrix} \tilde{\mathbf{X}}(z) & \tilde{\mathbf{W}}(z) \\ \tilde{\mathbf{Y}}(z) & -\tilde{\mathbf{Z}}(z) \end{bmatrix} = [I\ 0]. \tag{4.73}$$

Two-sided realizations play a role in certain design problems ([R-2]) but will not be used in this volume.

Exercises

(1) Show that $G(r; R)$ is a group. Can you extend this to a general ring R?

(2) Verify (4.7) in detail. State and prove the analog of Lemma 4.3 for a row.

(3) Prove Proposition 4.12.

(4) Prove the analogs of Proposition 4.14 and Theorem 4.15 for row operations.

(5) Prove Corollary 4.16 and extend it to the case $q < r$.

(6) Prove the Binet-Cauchy formula.

(7) Prove Proposition 4.44 and Corollary 4.45.

(8) Let $\partial_1 = 3$, $\partial_2 = 2$ and $n = 5$. Let $p = 2$, $m = 2$. Work out the development from (4.55) on for this case.

(9) Show that $g \cdot x_1$ is in controllable companion form for g given by (4.66).

(10) Show that $x = \oplus \sum x_i$ is a minimal realization of $\mathcal{H}_{\mathbf{G}}$ and that $\mathcal{H}_{\mathbf{G}} = \oplus \sum \mathcal{H}_{\epsilon_i/\psi_i}$ ((4.68)).

(11) Prove that $\rho(\mathcal{H}_{\mathbf{F}}) = \rho(\mathcal{H}_{\mathbf{G}})$.

5

Projective Space

We saw in I.23 and in Chapter 2 that the treatment of multivariable systems requires projective algebraic geometry. We begin the development with a study of projective space.

Let $\mathbf{V}$ be a vector space over k. As usual, if $A \subset \mathbf{V}$, we let $\operatorname{sp}[A]$ denote the *span of* A.

Definition 5.1 Let $\mathbb{P}(\mathbf{V}) = \{\operatorname{sp}[v] \colon v \neq 0,\ v \in \mathbf{V}\}$ be the set of 1-dimensional subspaces of $\mathbf{V}$. $\mathbb{P}(\mathbf{V})$ is the *projective space of* $\mathbf{V}$.

We can also view $\mathbb{P}(\mathbf{V})$ as $\mathbf{V} - \{0\}$ modulo the equivalence relation $v \equiv w$ if and only if $v = \lambda w$ for some $\lambda \in k^*$ (since $\operatorname{sp}[v] = \operatorname{sp}[w]$ if and only if $v = \lambda w$, $\lambda \neq 0$). The *dimension of* $\mathbb{P}(\mathbf{V})$ is $\dim \mathbf{V} - 1$. If $\mathbf{V} = k^{N+1}$, then we write $\mathbb{P}_k^N$ or $\mathbb{P}^N(k)$ in place of $\mathbb{P}(\mathbf{V})$. If $\mathbf{W}$ is a subspace of $\mathbf{V}$, then $\mathbb{P}(\mathbf{W}) = \{\operatorname{sp}[w] \colon w \neq 0,\ w \in \mathbf{W}\}$ may be viewed as a subspace of $\mathbb{P}(\mathbf{V})$. If we let $\pi \colon \mathbf{V} - \{0\} \to \mathbb{P}(\mathbf{V})$ be the natural map (i.e., $\pi(v) = \operatorname{sp}[v]$), then $\mathbb{P}(\mathbf{W}) = \pi(\mathbf{W})$. This leads us to the following:

Definition 5.2 A (projective) *linear subspace* L of $\mathbb{P}(\mathbf{V})$ is the image under π of a subspace $\mathbf{W}$ of $\mathbf{V}$ (i.e., $L = \pi(\mathbf{W})$).

If L_1 and L_2 are linear subspaces of $\mathbb{P}(\mathbf{V})$, then $L_1 \cap L_2$ is a linear subspace of $\mathbb{P}(\mathbf{V})$ and there is a smallest linear subspace $L_1 \vee L_2$ of $\mathbb{P}(\mathbf{V})$ containing $L_1 \cup L_2$. $L_1 \vee L_2$ is called the *join of* L_1 *and* L_2 and is often written $J(L_1, L_2)$.

Proposition 5.3 *Let* L_1, L_2 *be linear subspaces of* $\mathbb{P}(\mathbf{V})$. *Then* $\dim L_1 + \dim L_2 = \dim L_1 \cap L_2 + \dim L_1 \vee L_2$.*

---

*Conventionally, $\dim \emptyset = -1$.

P. Falb, *Methods of Algebraic Geometry in Control Theory: Part II*,
Modern Birkhäuser Classics, https://doi.org/10.1007/978-3-319-96574-1_5

Proof. Let $L_1 = \pi(\mathbf{W}_1)$, $L_2 = \pi(\mathbf{W}_2)$. Then $\dim L_1 + \dim L_2 = \dim \mathbf{W}_1 - 1 + \dim \mathbf{W}_2 - 1 = \dim(\mathbf{W}_1 \cap \mathbf{W}_2) - 1 + \dim(\mathbf{W}_1 + \mathbf{W}_2) - 1 = \dim L_1 \cap L_2 + \dim L_1 \vee L_2$ since $\pi(\mathbf{W}_1 \cap \mathbf{W}_2) = L_1 \cap L_2$ and $\pi(\mathbf{W}_1 + \mathbf{W}_2) = L_1 \vee L_2$.

Corollary 5.4 *If* $\dim L_1 + \dim L_2 \geq N = \dim \mathbb{P}(\mathbf{V})$, *then* $L_1 \cap L_2$ *is a linear subspace of dimension at least* $\dim L_1 + \dim L_2 - N \geq 0$ *and is not empty.*

Corollary 5.5 *Any two lines in* $\mathbb{P}_k^2$ *intersect* (cf. I.23).

Corollary 5.6 *A hyperplane* L *(i.e.,* $\dim L = N - 1$*) and any line in* $\mathbb{P}_k^N$ *intersect.*

Corollary 5.7 *If* $L_1 = \{P\}$ *(i.e.,* $\dim L_1 = 0$*),* $\dim L_2 = r$ *and* $P \notin L_2$*, then* $\dim L_1 \vee L_2 = r + 1$.

In effect, linear subspaces of $\mathbb{P}(\mathbf{V})$ correspond to vector subspaces of $\mathbf{V}$ and the dimension is one less. Thus: a *point* P of $\mathbb{P}(\mathbf{V})$ corresponds to a one-dimensional subspace of $\mathbf{V}$; a *line* $P \vee Q = J(P, Q)$ corresponds to a two-dimensional subspace of $\mathbf{V}$, etc. In this (limited) sense, the study of projective space may be viewed as the study of subspaces of a vector space and concepts such as independence can be translated from one context into the other.

Suppose that $\mathbf{V}$ is a vector space over k of dimension $N+1$ and let $e_0, \ldots, e_N$ be a basis of V. If $P = \mathrm{sp}[v]$ is a point of $\mathbb{P}(\mathbf{V})$ with $v = \xi_0 e_0 + \cdots + \xi_N e_N$, then $(\xi_0, \xi_1, \ldots, \xi_N)$ are called *homogeneous coordinates* of $P = \boldsymbol{\xi} = (\xi)$. Note that $\boldsymbol{\xi} \neq 0$ and that if $P = \mathrm{sp}[w]$, then $w = \lambda v$, $\lambda \neq 0$, and $(\lambda\xi_0, \ldots, \lambda\xi_N)$ are also homogeneous coordinates of $\mathbf{P}$ and conversely. If $g\colon \mathbf{V} \to \mathbf{V}_1$ is an injective linear map of $\mathbf{V}$ into $\mathbf{V}_1$, then g defines a map $\mathbb{P}(g)\colon \mathbb{P}(\mathbf{V}) \to \mathbb{P}(\mathbf{V}_1)$ by

$$\mathbb{P}(g)(P = \mathrm{sp}[v]) = \mathrm{sp}[g \cdot v]. \tag{5.8}$$

The map $\mathbb{P}(g)$ is called a *projective map*, and a *projective transformation* if $\dim \mathbf{V} = \dim \mathbf{V}_1$. We observe that if $h\colon \mathbf{V}_1 \to \mathbf{V}_2$ is also an injective linear map, then

$$\mathbb{P}(h \circ g) = \mathbb{P}(h) \circ \mathbb{P}(g) \tag{5.9}$$

and $\mathbb{P}(I) = I$.

Lemma 5.10 *Suppose* $\dim \mathbf{V} \geq 2$ *(i.e.,* $N \geq 1$ *and* $\dim \mathbb{P}(\mathbf{V}) \geq 1$*). Then two injective linear maps* g, h *of* $\mathbf{V}$ *into* $\mathbf{V}_1$ *have* $\mathbb{P}(g) = \mathbb{P}(h)$ *if and only if there is an* $\alpha \in k$ *with* $h = \alpha g$ *(such an* $\alpha \in k^*$*).*

Proof. If $h = \alpha g$, then h is linear and $\mathrm{sp}[hv] = \mathrm{sp}[\alpha g v] = \mathrm{sp}[gv]$ if $\alpha \neq 0$. Since g is injective, h will be injective only when $\alpha \neq 0$. On the other hand, if

$\mathbb{P}(g) = \mathbb{P}(h)$, then, for any $v \neq 0$, there is an $\alpha(v)$ such that $h(v) = \alpha(v)g(v)$. If v_0, v_1 are independent, then $g(v_0)$, $g(v_1)$ are independent (since g injective implies $\xi_0 g(v_0) + \xi_1(v_1) = g(\xi_0 v_0 + \xi_1 v_1) = 0$ means $\xi_0 v_0 + \xi_1 v_1 = 0$) and so, $\alpha(u_1+u_2)g(u_1) + \alpha(u_1+u_2)g(u_2) = \alpha(u_1+u_2)g(u_1+u_2) = h(u_1+u_2) = h(u_1) + h(u_2) = \alpha(u_1)h(u_1) + \alpha(u_2)h(u_2)$ for all u_1, u_2. Thus, $\alpha(u_1) = \alpha(u_1 + u_2) = \alpha(u_2)$ (when u_i independent). Hence $\alpha(v) = \alpha$ is constant for all $v \neq 0$.

Corollary 5.11 *The projective transformations of* $\mathbb{P}(\mathbf{V})$ *into* $\mathbb{P}(\mathbf{V})$ *form a group, denoted* $\mathrm{PGL}(\mathbf{V})$, *and called the projective (linear) group of* $\mathbf{V}$ *or* $\mathbb{P}(\mathbf{V})$. *Note that* $\mathrm{PGL}(\mathbf{V}) = \mathrm{GL}(\mathbf{V}) \setminus \{k^* I\}$.

For example, if $\alpha\colon \mathbb{P}_k^N \to \mathbb{P}_k^N$ is a projective transformation, then there is a $g_\alpha \in \mathrm{GL}(N+1, k)$ with $\alpha(\mathrm{sp}[v]) = \mathrm{sp}[g_\alpha v]$ for all $v \neq 0$ in k^{N+1}. If $\mathrm{sp}[g_\alpha v] = \mathrm{sp}[g'_\alpha v]$ for all $v \neq 0$, then $g_\alpha g_\alpha^{'-1} = \lambda I$, $\lambda \neq 0$ and conversely. Hence, $\mathrm{PGL}(N+1, k) = \mathrm{GL}(N+1,k)/\{k^* I\}$ via $g \equiv g'$ if $gg^{'-1} = \lambda I$, $\lambda \neq 0$. In essence, a projective transformation of $\mathbb{P}_k^N$ (or $\mathbb{P}(\mathbf{V})$) corresponds to a change of basis in k^{N+1} (or $\mathbf{V}$) modulo the center of $\mathrm{GL}(N+1,k)$ (or $\mathrm{GL}(\mathbf{V})$).

Definition 5.12 A subset $A \subset \mathbb{P}(\mathbf{V})$ is in *general position or is independent* if $A = \pi(A_1)$ with A_1 linearly independent in $\mathbf{V}$.

If $e_0, \ldots, e_N$ is a basis of V and $\pi(e_0) = P_0, \ldots, \pi(e_N) = P_N$, then $P_0, \ldots, P_N$ are in general position. If $\lambda_0, \ldots, \lambda_N$ are all $\neq 0$, then $\pi(\lambda_i e_i) = P_i$, $i = 0, 1, \ldots, N$ and so, $P_0, \ldots, P_N$ are not sufficient to determine a basis of $\mathbf{V}$. However, if $P_{N+1} = \pi(e_0 + \cdots + e_N)$, then $P_0, \ldots, P_N$, P_{N+1} do determine $e_0, \ldots, e_N$ since $\pi(e_0 + \cdots + e_N) = \pi(\lambda_0 e_0 + \cdots + \lambda_N e_N)$ implies $\lambda_0 = \lambda_1 = \cdots = \lambda_N = \lambda \neq 0$. In other words, the *projective coordinate system* on $\mathbb{P}(\mathbf{V})$ is uniquely determined by the $N+2$ points with homogeneous coordinates

$$(1,0,\ldots,0), (0,1,\ldots,0), \ldots, (0,\ldots,0,1), (1,1,\ldots,1).$$

Any subset of $N+1$ of these points is in general position. Conversely, if $P_0, \ldots, P_{N+1}$ are points in $\mathbb{P}(\mathbf{V})$ such that all subsets of $N+1$ points are in general position, then there is a projective coordinate system with $P_0 = (1,0,\ldots,0), \ldots, P_N = (0,\ldots,0,1)$, $P_{N+1} = (1,1,\ldots,1)$.

Suppose that $\{P^0, P^1, \ldots, P^r\}$ is a set of $r+1$ points in general position in $\mathbb{P}(\mathbf{V})$. Let $P^i = \mathrm{sp}[v^i] = (\xi_0^i, \ldots, \xi_N^i)$. Then we can say that $P = \sum_{i=0}^{r} c_i P^i$ is an element of $\mathrm{sp}[P^0, \ldots, P^r]$ if $P = \mathrm{sp}[\sum c_i v^i] = \left(\sum_{i=0}^{r} c_i \xi_j^i\right)$, c_i not all 0. This makes sense since $P = \mathrm{sp}[\sum c_i v^i]$ and $\sum c_i v^i = 0$ would imply $c_i = 0$, $i = 0, \ldots, r$ as the v^i are linearly independent in $\mathbf{V}$. Moreover, $\lambda w = \sum c_i \lambda_i v^i$ if and only if $w \in \mathrm{sp}[v^0, \ldots, v^r]$ ($\lambda \neq 0$). In other words, $L_r = \mathrm{sp}[P^0, \ldots, P^r]$

is a linear subspace of $\mathbb{P}(\mathbf{V})$ and, clearly, $\dim L_r = r$. Conversely, any linear subspace L_r of $\mathbb{P}(\mathbf{V})$ with $\dim L_r = r$ is the span of $r+1$ points in general position. We have:

Proposition 5.13 (*Incidence*)[†] (i) *if $L_r \subseteq L_s$ and $L_s \subseteq L_r$, then $L_r = L_s$;* (ii) *if $L_r \subseteq L_s$ and $L_s \subseteq L_t$, then $L_r \subseteq L_t$;* (iii) *$r+1$ points are not in general position if there is an L_s with $s < r$ containing them;* (iv) *any line L_1 contains at least three points;* (v) *if $P^0, \ldots, P^r$ are in general position, then there is an L_r with $P^j \in L_r$;* (vi) *any L_r contains $r+1$ points in general position;* (vii) *if $P^0, \ldots, P^r \in L_s$ and $P^0, \ldots, P^r \in L_r$, then $L_r \subset L_s$;* (viii) *If $L_r = \text{sp}[P^0, \ldots, P^r]$ and $L_s = \text{sp}[Q^0, \ldots, Q^s]$, then $\{P^0, \ldots, P^r, Q^0, \ldots, Q^s\}$ dependent implies $L_r \cap L_s \neq \emptyset$; and,* (ix) *if $\dim \mathbf{V} = N+1$, then, in $\mathbb{P}(\mathbf{V})$, there are $N+1$ points in general position and any $M > N+1$ points are dependent.*

Proof. See Exercise 3.

The conditions of the proposition can be used to develop projective geometry synthetically. Moreover, this classical view can be used to connect projective geometry with matrices, subspaces and Grassmannians.

If $\{P^0, \ldots, P^r\}$ are in general position and $L_r = \text{sp}[P^0, \ldots, P^r]$, then L_r is itself (isomorphic to) $\mathbb{P}_k^r$ since $P = \left(\sum_{i=0}^{r} c_i \xi_j^i\right) = \left(\sum_{i=0}^{r} \lambda c_i \xi_j^i\right)$, $\lambda \neq 0$, where $P^i = (\xi_0^i, \ldots, \xi_N^i)$. The isomorphism is given by $P \to (c_0, \ldots, c_r)$.[‡] Thus, we can and do view $L_r = \mathbb{P}_k^r$ as a projective subspace of $\mathbb{P}_k^N$. In fact, it is often useful to consider the set of all $\mathbb{P}_k^r \subset \mathbb{P}_k^N$ (it is a version of $\text{Gr}(r+1, N+1)$).

Now, suppose that $P^i = (\xi_j^i)$, $i = 0, 1, \ldots, N-1$ are in general position and so determine a $\mathbb{P}_k^{N-1}$ i.e., a hyperplane. Let $\mathbf{A}$ be the matrix given by

$$\mathbf{A} = (\xi_j^i) = \begin{pmatrix} P^0 \\ P^1 \\ \vdots \\ P^{N-1} \end{pmatrix} \tag{5.14}$$

where $i = 0, 1, \ldots, N-1$, $j = 0, \ldots, N$. $\mathbf{A}$ is an $N \times N+1$ matrix of rank N (since the P^i are in general position) and so

---

[†] We use L_r to denote a linear subspace of dimension r.

[‡] All are homogeneous coordinates. Perhaps, we should be pedantic and write $(\xi_0 : \xi_j : \cdots : \xi_N)$ or $\langle \xi_0, \ldots, \xi_N \rangle$ or $\cdots$ for homogeneous coordinates but we shall not and shall leave it to the reader to understand when we are using them.

$$\mathbf{A}u = 0, \quad u = \begin{pmatrix} u^0 \\ \vdots \\ u^N \end{pmatrix} \tag{5.15}$$

has a nonzero solution $\boldsymbol{\eta} = (\eta^0, \ldots, \eta^N)$. Since $\dim \operatorname{Ker} \mathbf{A} = 1$, all nonzero solutions of (5.15) are of the form $\lambda\boldsymbol{\eta}$, $\lambda \neq 0$. The equation (5.15) may be written as

$$\sum_{j=0}^{N} u^j \xi_j^i = 0 \tag{5.16}$$

where $i = 0, 1, \ldots, N-1$. Thus, $\mathbb{P}_k^{N-1}$ corresponds to a point $\boldsymbol{\eta}$ in a projective space $(\mathbb{P}_k^N)^*$. More precisely, if $\boldsymbol{\eta} = (\eta^0, \ldots, \eta^N) \in (\mathbb{P}_k^N)^*$, then the equation

$$\sum_{j=0}^{N} \eta^j Y_j = 0 \tag{5.17}$$

on $\mathbb{P}_k^N$ defines an L_{N-1} and any L_{N-1} is defined by such an equation. The projective space $(\mathbb{P}_k^N)^*$ is called the *dual space* of $\mathbb{P}_k^N$. Note that $\mathbb{P}(\mathbf{V})^* = \mathbb{P}(\mathbf{V}^*)$.

Definition 5.18 A point $\boldsymbol{\xi} \in \mathbb{P}_k^N$ and a hyperplane $\boldsymbol{\eta} \in (\mathbb{P}_k^N)^*$ are *incident* if $\sum_{j=0}^{N} \eta^j \xi_j = 0$ (i.e., if the points lies on the hyperplane or if, equivalently, the hyperplane passes through the point).

Proposition 5.19 *The hyperplane L_{N-1} of $\mathbb{P}_k^N$ which contain a given linear subspace L_r of $\mathbb{P}_k^N$ form an L_{N-r-1}^* in $(\mathbb{P}_k^N)^*$. In other words, $\{L_{N-1}: L_r \subset L_{N-1}\} = L_{N-r-1}^*$ in $(\mathbb{P}_k^N)^*$.*

Proof. Let $P^i = (\xi_0^i, \ldots, \xi_N^i)$, $i = 0, 1, \ldots, r$ be $r+1$ points of L_r in general position. Then L_{N-1} contains L_r if and only if the P^i all satisfy the equation

$$\sum_{j=0}^{N} \eta^j Y_j = 0 \tag{5.20}$$

of L_{N-1} $(= (\eta^0, \ldots, \eta^N)$ as point of $(\mathbb{P}_k^N)^*)$. If $\mathbf{A} = (\xi_j^i)$, then A is an $r+1 \times N \times 1$ matrix of rank $r+1$, and the hyperplane L_{N-1} $(= \boldsymbol{\eta})$ contains all the P^i if and only if $\boldsymbol{\eta} \in \operatorname{Ker} \mathbf{A}$. But $\dim \operatorname{Ker} \mathbf{A} = N - r$ and so, $\boldsymbol{\eta} \in L_{N-r+1}^*$ $(= \operatorname{sp}[\operatorname{Ker} \mathbf{A}])$.

Corollary 5.21 *The hyperplanes L_{N-1}^* of $(\mathbb{P}_k^N)^*$ which contain a given linear subspace L_{N-r-1}^* form an $L_r^{**} \simeq L_r$ in $(\mathbb{P}_k^N)^{**} \simeq \mathbb{P}_k^N$.*

Example 5.22 Consider $\mathbb{P}^N$ with (homogeneous) coordinates $(X_0, \ldots, X_N)$ and $\mathbb{P}^{N-1}$ with (homogeneous) coordinates $(Y_0, \ldots, Y_{N-1})$. Let $\boldsymbol{\xi}_\infty = (0, \ldots, 0, 1) \in \mathbb{P}^N$ and consider the map $\pi\colon \mathbb{P}^N - \{\boldsymbol{\xi}_\infty\} \to \mathbb{P}^{N-1}$ given by

$$\pi(\xi_0, \ldots, \xi_{N-1}, \xi_N) = (\xi_0, \ldots, \xi_{N-1}). \tag{5.23}$$

π is called the *projection with center* $\boldsymbol{\xi}_\infty$. The map π is clearly surjective. If $\boldsymbol{\xi} \neq \boldsymbol{\xi}_\infty$ then $\pi^{-1}(\pi(\boldsymbol{\xi})) = J(\boldsymbol{\xi}, \boldsymbol{\xi}_\infty)$ is the line joining $\boldsymbol{\xi}$ and $\boldsymbol{\xi}_\infty$. We can view this example in another way. In $\mathbb{P}^N$, $\boldsymbol{\xi}_\infty$ is an L_0 defined by the N independent forms $X_0 = 0, \ldots, X_{N-1} = 0$ and $\mathbb{P}^{N-1} \simeq L_{N-1}$ is the hyperplane defined by the form $X_N = 0$. If $\boldsymbol{\xi} = (\xi_0, \ldots, \xi_N) \neq \boldsymbol{\xi}_\infty$, then $J(\boldsymbol{\xi}, \boldsymbol{\xi}_\infty)$ is an L_1 (i.e., a line) which meets L_{N-1} in a point $Q = (q_0, \ldots, q_N)$ (since $J(\boldsymbol{\xi}_\infty, L_{N-1}) = \mathbb{P}^N$). But $J(\boldsymbol{\xi}, \boldsymbol{\xi}_\infty) = (\lambda\xi_i + \mu\xi_{\infty,i})$. Since $\xi_{\infty,0} = 0, \ldots, \xi_{\infty,N-1} = 0$ and $q_N = 0$ as $Q \in L_{N-1}$, we have $Q = \lambda(\xi_0, \ldots, \xi_{N-1}, 0)$. In fact, if L_{N-1} is any hyperplane and P is a point with $P \notin L_{N-1}$, then the *projection of* $\mathbb{P}^N$ *on* L_{N-1} *with center* P is the map $\pi_p\colon \mathbb{P}^N - \{P\} \to L_{N-1}$ defined as follows: if $Q \neq P$, then the line $J(P, Q)$ meets L_{N-1} in a point R which is $\pi_p(Q)$.

Example 5.24 (Generalization of Example 5.22). Let L_r, $0 \leq r \leq N-1$, be a linear subspace of $\mathbb{P}^N$ of dimension r and let L_{N-r-1} be a complimentary linear subspace i.e., $L_r \cap L_{N-r-1}$ is empty so that $J(L_r, L_{N-r-1}) = \mathbb{P}^N$. If $Q \notin L_{N-r-1}$, then $J(Q, L_{N-r-1}) = L_{N-r}$ and $L_{N-r} \cap L_r$ is a linear subspace of dimension 0 i.e., a point Q' (since $\dim L_{N-r} \cap L_r = \dim L_{N-r} + \dim L_r - \dim J(L_{N-r}, L_r) = N - N = 0$ as $\mathbb{P}^N = J(L_r, L_{N-r-1}) \subseteq J(L_r, L_{N-r}) \subseteq \mathbb{P}^N$). We can thus define a map $\pi_{L_{N-r-1}}\colon \mathbb{P}^N - L_{N-r-1} \to L_r \simeq \mathbb{P}^r$ by setting

$$\pi_{L_{N-r-1}}(Q) = Q' = J(Q, L_{N-r-1}) \cap L_r. \tag{5.25}$$

This surjective map is called the *projection of* $\mathbb{P}^N$ *on* L_r *with center* L_{N-r-1}. If $P^0, \ldots, P^r$ are in general position in L_r and $P^{r+1}, \ldots, P^N$ are in general position in L_{N-r-1} so that $P^0, \ldots, P^r, P^{r+1}, \ldots, P^N$ can be viewed as reference vertices, then L_r is given by $X_{r+1} = 0, \ldots, X_N = 0$ and L_{N-r-1} is given by $X_0 = 0, \ldots, X_r = 0$. If $Q = (q_i)$, then

$$J(Q, L_{N-r-1}) = \{(\lambda q_i + \mu\xi_i)\colon \boldsymbol{\xi} = (\xi_i) \in L_{N-r-1}\}.$$

Since $\xi_0, \ldots, \xi_r$ are all zero and $\pi_{L_{N-r-1}}(Q) = Q' \in L_r$, we have $Q' = \lambda(q_0, \ldots, q_r, 0, \ldots, 0)$. In another way, if $\ell_i = \sum_{j=0}^{N} a_{ij}X_j$, $i = 0, 1, \ldots, r$ are $r+1$ independent linear forms, then $\ell_0 = 0, \ldots, \ell_r = 0$ defines an $N - r - 1$ dimensional linear subspace L of $\mathbb{P}^N$ and the *projection* $\pi_L\colon \mathbb{P}^N - L \to \mathbb{P}^r$ *with center* L is given by

$$\pi_L(\xi_0, \ldots, \xi_N) = \left(\sum_{j=0}^{N} a_{0j}\xi_j, \ldots, \sum_{j=0}^{N} a_{rj}\xi_j\right). \tag{5.26}$$

π_L is clearly surjective.

Example 5.27 (Intrinsic form of Example 5.24). Let $\mathbb{P}^N = \mathbb{P}(\mathbf{V})$ and let $L = L_{N-r-1} = \mathbb{P}(\mathbf{W})$ correspond to the $N-r$ dimensional subspace $\mathbf{W}$ of $\mathbf{V}$. If $\mathbf{U} \subset \mathbf{V}$ is a complimentary subspace (i.e., $\mathbf{W} \oplus \mathbf{U} = \mathbf{V}$), then $\mathbb{P}(\mathbf{U}) = L_r$ and π_L is just the map which corresponds to the natural projection $\mathbf{W} \oplus \mathbf{U} \to \mathbf{U}$. This also shows that π_L depends (to within a projective transformation) only on L and not on the choice of L_r. The connection between subspaces of $\mathbf{V}$ and linear spaces in $\mathbb{P}^N$ is also illustrated by this example.

Projections play an important role in projective algebraic geometry as we shall see in the sequel.

Example 5.28 Consider $\mathbb{P}^2$ and let $P = \boldsymbol{\xi}_\infty = (0,0,1)$. Let $L \simeq \mathbb{P}^1$ be the hyperplane $X_2 = 0$. Then, as we have seen, $\pi_P(\boldsymbol{\xi}) = (\xi_0, \xi_1, 0)$ if $\boldsymbol{\xi} = (\xi_0, \xi_1, \xi_2) \neq \boldsymbol{\xi}_\infty$. Let $\tilde{L}$ be the hyperplane $X_0 + X_1 + X_2 = 0$ so that $\boldsymbol{\xi}_\infty \notin \tilde{L}$. Then $\pi_P(\tilde{L}) = \mathbb{P}^1$. For, if $Q = (q_0, q_1, 0) \in \mathbb{P}^1$, then $J(Q, \boldsymbol{\xi}_\infty) = (\lambda q_0, \lambda q_1, \mu)$ and $J(Q, \boldsymbol{\xi}_\infty) \cap \tilde{L} = (\lambda q_0, \lambda q_1, -\lambda(q_0 + q_1))$. Note that "above" each point Q of $\mathbb{P}^1$ there is one point of $\tilde{L}$ i.e., $\pi_P^{-1}(Q) \cap \tilde{L}$ contains a single point.

Example 5.29 Consider $\mathbb{P}^2$, $\boldsymbol{\xi}_\infty = (0,0,1)$ and $\mathbb{P}^1 = L$, the hyperplane $X_2 = 0$. Let

$$V = \{\boldsymbol{\xi} = (\xi_0, \xi_1, \xi_2) \colon \xi_2^2 - \xi_0\xi_1 = 0\} = V(X_2^2 - X_0X_1)$$

(note that if (ξ_0, ξ_1, ξ_2) is such that $\xi_2^2 - \xi_0\xi_1 = 0$ then $(\lambda\xi_0, \lambda\xi_1, \lambda\xi_2)$ also satisfies the equation $(\lambda\xi_2)^2 - (\lambda\xi_0)(\lambda\xi_1) = 0$). Since $\boldsymbol{\xi}_\infty \notin V$, $\pi_{\boldsymbol{\xi}_\infty}(V) = \tilde{V}$ is defined. If $Q = (q_0, q_1, 0) \in L$, then $J(Q, \boldsymbol{\xi}_\infty) = (\lambda q_0, \lambda q_1, \mu)$ and $J(Q, \boldsymbol{\xi}_\infty) \cap V = \{(q_0, q_1, 0) \colon q_0 q_1 = 0\}$ or $\{(q_0, q_1, \mu) \colon q_0 q_1 \neq 0 \text{ and } \mu^2 = q_0 q_1\}$. Hence $\tilde{V} = \mathbb{P}^1$ and "above" the points of $\mathbb{P}^1 - \{X_0X_1 = 0\}$, there are two points of V while "above" the points of $L \cap \{X_0X_1 = 0\}$, there is only one point of V. In other words, $L \cap \{X_0X_1 = 0\} = \{(1,0,0), (0,1,0)\}$ and $(1,0,0) \in V$ lies "above" the point $(1,0,0)$ (similarly for $(0,1,0)$) i.e., $\pi_{\boldsymbol{\xi}}((1,0,0) \in V) = (1,0,0) \in \tilde{V}$. "Above" the point $(1,1,0) \in L$, lie the points $(1,1,1)$ and $(1,1,-1)$ of V. (This is relevant to the notion of "degree").

Let H_0 be the hyperplane (linear subspace) $X_0 = 0$ in $\mathbb{P}^N$ (i.e., $H_0 = \{\boldsymbol{\xi} \in \mathbb{P}^N \colon \xi_0 = 0\}$) and let $U_0 = \mathbb{P}^N - H_0$. If $P = \boldsymbol{\xi} = (\xi_0, \dots, \xi_N) \in U_0$, then $\xi_0 \neq 0$ and $P = (1, \xi_1/\xi_0, \dots, \xi_N/\xi_0) = (1, \eta_1, \dots, \eta_N) = (1, \boldsymbol{\eta})$ where $\eta_i = \xi_i/\xi_0$, $i = 1, \dots, N$. The η_i are uniquely determined by P (i.e., do not depend on the choice of homogeneous coordinates). The map $\psi_0 \colon U_0 \to \mathbb{A}^N$ given by

$$\psi_0(P) = (\xi_1/\xi_0, \dots, \xi_N/\xi_0) = (\eta_1, \dots, \eta_N) \tag{5.30}$$

is clearly bijective. Similarly, if H_i is the hyerplane $X_i = 0$ and $U_i = \mathbb{P}^N - H_i$, then the map $\psi_i \colon U_i \to \mathbb{A}^N$ given by

$$\psi_i(\xi_0, \dots, \xi_N) = (\xi_0/\xi_i, \dots, \xi_{i-1}/\xi_i, \xi_{i+1}/\xi_i, \dots, \xi_N/\xi_i) \tag{5.31}$$

is also bijective. Moreover, $\mathbb{P}^N = \bigcup_{i=0}^{N} U_i$ so that $\mathbb{P}^N$ is covered by affine spaces $\mathbb{A}^N$. Since, for example, if $P = (\xi_0, \ldots, \xi_N) \in U_0$, the $\eta_i = \xi_i/\xi_0$ are uniquely determined, these are called *affine coordinates of* P (with respect to H_0 or U_0). Thus, locally at least, $\mathbb{P}^N$ looks like $\mathbb{A}^N$. We also often refer to (say) H_0 as the *hyerplane at infinity.*

Exercises

(1) Show that if $L_1 = \pi(\mathbf{W}_1)$, $L_2 = \pi(\mathbf{W}_2)$, then $L_1 \cap L_2 = \pi(\mathbf{W}_1 \cap \mathbf{W}_2)$ and $L_1 \vee L_2 = \pi(\mathbf{W}_1 + \mathbf{W}_2)$.

(2) Show that if $P_0, \ldots, P_{N+1}$ are points in $\mathbb{P}_k^N$ such that all subsets of $N+1$ points are in general position, then there is a projective coordinate system with $P_0 = (1, 0, \ldots, 0), \ldots, P_N = (0, \ldots, 0, 1)$, $P_{N+1} = (1, 1, \ldots, 1)$.

(3) Prove Proposition 5.13.

(4) Show that $\mathbb{P}(\mathbf{V})^* = \mathbb{P}(\mathbf{V}^*)$ where $\mathbf{V}^*$ is the dual space of $\mathbf{V}$.

(5) Show that $(\mathbb{P}_k^N)^{**} \simeq \mathbb{P}_k^N$.

(6) Show that if L_r, $0 \leq r \leq N-1$, is a linear subspace of $\mathbb{P}^N$, then there are L_{N-r-1} such that $L_r \cap L_{N-r-1} = \emptyset$ and, for any such L_{N-r-1}, $J(L_r, L_{N-r-1}) = \mathbb{P}^N$.

6

Projective Algebraic Geometry I: Basic Concepts

Let k be an algebraically closed field and let $\mathbb{P}^N = \mathbb{P}_k^N$ be projective N-space over k. We let $(Y_0, \ldots, Y_N)$ be the coordinate functions on $\mathbb{P}^N$.

Definition 6.1 A polynomial $F(Y)$ in $k[Y_0, \ldots, Y_N]$ *vanishes at* $\boldsymbol{\xi} \in \mathbb{P}^N$ if $F(\xi_0, \ldots, \xi_N) = 0$ for *all* sets $(\xi_0, \ldots, \xi_N)$ of homogeneous coordinates of $\boldsymbol{\xi}$.

Remark 6.2 If $F = F_0 + F_1 + \cdots + F_r$, F_i homogeneous of degree i, then $F(\lambda\boldsymbol{\xi}) = F_0(\boldsymbol{\xi}) + \lambda F_1(\boldsymbol{\xi}) + \cdots + \lambda^r F_r(\boldsymbol{\xi}) = 0$ for all $\lambda \in k^*$ if and only if $F_i(\boldsymbol{\xi}) = 0$ $i = 0, 1, \ldots, r$ (since k is infinite). In particular, $F_0(\boldsymbol{\xi}) = 0$ means $F_0 \equiv 0$.

Definition 6.3 $V \subset \mathbb{P}^N$ is a *projective algebraic set* if V is the set of common zeros of a family of polynomials in $S = k[Y_0, \ldots, Y_N]$ (or, equivalently, the set of common zeros of a family of homogeneous elements of S in view of Remark 6.1).

We want to view $S = k[Y_0, \ldots, Y_N] = k[Y]$ as a graded ring.

Definition 6.4 A *grading* of a ring R is a family $\{R_q\}$, $q \in \mathbb{Z}$, of subgroups of the additive group of R such that (i) $R = \oplus \sum R_q$; and, (ii) $R_q R_{q'} \subset R_{q+q'}$. R is *positively graded* if $R_q = 0$ for $q < 0$. The elements of R_q are called *homogeneous of degree* q. An ideal $\mathfrak{a}$ of R is *homogeneous* if $F = \sum F_q$, $F \in \mathfrak{a}$, $F_q \in R_q$ implies all homogeneous components F_q of F are also in $\mathfrak{a}$. An R-module M is a *graded R-module* if (i) $M = \oplus \sum M_q$; and, (ii) $R_d \cdot M_q \subset M_{q+d}$.

The polynomial ring $S = k[Y_0, \ldots, Y_N]$ is a (positively) graded ring if we let S_q be the set of all k-linear combinations of monomials of degree q.

P. Falb, *Methods of Algebraic Geometry in Control Theory: Part II*,
Modern Birkhäuser Classics, https://doi.org/10.1007/978-3-319-96574-1_6

Example 6.5 Let $N = 2$ so that $S = k[Y_0, Y_1, Y_2]$. Then $S_0 = \mathrm{sp}_k[1]$, $S_1 = \mathrm{sp}_k[Y_0, Y_1, Y_2]$, $S_2 = \mathrm{sp}_k[Y_0^2, Y_0Y_1, Y_0Y_2, Y_1^2, Y_1Y_2, Y_2^2], \ldots$. In fact, $\dim_k S_q = \binom{q+2}{q} = (q+2)!/2!q! = (q+2)(q+1)/2$ since (abusing notation) S_q is spanned by $Y_0^q, Y_0^{q-1}S_1[Y_1, Y_2], Y_0^{q-2}S_2[Y_1, Y_2], \ldots, S_q[Y_1, Y_2]$. If $(\alpha_0, \alpha_1, \alpha_2)$ is a given element of $\mathbb{Z}^3$, then we can also grade S by setting $S_q^\alpha = \mathrm{sp}_k[Y_0^{t_0}Y_1^{t_1}Y_2^{t_2}\colon \alpha_0 t_0 + \alpha_1 t_1 + \alpha_2 t_2 = q]$.

Example 6.6 Consider $S = k[Y_0, \ldots, Y_N]$. Then $S_0 = \mathrm{sp}_k[1]$, $S_1 = \mathrm{sp}_k[Y_0, \ldots, Y_N], \ldots$. We have $\dim_k S_q = \binom{q+N}{q} = \binom{q+N}{N}$ since (abuse of notation) S_q is "spanned" by $Y_0^q, Y_0^{q-1}S_1[Y_1, \ldots, Y_N], \ldots, S_q[Y_1, \ldots, Y_N]$. As in the previous example, if $(\alpha_0, \ldots, \alpha_N) \in \mathbb{Z}^{N+1}$, then we can also grade S by setting $S_q^\alpha = \mathrm{sp}_k[Y_0^{t_0}Y_1^{t_1}\cdots Y_N^{t_N}\colon \sum \alpha_j t_j = q]$. We note also that

$$\begin{aligned} \dim_k S_q = \binom{q+N}{q} &= \frac{(q+N)\cdots(q+1)}{N!} \\ &= 1 \cdot \frac{q^N}{N!} + \cdots = h(q) \end{aligned} \tag{6.7}$$

is a polynomial in q of degree N.

We return now to graded rings.

Lemma 6.8 *Let R be a graded ring and let $\mathfrak{a}$ be an ideal in R. Then the following are equivalent:* (a) *$\mathfrak{a}$ is homogeneous;* (b) *$\mathfrak{a}$ has a basis of homogeneous elements; and,* (c) *$R/\mathfrak{a}$ is a graded ring with grading $(R/\mathfrak{a})_q = (R_q + \mathfrak{a})/\mathfrak{a}$. In fact, $R/\mathfrak{a}$ is an R-module under $r_1 \cdot \bar{r}_2 = \overline{r_1 r_2}$.*

Proof. (a) $\Rightarrow$ (b) If $f^{(\alpha)}$ is a basis of $\mathfrak{a}$, then the homogeneous components of $f_i^{(\alpha)}$ are in $\mathfrak{a}$ and give a basis. (b) $\Rightarrow$ (a) If $f^{(\alpha)}$ is a basis of homogeneous elements and $f \in \mathfrak{a}$, then $f = \sum r_\alpha f^{(\alpha)}$, $r_\alpha \in R$. If $r_\alpha = \sum_q r_{q,\alpha}$ with $r_{q,\alpha}$ homogeneous of degree q, then $f = \sum_{\alpha,q} f_{q,\alpha} f^{(\alpha)} = \sum_{q'} f_{q'}$ where $f_{q'} = \sum_{q+q_\alpha} r_{q,\alpha'} f^{(\alpha)}$ where $q_\alpha = \text{degree } f^{(\alpha)}$. Hence, $f_{q'} \in \mathfrak{a}$ for all q' and $\mathfrak{a}$ is homogeneous. (a) $\Rightarrow$ (c) It is clear that $(R/\mathfrak{a}) = \sum (R/\mathfrak{a})_q$ and that $(R/\mathfrak{a})_q (R/\mathfrak{a})_{q'} \subset (R/\mathfrak{a})_{q+q'}$. Thus it is enough to show the sum is direct. So suppose that $\sum \bar{r}_q = 0$ and $\bar{r}_q = r_q + \mathfrak{a}$. Then $\sum r_q \in \mathfrak{a}$ and, since $\mathfrak{a}$ is homogeneous, $r_q \in \mathfrak{a}$ (all q) and $\bar{r}_q = 0$. (c) $\Rightarrow$ (a) If $\sum r_q \in \mathfrak{a}$, then $\sum \bar{r}_q = 0$ and $\bar{r}_q = 0$ so that $r_q \in \mathfrak{a}$.

Corollary 6.9 *If $\mathfrak{a}, \mathfrak{b}$ are homogeneous ideals, then $\mathfrak{a} + \mathfrak{b}$, $\mathfrak{a}\mathfrak{b}$, $\mathfrak{a} \cap \mathfrak{b}$, and $\sqrt{\mathfrak{a}}$ are also homogeneous ideals.*

Proof. Trivial for $\mathfrak{a}+\mathfrak{b}$, $\mathfrak{ab}$, and $\mathfrak{a}\cap\mathfrak{b}$. Let $f \in \sqrt{\mathfrak{a}}$ with $f = f_{s+1} + \cdots$, f_i homogeneous. Then $f^{\nu} = f_s^{\nu} + \cdots$ terms of degree $> s\nu$. Since $f^{\nu} \in \mathfrak{a}$, $f_s^{\nu} \in \mathfrak{a}$ and $f_s \in \sqrt{\mathfrak{a}}$. Apply induction to $f - f_s$ to get the result.

Lemma 6.10 *If $\mathfrak{p}$ is a prime ideal in R, a graded ring, and $\mathfrak{p}_h$ is the ideal generated by the homogeneous elements of $\mathfrak{p}$, then $\mathfrak{p}_h$ is prime.*

Proof. We may suppose that R is positively graded. Let $f = f_0 + \cdots + f_r$, $g = g_0 + \cdots + g_s$ with $fg \in \mathfrak{p}_h$. If both f and g are not in $\mathfrak{p}_h$, then there is a largest i with $f_i \notin \mathfrak{p}_h$ and a largest j with $g_j \notin \mathfrak{p}_h$. But $(fg) = \sum (fg)_q$ and $(fg)_{i+j} = \sum_{\alpha+\beta=i+j} f_\alpha g_\beta$ taking $q = i+j$. Since $\mathfrak{p}_h$ is a homogeneous ideal, $(fg)_{i+j} \in \mathfrak{p}_h$. But all summands $f_\alpha g_\beta$ except perhaps $f_i g_j$ are in $\mathfrak{p}_h$ so that $f_i g_j$ is in $\mathfrak{p}_h \subset \mathfrak{p}$. Since $\mathfrak{p}$ is prime, (say) $f_i \in \mathfrak{p}$ so that $f_i \in \mathfrak{p}_h$ which is a contradiction.

Corollary 6.11 *A homogeneous ideal $\mathfrak{p}$ is prime if and only if $f \notin \mathfrak{p}$, $g \notin \mathfrak{p}$ implies $fg \notin \mathfrak{p}$ for homogeneous f, g.*

Example 6.12 Let $S = k[Y_0, \ldots, Y_N]$ and let F be a homogeneous element of degree d and $\mathfrak{a} = (F)$. Then $\mathfrak{a}$ is a homogeneous ideal and F defines an S-module homomorphism of $S_{q-d} \to S_q$ by sending $G \to F \cdot G$. The sequence

$$0 \longrightarrow S_{q-d} \xrightarrow{F} S_q \longrightarrow [S/\mathfrak{a}]_q \longrightarrow 0 \tag{6.13}$$

is exact (note $S/\mathfrak{a} = S/(F)$ is a graded S-module). It follows that $\dim_k S_q = \dim_k S_{q-d} + \dim_k [S/(F)]_q$ and hence, that

$$\begin{aligned}\dim_k [S/(F)]_q &= \binom{q+N}{q} - \binom{q+N-d}{q-d} \\ &= \binom{q+N}{N} - \binom{q-d+N}{N} \\ &= \frac{dq^{N-1}}{(N-1)!} + \cdots = h_F(q)\end{aligned}$$

where $h_F(q)$ is a polynomial of degree $N-1$ for $q \geq d$.

We can generalize the example as follows:

Proposition 6.14 *Let $\mathfrak{a}$ be a homogeneous ideal in $S = k[Y_0, \ldots, Y_N]$. Then there is a polynomial $h_\mathfrak{a}(q)$ of degree less than or equal to N such that* $\dim_k [S/\mathfrak{a}]_q = h_\mathfrak{a}(q)$ *for q sufficiently large. $h_\mathfrak{a}(q)$ is called the Hilbert polynomial of $\mathfrak{a}$.* (See Appendix A.)

Proof. Let $S^1 = k[Y_0, \ldots, Y_{N-1}]$ so that $S = S^1[Y_N]$. We use induction on N assuming the result for $N-1$. Let $(S/\mathfrak{a})' = \{\bar{r} \in S/\mathfrak{a}\colon Y_N\bar{r} = 0\}$ and $(S/\mathfrak{a})'' = (S/\mathfrak{a})/Y_N(S/\mathfrak{a})$. These are finitely generated graded S-modules which are annihilated by Y_N. Thus we have the exact sequence of S^1-modules

$$0 \longrightarrow (S/\mathfrak{a})'_q \longrightarrow (S/\mathfrak{a})_q \xrightarrow{Y_N} (S/\mathfrak{a})_{q+1} \longrightarrow (S/\mathfrak{a})''_{q+1} \longrightarrow 0. \tag{6.15}$$

By induction, there are polynomials $h'(q)$, $h''(q)$ (for large q) of degree $\leq N-1$ with $h'(q) = \dim_k(S/\mathfrak{a})'_q$ and $h''(q) = \dim_k(S/\mathfrak{a})''_q$. In view of (6.15),

$$\dim_k(S/\mathfrak{a})_{q+1} - \dim_k(S/\mathfrak{a})_q = h''(q+1) - h'(q). \tag{6.16}$$

It follows that there is a polynomial $h(q)$ of degree $\leq N$ with $h(q+1) - h(q) \equiv h''(q+1) - h'(q)$ and hence that

$$\dim_k(S/\mathfrak{a})_{q+1} - h(q+1) = \dim_k(S/\mathfrak{a})_q - h(q) \tag{6.17}$$

which means that $\dim_k(S/\mathfrak{a})_q = h(q) + c$, some constant c, for q large.

Now let us develop some of the elementary concepts such as the ideal of an algebraic set, the Zariski topology, irreducibility, and the Hilbert Nullstellensatz. We begin with an example.

Example 6.18 Let $N = 3$. Any linear space in $\mathbb{P}^3$ is a projective algebraic set as it is the set of common zeros of linear forms. Let $P = (\xi_0, \xi_1, \xi_2, \xi_3) \neq Q = (\eta_0, \eta_1, \eta_2, \eta_3)$ be points of $\mathbb{P}^3$. Then there is a linear form ℓ with $\ell(P) = 0$ and $\ell(Q) \neq 0$. Thus any finite set of points in $\mathbb{P}^3$ is a projective algebraic set. Let $F(Y_0, Y_1, Y_2, Y_3)$ be a homogeneous polynomial and let $V(F) = \{\boldsymbol{\xi} \in \mathbb{P}^3\colon F(\boldsymbol{\xi}) = 0\}$ (i.e., $F(\xi_0, \xi_1, \xi_2, \xi_3) = 0$ for all sets of homogeneous coordinates of $\boldsymbol{\xi}$). $V(F)$ is a hypersurface in $\mathbb{P}^3$. Consider the quadratic polynomials

$$Q_0(Y) = Y_1^2 - Y_0Y_2$$
$$Q_1(Y) = Y_1Y_2 - Y_0Y_3$$
$$Q_2(Y) = Y_2^2 - Y_1Y_3$$

and let $C = V(Q_0) \cap V(Q_1) \cap V(Q_2)$. C is an algebraic set called the *twisted cubic curve.* C can also be defined as the image of the map $\psi\colon \mathbb{P}^1 \to \mathbb{P}^3$ given by $\psi(X_0, X_1) = (X_0^3, X_0^2X_1, X_0X_1^2, X_1^3)$.

Definition 6.19 If $W \subset \mathbb{P}^N$, then $I_h(W) = \{f\colon f(W) = 0\}$ is the *ideal of* W (note $I_h(W)$ is a homogeneous ideal). If $\mathfrak{a}_h$ is a homogeneous ideal, then $V(\mathfrak{a}_h) = \{\boldsymbol{\xi}\colon f(\boldsymbol{\xi}) = 0 \text{ all } f \in \mathfrak{a}_h\}$ is the *algebraic (or zero) set of* $\mathfrak{a}_h$.

We observe that $\emptyset = V(S)$ and that $\mathbb{P}^N = V((0))$ so that both $\emptyset$ and $\mathbb{P}^N$ are algebraic sets.

Example 6.20 Let $P_1, \dots, P_d$ be a finite set of points in $\mathbb{P}^N$ not lying on the hyperplane H_0 at infinity. Let $\mathfrak{a} = I_h(\{P_1, \dots, P_d\})$. $\mathfrak{a}$ is a homogeneous ideal and $S/\mathfrak{a}$ is a graded S-module. Define an S-module homomorphism ψ of S into $\oplus \sum_{i=1}^{d} k$ $(= k^d)$ by setting $\psi(F) = \oplus \sum_{i=1}^{d} F(P_i)/Y_0^d$. Then the sequence

$$0 \longrightarrow \mathfrak{a}_q \longrightarrow S_q \xrightarrow{\psi} k^d$$

is exact. If $q > d$, then there are F_i homogeneous of degree q with $F_i(P_i) \neq 0$ and $F_i(P_j) = 0$, $i \neq j$. It follows that ψ is surjective for $q \geq d$ and hence that $\dim_k [S/\mathfrak{a}]_q = d \cdot t^0/0!$ is a polynomial $h_{\mathfrak{a}}(q)$ of degree 0.

Proposition 6.21 *Let $\mathfrak{a}, \mathfrak{a}_i$ be homogeneous ideals in $S = k[Y_0, \dots, Y_N]$ and let W, W_j be subsets of $\mathbb{P}_k^N$. Then*

(1) *if $\mathfrak{a} \subset \mathfrak{a}_1$, then $V(\mathfrak{a}) \supset V(\mathfrak{a}_1)$;*

(2) *if $W \subset W_1$, then $I_h(W) \supset I_h(W_1)$;*

(3) $V(\sum \mathfrak{a}_i) = \cap V(\mathfrak{a}_i)$;

(4) $I_h(\cup W_j) = \cap I_h(W_j)$;

(5) $V(\mathfrak{a} \cap \mathfrak{a}_1) = V(\mathfrak{a}) \cup V(\mathfrak{a}_1)$;

(6) $V(I_h(W)) \supset W$ *and* $I_h(V(\mathfrak{a})) \supset \mathfrak{a}$;

(7) $V(\mathfrak{a}) = V(\sqrt{\mathfrak{a}})$;

(8) *if $\sqrt{\mathfrak{a}} = \mathfrak{p}_1 \cap \cdots \cap \mathfrak{p}_r$, $\mathfrak{p}_i$ prime (necessarily homogeneous), then $V(\mathfrak{a}) = V(\mathfrak{p}_1) \cup \cdots \cup V(\mathfrak{p}_r)$.* [Cf. I.4.9, etc.]

Thus we can define the *Zariski-topology on* $\mathbb{P}_k^N$ by using the $V(\mathfrak{a})$ as the closed sets. Since $\mathbb{P}^N$ becomes a topological space, the notion of *irreducibility* (I.6) and its properties (e.g. I.6.3, I.6.4, I.6.10) applies to algebraic sets in $\mathbb{P}^N$. Let $\mathfrak{a}_s$ be the homogeneous ideal in S consisting of polynomials with terms of at least degree s i.e., $\mathfrak{a}_s = \sum_{q \geq s} S_q$. The maximal ideal $\mathfrak{a}_1 = (Y_0, \dots, Y_N) = \sum_{q \geq 1} S_q$ is called the *irrelevant ideal.* We observe that $\mathfrak{a}_1$ is a proper ideal i.e., $\mathfrak{a}_1 < S$ but that $V(\mathfrak{a}_1) = \emptyset$. Thus, we must modify the Nullstellensatz (I.5.9) in the projective case.

Lemma 6.22 *A proper homogeneous ideal $\mathfrak{a}$ has $V(\mathfrak{a})$ empty if and only if $\mathfrak{a}_s \subset \mathfrak{a}$ for some $s > 0$ i.e., if and only if $\sqrt{\mathfrak{a}} = (Y_0, \dots, Y_N)$.*

Proof. If $\mathfrak{a}_s \subset \mathfrak{a}$, then $V(\mathfrak{a}) \subset V(\mathfrak{a}_s) = \emptyset$ as $Y_i^s \in \mathfrak{a}_s$, $i = 0, 1, \ldots, N$. Suppose, on the other hand, $V(\mathfrak{a}) = \emptyset$ and $\mathfrak{a} = (f_1, \ldots, f_\nu)$ with f_i homogeneous of degree $d_i = d$ (we may assume all of same degree as

$$V(\mathfrak{a}) = V(f_1, \ldots, f_\nu) = V\big(Y_0^{d-d_1} f_1, Y_1^{d-d_1} f_1, \ldots, Y_N^{d-d_1} f_1, \ldots, Y_N^{d-d_\nu} f_\nu\big).$$

Let $X_j = Y_j/Y_0$ and $\tilde{f}_i(X_1, \ldots, X_N) = f_i(1, X_1, \ldots, X_N)$. Then the $\tilde{f}_i$, $i = 1, \ldots, \nu$ do not have a common zero in $\mathbb{A}_k^N$ (if all $\tilde{f}_i(\xi_1, \ldots, \xi_N) \in V(\mathfrak{a})$), then $f_i(1, \boldsymbol{\xi}_1, \ldots, \boldsymbol{\xi}_N) = 0$ all i and $(1, \xi_1, \ldots, \xi_N) \in V(\mathfrak{a})$) and so $V(\tilde{f}_1, \ldots, \tilde{f}_\nu) = \emptyset$ in $\mathbb{A}_k^N$. Hence, by the affine Nullstellensatz, there are $\tilde{g}_1(X_1, \ldots, X_N), \ldots, \tilde{g}_\nu(X_1, \ldots, X_N)$ such that $\sum \tilde{g}_i \tilde{f}_i = 1$. It follows that $Y_0^{r_0} \in \mathfrak{a}$ for some r_0 (multiplying by Y_0 to a suitable power). Similarly, $Y_i^{r_i} \in \mathfrak{a}$ for $i = 1, \ldots, N$. Thus, for large s (e.g. $s = (\nu-1)(d+1)+1)$, $m = \max(r_i)$), in every monomial $Y_0^{t_0} \cdots Y_N^{t_N}$, $\sum t_i \geq s$, there is a $t_i \geq m \geq r_i$ so that $\mathfrak{a}_s \subset \mathfrak{a}$.

Let $V = V(\mathfrak{a})$ be a projective algebraic set and let $C(V) = V_a(\mathfrak{a}) = \{\boldsymbol{\xi} \in \mathbb{A}_k^{N+1} \colon f(\boldsymbol{\xi}) = 0$ for all $f \in \mathfrak{a}\}$ be the *affine cone over* V. Since $\mathfrak{a}$ is a homogeneous ideal, $\boldsymbol{\xi} \in C(V)$ implies $t\boldsymbol{\xi} \in C(V)$ for all t in k. Thus, $C(V)$ is a union of lines in $\mathbb{A}^{N+1}$ containing 0 i.e., $C(V)$ is, indeed, a cone. Moreover, if $\boldsymbol{\xi} = (\xi_0, \ldots, \xi_N) \neq 0$ is an element of $C(V)$, then the point $P_{\boldsymbol{\xi}}$ with homogeneous coordinates $(\xi_0, \ldots, \xi_N)$ is in V and conversely.

Theorem 6.23 (Projective Hilbert Nullstellensatz) *If* $\mathfrak{a}$ *is a homogeneous ideal with* $\mathfrak{a}_1 = (Y_0, \ldots, Y_N) \not\subset \sqrt{\mathfrak{a}}$, *then* $V(\mathfrak{a})$ *is nonempty and* $I_h(V(\mathfrak{a})) = \sqrt{\mathfrak{a}}$.

Proof. Since $\mathfrak{a}_1 \not\subset \sqrt{\mathfrak{a}}$, the affine cone $C(V(\mathfrak{a}))$ is neither empty nor 0 alone so that $V = V(\mathfrak{a}) \neq \emptyset$. Since $V \neq \emptyset$, $I_h(V) \subset I(C(V))$. If $f \in I(C(V))$, then $f(t\boldsymbol{\xi}) = 0$ for $\boldsymbol{\xi} \in V$ and all $t \in k$. If $f = f_0 + f_1 + \cdots + f_r$, then $f(t\boldsymbol{\xi}) = f_0(\boldsymbol{\xi}) + \cdots + t^r f_r(\boldsymbol{\xi})$ and so, $I_h(V) = I(C(V))$. The result follows by the affine Nullstellensatz.

Corollary 6.24 *If* $W \subset \mathbb{P}^N$, *then* $V(I_h(W)) = \bar{W}$ *(the closure of* W*).*

Corollary 6.25 *The map* $V \to I_h(V)$ *is a bijection between the set of all nonempty projective algebraic sets in* $\mathbb{P}^N$ *and the set of all homogeneous ideals* $\mathfrak{a} < \mathfrak{a}_1 = (Y_0, \ldots, Y_N)$ *with* $\sqrt{\mathfrak{a}} = \mathfrak{a}$. *For any homogeneous* $\mathfrak{a} \neq S$, $I_h(V(\mathfrak{a})) = \sqrt{\mathfrak{a}}$. *Irreducible algebraic sets correspond to homogeneous prime ideals.*

Proof. Only the last assertion remains to be proved. If $\mathfrak{p}$ is a homogeneous prime ideal, then $V_a(\mathfrak{p})$ is irreducible (in $\mathbb{A}^{N+1}$) and hence, $V(\mathfrak{p})$ is irreducible. Conversely, if V is irreducible, then $C(V)$ is irreducible. But $I_h(V) = I(C(V))$ and so $I_h(V)$ is prime.

Definition 6.26 A *projective variety* is an irreducible projective algebraic set in $\mathbb{P}^N$. An open subset of a projective variety is called a *quasi-projective variety.*

Definition 6.27 Let $V \subset \mathbb{P}^N$ be a projective algebraic set and let F be a homogeneous element of $S = k[Y_0, \ldots, Y_N]$ with $F \notin I_h(V)$. The set $V_F = \{v \in V\colon F(v) \neq 0\}$ is called a *principal open subset of* V. [Note that $V_F = \{\mathbb{P}^N - V(F)\} \cap V$ is open in V.]

Proposition 6.28 *The principal open sets* $(\mathbb{P}^N)_F$*, where* F *is a homogeneous element of* $k[Y_0, \ldots, Y_N]$*, form a base for the Zariski topology.*

Proof. If U is open, then $U = \mathbb{P}^N - V$ with V a projective algebraic set. $I_h(V) = (F_1, \ldots, F_r)$, F_i homogeneous, and $V = \cap V(F_i)$. It follows that $U = \bigcup_{i=1}^{r} (\mathbb{P}^N)_{F_i}$.

Corollary 6.29 *The principal open sets* V_F *form a basis for the (induced) Zariski topology on* V.

Definition 6.30 If $V \subset \mathbb{P}^N$ is a projective (or quasi-projective) variety, then $S(V) = S/I_h(V) = k[Y_0, \ldots, Y_N]/I_h(V)$ is called the *homogeneous coordinate ring of* V.

Since $I_h(V)$ is a homogeneous ideal, $S(V)$ is naturally a graded ring. By Lemma 6.8 we have

$$\begin{aligned} S(V) &= S_0(V) + S_1(V) + \cdots \\ &= (S_0 + I_h(V))/I_h(V) + (S_1 + I_h(V))/I_h(V) + \cdots . \end{aligned} \tag{6.31}$$

We note that $S(V)$ is an integral domain (as V is irreducible) and that elements of $S(V)$ are not, in general, functions on V. In fact, we must use a different approach to defining functions on V. We note also that, since $I_h(V) = I(C(V))$, $S(V) = k[Y_0, \ldots, Y_N]/I(C(V)) = k[C(V)]$ is the affine coordinate ring of the affine cone over V.

Let $H_i = V(Y_i)$, $i = 0, 1, \ldots, N$ and let $U_i = \mathbb{P}^N - H_i = (\mathbb{P}^N)_{Y_i}$. The U_i are open and cover $\mathbb{P}^N$. We define maps $\psi_i\colon U_i \to \mathbb{A}_i^N$ (cf. (5.31)) via

$$\psi_i(\xi_0, \ldots, \xi_N) = (\xi_0/\xi_i, \ldots, \xi_{i-1}/\xi_i, \xi_{i+1}/\xi_i, \ldots, \xi_N/\xi_i). \tag{6.32}$$

Since $\xi_i \neq 0$ on U_i, the map ψ_i is well-defined and is clearly bijective. We have

Proposition 6.33 *The map* ψ_i *is a homeomorphism.*

Proof. We need only show that closed sets of U_i are identified with closed sets in $\mathbb{A}_i^N \simeq \mathbb{A}^N$. We may assume $i = 0$. If V is closed in U_0, then $V = U_0 \cap V_1$, V_1

closed in $\mathbb{P}^N$. Then $V_1 = V(f_1, \dots, f_\nu)$, f_i homogeneous of degree d_i. Consider the map $\tilde{}: S \to R$ ($S = k[Y_0, \dots, Y_N]$, $R = k[X_1, \dots, X_N]$) given by

$$f(Y_0, \dots, Y_N) \longrightarrow \tilde{f}(X_1, \dots, X_N) = f(1, X_1, \dots, X_N). \tag{6.34}$$

We claim that $\psi_0(V) = V(\tilde{f}_1, \dots, \tilde{f}_\nu)$. If $\boldsymbol{\eta} \in V(\tilde{f}_1, \dots, \tilde{f}_\nu)$, then $(1, \boldsymbol{\eta}) \in U_0 \cap V_1 = V$ so that $V(\tilde{f}_1, \dots, \tilde{f}_\nu) \subset \psi_0(V)$. On the other hand, if $\boldsymbol{\xi} \in U_0 \cap V_1$, then $\xi_0 \neq 0$ and $\xi_0^{-1}\boldsymbol{\xi} = (1, \xi_1/\xi_0, \dots, \xi_N/\xi_0) \in U_0 \cap V_1$ so that

$$0 = f_i(1, \xi_1/\xi_0, \dots, \xi_N/\xi_0) = \tilde{f}_i(\xi_1/\xi_0, \dots, \xi_N/\xi_0)$$

and $\psi_0(\boldsymbol{\xi}) \in V(\tilde{f}_1, \dots, \tilde{f}_\nu)$. Conversely, if W is closed in $\mathbb{A}^N$ and $I(W) = (g_1(X_1, \dots, X_N), \dots, g_\mu(X_1, \dots, X_N))$ with g_j of degree d_j, then $h_j(Y_0, \dots, Y_N) = Y_0^{d_j} g_j(Y_1/Y_0, \dots, Y_N/Y_0)$ is homogeneous of degree d_j and $V(h_1, \dots, h_\mu)$ is closed in $\mathbb{P}^N$. We claim that $\psi_0^{-1}(W) = U_0 \cap V(h_1, \dots, h_\mu)$. Clearly, $U_0 \cap V(h_1, \dots, h_\mu) \subset \psi_0^{-1}(W)$. If $\boldsymbol{\xi} \in \psi_0^{-1}(W)$, then $\boldsymbol{\xi} \in U_0$ (i.e., $\xi_0 \neq 0$) and $h_j(\boldsymbol{\xi}) = \xi_0^{d_j} g_j(\xi_1/\xi_0, \dots, \xi_N/\xi_0) = \xi_0^{d_j}(g_j \circ \psi_0)(\boldsymbol{\xi})$. But $\psi_0(\boldsymbol{\xi}) \in W$ so that $(g_j \circ \psi_0)(\boldsymbol{\xi}) = 0$, $j = 1, \dots, \mu$ and a fortiori $h_j(\boldsymbol{\xi}) = 0$, $j = 1, \dots, \mu$. Thus, $\psi_0^{-1}(W) \subset U_0 \cap V(h_1, \dots, h_\mu)$. Since ψ_0, ψ_0^{-1} are closed maps, ψ_0 is a homeomorphism.

Corollary 6.35 *If V is a quasi-projective variety, then $V = \bigcup_{i=0}^{N} (V \cap U_i)$ and each $V \cap U_i$ is homeomorphic to a quasi-affine variety.*

Corollary 6.36 *If $\mathfrak{p}_h = I_h(V)$ is a homogeneous prime ideal, then $V - V \cap H_0 = V(\mathfrak{p}_h^*)$ where $\mathfrak{p}_h^* = \{f(1, X_1, \dots, X_N) = \tilde{f}(X_1, \dots, X_N) : f \in \mathfrak{p}_h\}$.*

Corollary 6.37 *If $\mathfrak{p}^*$ is a prime ideal in $R = k[X_1, \dots, X_N]$ with $V(\mathfrak{p}^*) = V^*$, then $\overline{V^*}$ in $\mathbb{P}^N$ is $V(\mathfrak{p})$ where $\mathfrak{p}$ is generated by*

$$f(Y_0, \dots, Y_N) = Y_0^d g(Y_1/Y_0, \dots, Y_N/Y_0), \quad g \in \mathfrak{p}^*.$$

The proposition allows us to carry over results and concepts from affine algebraic geometry to projective algebraic geometry. We can view Corollary 6.35 as leading to the following:

Lemma 6.38 *Every point $\boldsymbol{\xi}$ of a projective (or quasi-projective) variety V has a neighborhood which is homeomorphic to an affine (or quasi-affine) variety.*

Proof. Either by Corollary 6.35 or by the following simple argument. Let $\boldsymbol{\xi} \in V$ and suppose $\boldsymbol{\xi} \in V \cap U_0 = W - W_1$ where W, W_1 are closed in U_0. Since $\boldsymbol{\xi} \in W$, there is an f in the affine coordinate ring $k[W]$ with $\boldsymbol{\xi} \in W_f$ and $f(W_1) = 0$. But W_f is an affine (or quasi-affine) variety.

Once we have introduced regular functions and morphisms for projective varieties (Chapter 7) we can change homeomorphic to isomorphic in the Lemma 6.38.

Example 6.39 Let $\ell_a = a_0Y_0 + a_1Y_1 + \cdots + a_3Y_3$ be a linear form on $\mathbb{P}^3$. Then $\mathbb{P}^3 - V(\ell_a) = U_a$ is open in $\mathbb{P}^3$ and we can define a map $\psi_a \colon U_a \to \mathbb{A}^3$ (e.g. by a suitable coordinate change) which is a homeomorphism. $V(\ell_a)$ is often called the *hyperplane at infinity* and we view $\mathbb{P}^3$ as $\mathbb{A}^3 \cup V(\ell_a)$ with the points of $V(\ell_a)$ being *points at infinity*.

Example 6.40 Consider the twisted cubic curve C of Example 6.18. Let $H_0 = V(Y_0)$. Then $C \cap U_0$ is mapped by ψ_0 into the curve C_0 in $\mathbb{A}^3$ given by $C_0 = V(\tilde{Q}_0) \cap V(\tilde{Q}_1) \cap V(\tilde{Q}_2)$ where

$$\begin{aligned} \tilde{Q}_0(X) &= X_1^2 - X_2 \\ \tilde{Q}_1(X) &= X_1X_2 - X_3 \\ \tilde{Q}_2(X) &= X_2^2 - X_1X_3 \end{aligned} \tag{6.41}$$

and $X = (X_1, X_2, X_3)$. C_0 is also the image of the map $\tilde{\psi} \colon \mathbb{A}^1 \to \mathbb{A}^3$ with $\tilde{\psi}(t) = (t, t^2, t^3)$.

Lemma 6.38 allows us to adopt a point of view which ultimately will prove most fundamental. We know that $\mathbb{P}^N = \bigcup_{i=0}^{N} U_i$ where $U_i = (\mathbb{P}^N)_{Y_i} = \mathbb{A}_i^N$ is affine N-space. Let $U_{ij} = U_i \cap U_j$ for all $i, j = 0, 1, \ldots, N$. Then $U_{ij} = (U_i)_{Y_j} = (U_j)_{Y_i}$ is open in U_i and in U_j. We let $\psi_{ji} \colon U_{ij} \to U_{ji}$ be given by

$$\begin{aligned} &\psi_{ji}(\xi_0, \ldots, \xi_{i-1}, 1, \xi_{i+1}, \ldots, \xi_N) \\ &\quad = (\xi_0/\xi_j, \ldots, \xi_{i-1}/\xi_j, 1/\xi_j, \xi_{i+1}/\xi_j, \ldots, \xi_N/\xi_j) \end{aligned} \tag{6.42}$$

and $\psi_{ii} = \mathbf{1}$ (the identity). The maps ψ_{ji} are all bijective maps of (quasi-) *affine* varieties. Moreover, since the affine coordinate ring $k[U_{ij}] = k[Y_\ell/Y_i, Y_i/Y_j] = k[Y_\ell/Y_j, Y_j/Y_i] = k[U_{ji}]$, the ψ_{ji} are isomorphisms of *affine* varieties with $\psi_{ji} \circ \psi_{ij} = \mathbf{1} = \psi_{ij} \circ \psi_{ji}$. If ψ_{ji}^{ℓ} is the restriction of ψ_{ji} to $U_{ij} \cap U_{i\ell} = (U_i \cap U_j) \cap (U_i \cap U_\ell) = U_i \cap U_j \cap U_\ell$, then ψ_{ji}^{ℓ} is an isomorphism of $U_{ij} \cap U_{i\ell}$ onto $U_{ji} \cap U_{j\ell}$ $(= U_j \cap U_i \cap U_\ell)$ and

$$\psi_{\ell i}^{j} = \psi_{\ell j}^{i} \circ \psi_{ji}^{\ell} \tag{6.43}$$

for all i, j, ℓ. Thus, $\mathbb{P}^N$ is obtained as a "patching" (or "glueing") of affine pieces in a compatible way. If $V \subset \mathbb{P}^N$ is a projective variety, then $V = \bigcup_{i=0}^{N} (V \cap U_i)$ and V can be obtained by a similar patching.

Example 6.44 Consider $\mathbb{P}^1 = (\mathbb{P}^1)_{Y_0} \cup (\mathbb{P}^1)_{Y_1}$. Let $U_0 = (\mathbb{P}^1)_{Y_0}$, $U_1 = (\mathbb{P}^1)_{Y_1}$ so that $U_0 \simeq \mathbb{A}^1$, $U_1 \simeq \mathbb{A}^1$. Then $U_{01} = \{(1, \xi_1) \colon \xi_1 \neq 0\}$, $U_{10} = \{(\xi_0, 1) \colon \xi_0 \neq 0\}$.

We have $\psi_{00} = \mathbf{1}$, $\psi_{11} = \mathbf{1}$, $\psi_{10}(1, \xi_1) = (1/\xi_1, 1)$ and $\psi_{01}(\xi_0, 1) = (1, 1/\xi_0)$. Thus, $(\psi_{10} \circ \psi_{01})(\xi_0, 1) = \psi_{10}(1, 1/\xi_0) = (1/(1/\xi_0), 1) = (\xi_0, 1)$ and $\psi_{01} \circ \psi_{10}(1, \xi_1) = \psi_{01}(1/\xi_1, 1) = (1, 1/(1/\xi_1)) = (1, \xi_1)$. Then $k[U_{01}] = k[Y_1/Y_0, Y_0/Y_1] = k[x_1, x_1^{-1}]$ and $k[U_{10}] = k[Y_0/Y_1, Y_1/Y_0] = k[x_0, x_0^{-1}]$ so that $\psi_{10}^*\colon k[U_{10}] \to k[U_{01}]$ is given by $\psi_{10}^*(x_0) = x_1^{-1}$ and $\psi_{10}^*(x_0^{-1}) = x_1$ [we leave it to the reader to determine what happens if we define the isomorphism by $x_0 \to x_1$, $x_0^{-1} \to x_1^{-1}$].

We can view the "glueing" process a bit more abstractly. Consider a finite set $U_0, \ldots, U_N$ of affine (or quasi-affine) varieties such that there are open sets U_{ij} of U_i and U_{ji} of U_j and an isomorphism $\psi_{ji}\colon U_{ij} \to U_{ji}$ for all i, j with the properties: (i) $U_{ii} = U_i$ and $\psi_{ii} = \mathbf{1}$; (ii) $\psi_{ji} \circ \psi_{ij} = \mathbf{1}$, $\psi_{ij} \circ \psi_{ji} = \mathbf{1}$; and, (iii) if ψ_{ji}^{ℓ} is the restriction of ψ_{ji} to $U_{ij} \cap U_{i\ell}$, then ψ_{ji}^{ℓ} is an isomorphism of $U_{ij} \cap U_{i\ell}$ onto $U_{ji} \cap U_{j\ell}$ and $\psi_{\ell i}^{j} = \psi_{\ell j}^{i} \circ \psi_{ji}^{\ell}$ for all i, j, ℓ. Then there is a space V and homeomorphisms $\psi_i\colon U_i \to V$ such that (a) $\psi_i(U_i)$ is open in V; (b) $V = \bigcup_{i=0}^{N} \psi_i(U_i)$; and, (c) $\psi_i(U_{ij}) = \psi_j(U_{ji}) = \psi_i(U_i) \cap \psi_j(U_j)$, $\psi_{ji} = \psi_j^{-1} \circ \psi_i$. If U is open in $\psi_i(U_i)$, then we can *define* $k[U]$ as the set $\{f \circ \psi_i^{-1}\colon f \in k[\psi_i^{-1}(U)]$ (i.e., $k[U] = (\psi_i^{-1})^*[k[\psi_i^{-1}(U)]]$. Thus we have an "abstract variety structure" obtained by "patching" the U_i. This, among other things, indicates that regular functions should (as in the affine case) be defined locally.

Now let us use Corollary 6.35 and Lemma 6.38 to help develop the notion of dimension and its properties. Since $\mathbb{P}^N$ is a topological space, we have, just as in I.16.1,

Definition 6.45 If $V \subset \mathbb{P}_k^N$ is a projective or quasi-projective variety, then the *dimension of* V, $\dim V$ (or $\dim_k V$) is the largest integer n such that there is a strict chain $V_0 < V_1 < \cdots < V_n \subset V$ of distinct nonempty closed irreducible subsets. If $V \subset \mathbb{P}_k^N$, then the *dimension of* V, $\dim V$, is the maximum of $\dim V_i$, where $V = \cup V_i$ is the decomposition of V into irreducible components.

Let $\ell_a = a_0 Y_0 + \cdots + a_N Y_N$ be a linear form on $\mathbb{P}^N$. Then $U_a = \mathbb{P}^N - V(\ell_a)$ is open in $\mathbb{P}^N$ and we can define a map $\psi_a\colon U_a \to \mathbb{A}^N$ (e.g. by a change of coordinates) which is a homeomorphism. $V(\ell_a)$ is called the *hyerplane at infinity* and $\mathbb{P}^N = \mathbb{A}^N \cup V(\ell_a)$. Any hyperplane ℓ_a in $(\mathbb{P}^N)^*$ can be chosen at infinity and we say that an algebraic set V does not *lie at infinity* if no irreducible component of V is entirely contained in $V(\ell_a)$.

Proposition 6.46 *Let V be a variety in $\mathbb{A}^N$ and let $\bar{V}$ be its projective closure in $\mathbb{P}^N = \mathbb{A}^N \cup V(\ell_a)$. Then the mapping $V \to \bar{V}$ is a bijection between the affine varieties in $\mathbb{A}^N$ and the projective varieties in $\mathbb{P}^N$ which do not lie at infinity.*

Proof. Since V is irreducible if and only if its closure $\bar{V}$ is irreducible, the map $V \to \bar{V}$ is an injection into the set of varieties not lying at infinity. If W is a projective variety not lying at infinity, then $W_a = W \cap U_a$ is a nonempty open set in W and hence is dense (as W is irreducible) so that $\bar{W}_a = W$. But W_a is an affine variety in U_a ($\simeq \mathbb{A}^N$) and so the map is surjective.

Corollary 6.47 *If V is an algebraic set in $\mathbb{A}^N$ not lying at infinity and $V = \cup V_i$ is its decomposition into irreducible components, then $\bar{V} = \cup \bar{V}_i$ is the decomposition of $\bar{V}$ into irreducible components.*

If $V \neq \emptyset$, $V < \mathbb{P}^N$, then there is a hyperplane $V(\ell_a)$ such that $V \not\subset V(\ell_a)$. For, if not, then $V \subset \bigcap_{i=0}^{N} H_i = \emptyset$. Thus, by a suitable change of coordinates, we may assume such a V does not lie in the hyperplane H_0 at infinity. If V is a finite set of points, then we can choose ℓ_a so that $V \cap V(\ell_a)$ is empty. Finally, we have:

Proposition 6.48 *Let $\alpha\colon \mathbb{P}_k^N \to \mathbb{P}_k^N$ be a projective transformation (i.e., an element of $\mathrm{PGL}(N+1,k)$) and let $V \subset \mathbb{P}_k^N$ be a projective algebraic set. Then $\alpha(V)$ is also a projective algebraic set and $I_h(\alpha(V)) = g \cdot I_h(V) = \alpha(I_h(V))$ for any $g \in \mathrm{GL}(N+1,k)$ with $\bar{g} = \alpha$.*

Proof. Let g be an element of $\mathrm{GL}(N+1,k)$ with $\bar{g} = \alpha$ and let $\mathbf{Z} = g \cdot \mathbf{Y}$ ($\mathbf{Y} = (Y_0, \ldots, Y_N)$, $\mathbf{Z} = (Z_0, \ldots, Z_N)$). If $F \in k[Y_0, \ldots, Y_N]$ is a homogeneous polynomial, then

$$G_F(\mathbf{Z}) = G_F(Z_0, \ldots, Z_N) = F(g^{-1}\mathbf{Z}) = F((g^{-1}\mathbf{Z})_0, \ldots, (g^{-1}\mathbf{Z})_N))$$

is also a homogeneous polynomial of the same degree. If $\boldsymbol{\eta} = \alpha(\boldsymbol{\xi})$, then $G_F(\boldsymbol{\eta}) = 0$ if and only if $F(\boldsymbol{\xi}) = 0$. Let $I_h(V) = \{F_s(\mathbf{Y})\}$. Then we claim $I_h(\alpha(V)) = \{G_{F_s}(\mathbf{Z})\}$. If $F_s \in I_h(V)$, then $F_s(\boldsymbol{\xi}) = 0$ all $\boldsymbol{\xi} \in V$ implies $G_{F_s}(\boldsymbol{\eta}) = 0$ all $\boldsymbol{\eta} \in \alpha(V)$. On the other hand, if $H(\boldsymbol{\eta}) = 0$ for all $\boldsymbol{\eta} \in \alpha(V)$, then $H(g \cdot \boldsymbol{\eta}) = 0$ all $\boldsymbol{\xi} \in V$ implies $H(g \cdot \mathbf{Y}) = F_H(\mathbf{Y}) \in I_h(V)$ and so, $H(\mathbf{Z}) = G_{F_H}(\mathbf{Z})$.

The proposition essentially says that the notion of a projective algebraic set or variety is invariant under coordinate change. We can now translate some simple properties of dimension from affine to projective varieties. We view $\mathbb{P}^N$ as $\mathbb{A}^N \cup H_0$ via the mapping ψ_0.

Theorem 6.49 (i) *Let V be a variety in $\mathbb{A}^N$ and let $\bar{V}$ be its projective closure. Then $\dim V = \dim \bar{V}$;* (ii) *If $W_0 < \cdots < W_m$ is a strict chain of projective varieties in $\mathbb{P}^N$, then it is contained in a maximal such chain and all maximal chains have length N; and,* (iii) *If $\mathfrak{p}_0 < \cdots < \mathfrak{p}_m$ is a strict chain of homogeneous prime ideals in $S = k[Y_0, \ldots, Y_N]$, then it is contained in a maximal such chain and all maximal chains have length $N+1$ (due to the irrelevant ideal).*

Proof. (i) Let $r = \dim V$. Then (I.16) there is a chain $V_0 < V_1 < \cdots < V_r = V < V_{r+1} < \cdots < V_N = \mathbb{A}^N$ of varieties. By Proposition 6.46, the projective closures form a chain $\bar{V}_0 < \cdots < \bar{V}_r = \bar{V} < \cdots < \bar{V}_N = \mathbb{P}^N$. If $\mathfrak{p}_i = I_h(\bar{V}_i)$, then $\mathfrak{p}_0 > \cdots > \mathfrak{p}_r = I_h(\bar{V}) > \cdots > \mathfrak{p}_N = (0)$ is a strict chain of homogeneous prime ideals in $k[Y_0, \ldots, Y_N]$ with $\mathfrak{p}_0 \neq (Y_0, \ldots, Y_N)$. But this chain is maximal since no chain can have length greater than $N + 1$. Hence $\dim \bar{V} = r$. As for (ii), let $W_0 < \cdots < W_m$ be a strict chain of projective varieties. We may assume $W_0 \not\subset H_0$ (the hyperplane at infinity). Then $W_i = \overline{W_i \cap U_0}$ and letting $V_i = W_i \cap U_0$, we note that the affine chain $V_0 < \cdots < V_m$ is contained in a maximal chain which has length N. Applying projective closure and Proposition 6.46, we get the result. (iii) is simply a translation of (ii) to homogeneous prime ideals.

Corollary 6.50 *If $V \subset \mathbb{P}^N$ is a projective variety, then $\dim V \leq N$ and $\dim V = N$ if and only if $V = \mathbb{P}^N$. If $C_a(V)$ is the affine cone of V, then $\dim C_a(V) = \dim V + 1$.*

Corollary 6.51 *If V has pure dimension r and W is a nonempty irreducible subvariety, then $\dim V = \dim W + \operatorname{codim}_V W$.*

Corollary 6.52 $\dim V = 0$ *if and only if V is a finite set.*

Proof. If V is finite, then we may assume $V \subset U_0$ and so, $\dim V = 0$ (as V affine). If $\dim V = 0$, then $V \cap U_i$, $i = 0, 1, \ldots, N$ is finite (or empty) so that $V = \cup(V \cap U_i)$ is finite.

Corollary 6.53 *A projective variety $V \subset \mathbb{P}_k^N$ has dimension $N - 1$ (or, in other words, codimension 1) if and only if $V = V(F)$ where F is a prime homogeneous element of $k[Y_0, \ldots, Y_N]$.*

Proof. If F is a prime homogeneous element of $S = k[Y_0, \ldots, Y_N]$, then (F) is a minimal prime homogeneous ideal and so, $\dim_k V(F) = \dim_k (F) = N - 1$. If $\dim_k V = N - 1$, then $I_h(V)$ is a minimal prime homogeneous ideal in S and hence, is principal with a homogeneous generator.

Corollary 6.54 *A projective algebraic set is a hypersurface if and only if all its irreducible components have codimension 1 i.e.. if and only if it is of pure dimension $N - 1$.*

Corollary 6.55 *Let V be a projective variety and let F be a form with $V \cap V(F) \neq \emptyset$ and $V \not\subset V(F)$. Then* $\dim V \cap V(F) = \dim V - 1$.

Proof. We apply Theorem I.16.54. Let ℓ_a be a linear form with $U_a = \mathbb{P}^N - V(\ell_a)$ such that $V \cap V(F) \cap U_a \neq \emptyset$ and $\psi_a(F) = F(\xi_0/\ell_a, \cdots, \xi_N/\ell_a) = f \neq 0$ is a nonunit in the affine coordinate ring $k[V \cap U_a]$ $(= k[\psi_a(V)] = k[V_a])$. Then, by I.16.54, for any irreducible component Z of $V(f) = \{v \in V_a\colon f(v) = 0\} = V_a \cap V(f) = (V \cap U_a) \cap (V(F) \cap U_a)$, we have $\dim Z = \dim V_a - 1$. But, by the theorem, $\dim V = \dim V_a$ and $\dim V \cap V(F) = \dim V \cap V(F) \cap U_a$. The corollary can also be established by considering affine cones. Thus, $\dim C_a(V) \cap C_a(V(F)) = \dim(C_a(V) \cap V_a(F))$ (i.e., $C_a(V(F)) = \{\boldsymbol{\xi} \in \mathbb{A}^{N+1}\colon F(\boldsymbol{\xi}) = 0\}$ as F a form) $= \dim C_a(V) - 1$. But $C_a(V \cap V(F)) = C_a(V) \cap C_a(V(F))$ so that $\dim V \cap V(F) + 1 = \dim C_a(V \cap V(F)) = \dim C_a(V) - 1 = \dim V$.

Corollary 6.55 can be used to develop results similar to Corollaries I.16.55–I.16.58. Since open sets are dense in irreducibles and since dimension can be checked on an affine piece (e.g. $V = \overline{V \cap U_0}$) and since closure is a *local property*, the assertion about the corollaries is plausible. For example, we have:

Lemma 6.56 *Closure is a local property. In other words, if $W \subset V$ and $V = \cup U_\alpha$, U_α open, then W is closed (in V) if and only if $W \cap U_\alpha$ is closed in U_α for all α.*

Proof. The implication if W is closed is trivial. Let $U_\alpha = V - Z_\alpha$ where Z_α is closed and let $W \cap U_\alpha = U_\alpha \cap X_\alpha$ where X_α is closed. We *claim* that $W = \cap(Z_\alpha \cup X_\alpha)$ (and hence that W is closed). If $\xi \in W$, then, for any α, either $\xi \in U_\alpha$ or $\xi \notin U_\alpha$. If $\xi \in U_\alpha$, then $\xi \in W \cap U_\alpha \subset X_\alpha$ so that $\xi \in Z_\alpha \cup X_\alpha$. If $\xi \notin U_\alpha$, then $\xi \in V - U_\alpha = Z_\alpha$ and, again, $\xi \in Z_\alpha \cup X_\alpha$. Thus, $W \subset \bigcap_\alpha (Z_\alpha \cup X_\alpha)$. On the other hand, if $\xi \in \bigcap_\alpha (Z_\alpha \cup X_\alpha)$, then $\xi \in Z_\alpha \cup X_\alpha$ for all α. Since $\xi \in V = \cup U_\alpha$, $\xi \in$ some U_{α_0} and $\xi \notin Z_{\alpha_0}$. It follows that $\xi \in U_{\alpha_0} \cap X_{\alpha_0} \subset W$ and we are done.

We leave to the reader the formulation and proof of the analogs of Corollaries I.16.55–I.16.58. We can also use Corollary 6.55 to prove the following:

Theorem 6.57 *Let V^r, W^s be varieties in $\mathbb{P}^N$ of dimensions r, s respectively. Then, every component Z of $V^r \cap W^s$ has dimension $\geq r+s-N$ (i.e., $\dim V^r \cap W^s \geq r+s-N$). If $r+s \geq N$, then $V^r \cap W^s$ is nonempty.*

Proof. Since $\mathbb{P}^N = \bigcup_{i=0}^{N} U_i$ and the U_i are affine spaces, we may suppose that V^r, W^s are affine varieties in $\mathbb{A}^N$. If $W^s = W^{N-1} = V(f)$ is a hypersurface, then the result follows by Corollary 6.55 (or, perhaps better, by Theorem I.16.54). Now, let Δ be the diagonal in $\mathbb{A}^N \times \mathbb{A}^N$ (i.e., $X_i - Y_i = 0$, $i = 1, \ldots, N$ define Δ). Δ is the intersection of exactly N hypersurfaces and Δ is isomorphic to $\mathbb{A}^N$. Moreover, $(V^r \times W^s) \cap \Delta$ is isomorphic to $V^r \cap W^s$.* Thus, $\dim V^r \cap W^s = \dim(V^r \times W^s) \cap \Delta = \dim Z^{r+s} \cap \Delta$ where $Z^{r+s} = V^r \times W^s$ has dimension

*Note all is affine here and so the results of Part I apply.

$r+s$. Repeatedly applying the case of a hypersurface, we get $\dim Z^{r+s} \cap \Delta \geq r+s+N-2N = r+s-N$ and the result follows. If $r+s \geq N$, consider the affine cones $C_a(V^r)$ and $C_a(W^s)$. By Corollary 6.50, $\dim C_a(V^r) = r+1$, $\dim C_a(W^s) = s+1$. Since $0 \in C_a(V^r) \cap C_a(W^s)$, $C_a(V^r) \cap C_a(W^s) \neq \emptyset$. But, by what we just proved, $\dim C_a(V^r) \cap C_a(W^s) \geq r+1+s+1-(N+1) = r+s+1-N > 0$ and so, there is a point $\boldsymbol{\xi} = (\xi_0, \ldots, \xi_N) \neq 0$ in $C_a(V^r) \cap C_a(W^s)$. But then $\boldsymbol{\xi}$ is in $V^r \cap W^s$ in $\mathbb{P}^N$.

Corollary 6.58 *If $\dim V = r$ and $s \geq N-r$ and L_s is a linear space of dimension s, then $V \cap L_s \neq \emptyset$.*

Proposition 6.59 *If $\dim V = r$, then there is a linear space L_{N-r-1} with $V \cap L_{N-r-1} = \emptyset$.*

Proof. If $\dim V = r < N$, then there is a linear form ℓ_0 which does not vanish on V and for which, $V_0 = V(\ell_0) \cap V \neq \emptyset$. Then $\dim V_0 = r-1$. By induction, there is an L_{N-r} with $V_0 \cap L_{N-r} = \emptyset$. If $L_{N-r} = V(\ell_1, \ldots, \ell_r)$, then $V(\ell_0, \ell_1, \ldots, \ell_{N-r}) = L_{N-r-1}$ and $L_{N-r-1} \cap V \subseteq V \cap V(\ell_0) \cap L_{N-r} = V_0 \cap L_{N-r} = \emptyset$.

We observe that $\dim V = r$ is the maximum $\{s$: there is an L_{N-s-1} which does not meet $V\}$.

Now let V be a projective variety with $\mathfrak{p} = I_h(V)$, a homogeneous prime ideal. In view of Proposition 6.14, $\dim_k S(V)_q = \dim_k [S/\mathfrak{p}]_q = h_{\mathfrak{p}}(q)$ is a polynomial for q large. Since $\dim C_a(V) = \dim V + 1$ and $\dim C_a(V) = \mathrm{tr} \cdot \deg_k S(V) = \dim_k \mathfrak{p}$ (I.16.48), we have

$$\dim_k V = \mathrm{tr} \cdot \deg_k S(V) - 1 \tag{6.60}$$

which leads to:

Proposition 6.61 *If V is a projective variety with $\mathfrak{p} = I_h(V)$, then the degree of the Hilbert polynomial $h_{\mathfrak{p}}(q)$ is the dimension of V, $\dim_k V$.*

Proof. We may assume that $V \cap H_0 \neq \emptyset$ and that $V \not\subset H_0$ (since $\mathfrak{p} \neq (Y_0, \ldots, Y_N)$). Consider the exact sequence

$$0 \longrightarrow S/\mathfrak{p} \xrightarrow{Y_0} S/\mathfrak{p} \longrightarrow (S/\mathfrak{p})/Y_0(S/\mathfrak{p}) \longrightarrow 0 \tag{6.62}$$

(multiplication by Y_0). Then

$$\dim_k [(S/\mathfrak{p})/Y_0(S/\mathfrak{p})]_q = \dim_k [S/\mathfrak{p}]_q - \dim_k [S/\mathfrak{p}]_{q-1}. \tag{6.63}$$

The S-module $M = (S/\mathfrak{p})/Y_0(S/\mathfrak{p})$ has an annihilator $\mathfrak{a}$ $(= \{r\colon rM = 0\})$ which is a homogeneous ideal since M is S-graded. Moreover, $V(\mathfrak{a}) = V \cap H_0$. Since $V \cap H_0 \neq \emptyset$ and $V \not\subset H_0$, $\dim V \cap H_0 = \dim V - 1$. Using induction on the dimension, we have

$$\begin{aligned} h_{\mathfrak{p}}(q) - h_{\mathfrak{p}}(q-1) = h_{\mathfrak{a}}(q) &= \frac{cq^{r-1}}{(r-1)!} + \cdots \\ &= \sum_{j=1}^{r} \alpha_j \binom{q}{r-j} \end{aligned} \tag{6.64}$$

where $r = \dim V$. It follows that

$$h_{\mathfrak{p}}(q) = \sum_{j=1}^{r+1} \beta_j \binom{q}{r+1-j} = d_{\mathfrak{p}} \frac{q^r}{r!} + \cdots \tag{6.65}$$

is of degree r.

Definition 6.66 If V is a projective variety with $\mathfrak{p} = I_h(V)$, then $h_{\mathfrak{p}}(q) = h_V(q)$ is the *Hilbert polynomial of* V and $d_{\mathfrak{p}} = d_V$ is called the *degree of* V.

We shall use the notion of degree in connection with intersections in the sequel (Chapter 18).

If $V = V(F)$ is a hypersurface with F a form of degree d, then, in view of Example 6.12, we have

$$h_V(q) = \frac{dq^{N-1}}{(N-1)!} + \cdots \tag{6.67}$$

so that V also has degree d. In particular, if $F = \ell$ is a linear form, then $V(F) = V(\ell)$ is a hyperplane and the degree is 1.

Example 6.68 Suppose that $(\ell_1, \ldots, \ell_r)$ are r independent linear forms which define a linear subspace $L_{N-r} = V(\ell_1, \ldots, \ell_r)$ in $\mathbb{P}^N$. Then L_{N-r} also has degree 1. For example, if $r = 2$ and we let $\mathfrak{a}_1 = (\ell_1)$, $\mathfrak{a}_2 = (\ell_1, \ell_2)$, $S_1 = S/\mathfrak{a}_1$, $S_2 = S/\mathfrak{a}_2$, $h_1(q) = \dim_k(S_1)_q$, $h_2(q) = \dim_k(S_2)_q$, then

$$h_2(q) = h_1(q) - h_1(q-1) \tag{6.69}$$

since multiplication by ℓ_2 gives an exact sequence

$$0 \longrightarrow (S_1)_{q-1} \xrightarrow{\ell_2} (S_1)_q \longrightarrow (S_2)_q \to 0. \tag{6.70}$$

It follows that

$$\begin{aligned} h_2(q) &= \binom{q+N}{N} - \binom{q+N-1}{N} - \binom{q+N-1}{N} + \binom{q+N-2}{N} \\ &= \binom{q+N-2}{N-2} = 1\frac{q^{N-2}}{(N-2)!} + \cdots \end{aligned} \tag{6.71}$$

and that degree L_2 = degree $V(\ell_1, \ell_2) = 1$. A similar argument proves the assertion for any linear subspace L_{N-r}.

Example 6.72 Suppose that F is an irreducible form of degree d and that $V = V(F)$ is the irreducible hypersurface defined by F. Let L_1 be a line which is not contained in V. Then $\dim(L_1 \cap V) = 0$ and $L_1 \cap V$ is a finite set. Let $s\boldsymbol{\xi} + t\boldsymbol{\eta}$ be the "equation" of L_1 ($L_1 = J(\boldsymbol{\xi}, \boldsymbol{\eta})$) and let $F(s\boldsymbol{\xi} + t\boldsymbol{\eta}) = G(s,t)$. Then $P = s_p\boldsymbol{\xi} + t_p\boldsymbol{\eta} \in L_1 \cap V$ if and only if $G(s_p, t_p) = 0$ i.e., if and only if (s_p, t_p) is a root of G. Since F is irreducible and so is not a pth power (p = characteristic of k), G is of degree d and has, if counted with proper multiplicity, d roots. In other words, the degree of the hypersurface V is the number of points in its intersection with a "general" line. (This can be generalized as we shall see in the sequel.) More precisely, since k is algebraically closed (therefore perfect), we can, by a change of coordinates, assume that $F(Y_0, \ldots, Y_N)$ is separable of degree d in Y_N. If L_1 is given by $Y_i = s\xi_i + t\eta_i$, $(s,t) \in \mathbb{P}^1$, then $L_1 \cap V = V(G(s,t))$ where $G(s,t) = F(s\xi_0 + t\eta_0, \ldots, s\xi_N + t\eta_N)$. If $\Delta_G(\boldsymbol{\xi}, \boldsymbol{\eta})$ is the discriminant, then $\Delta_G(\boldsymbol{\xi}, \boldsymbol{\eta}) \not\equiv 0$ since $F(0, \ldots, 0, t) \not\equiv 0$. Thus, the lines $J(\boldsymbol{\xi}, \boldsymbol{\eta})$ where $\Delta_G(\boldsymbol{\xi}, \boldsymbol{\eta}) \neq 0$ give d distinct roots i.e., d = degree F = degree $V(F)$ points of intersection.

Exercises

(1) Let $S = k[Y_0, \ldots, Y_N]$. Show that $\dim_k S_q = \binom{q+N}{q} = \binom{q+N}{N}$.

(2) Show that if P, Q are distinct points of $\mathbb{P}^N$, then there is a linear form ℓ with $\ell(P) = 0$ and $\ell(Q) \neq 0$.

(3) Show that the twisted cubic is indeed the image of the map $\psi(X_0, X_1) = (X_0^3, X_0^2X_1, X_0X_1^2, X_1^3)$.

(4) Prove Proposition 6.21. Show that the $V(\mathfrak{a})$ define a topology on $\mathbb{P}^N$.

(5) Determine what happens in Example 6.44 if the isomorphism is given by $x_0 \to x_1$, $x_0^{-1} \to x_1^{-1}$.

(6) Formulate and prove the projective versions of I.16.55–I.16.58.

(7) Show that the diagonal Δ in $\mathbb{A}^N \times \mathbb{A}^N$ is isomorphic to $\mathbb{A}^N$ and that $(V^r \times W^s) \cap \Delta$ is isomorphic to $V^r \cap W^s$.

(8) Show how (6.64) implies (6.65).

7

Projective Algebraic Geometry II: Regular Functions, Local Rings, Morphisms

Let $V \subset \mathbb{P}_k^N$ be a quasi-projective variety and let $\mathfrak{p} = I_h(V)$ be the homogeneous prime ideal of V. Let $S = k[Y_0, \ldots, Y_N]$ and let $S(V) = S/\mathfrak{p}$ be the homogeneous coordinate ring of V. $S(V)$ is a graded integral domain.

Definition 7.1 If R is a graded ring and M is a multiplicative set of homogeneous elements, then, for $r \in R_q$, $m \in M$, define the *degree of r/m* to be $q - \deg m = \deg r - \deg m$.

Proposition 7.2 *Let R be a graded integral domain.* (i) *If M is a multiplicative set of homogeneous elements, then R_M is a graded integral domain;* (ii) *if $\mathfrak{p}$ is a homogeneous prime ideal in R and $M_{\mathfrak{p}} = \{m\colon m \notin \mathfrak{p},\ m \text{ homogeneous}\}$ (a multiplicative set), then, letting $R_{\mathfrak{p}} = R_{M_{\mathfrak{p}}}$, the set $(R_{\mathfrak{p}})_0$ of elements of degree 0 is a local ring with maximal ideal $\mathfrak{p}R_{\mathfrak{p}} \cap (R_{\mathfrak{p}})_0$ and hence, $(R_{(0)})_0$ is a field in $Q(R)$ (the quotient field of R);* (iii) *if $\mathfrak{p}$ is a homogeneous prime ideal in R, and $\mathfrak{p}'$ is a homogeneous prime ideal with $\mathfrak{p} \subset \mathfrak{p}'$, then $(R/\mathfrak{p})_{\mathfrak{p}'/\mathfrak{p}} \simeq R_{\mathfrak{p}'}/\mathfrak{p}R_{\mathfrak{p}'}$ and $[(R/\mathfrak{p})_{\mathfrak{p}'/\mathfrak{p}}]_0 \simeq (R_{\mathfrak{p}'}/\mathfrak{p}R_{\mathfrak{p}'})_0$; and* (iv) *if F is homogeneous in R and $M_F = \{F^t\}$, then, letting $R_F = R_{M_F}$, $(R/\mathfrak{p})_{(F/\mathfrak{p})} \simeq R_F/\mathfrak{p}R_F$ and $[(R/\mathfrak{p})_{F/\mathfrak{p}}]_0 \simeq (R_F/\mathfrak{p}R_F)_0$ for $\mathfrak{p}$ a homogeneous prime ideal with $F \notin \mathfrak{p}$.*

Proof. (i) Clearly, $R_M = \oplus \sum (R_M)_q$. Now we claim that $(R_M)_q (R_M)_{q'} \subseteq (R_M)_{q+q'}$. If $\deg r/m = q = \deg r - \deg m$ and $\deg r'/m' = q' = \deg r' - \deg m'$, then $(r/m)(r'/m') = rr'/mm'$ and $\deg rr' - \deg mm' = \deg r + \deg r' - \deg m - \deg m' = q + q'$.

(ii) If $r/m \in (R_{\mathfrak{p}})_0$ and $r/m \notin \mathfrak{p}R_{\mathfrak{p}}$, then $r \notin \mathfrak{p}$ and $m/r \in (R_{\mathfrak{p}})_0$ so that $\mathfrak{p}R_{\mathfrak{p}} \cap (R_{\mathfrak{p}})_0$ is the ideal of non-units in $(R_{\mathfrak{p}})_0$. Since $(0) \cap (R_{\mathfrak{p}})_0 = (0)$, $(R_{(0)})_0$ is a field.

P. Falb, *Methods of Algebraic Geometry in Control Theory: Part II*,
Modern Birkhäuser Classics, https://doi.org/10.1007/978-3-319-96574-1_7

(iii) Consider the map $\varphi\colon (R/\mathfrak{p})_{\mathfrak{p}'/\mathfrak{p}} \to R_{\mathfrak{p}'}/\mathfrak{p}R_{\mathfrak{p}'}$ given by

$$\varphi(\bar{r}/\bar{m}) = \varphi(r+\mathfrak{p}/m+\mathfrak{p}) = r/m + \mathfrak{p}R_{\mathfrak{p}'} = \overline{(r/m)}. \tag{7.3}$$

φ is clearly a surjective homomorphism. If $\varphi(\bar{r}/\bar{m}) = 0$ then $r/m \in \mathfrak{p}R_{\mathfrak{p}'}$ so there are $p \in \mathfrak{p}$, $m' \in R - \mathfrak{p}'$, $m'' \in R - \mathfrak{p}'$ with $m''(m'r - pm) = 0$. But then

$$\bar{r}/\bar{m} = \overline{m''m'r}/\overline{m''m'm} = \overline{m''pm}/\overline{m''m'm} = 0.$$

Thus φ is injective. If $\deg \bar{r}/\bar{m} = 0$, then $\deg \bar{r} = \deg \bar{m} = q$ so that $r = r_1 + p_1$, $r_1 \in R_q$, $p_1 \in \mathfrak{p}$ and $m = m_1 + p_2$, $m_1 \in R_1$, $p_2 \in \mathfrak{p}$. We claim that $\varphi(\bar{r}/\bar{m}) = r_1/m_1 + \mathfrak{p}R_{\mathfrak{p}'}$ which implies $\deg\overline{(r/m)} = \deg\overline{(r_1/m_1)} = 0$. But $r_1/m_1 - r/m = (mr_1 - m_1 r)/m_1 m = (m_1r_1 + p_2r_1 - m_1r_1 - m_1p_1)/m_1 m \in \mathfrak{p}R_{\mathfrak{p}'}$ (note $m, m_1 \in R - \mathfrak{p}'$). On the other hand, if $\deg\overline{(r/m)} = 0$, then clearly $\deg \bar{r}/\bar{m} = 0$.

(iv) Consider here the map $\psi\colon (R_{\mathfrak{p}})_{F/\mathfrak{p}} \to R_F/\mathfrak{p}R_F$ given by

$$\psi(\bar{r}/\bar{F}^t) = r/F^t + \mathfrak{p}R_F = \overline{(r/F^t)}. \tag{7.4}$$

ψ is obviously a surjective homomorphism. If $\psi(\bar{r}/\bar{F}^t) = 0$ then $r/F^t \in \mathfrak{p}R_F$ and there is a $p \in \mathfrak{p}$ with $r/F^t = p/F^{t'}$ i.e., $F^{t''}(rF^{t'} - pF^t) = 0$. But then $\bar{r}/\bar{F}^t = \overline{rF}^{t''+t'}/\bar{F}^{t+t'+t''} = \overline{F_p^{t+t''}}/\bar{F}^{t+t'+t''} = 0$ so that $rF^{t''+t} \in \mathfrak{p}$. Since $F \notin \mathfrak{p}$, $r \in \mathfrak{p}$ and $\bar{r}/\bar{F}^t = 0$. The rest is proved similarly to that part of (iii).

We use the proposition to define the local ring of a point (or later of a subvariety). If $\boldsymbol{\xi} = (\xi_0, \ldots, \xi_N) \in \mathbb{P}_k^N$, then $M_{\boldsymbol{\xi}} = I_h(\boldsymbol{\xi})$ is a homogeneous prime ideal in S and $\mathfrak{m}_{\boldsymbol{\xi}} = M_{\boldsymbol{\xi}}/\mathfrak{p}$ is a homogeneous prime ideal in $S(V)$ when $\boldsymbol{\xi} \in V$.

Definition 7.5 (Local Ring 1) Let

$$\begin{aligned}\mathfrak{o}^1_{\boldsymbol{\xi},V} &= (S(V)_{\mathfrak{m}_{\boldsymbol{\xi}}})_0 = [(S/\mathfrak{p})_{M_{\boldsymbol{\xi}}/\mathfrak{p}}]_0 \simeq [(S_{M_{\boldsymbol{\xi}}})/\mathfrak{p}S_{M_{\boldsymbol{\xi}}}]_0 \\ &= \{F/G\colon F, G \in S_q,\ G(\boldsymbol{\xi}) \neq 0\}/\mathrm{ideal}\{F/G\colon F, G \in S_q, F \in \mathfrak{p}_q, G(\boldsymbol{\xi}) \neq 0\}.\end{aligned}$$

Then $\mathfrak{o}^1_{\boldsymbol{\xi},V}$ is a local ring called the *local ring of $\boldsymbol{\xi}$ on V*.

We observe that $(S(V)_{(0)})_0$ is a field $k(V)$ which we call the *function field of V*. Note that $\mathfrak{o}^1_{\boldsymbol{\xi},V} \subset k(V)$. In fact, $\mathfrak{o}^1_{\boldsymbol{\xi},V} = \{f/g \in k(V)\colon g(\boldsymbol{\xi}) \neq 0\}$ and $\mathfrak{m}^1_{\boldsymbol{\xi},V} = \{f/g \in k(V)\colon f(\boldsymbol{\xi}) = 0,\ g(\boldsymbol{\xi}) \neq 0\}$ is the maximal ideal.

Definition 7.6 A map $h\colon V \to k$ is *regular* at $\boldsymbol{\xi}$ if there is an open neighborhood U of $\boldsymbol{\xi}$ in V with $h = F/G$, $F, G \in S_q$, $G(\boldsymbol{\xi}) \neq 0$, $G \neq 0$ on U. h is *regular on a subset* V_1 of V if h is regular at all $\boldsymbol{\xi} \in V_1$.

Proposition 7.7 *If h is regular, then h is continuous.*

Proof. If $h = F/G$ on U, then $h^{-1}(a) \cap U = \{\boldsymbol{\xi}\colon F(\boldsymbol{\xi})/G(\boldsymbol{\xi}) = a\} = \{\boldsymbol{\xi}\colon (F - aG)(\boldsymbol{\xi}) = 0\} = \{\boldsymbol{\xi}\colon \boldsymbol{\xi} \in V(F - aG) \cap U\}$ which is closed in U.

Corollary 7.8 *If $h_1 = h_2$ on U open in V, then $h_1 = h_2$ on V (for $\{\boldsymbol{\xi}\colon (h_1 - h_2)(\boldsymbol{\xi}) = 0\}$ is closed and dense).*

Corollary 7.9 *If U_1, U_2 are open in V, h_1 is regular on U_1, h_2 is regular on U_2 and there is an open $U \subset U_1 \cap U_2$ with $h_1 = h_2$ on U, then $h_1 = h_2$ on $U_1 \cap U_2$.*

Corollary 7.10 *If U is open in V and h is regular on U, then there is a unique maximal open set U' with $U \subset U'$ such that h extends to a regular function h' on U'.*

The function h' is called a *rational function on* V (note that h' need not be defined at all points of V), U' is called its *domain of definition* and $V - U'$ is called its *pole set.* If $f/g \in k(V)$, then $f/g = \bar{F}/\bar{G}$, $F, G \in S_q$, $G \notin I_h(V)$ and $\bar{F}/\bar{G}$ is regular on $V - V(G) \cap V$. In other words, elements of $k(V)$ are rational functions on V. Conversely, if h is rational with domain of definition U, then, for $\boldsymbol{\xi} \in U$, $h = \bar{F}/\bar{G}$ with $\bar{G}(\boldsymbol{\xi}) \neq 0$ (F, G may depend on $\boldsymbol{\xi}$) and $\bar{F}/\bar{G}$ is in $k(V)$.

Clearly $\mathfrak{o}^1_{\boldsymbol{\xi},V}$ can be viewed as the ring of regular functions at $\boldsymbol{\xi}$.

Example 7.11 Let $V = \mathbb{P}^1_k$ with coordinates (x_0, x_1) and set $h(\xi_0, \xi_1) = \xi_0/\xi_1$ or $h(x_0, x_1) = x_0/x_1$. Then h is a rational function on $\mathbb{P}^1$ with domain of definition $(\mathbb{P}^1)_{x_1} = U_1$. Note $k(\mathbb{P}^1) = k(x_0/x_1) = k(x_1/x_0)$.

Definition 7.12 (Local Ring 2). Let $\ell_a = \sum a_i Y_i$ be a linear form and let $H_a = V(\ell_a)$ be a hyperplane. If $\ell_a \notin \mathfrak{p} = I_h(V)$, then $V_a = V - V \cap H_a = V_{\ell_a}$ is homeomorphic to an affine variety. Let $\mathfrak{o}^2_{\boldsymbol{\xi},V} = \mathfrak{o}_{\boldsymbol{\xi},V_a}$ if $\boldsymbol{\xi} \in V_a$ be the local ring of $\boldsymbol{\xi}$ on V_a and call $\mathfrak{o}^2_{\boldsymbol{\xi},V}$ the *local ring of* $\boldsymbol{\xi}$ *on* V. The *affine coordinate ring,* $k[V_a]$ *of* V_a is $[(S_{\ell_a})_{\mathfrak{p}S_{\ell_a}}]_0 = \{F/\ell_a^q\colon F \in S_q\}/\ \text{ideal}\{F/\ell_a^q\colon F \in \mathfrak{p}_q\}$.

Since V_a is open in V, and, as we shall shortly show, affine, we can take an affine representative of an element of $\mathfrak{o}^1_{\boldsymbol{\xi},V}$ so that $\mathfrak{o}^1_{\boldsymbol{\xi},V} \simeq \mathfrak{o}^2_{\boldsymbol{\xi},V}$ (see Part I). For example, let $\ell_a = Y_0$ so that $\xi_0 \neq 0$. Then map

$$F(\xi_0, \xi_1, \ldots, \xi_N)/G(\xi_0, \ldots, \xi_N)$$
$$\to F(1, \xi_1/\xi_0, \ldots, \xi_N/\xi_0)/G(1, \xi_1/\xi_0, \ldots, \xi_N/\xi_0).$$

In view of the corollaries, the two definitions of the local ring are equivalent to: $h\colon V \to k$ is *regular at* $\boldsymbol{\xi}$ if $h\colon V \cap U_i \to k$ ($U_i = \mathbb{P}^N - H_i = \mathbb{P}^N_{Y_i}$) is regular at $\boldsymbol{\xi}$ for some i. If U is open in V, denote by $\mathfrak{o}_V(U)$, the *ring of all regular* *functions on U*. Since h is regular at $\boldsymbol{\xi}$ if and only if $h \in \mathfrak{o}^1_{\boldsymbol{\xi},V}$ ($\simeq \mathfrak{o}^2_{\boldsymbol{\xi},V}$), we have $\mathfrak{o}_V(U) = \bigcap_{\boldsymbol{\xi} \in U} \mathfrak{o}^1_{\boldsymbol{\xi},V}$.

Definition 7.13 (Local Ring 3) If $\boldsymbol{\xi} \in U, U'$ and $h \in \mathfrak{o}_V(U)$, $h' \in \mathfrak{o}_V(U')$, then (h, U) is *equivalent to* (h', U') *at* $\boldsymbol{\xi}$ if $h = h'$ on $U \cap U'$ (this is an equivalence relation). Let $\mathfrak{o}^3_{\boldsymbol{\xi},V}$ be the set of equivalence classes of (h, U) at $\boldsymbol{\xi}$. Then $\mathfrak{o}^3_{\boldsymbol{\xi},V}$ is a local ring called the *local ring of* $\boldsymbol{\xi}$ *on* V.

We will, of course, show that all these definitions are essentially the same and shall write $\mathfrak{o}_{\boldsymbol{\xi},V}$ for the *local ring of* $\boldsymbol{\xi}$ *on* V. Note that if $(h, U) \in \mathfrak{o}^3_{\boldsymbol{\xi},V}$ and $\boldsymbol{\xi} \in U_i$, then $(h, U \cap U_i) \equiv (h, U)$ is also in $\mathfrak{o}^3_{\boldsymbol{\xi},V}$ and so, if we show that $V \cap \mathbb{P}^N_{Y_i}$ is an affine variety (I.22), then $\mathfrak{o}^3_{\boldsymbol{\xi},V} \simeq \mathfrak{o}^2_{\boldsymbol{\xi},V} \simeq \mathfrak{o}^1_{\boldsymbol{\xi},V}$. We recall (I.22):

Definition 7.14 A *sheaf* is a composite concept consisting of a topological space X, and, for each non-empty open subset U of X, a k-algebra $\mathfrak{o}_X(U)$, and, a family of k-homomorphisms $\rho_{U,V} \colon \mathfrak{o}_X(U) \to \mathfrak{o}_X(V)$ for $U \supset V$, such that (a) $\rho_{U,U}$ = identity; (b) if $W \subset V \subset U$, then $\rho_{U,V} \circ \rho_{V,W} = \rho_{U,W}$; (c) if $U = \cup U_\alpha$, $f_1, f_2 \in \mathfrak{o}_X(U)$ and $\rho_{U,U_\alpha}(f_1) = \rho_{U,U_\alpha}(f_2)$ for all α, then $f_1 = f_2$; and, (c) if $U = \cup U_\alpha$ and $f_\alpha \in \mathfrak{o}_X(U_\alpha)$ with $\rho_{U_\alpha, U_\alpha \cap U_\beta}(f_\alpha) = \rho_{U_\beta, U_\alpha \cap U_\beta}(f_\beta)$ for all α, β, then there is a (unique) f in $\mathfrak{o}_X(U)$ with $\rho_{U,U_\alpha}(f) = f_\alpha$.

Proposition 7.15 *For all i, $0 \le i \le N$, $(V \cap \mathbb{P}^N_{Y_i}, \mathfrak{o}_V \mid V \cap \mathbb{P}^N_{Y_i})$ is an affine variety where $\mathfrak{o}_V$ is the sheaf $\mathfrak{o}_V(U)$, U open in V.*

Proof. Enough to treat $i = 0$. Let $F \in \mathfrak{p}_q$ and set

$$F/Y_0^a = F^*(Y_1/Y_0, \ldots, Y_N/Y_0) = F^*(X_1, \ldots, X_N).$$

Let $\mathfrak{p}^* = \text{ideal}\{F^* \colon F \in \mathfrak{p}\}$. We map $k[X_1, \ldots, X_N] \to k(V)$ via $X_i \to Y_i/Y_0 \mid V$. The kernel is $\mathfrak{p}^*$ and $\mathfrak{p}^*$ is prime. Now the map $\psi_0 \colon V \cap \mathbb{P}^N_{Y_0} \to V^* = \bar{V}(\mathfrak{p}^*) \subset \mathbb{A}^N_k$ given by

$$\psi_0(\xi_0, \ldots, \xi_N) = (\xi_1/\xi_0, \ldots, \xi_N/\xi_0) \tag{7.16}$$

is a homeomorphism (Proposition 6.33). If $\boldsymbol{\xi} \in V \cap \mathbb{P}^N_{Y_0}$, then $\mathfrak{o}_{\boldsymbol{\xi},V} = \{F/G \colon F, G \in S(V)_q,\ G(\boldsymbol{\xi}) \neq 0\}$. Let $R^* = k[V^*] = k[X_1, \ldots, X_N]/\mathfrak{p}^*$ and let $\mathfrak{o}_{\psi_0(\boldsymbol{\xi}),V^*} = \{f/g \colon f, g \in R^*,\ g(\psi_0(\boldsymbol{\xi})) \neq 0\}$. There is a map $\psi^* \colon R^* \to k(V)$ given by $\psi^*(\bar{X}_i) = \overline{Y_i/Y_0}$ which extends to an isomorphism of $k(V^*)$ into $k(V)$. We claim that

$$\psi^*(\mathfrak{o}_{\psi_0(\boldsymbol{\xi}),V^*}) = \mathfrak{o}_{\boldsymbol{\xi},V}. \tag{7.17}$$

If $F, G \in S(V)_q$ with $G(\boldsymbol{\xi}) \neq 0$, then $F/G = (F/Y_0^a)/(G/Y_0^a)$ in $k(V)$ and F/Y_0^a, G/Y_0^a come from f, g in R^* with $g(\psi_0(\boldsymbol{\xi})) \neq 0$. Thus (7.17) holds and the sheaves are the same.

Corollary 7.18 $\mathbb{P}^N_{Y_i} \simeq \mathbb{A}^N$ *as affine varieties.*

Corollary 7.19 $V_i = V \cap \mathbb{P}^N_{Y_i}$ *is an affine variety and* $k[V_i] \simeq (S(V)_{Y_i})_0$.

Proof. Let $\psi_i \colon V_i \to \mathbb{A}^N_k$ and $\tilde{\psi}_i \colon k[V_i] \to (S(V)_{Y_i})_0$ give a natural isomorphism. In other words, the map $k[X_1, \ldots, X_N] \to (k[Y_0, \ldots, Y_N]_{Y_i})_0$ via $f(X_1, \ldots, X_N) \to f(Y_0/Y_i, \ldots, Y_N/Y_i)$ sends $I(V_i)$ into $I(V)S_{Y_i}$ and then apply the proposition.

Corollary 7.20 $k(V_i) \simeq (S(V)_{(0)})_0 = k(V)$.

Example 7.21 Consider $\mathbb{P}^N_k$ and $U_i = (\mathbb{P}^N_k)_{Y_i}$. If $\xi \in U_i$, then $\mathfrak{o}_{\xi,\mathbb{P}^N} = \mathfrak{o}_{\xi,U_i}$. Note that if $\xi \in U_i \cap U_j$, then $\mathfrak{o}_{\xi,\mathbb{P}^N} = \mathfrak{o}_{\xi,U_i} = \mathfrak{o}_{\xi,U_j}$ since $(S_{Y_i})_0[Y_i, Y_i^{-1}] \simeq (S_{Y_j})_0[Y_j, Y_j^{-1}]$. For U open in $\mathbb{P}^N$, $\mathfrak{o}_{\mathbb{P}^N}(U) = \bigcap_{\xi \in U} \mathfrak{o}_{\xi,\mathbb{P}^N}$. Then $(\mathbb{P}^N, \mathfrak{o}_{\mathbb{P}^N})$ with the natural restriction maps is a sheaf. For instance, if $f \in \bigcap_{\xi \in W} \mathfrak{o}_{\xi,\mathbb{P}^N}$, then $f \in \bigcap_{\xi \in U} \mathfrak{o}_{\xi,\mathbb{P}^N}$ if $W \supset U$. Similarly, if $V \subset \mathbb{P}^N$, then $(\mathbb{P}^N, \mathfrak{o}_{\mathbb{P}^N})$ induces a sheaf on V via $V = \cup V_i$, $V_i = V \cap U_i$, and $\mathfrak{o}_{\mathbb{P}^N} \mid V_i$.

Corollary 7.22 *Let* $\mathfrak{m}_\xi$ *be the homogeneous ideal in* $S(V)$ *of elements vanishing at* ξ. *Then* $\mathfrak{o}_{\xi,V} = (S(V)_{\mathfrak{m}_\xi})_0$.

Proof. Say $\xi \in V_0$ so that $\mathfrak{o}_{\xi,V} = \mathfrak{o}_{\xi,V_0} \simeq k[V_0]_{\tilde{\mathfrak{m}}_\xi}$ where $\tilde{\mathfrak{m}}_\xi$ is ideal of ξ in $k[V_0]$. But as in the proposition, $\psi^*(\tilde{\mathfrak{m}}_\xi) = \mathfrak{m}_\xi(S(V)_{Y_0})_0$ and hence, since $Y_0 \notin \mathfrak{m}_\xi$, $k[V_0]_{\tilde{\mathfrak{m}}_\xi} \simeq (S(V)_{\mathfrak{m}_\xi})_0$.

Proposition 7.23 $\mathfrak{o}_V(V) = k$ *and hence,* $\mathfrak{o}_{\mathbb{P}^N}(\mathbb{P}^N) = k$.

Proof. ([H-2]) If $f \in \mathfrak{o}_V(V)$, then $f \in \mathfrak{o}_{V_i}(V_i) = k[V_i]$ for $i = 0, \ldots, N$. Thus $f = F_i/Y_i^{q_i}$, $F_i \in S_{q_i}$ for $i = 0, \ldots, N$. We may assume $q_i = q$ for all i. Take $r \geq Nq$ so that $S(V)_r f \subseteq S(V)_r$ and hence $S(V)_r f^\mu \subseteq S(V)_r$ all $\mu > 0$. In other words, $S(V)[f]$ is a finite $S(V)$ module and so, (I.16.19), f is integral over $S(V)$ with $f^\nu + a_1 f^{\nu-1} + \cdots + a_\nu = 0$. Since $\deg f = 0$, the a_i are in k and f is algebraic over k (hence in k as k is algebraically closed).

Corollary 7.24 *Every nonconstant rational function* h *has a nonempty pole set and a nonempty zero set.*

Proof. If h has domain of definition U and $V - U = \emptyset$, then h would be in $\mathfrak{o}_V(V) = k$. The zero set is nonempty since $1/h$ has a nonempty pole set.

Corollary 7.25 *If h is a nonconstant rational function with domain U, then* $\dim(V - U) \leq \dim V - 1$.

Let us view $\mathbb{P}^N$ as $\mathbb{A}^N \cup H_0$ with $H_0 = V(Y_0)$. Let W be an affine variety in $\mathbb{A}^N$ and let $\bar{W}$ be its projective closure. If U is open in W, then $\mathfrak{o}_W(U)$ and $\mathfrak{o}_{\bar{W}}(U)$ are the same. If $h = f/g$ is in $\mathfrak{o}_W(U)$ so that $h = F/G$ with $F, G \in k[X_1, \ldots, X_N]$, then $h = \tilde{F}/\tilde{G}$ in $\mathfrak{o}_{\bar{W}}(U)$ where $\tilde{F} = Y_0^q F(Y_1/Y_0, \ldots, Y_N/Y_0)$, $\tilde{G} = Y_0^q G(Y_1/Y_0, \ldots, Y_N/Y_0)$ and $q = \max(\deg F, \deg G)$. Similarly, if

$$\tilde{F}(Y_0, \ldots, Y_N)/\tilde{G}(Y_0, \ldots, Y_N) \in \mathfrak{o}_{\bar{W}}(U),$$

then $\tilde{F}(1, X_1, \ldots, X_N)/\tilde{G}(1, X_1, \ldots, X_N)$ is in $\mathfrak{o}_W(U)$.

We recall (Part I) that if W is an affine variety and $f \in k[W]$, then $V(f) = \{\xi \in W\colon f(\xi) = 0\}$ is closed and $V(f) \neq W$ unless $f = 0$ (by the Nullstellensatz). We have:

Proposition 7.26 *If U is open in V and $f \in \mathfrak{o}_V(U)$, then $f(\xi) = 0$ for all $\xi \in U$ implies that $f = 0$.*

Proof. Since $U' \subset U$ means $\mathfrak{o}_V(U) \subset \mathfrak{o}_V(U')$, we can make U smaller if necessary. We may also assume that V is affine (as $V = \cup(V \cap \mathbb{P}^N_{Y_i})$). Then $f = f_1/g_1$ with f_1, g_1 in $k[V]$. We work on V_{g_1} and U_{g_1} so we may assume that $g_1(\boldsymbol{\xi}) \neq 0$ all $\boldsymbol{\xi}$. Thus, $f_1(\boldsymbol{\xi}) = 0$ on U_{g_1} which is dense in V and so $f_1 = 0$ on V i.e., $f = 0$.

We can now define the notion of a morphism.

Definition 7.27 A map $\psi\colon V \to W$ of projective varieties is a *morphism* if (i) ψ is continuous; (ii) if, for all open U in W, $f \in \mathfrak{o}_W(U)$, then $\psi^*(f) = f \circ \psi \in \mathfrak{o}_V(\psi^{-1}(U))$. ψ^* is called the *comorphism* and is a ring homomorphism. A bijective morphism $\psi\colon V \to W$ is an *isomorphism* if there is a morphism $\varphi\colon W \to V$ with $\psi \circ \varphi = 1_W$ and $\varphi \circ \psi = 1_V$.

Theorem 7.27 (Affine Criterion for Morphisms) *Let $\psi\colon V \to W$. Suppose $W = \cup W_j$, W_j open and affine. If $V = \cup V_j$, V_j open and* (1) *$\psi(V_j) \subset W_j$ and* (2) *$\psi^*(\mathfrak{o}_W(W_j)) \subset \mathfrak{o}_V(V_j)$ for all j, then ψ is a morphism.*

Proof. We may assume that the V_j are also affine for if $U_j \subset V_j$, U_j affine, then $\psi(U_j) \subset \psi(V_j) \subset W_j$ and $\mathfrak{o}_V(V_j) \subset \mathfrak{o}_V(U_j)$. So let $\psi_j = \psi \mid V_j$. Then ψ_j is a morphism of affine varieties since $\psi_j^*\colon \mathfrak{o}_W(W_j) = k[W_j] \to \mathfrak{o}_V(V_j) = k[V_j]$ is, in fact, a k-homomorphism and (I.7.16) is induced by a morphism φ_j. Since ψ_j^* and φ_j determine each other, $\psi_j = \varphi_j$. Thus all ψ_j and hence ψ are continuous. Let U be open in W and set $\tilde{U} = \psi^{-1}(U)$. If $f \in \mathfrak{o}_W(U)$, then $\psi^*(f) = f \circ \psi \in$

$\mathfrak{o}_V(\psi^{-1}(U \cap W_j))$. But $\psi^{-1}(U \cap W_j) \supset \tilde{U} \cap V_j$ and so, $\psi^*(f) \in \mathfrak{o}_V(\tilde{U} \cap V_j)$ for all j. But $\tilde{U} = \cup \tilde{U} \cap V_j$ and the $\mathfrak{o}_v(O)$, O open in V, form a sheaf, so that $\psi^*(f) \in \mathfrak{o}_V(\tilde{U})$.

Observe that the theorem provides a basis for constructing morphisms by "patching". More precisely, let $V = \cup V_j$, $W = \cup W_j$ where the V_j, W_j are open and affine and let $\varphi_j \colon V_j \to W_j$ be (affine) morphisms such that $\varphi_i = \varphi_j$ on $V_i \cap V_j$. If $\psi \colon V \to W$ is given by $\psi(\boldsymbol{\xi}) = \varphi_j(\boldsymbol{\xi})$ for $\boldsymbol{\xi} \in V_j$, then ψ is a morphism.

Suppose that $\psi \colon V \to W$ is a map of quasi-projective varieties with $V \subset \mathbb{P}^N$ and $W \subset \mathbb{P}^M$. Then ψ is *regular* (i.e., a morphism) if, for $\boldsymbol{\xi} \in V$, $\psi(\boldsymbol{\xi}) = \boldsymbol{\eta} \in W$, and, for each $\mathbb{A}_i^M$ $(= \mathbb{P}^M_{Y_i})$ containing $\boldsymbol{\eta}$, there is an open (affine) neighborhood U of $\boldsymbol{\xi}$ with $\psi(U) \subset \mathbb{A}_i^M$ and $\psi \colon U \to \mathbb{A}_i^M$ regular. Now, say $\psi(\boldsymbol{\xi}) \in \mathbb{A}_0^M$, $\psi = (\psi_1^0, \ldots, \psi_M^0)$ where $\psi_i^0 = F_i/G_i$ with F_i, G_i forms of the same degree in $k[Y_0, \ldots, Y_N]$ and $G_i(\boldsymbol{\xi}) \neq 0$. We can write $\psi_i^0 = \tilde{F}_i/\tilde{G}$ where $\tilde{F}_i, \tilde{G}$ are forms of the same degree and $\tilde{G}(\boldsymbol{\xi}) \neq 0$. Thus the morphism or regular map ψ is given concretely by

$$\psi = (\psi_0, \ldots, \psi_M) \tag{7.29}$$

where the ψ_i are forms of like degree in $S = k[Y_0, \ldots, Y_N]$. The representation is *not unique*. In fact, $\psi = (\psi_0, \ldots, \psi_M) = \psi' = (\psi_0', \ldots, \psi_M')$ if and only if

$$\psi_i \psi_j' = \psi_i' \psi_j \tag{7.30}$$

for all $i, j = 0, 1, \ldots, M$.

Example 7.31 Let $V \subset \mathbb{P}^2$ be $V(Y_2^2 - (Y_0 + Y_1)(Y_0 - Y_1))$ and let $\psi(Y_0, Y_1, Y_2) = (Y_0 - Y_1, Y_2)$, $\psi \colon \mathbb{P}^2 \to \mathbb{P}^1$. Then $(1,1,0) \in V$ but ψ is not defined there. But $\psi'(Y_0, Y_1, Y_2) = (Y_2, Y_1 + Y_0)$ is defined at $(1,1,0)$. So $\psi_0 = Y_0 - Y_1$, $\psi_1 = Y_2$, $\psi_0' = Y_2$, $\psi_1' = Y_1 + y_0$ and $\psi_0 \psi_1' = (Y_0 - Y_1)(Y_0 + Y_1) = Y_2^2 = \psi_0' \psi_1$ *on* V. Both ψ and ψ' define the same map on V.

This nonuniqueness is quite different from the affine case and indicates why ultimately, rational (not regular) maps play a critical role in projective algebraic geometry.

Example 7.32 Consider $\mathbb{P}_k^2$ and let $\boldsymbol{\xi}_\infty = (0,0,1)$. Then the projection $\pi \colon \mathbb{P}_k^2 - \{\boldsymbol{\xi}_\infty\} \to \mathbb{P}_k^1$ given by $\pi(y_0, y_1, y_2) = (y_0, y_1)$ is a regular map (note π is not defined at $\boldsymbol{\xi}_\infty$). If $(\xi_0, \xi_1) \in \mathbb{P}_k^1$, then $\pi^{-1}(\xi_0, \xi_1) = J((\xi_0, \xi_1, 0), \boldsymbol{\xi}_\infty) - \{\boldsymbol{\xi}_\infty\} = L_\xi - \boldsymbol{\xi}_\infty$ where L_ξ is the line joining $\xi = (\xi_0, \xi_1, 0)$ and $\boldsymbol{\xi}_\infty$. If $V \subset \mathbb{P}_k^2$ is a variety and $\boldsymbol{\xi}_\infty \notin V$, then $\pi(V) = \tilde{V} \subset \mathbb{P}_k^1$. If $(\tilde{\xi}_0, \tilde{\xi}_1) \in \tilde{V}$, then $\pi^{-1}(\tilde{\xi}_0, \tilde{\xi}_1) \cap V < L_\xi$ (otherwise $L_\xi - \boldsymbol{\xi}_\infty \subset V$ and V closed implies $\boldsymbol{\xi}_\infty \in V$). Thus $\pi^{-1}(\tilde{\xi}_0, \tilde{\xi}_1) \cap V$ is a finite set (dim 0) and π is a "finite" regular map.

Example 7.33 Consider $\mathbb{P}_k^N$ and let $\boldsymbol{\xi}_\infty = (0, \ldots, 0, 1)$. Then the projection $\pi \colon \mathbb{P}_k^N - \{\boldsymbol{\xi}_\infty\} \to \mathbb{P}_k^{N-1}$ given by $\pi(Y_0, \ldots, Y_N) = (Y_0, \ldots, Y_{N-1})$ is regular and, as in the previous example, $\pi^{-1}(\xi_0,\dots,\xi_{N-1}) = L_\xi - \{\boldsymbol{\xi}_\infty\}$ where L_ξ is the line joining $\boldsymbol{\xi}_\infty$ and $(\xi_0,\dots,\xi_{N-1},0)$. Then $\pi^{-1}(\xi_0,\dots,\xi_{N-1})\cap V < L_\xi$ so that $\dim \pi^{-1}(\xi_0,\dots,\xi_{N-1})\cap V < 1$ and map is "finite".

Example 7.34 Let $M_0,\dots,M_n$ be all the monomials of degree q in $Y_0,\dots,Y_m$ with $n = \binom{m+q}{q} - 1$. Let $p_{i_0 i_1\cdots i_m}$ where $i_0+\cdots+i_m = q$ be the coordinates in $\mathbb{P}^n$ and consider the map $v_q\colon \mathbb{P}^m \to \mathbb{P}^n$ given by

$$v_q(\xi_0,\dots,\xi_n) = (\xi_0^{i_0}\cdots\xi_m^{i_m}) = (p_{i_0 i_1\cdots i_m}). \tag{7.35}$$

v_q is a regular map into $\mathbb{P}^n$ since not all ξ_i^q are zero. The map v_q is called the *Veronese mapping.* We observe that $v_q(\mathbb{P}^m)$ is a variety. Since, on $v_q(\mathbb{P}^m)$,

$$p_{i_0\cdots i_m}p_{j_0\cdots j_m} = p_{k_0\cdots k_m}p_{\ell_0\cdots\ell_m} \tag{7.36}$$

if $i_0+j_0 = k_0+\ell_0,\dots,i_m+j_m = k_m+\ell_m$. Conversely, any solution of these equations has some $p_{0\cdots q\cdots 0} \neq 0$, say $p_{q0\cdots 0}\neq 0$, in which case we can take $\xi_0 = p_{q0\cdots 0}$, $\xi_1 = p_{q-1\,1\,0\cdots 0}$, $\dots$, $\xi_m = p_{q-1 0\cdots 0 1}$ to get an inverse of v_q. Note also that the map $v_q^*\colon k[Z_0,\dots,Z_n]\to k[Y_0,\dots,Y_m]$ given by

$$v_q^*(Z_i) = M_i(Y) \tag{7.37}$$

is a homomorphism with $\mathfrak{a} = \operatorname{Ker} v_q^*$. Since v_q^* maps onto the graded integral domain $\sum_{t=0} k[Y_0,\dots,Y_m]_{qt}$, $\mathfrak{a}$ is a homogeneous prime ideal and $V(\mathfrak{A}) = v_q(\mathbb{P}^m)$. Thus $v_q(\mathbb{P}^m)$ is indeed a projective variety called the *Veronese variety.* v_q is, in fact, an isomorphism between $\mathbb{P}^m$ and $V(\mathfrak{a})$. Let $\beta^i = (\beta_0^i,\dots,\beta_m^i)$ where $\beta_i^i = q$ and $\beta_j^i = 0$ for $i\neq j$ and let $W_i = \mathbb{P}^{n_i}_\beta$ (i.e., $\mathbb{P}^n_{p_{\beta_0^i\cdots\beta_m^i}}$). Then $v_q(\mathbb{P}^m_{Y_i})\subset W_i$ and if $\boldsymbol{\xi}\notin U_i = \mathbb{P}^m_{Y_i}$, then $v_q(\boldsymbol{\xi})\notin W_i$ so that $v_q^{-1}(W_i) = U_i$. If $\varphi_i\colon W_i\to\mathbb{P}^m$ is given by $\varphi_i(p_{\beta^i}) = Y_i$ and, for $j\neq i$, as above, then $\varphi_i\circ v_q = I$ on U_i. For example, $\varphi_0(p_{q0\cdots 0}) = Y_0$, $\varphi_0(p_{q-1\,1\,0\cdots 0}) = Y_1,\dots,\varphi_0(p_{q-1\,0\cdots 0\,1}) = Y_m$ on U_0 (i.e., the φ_i give an inverse morphism for v_q). Let $N_{m,q} = \binom{m+q}{q} - 1$ and let $v_q\colon\mathbb{P}^m\to\mathbb{P}^{N_{m,q}}$ be the Veronese mapping. If $F = \sum a_{i_0\cdots i_m}Y_0^{i_0}\cdots Y_m^{i_m}$ is a form of degree q, $i_0+\cdots+i_m = q$, then $H_F = V(F) = \{\boldsymbol{\xi}\colon F(\boldsymbol{\xi}) = 0\}$ is a hypersurface in $\mathbb{P}^m$ and

$$v_q(H_F) = v_q(\mathbb{P}^m)\cap L_F \tag{7.38}$$

where $L_F = V(\sum a_{i_0\cdots i_m}p_{i_0\cdots i_m})$ is a hyperplane in $\mathbb{P}^{N_{m,q}}$. In other words, hypersurfaces go into intersections of the Veronese variety with hyperplanes i.e., hyperplane sections of the Veronese variety. Let $m = 2$, $q = 2$. Then $v_2(Y_0,Y_1,Y_2) = (Y_0^2, Y_0Y_1, Y_0Y_2, Y_1^2, Y_1Y_2, Y_2^2)$. If $z_0,\dots,z_5$ are the coordinates on $\mathbb{P}^{N_{2,2}} = \mathbb{P}^5$, then $v_q(\mathbb{P}^2)$ is defined by the equations $z_0z_3 - z_1^2 = 0$, $z_1z_4 - z_2z_3 = 0$, $z_0z_4 - z_1z_2 = 0$, $z_0z_5 - z_2^2 = 0$, $z_1z_5 - z_2z_4 = 0$, $z_3z_5 - z_4^2 = 0$. These may be viewed as equivalent to requiring

$$\operatorname{rank}\begin{bmatrix} z_0 & z_1 & z_2 \\ z_1 & z_3 & z_4 \\ z_2 & z_4 & z_5 \end{bmatrix} \leq 1. \tag{7.39}$$

The hypersurface $Y_0^2 + Y_1^2 + Y_2^2 = 0$ is mapped by v_2 into $v_2(\mathbb{P}^2) \cap V(z_0 + z_3 + z_5)$. Observe also that the hypersurface $Y_1^2 - Y_0Y_2 = 0$ is mapped by v_2 into $v_2(\mathbb{P}^2) \cap V(z_3 - z_2)$ and, in view of (7.39), this is isomorphic to the Hankel variety $H(1,k)$.

We can also treat subvarieties. More precisely, we have:

Definition 7.40 Let W be a (nonempty) irreducible subvariety of V. Consider pairs (f_1, U_1), (f_2, U_2) with U_1, U_2 open in V, $U_1 \cap W \neq \emptyset$, $U_2 \cap W \neq \emptyset$, and $f_1 \in \mathfrak{o}_V(U_1)$, $f_2 \in \mathfrak{o}_V(U_2)$. Call two such pairs *equivalent* if there is a U open in V with $U \subset U_1 \cap U_2$, $U \cap W \neq \emptyset$ and $f_1 = f_2$ on U. The set of equivalence classes $\mathfrak{o}_{W,V}$ is called the *local ring of* W *on* V.

$\mathfrak{o}_{W,V}$ is, in fact, a local ring with maximal ideal $\mathfrak{m}_{W,V} = \{(f,U)\colon f = 0$ on $U \cap W\}$. If V is affine, then $\mathfrak{o}_{W,V} \simeq \mathfrak{o}_{\bar{W},\bar{V}}$ where $\bar{W}, \bar{V}$ are the projective closures of W, V respectively. If V is affine with coordinate ring $k[V]$ and $\mathfrak{p}_W$ is the ideal of W in $k[V]$ (i.e., $\mathfrak{p}_W = I(W)/I(V)$), then $\mathfrak{o}_{W,V} \simeq k[V]_{\mathfrak{p}_W}$ Since prime ideals in R_M, (R a ring, M a multiplicative set) are of the form $\mathfrak{q}$, $\mathfrak{q}$ prime in R, $\mathfrak{q} \cap M = \emptyset$, the prime ideals in $\mathfrak{o}_{W,V}$ correspond to irreducible subvarieties of V which contain W (if V is reducible, then minimal prime ideals correspond to the irreducible components V_i of V with $W \subset V_i$). If W is irreducible, then $\mathfrak{o}_{W,V}$ is an integral domain. Similarly, in the projective case, $\mathfrak{o}_{W,V} = (S(V)_{\mathfrak{p}_W})_0$ where $\mathfrak{p}_W = I_h(W)/I_h(V)$. In other words, $\mathfrak{o}_{W,V} = \{F/G\colon F, G \in S_q, G \neq 0$ on $W\}/\text{ideal}\{F/G\colon F, G \in S_q, G \neq 0$ on W, $F \in I_h(V)_q\}$. Note that $\mathfrak{o}_{W,V}/\mathfrak{m}_{W,V}$ is a field which need not be algebraically closed.

Proposition 7.41 *If V is affine, $W \subset V$, $\dim V = r$ and $\dim W = s$, then* $\operatorname{tr}\deg_k \mathfrak{o}_{W,V}/\mathfrak{m}_{W,V} = s$.

Proof. Let $R^* = \mathfrak{o}_{W,\mathbb{A}^N} = k[X_1, \ldots, X_N]_{I(W)}$ and let $\mathfrak{p} = I(V)$, $\mathfrak{p}^* = \mathfrak{p}R^*$. Let $\mathfrak{m}^* = \mathfrak{m}_{W,\mathbb{A}^N} = I(W)R^*$. Then $\mathfrak{o}_{W,V} \simeq R^*/\mathfrak{p}^*$ and $\mathfrak{m}_{W,V} \simeq \mathfrak{m}^*/\mathfrak{p}^*$. It follows that

$$\mathfrak{o}_{W,V}/\mathfrak{m}_{W,V} = (R^*/\mathfrak{p}^*)/(\mathfrak{m}^*/\mathfrak{p}^*) \simeq R^*/\mathfrak{m}^* = \mathfrak{o}_{W,\mathbb{A}^N}/\mathfrak{m}_{W,\mathbb{A}^N}. \tag{7.42}$$

But $\mathfrak{o}_{W,\mathbb{A}^N}/\mathfrak{m}_{W,\mathbb{A}^N} = k[X_1, \ldots, X_N]_{I(W)}/I(W)k[X_1, \ldots, X_N]_{I(W)} = k(W)$, the function field of W (i.e., the quotient field of $k[W] = k[X_1, \ldots, X_N]/I(W)$).

Definition 7.43 If R is a ring, then the *Krull dimension of* R, $K\operatorname{Dim} R$, is the supremum of the heights of prime ideals in R (see I.16.15).

Corollary 7.44 $\mathfrak{o}_{W,V}/\mathfrak{m}_{W,V} = k(W)$ *and* $K\operatorname{Dim}\mathfrak{o}_{W,V} = \dim V - \dim W = r - s = \operatorname{codim}_V W$. *In particular, if* $W = \boldsymbol{\xi}$ *is a point, then* $K\operatorname{Dim}\mathfrak{o}_{\boldsymbol{\xi},V} = \dim V$.

Proof. By virtue of Theorem I.16.46,

$$\begin{aligned}
K\operatorname{Dim}\mathfrak{o}_{W,V} + h(\mathfrak{p}^*) &= K\operatorname{Dim}\mathfrak{o}_{W,\mathbf{A}^N} \\
&= K\operatorname{Dim} k(W) + h(\mathfrak{m}_{W,\mathbf{A}^N}) \\
&= N - d(\mathfrak{m}_{W,\mathbf{A}^N}) = N - \dim_k \mathfrak{m}_{W,\mathbf{A}^N} \\
&= N - s.
\end{aligned}$$

Similarly, $h(\mathfrak{p}^*) = N - r$ and so, $K\operatorname{Dim}\mathfrak{o}_{W,V} = N - s - (N - r) = r - s$. (Here $h(\cdot)$ is height and $d(\cdot)$ is depth.)

We shall relate the proposition and corollary to nonsingularity in the sequel (Chapter 15). For the moment, however, let us recall that (I.20.40) if $\boldsymbol{\xi}$ is a point of the affine variety V and $T_{V,\boldsymbol{\xi}}$ is the tangent space to V at $\boldsymbol{\xi}$, then $\dim_k T_{V,\boldsymbol{\xi}} = \dim_k(\mathfrak{m}_{\boldsymbol{\xi}}/\mathfrak{m}_{\boldsymbol{\xi}}^2)^* = \dim_k(\mathfrak{m}_{\boldsymbol{\xi}}/\mathfrak{m}_{\boldsymbol{\xi}}^2) \geq \dim V = K\operatorname{Dim}\mathfrak{o}_{\boldsymbol{\xi},V}$ and so, $\boldsymbol{\xi}$ is nonsingular if and only if $\dim_k(\mathfrak{m}_{\boldsymbol{\xi}}/\mathfrak{m}_{\boldsymbol{\xi}}^2) = K\operatorname{Dim}\mathfrak{o}_{\boldsymbol{\xi},V}$ $(= \dim V = r)$. If $(\mathfrak{o}, \mathfrak{m})$ is any Noetherian local ring, we call $\dim_k(\mathfrak{m}/\mathfrak{m}^2)$ (where $k = \mathfrak{o}/\mathfrak{m}$) the *embedding dimension*, $E\operatorname{Dim}\mathfrak{o}$, *of* $\mathfrak{o}$. We say that $(\mathfrak{o}, \mathfrak{m})$ is *regular* or that $\mathfrak{o}$ is a *regular local ring* if $E\operatorname{Dim}\mathfrak{o} = K\operatorname{Dim}\mathfrak{o}$ (see Appendix A).

Lemma 7.45 (Nakayama) *If* $(\mathfrak{o}, \mathfrak{m})$ *is a local ring,* $\mathfrak{a}$ *is an ideal in* $\mathfrak{o}$, N, M *are* $\mathfrak{o}$*-modules with* $M \subset N$, *and* N/M *is finitely generated, then* $N = M + \mathfrak{a}N$ *implies* $N = M$.

Proof. Let $n_1, \ldots, n_\nu$ be a minimal set of generators of N/M. Then $N/M = \mathfrak{a}N/M$ so that $n_\nu = \sum_{i=1}^{\nu} a_i n_i$ with $a_i \in \mathfrak{a}$. In particular, $a_\nu \in \mathfrak{a}$ and $1 - a_\nu \notin \mathfrak{m}$ i.e., is a unit in $\mathfrak{o}$. It follows that $n_\nu \in (n_1, \ldots, n_{\nu-1})$ which contradicts minimality.

Corollary 7.46 *If* N *is a finite* $\mathfrak{o}$*-module, then* $N = (n_1, \ldots, n_r)$ *if and only if* $N/\mathfrak{m}N = \operatorname{span}_k\{\bar{n}_1, \ldots, \bar{n}_r\}$ *where* $k = \mathfrak{o}/\mathfrak{m}$.

Proof. If $N/\mathfrak{m}N = \operatorname{span}_k\{\bar{n}_1, \ldots, \bar{n}_r\}$, then $N = (n_1, \ldots, n_r) + \mathfrak{m}N$ and apply the lemma. The other direction is obvious.

Corollary 7.47 *If* N *is a finite* $\mathfrak{o}$*-module, then:* (1) $n_1, \ldots, n_r$ *is a minimal set of generators of* N *if and only if* $\bar{n}_1, \ldots, \bar{n}_r$ *is a* k*-basis of* $N/\mathfrak{m}N$; (2)

$n_1, \ldots, n_r$ a minimal set of generators and $\sum r_i n_i = 0$ imply $r_i \in \mathfrak{m}$ for all i; (3) any set of generators contains a minimal set; and, (4) $n_1, \ldots, n_t$ extends to a minimal set of generators if and only if $\bar{n}_1, \ldots, \bar{n}_t$ are linearly independent.

Corollary 7.48 *The following are equivalent: (a) $\boldsymbol{\xi}$ is a simple point of V (I.20); (b) $(\mathfrak{o}_{\boldsymbol{\xi},V}, \mathfrak{m}_{\boldsymbol{\xi},V})$ is a regular local ring; and, (c) $\mathfrak{m}_{\boldsymbol{\xi},V}$ has $r = \dim V$ generators.*

Corollary 7.49 *If $(\mathfrak{o}, \mathfrak{m})$ is a local domain, then $\mathfrak{m}^\nu \neq \mathfrak{m}^{\nu+1}$ for all $\nu \geq 0$.*

Proof. If $\mathfrak{m}^\nu = \mathfrak{m}^{\nu+1} = \mathfrak{m}\mathfrak{m}^\nu$, then $\mathfrak{m}^\nu = (0)$ by Nakayama Lemma.

Of course, since nonsingularity is a local property, Corollary 7.48 will apply when V is projective.

Let W be an irreducible subvariety of V with $s = \dim W < \dim V = r$. Let Z be a subvariety of V. We say that $f_1, \ldots, f_t$ are *local equations of Z near W* if there is an open, affine U with $W \cap U \neq \emptyset$ such that $Z \cap U = Z' \neq \emptyset$ and $I(Z') = (f_1, \ldots, f_t)$ in $k[U]$. If $\mathfrak{o} = \mathfrak{o}_{W,V}$, then let $\mathfrak{a}_Z = \{f \in \mathfrak{o}$: there is an open, affine U' with $U' \cap W \neq 0$ and $f(Z \cap U') = 0\}$. Clearly $\mathfrak{a}_Z$ is an ideal in $\mathfrak{o}$. If V is affine, then $\mathfrak{a}_Z = \{f_1/g_1\colon f_1, g_1 \in k[V],\ g_1 \notin \mathfrak{p}_W,\ f_1 \in I(Z)$ (in $k[V]$)$\}$.

Proposition 7.50 *$f_1, \ldots, f_t$ are local equations of Z near W if and only if $\mathfrak{a}_Z = (f_1, \ldots, f_t)$.*

Proof. If $I(Z') = (f_1, \ldots, f_t)$ in $k[U]$, then $f_1, \ldots, f_t \in \mathfrak{o}$ and $f_i(Z \cap U) = 0$ so that $f_i \in \mathfrak{a}_Z$. If $g \in \mathfrak{a}_z$, then $g = g_1/g_2$, $g_i \in k[U \cap U'] \simeq k[U]$. But $g_1 = 0$ on $Z \cap U' \cap U$ so that $g_1 \in I(Z')$. Since $g_2 \neq 0$ on U, $g_2 \neq 0$ on $W \cap U$. In other words, $g \in (f_1, \ldots, f_t)$ in $\mathfrak{o}$ and $\mathfrak{a}_Z = (f_1, \ldots, f_t)$. Conversely, suppose that $\mathfrak{a}_Z = (f_1, \ldots, f_t)$ in $\mathfrak{o}$ and let $I(Z') = I(Z) = (g_1, \ldots, g_s)$ in $k[V]$ (may assume affine as local property). Then the g_i are in $\mathfrak{a}_Z$ and $g_i = \sum \alpha_{ij} f_j$, $\alpha_{ij} \in \mathfrak{o}$. We may suppose the g_i and α_{ij} are regular in some principal affine neighborhood $U = V_h$ of W in V so that $k[U] = k[V]_h$. Then $(g_1, \ldots, g_s) = I(Z) \cdot k[U] \subset (f_1, \ldots, f_t) = \mathfrak{a}_Z$. Let $Z' = Z \cap U$. It will be enough to show that $I(Z') = I(Z) \cdot k[U]$ (for $\mathfrak{a}_Z = I(Z') \cdot k[U]$ and then the $f_i \in I(Z')$ are local equations of Z near W). Clearly $I(Z) \cdot k[U] \subset I(Z')$. If $f \in I(Z')$, then $f \in k[U]$ and $f = g/h^\nu$, $g \in k[V]$. It follows that $h^\nu f = g \in I(Z)$. But $1/h^\nu \in k[U]$ and so, $f \in I(Z) \cdot k[U]$.

We shall use the notion of local equation in connection with divisors (Chapter 17) and where $W = \{\boldsymbol{\xi}\}$ is a point.

Proposition 7.51 *Suppose that $\mathfrak{o} = \mathfrak{o}_{W,V}$ is a regular local ring (and hence a UFD, Appendix A). If Γ is an irreducible subvariety of V with* $\operatorname{codim}_V \Gamma = 1$, *then Γ has one local equation near W.*

Proof. Since there is an $f \in \mathfrak{o}$ with $f \in \mathfrak{p}_\Gamma$ (in $k[V]$), there is a $g \in \mathfrak{p}_\Gamma$, g a prime element in the UFD $\mathfrak{o}$. We may assume that $g \in k[V]$. We claim that $\mathfrak{a}_{W,\Gamma} = (g)$. Consider $V(g) = \Gamma \cup Z$. If $W \cap \Gamma = \emptyset$ so that $W \subset Z$, then there are f_1, f_2 with $f_1 f_2 = 0$ on $V(g)$ and $f_1 \neq 0$, $f_2 \neq 0$ on $V(g)$. But then $(f_1 f_2)^t$, for some t, is divisible by g in $k[V]$ and hence in $\mathfrak{o}$. Since $\mathfrak{o}$ is a UFD and g is prime, $g \mid f_1$ (say) and $f_1 = 0$ on $V(g)$ (a contradiction). Thus $W \cap \Gamma \neq \emptyset$. By taking a neighborhood U of W, we may suppose that $\Gamma = V(g)$. In other words, $\mathfrak{a}_{W,\Gamma} = (g)$.

Example 7.52 Let $C \subset \mathbb{A}^4$ be $V(f_1, f_2, f_3)$ where

$$
\begin{aligned}
f_1(x_1, x_2, x_3, x_4) &= (x_3 + 1)x_1 + x_3 \\
f_2(x_1, x_2, x_3, x_4) &= (x_3 + 1)x_2 + x_3(x_3 + 2) \\
f_3(x_1, x_2, x_3, x_4) &= (x_3 + 1)x_4 - x_3^2(x_3 + 2).
\end{aligned}
$$

C is a "curve" in $\mathbb{A}^4$ i.e., $\dim C = 1$. Let $\Gamma_1 = P_1 = (0,0,0,0)$ and $\Gamma_2 = P_2 = (-2,0,-2,0)$. P_1 and P_2 are points of C. Let $\mathfrak{m}_1 = \mathfrak{m}_{P_1,C}$ and $\mathfrak{m}_2 = \mathfrak{m}_{P_2,C}$. Let $k[C] = k[x_1, x_2, x_3, x_4]/I(C)$ and let $^-$ denote the $I(C)$-residue. Then $\bar{x}_3 \in \mathfrak{m}_1$, $\bar{x}_3 + 1 \notin \mathfrak{m}_1$ and $\bar{x}_3 + 2 \notin \mathfrak{m}_1$ so that $\bar{x}_4 \in \mathfrak{m}_1^2$ and $\bar{x}_4$ is a local equation of $Z = V(x_4)$ at P_1. Now $\bar{x}_3 \notin \mathfrak{m}_2$, $\bar{x}_3 + 1 \notin \mathfrak{m}_2$, and $\bar{x}_3 + 2 \in \mathfrak{m}_2$ (otherwise $1 \in \mathfrak{m}_2$) so that $\bar{x}_4 = \varepsilon(\bar{x}_3 + 2)$ with ε a unit in $\mathfrak{o}_{P_2,C}$. Then $\bar{x}_4$ is a local equation of Z at P_2 but $\bar{x}_3 + 2$ is also a local equation of Z at P_2.

Exercises

(1) Show that if h is regular on U, then the domain of definition of h is well-defined. In $\mathbb{P}^2$, consider the rational function $h = (x_0 + x_1)/(x_1 + x_2)$. What is the domain of definition of h?

(2) Show that the map $\psi^* : R^* \to k(V)$ given by $\psi^*(\bar{X}_i) = (\overline{Y_i/Y_0})$ (Proposition 7.15) is indeed extendable to an isomorphism of $k(V^*)$ *into* $k(V)$.

(3) Prove Corollary 7.20.

(4) Show that $(\mathbb{P}^N, \mathfrak{o}_{\mathbb{P}^N})$ is a sheaf and that if $V \subset \mathbb{P}^N$, then $(\mathbb{P}^N, \mathfrak{o}_{\mathbb{P}^N})$ induces a sheaf on V.

(5) Prove that $v_q(\mathbb{P}^m)$ is isomorphic to $V(\mathfrak{a})$ [Example 7.34].

(6) Show that the algebraic set C of Example 7.52 is in fact a curve. [*Hint*: consider the map $\psi\colon \mathbb{P}^1 \to \mathbb{P}^5$ with $\psi(s,t) = (\psi_0, \dots, \psi_5)$ where $\psi_0(s,t) = s^2t$, $\psi_1(s,t) = -s^2(t+s)$, $\psi_2(s,t) = s(t+s)(t-s)$, $\psi_3(s,t) = -st(t+s)$, $\psi_4(s,t) = 0$, $\psi_5(s,t) = (t+s)^2(t-s)$. Let $\mathbb{P}^4 = V(Y_4)$. Then $\psi(\mathbb{P}^1) \subset \mathbb{P}^4$. Note $\psi(0,1) = (0,0,0,0,1)$ (in $\mathbb{P}^4$) and $\psi(1,t) = (t, -(t+1), (t+1)(t-1), -t(t+1), (t+1)^2(t-1))$. Consider the appropriate map into $(\mathbb{P}^4)_{Y_0}$.]

8

Exterior Algebra and Grassmannians

We saw in Chapter 1 that a $p \times 1$ system could be represented by a morphism form $\mathbb{P}^1$ into $\mathbb{P}^p$ and we indicated in I.Chapter 23 how a 2×2 system could be viewed as a morphism of $\mathbb{P}^1$ into a projective variety $\mathrm{Gr}(2,4)$ representing the two-dimensional subspaces of 4-space. Here we wish to develop the ideas we need to represent an $\mathbf{F}(z) \in \mathbf{Rat}(n; p, m)$ by a morphism of $\mathbb{P}^1$ into the projective variety representing the m-dimensional subspaces of $m + p$-space.

Example 8.1 Let $\mathbf{F}(z) \in \mathbf{Rat}(3; 2, 2)$ be given by

$$\mathbf{F}(z) = \begin{bmatrix} 1/z+1 & 1/(z+1)(z-1) \\ 0 & 1/(z+1)(z-1) \end{bmatrix}.$$

If

$$\mathbf{P}(z) = \begin{bmatrix} 1 & -1 \\ z & 0 \end{bmatrix}, \quad \mathbf{Q}(z) = \begin{bmatrix} 0 & -(z+1) \\ (z+1)(z-1) & 0 \end{bmatrix}$$

then $\mathbf{Q}(z)$ is column proper with $\partial_1(\mathbf{Q}) = 2 > \partial_2(\mathbf{Q}) = 1$, $\det \mathbf{Q}(z) = z^3 + z^2 - z - 1$ monic of degree 3, and $(\mathbf{P}(z), \mathbf{Q}(z))$ is a minimal (i.e., coprime) realization of $\mathbf{F}(z)$. Let $M_{\mathbf{F}}(z)$ be the 4×2 matrix given by

$$M_{\mathbf{F}}(z) = \begin{bmatrix} \mathbf{P}(z) \\ \mathbf{Q}(z) \end{bmatrix} = \begin{bmatrix} 1 & -1 \\ z & 0 \\ 0 & -(z+1) \\ z^2-1 & 0 \end{bmatrix}.$$

P. Falb, *Methods of Algebraic Geometry in Control Theory: Part II*,
Modern Birkhäuser Classics, https://doi.org/10.1007/978-3-319-96574-1_8

Let $\boldsymbol{\alpha} = (\alpha_1, \alpha_2)$, $1 \leq \alpha_1 < \alpha_2 \leq 4$, be a row-index and let $\psi_{\mathbf{F},\boldsymbol{\alpha}}(z) = \det M_{\mathbf{F}}^{\alpha_1,\alpha_2}(z)$. Then,

$$\begin{aligned}(\psi_{\mathbf{F},\boldsymbol{\alpha}}(z)) = (&\psi_{\mathbf{F},(1,2)}(z), \psi_{\mathbf{F},(1,3)}(z), \psi_{\mathbf{F},(1,4)}(z),\\ &\psi_{\mathbf{F},(2,3)}(z), \psi_{\mathbf{F},(2,4)}(z), \psi_{\mathbf{F},(3,4)}(z))\\ &= (z, -(z+1), z^2-1, -z(z+1), 0, z^3+z^2-z-1) = \psi_{\mathbf{F}}(z) \in \mathbb{P}^5.\end{aligned}$$

Letting $z = x_1/x_0$ and homogenizing, we have

$$\begin{aligned}\psi_{\mathbf{F}}(x_0, x_1) = (&x_0^2 x_1, -x_0^2(x_1+x_0), x_0(x_1^2 - x_0^2), -x_0 x_1(x_0+x_1),\\ &0, x_1^3 + x_1^2 x_0 - x_1 x_0^2 - x_0^3)\end{aligned}$$

so that $\psi_{\mathbf{F}}\colon \mathbb{P}^1 \to \mathbb{P}^5$ is a regular map. We note that $\psi_{\mathbf{F}}(0,1) = (0, \ldots, 0, 1)$, that $\psi_{\mathbf{F}}(\mathbb{P}^1) \subset V(Y_4) = \mathbb{P}^4$, and that $\psi_{\mathbf{F}}(\mathbb{P}^1) \subset V(Y_0 Y_5 + Y_2 Y_3 - Y_1 Y_4) = \mathrm{Gr}(2,4)$. In other words, $\psi_{\mathbf{F}}(z)$ is, for every z, a 4×2 matrix of rank 2 whose columns span a two-dimensional subspace of 4-space. The set of all such is a projective variety which can also be viewed as the set of all lines in $\mathbb{P}^3$.

Example 8.2 Let V be a k-vector space with $\dim V = 4$ and let ϵ_1, ϵ_2, ϵ_3, ϵ_4 be the standard basis of V. If $w \in V$, then $w = \sum_{i=1}^{4} a^i \epsilon_i$. Let w_1, w_2 be elements of V with $w_j = \sum a_j^i \epsilon_i$ for $j = 1, 2$. Set $W = \mathrm{span}[w_1\ w_2]$. Then $\dim W = 2$ if and only if $\rho(A_W) = 2$ where A_W is the 4×2 matrix (a_j^i). Conversely, if $B = (b_j^i)$ is a 4×2 matrix of rank 2 and $\omega_j = \sum b_j^i \epsilon_j$, $j = 1, 2$, then $\omega = \mathrm{span}[\omega_1\ \omega_2]$ is two dimensional. Thus there is a natural association between two-dimensional subspaces of V and matrices in $M_*(4,2)$. When do two matrices correspond to the same subspace? If $W = \mathrm{span}[w_1\ w_2] = \mathrm{span}[w_1'\ w_2'] = W'$, then $w_1' = \sum_{j=1}^{2} g_{1j} w_j$, $w_2' = \sum_{j=1}^{2} g_{2j} w_j$ and $A_{W'} = A_W g$. Conversely, if $A_{W'} = A_W g$ for some $g \in \mathrm{GL}(2; k)$, then $W' = W$. Let $\mathrm{Gr}(2,4) = \{W\colon W$ is a two-dimensional subspace of $V\}$. Then $\mathrm{Gr}(2,4) = M_*(4,2)/G$ where $G = \mathrm{GL}(2;k)$ acts on the right. Let $\mathbf{i} = (i_1\ i_2)$, $1 \leq i_1 < i_2 \leq 4$, and let $\pi_W(\mathbf{i}) = \det\bigl[A_W\bigr|_{1\ 2}^{i_1\ i_2}\bigr]$. If $\sigma = (1\ 2)$ is the nonidentity permutation, then $\pi_W(\sigma\mathbf{i}) = (\mathrm{sign}\,\sigma)\pi_W(\mathbf{i}) = (-1)\pi_W(\mathbf{i})$ (this is a fancy way of saying that $\det\bigl[A_W\bigr|_{1\ 2}^{i_2\ i_1}\bigr] = (-1)\det\bigl[A_W\bigr|_{1\ 2}^{i_1\ i_2}\bigr]$ in this case). We also note that

$$\pi_W(j, i_1)\pi_W(i_2, i_3) - \pi_W(j, i_2)\pi_W(i_1\ i_3) + \pi_W(j, i_3)\pi_W(i_1\ i_2) = 0 \qquad (8.3)$$

for $j = 1, 2, 3, 4$, $1 \leq i_1 < i_2 < i_3 \leq 4$, and the only nontrivial version of this equation is

$$\pi_W(1\ 2)\pi_W(3\ 4) - \pi_W(1\ 3)\pi_W(2\ 4) + \pi_W(1\ 4)\pi_W(2\ 3) = 0$$

[which is the equation of Gr(2, 4)]. Let

$$\boldsymbol{\xi}_W = (\pi_W(1\ 2), \pi_W(1\ 3), \pi_W(1\ 4), \pi_W(2\ 3), \pi_W(2\ 4), \pi_W(3\ 4))$$

and note that if $W \in \mathrm{Gr}(2,4)$, then $\boldsymbol{\xi}_W \in \mathbb{P}^5$ and $\boldsymbol{\xi}_W \in V(Y_0Y_5 - Y_1Y_4 + Y_2Y_3)$. Suppose that $\boldsymbol{\xi} \in V(Y_0Y_5 - Y_1Y_4 + Y_2Y_3)$ with $\xi_0 \neq 0$. Then $\boldsymbol{\xi} = (1, \xi_1, \xi_2, \xi_3, \xi_4, \xi_5)$ with $\xi_5 = \xi_1\xi_4 - \xi_2\xi_3$. If A is given by

$$A = \begin{bmatrix} 1 & 0 \\ 0 & 1 \\ -\xi_3 & \xi_1 \\ -\xi_4 & \xi_2 \end{bmatrix}$$

then $(\pi_W(\mathbf{i})) = \boldsymbol{\xi}_W$ where $W = \mathrm{span}[A_1\ A_2]$ (the span of the columns of A).

We now turn our attention to developing the ideas of Example 8.2 in the general situation. We let k be a field (which, in much of this chapter, need not be algebraically closed) and V be a k-vector space with $\dim V = n$.

Definition 8.4 The *set of p-vectors*, $\Lambda^p V$, $p = 0, 1, \ldots$ is defined as follows: (i) $\Lambda^0 V = k$; (ii) $\Lambda^1 V = V$; and, (iii) $\Lambda^p V$ is the set of formal sums $\sum a_w w_1 \wedge \cdots \wedge w_p$ where $w_i \in V$ subject to the conditions:

(a) $w_1 \wedge \cdots \wedge (aw_i + bw_i') \wedge \cdots \wedge w_p = a(w_1 \wedge \cdots \wedge w_p) + b(w_1 \wedge \cdots \wedge w_i' \wedge \cdots \wedge w_p)$;

(b) $w_{\sigma(1)} \wedge \cdots \wedge w_{\sigma(p)} = (\mathrm{sign}\,\sigma) w_1 \wedge \cdots \wedge w_p$ if $\sigma \in S_p$ (the symmetric group); and,

(c) $w_1 \wedge \cdots \wedge w_p \neq 0$ if the w_i are linearly independent.

[(a) is called *multilinearity* and (b) is called *skew-symmetry*.]

We note that (char $k \neq 2$) $w_1 \wedge \cdots \wedge w_p = 0$ if $w_i = w_j$ for some $i \neq j$. Thus, if $w_1, \ldots, w_p$ are dependent, then $w_1 \wedge \cdots \wedge w_p = 0$.

Example 8.5 Let $n = 3$ and let v_1, v_2, v_3 be a basis of V. Then $\Lambda^0 V = k$, $\Lambda^1 V = V$, $\Lambda^2 V = \{\sum a(v \wedge w)\}$ and $\Lambda^3 V = \{\sum b(v \wedge w \wedge x)\}$. If $v = \sum_{i=1}^{3} a^i v_i$ and $w = \sum_{j=1}^{3} b^j v_j$, then $v \wedge w = \sum_{i,j} a^i b^i (v_i \wedge v_j)$ by multilinearity. But $v_i \wedge v_i = 0$ and $v_j \wedge v_i = -v_i \wedge v_j$ for $i < j$ by skew-symmetry. Thus $v \wedge w = \sum_{i<j} (a^i b^j - b^i a^j) v_i \wedge v_j$ and $v_1 \wedge v_2$, $v_1 \wedge v_3$, $v_2 \wedge v_3$ form a basis for $\Lambda^2 V$. Similarly, $v_1 \wedge v_2 \wedge v_3$ is a basis of $\Lambda^3 V$. If $A_{v,w}$ is the matrix given by

$$A_{v,w} = \begin{bmatrix} a^1 & b^1 \\ a^2 & b^2 \\ a^3 & b^3 \end{bmatrix}$$

and $\alpha = (i, j)$, $1 \leq i < j \leq 3$, then

$$v \wedge w = \sum_{\alpha} \det\left[A_{v,w}\Big|_{1\ 2}^{\alpha}\right] v_i \wedge v_j = \sum_{\alpha} \det\left[A_{v,w} \mid \alpha\right] \hat{v}_{\alpha}$$

where $\hat{v}_{\alpha} = v_i \wedge v_j$. Thus p-vectors relate to $n \times p$ (or $p \times n$) matrices. If we call a 2-vector z *decomposable* if z is of the form $z = v \wedge w$, then z is decomposable implies there is a 3×2 matrix A_z with $z = \sum_{\alpha} \det[A_z \mid \alpha]\hat{v}_{\alpha}$. Conversely, if $A \in M(3,2)$, $A = (a_j^i)$, and $z = \sum_{\alpha} \det[A \mid \alpha]\hat{v}_{\alpha}$, then $z = (\sum a_1^i v_i) \wedge (\sum a_2^i v_i)$ is decomposable.

We observe that, in general, if $v_1, \ldots, v_n$ is a basis of V, then $v_{i_1} \wedge \cdots \wedge v_{i_p}$, $1 \leq i_1 < i_2 < \cdots < i_p \leq n$, is a basis of $\Lambda^p V$ so that $\dim \Lambda^p V = \binom{n}{p} = n!/p!(n-p)!$. If $\alpha = (i_1, \ldots, i_p)$, then we write $\hat{v}_{\alpha}$ for $v_{i_1} \wedge \cdots \wedge v_{i_p}$.

Definition 8.6 An element x of $\Lambda^p V$ is *decomposable* if $x = w_1 \wedge \cdots \wedge w_p$ for some $w_i \in V$.

Proposition 8.7 *An element x of $\Lambda^p V$ is decomposable if and only if there is an $n \times p$ matrix A with*

$$x = \sum_{\alpha} \det\left[A\Big|_{1\cdots p}^{i_1 \cdots i_p}\right] v_{i_1} \wedge \cdots \wedge v_{i_p} = \sum_{\alpha} \det[A \mid \alpha]\hat{v}_{\alpha}. \tag{8.8}$$

Proof. If $x = w_1 \wedge \cdots \wedge w_p$, then

$$w_j = \sum_{i=1}^{n} a_j^i v_i \tag{8.9}$$

for $j = 1, \ldots, p$, and $w_1 \wedge \cdots \wedge w_p = \sum_{\alpha} \det[A \mid \alpha]\hat{v}_{\alpha}$ where $A = (a_j^i)$. Conversely, if $x = \sum \det[A \mid \alpha]\hat{v}_{\alpha}$ and we define the w_j by (8.9), then $x = w_1 \wedge \cdots \wedge w_p$.

Corollary 8.10 *$x \neq 0$ is decomposable if and only if $x = w_1 \wedge \cdots \wedge w_p$ where the w_j are linearly independent (or, equivalently, $\dim W = p$ where $W = \operatorname{span}[w_1 \cdots w_p]$).*

Thus, decomposable p-vectors correspond to p-dimensional subspaces of V.

Let $f \colon \Lambda^p V \to X$ be a map (X a k-vector space). Then f is linear if and only if there is a map $g \colon V^p \to X$ with g linear in each variable and g alternating (i.e., $g = 0$ when two arguments are equal and g changes sign when two arguments are interchanged). This leads to:

Definition 8.11 Let $x = \otimes^p V$ (the p-fold tensor product of V, I.Appendix A). x is *skew-symmetric* (or *alternating*) if $\sigma(x) = (\text{sign } \sigma)x$ for all permutations σ in S_p.

Theorem 8.12 (i) *The set $\Lambda^p V$ (sic!) of skew-symmetric elements of $\otimes^p V$ is a subspace of $\otimes^p V$;* (ii) *if $w_1, \ldots, w_p \in V$, then*

$$\sum_{\sigma \in S_p} (\text{sign } \sigma)(w_{\sigma(1)} \otimes \cdots \otimes w_{\sigma(p)})$$

is in $\Lambda^p V$ and we denote it by $w_1 \wedge \cdots \wedge w_p$; (iii) *$w_1 \wedge \cdots \wedge w_p$ is multilinear and alternating;* (iv) *$w_1 \wedge \cdots \wedge w_p \neq 0$ if and only if the w_i are linearly independent; and,* (v) *if $v_1, \ldots, v_n$ is a basis of V, then $v_{i_1} \wedge \cdots \wedge v_{i_p}$, $1 \leq i_1 < \cdots < i_p \leq n$ is a basis of $\Lambda^p V$.*

Proof. (i) The map $\varphi_\sigma \colon \otimes^p V \to \otimes^p V$ given by $\varphi_\sigma(x) = (\text{sign } \sigma)x$ for σ in S_p is a linear map and so, $\Lambda^p V$ is a subspace (e.g. $\varphi_\sigma(x+y) = \varphi_\sigma(x) + \varphi_\sigma(y)$ so $x + y \in \Lambda^p V$ if $x, y \in \Lambda^p V$).

(ii) Let $\tau \in S_p$. Then

$$\begin{aligned}
&\varphi_\tau \sum_\sigma (\text{sign } \sigma)(w_{\sigma(1)} \otimes \cdots \otimes w_{\sigma(p)}) \\
&= \sum_\sigma (\text{sign } \sigma)(w_{\tau\sigma(1)} \otimes \cdots \otimes w_{\tau\sigma(p)}) \\
&= \sum_{\tau^{-1}\sigma'} (\text{sign } \tau^{-1}\sigma')(w_{\sigma'(1)} \otimes \cdots \otimes w_{\sigma'(p)}) \\
&= (\text{sign } \tau) \sum_{\sigma'} (\text{sign } \sigma')(w_{\sigma'(1)} \otimes \cdots \otimes w_{\sigma'(p)})
\end{aligned}$$

as $\text{sign } \tau^{-1}\sigma' = (\text{sign } \tau^{-1})(\text{sign } \sigma')$, $\text{sign } \tau = \text{sign } \tau^{-1}$, and $\sigma' \to \tau^{-1}\sigma'$ is a bijection of S_p.

(iii) Multilinearity is obvious and skew-symmetry follows by applying φ_σ with σ a transposition (I.Appendix C).

(iv) If $w_1, \ldots, w_p$ dependent, then $w_1 \wedge \cdots \wedge w_p = 0$ by the observation that (say) $w_2 \wedge w_2 \wedge \cdots w_j = 0$. If $w_1, \ldots, w_p$ are independent, then extend to a basis of V. But then the elements $w_{\sigma(i)} \otimes \cdots \otimes w_{\sigma(p)}$ are distinct and are part of a basis of $\otimes^p V$ so that $w_1 \wedge \cdots \wedge w_p \neq 0$.

(v) If the $v_{i_1} \wedge \cdots \wedge v_{i_p}$ are linearly independent, then they will form a basis of $\Lambda^p V$ since any $x \in \Lambda^p V$ can be written as a sum of the form $\sum a_{j_1 \cdots j_p} v_{j1} \otimes \cdots \otimes v_{j_p}$ with $a_{j\sigma(1) \cdots j\sigma(p)} = (\text{sign } \sigma)(a_{j_1 \cdots j_p})$ and hence (by induction) as a sum $\sum b_{j_1 \cdots j_p} v_{j_1} \wedge \cdots \wedge v_{j_p}$. But since the $v_{i_1} \wedge \cdots \wedge v_{i_p}$ involve different sets of basis vectors of $\otimes^p V$, they are independent.

Thus, p-vectors can be defined in terms of the tensor product.

Proposition 8.13 *Let $x \in \Lambda^p V$ and let $v \in V$, $v \neq 0$. Let $[v] \oplus W = V$ and let $w_1, \ldots, w_{n-1}$ be a basis of W.* (i) *if $x = v \wedge y$ with $y \in \Lambda^{p-1}V$, then $x \wedge v = 0$ (as an element of $\Lambda^{p+1}V$);* (ii) *if $z \in \Lambda^p W$ and $v \wedge z = 0$, then $z = 0$;* (iii) *the set $\{z \wedge v : z \in \Lambda^{p-1}W\}$ is a subspace of $\Lambda^p V$ written as $(\Lambda^{p-1}W) \wedge v$ and $\Lambda^p V = [(\Lambda^{p-1}W) \wedge v] \oplus \Lambda^p W$; and,* (iv) *if $x \in \Lambda^p V$ and $x \wedge v = 0$, then $x = y \wedge v$ with $y \in \Lambda^{p-1}W$.*

Proof. (i) $x \wedge v = v \wedge y \wedge v = 0$. (ii) Let $z = \sum a_{\alpha}\hat{w}_{\alpha}$. Then $0 = v \wedge z = \sum a_{\alpha}\hat{w}_{\alpha} \wedge v$. But the $\hat{w}_{\alpha} \wedge v$ are linearly independent as basis elements of $\Lambda^{p+1}V$ $(p < n)$ and so $a_{\alpha} = 0$. If $p = n$, then $\Lambda^n W = 0$ as $\dim W = n-1$. (iii) Clearly a subspace. $(\Lambda^{p-1}W) \wedge v$ is spanned by the elements $\hat{w}_{\alpha} \wedge v$ for $\alpha = (i_1, \ldots, i_{p-1})$. But $\Lambda^p W$ has basis $\hat{w}_{\beta}$ with $\beta = (j_1, \ldots, j_p)$ and so the sets $\hat{w}_{\alpha} \wedge v$ and $\hat{w}_{\beta}$ are disjoint. Since $\dim(\Lambda^{p-1}W) \wedge v = \binom{n-1}{p-1}$ and $\dim \Lambda^p W = \binom{n-1}{p}$, we have

$$\dim[(\Lambda^{p-1}W) \wedge v] \oplus \Lambda^p W = \binom{n-1}{p-1} + \binom{n-1}{p} = \binom{n}{p} = \dim \Lambda^p V.$$

(iv) If $x \in \Lambda^p V$, then $x = y \wedge v + z$ where $y \in \Lambda^{p-1}W$ and $z \in \Lambda^p W$. If $x \wedge v = 0$, then $z \wedge v = 0$ and $z = 0$ by (ii).

Corollary 8.14 *If $x \in \Lambda^p V$ and $M_x = \{v \mid v \wedge x = 0\}$, then $v \in M_x$ if and only if $x = v \wedge y$ with y in $\Lambda^{p-1}W$.*

Corollary 8.15 *If $x \in \Lambda^p V$, $x \neq 0$, $x = v \wedge y$ with $v \in M_x$ and $y \in \Lambda^{p-1}W$ (note $[v] \oplus W = V$), then $M_y \subset M_x$ and $v \notin M_y$ so that* $\dim M_y < \dim M_x$.

Proof. Clearly $M_y \subset M_x$. If $v \in M_y$, then $v \wedge y = 0$ contradicts $x \neq 0$.

Corollary 8.16 $\dim M_x \leq p$.

Proof. By Corollary 8.15, $x = v \wedge y$ with $y \in \Lambda^{p-1}W$ and $v \notin M_y$. Then use induction.

Corollary 8.17 *If $x \in \Lambda^p V$, $x \neq 0$, is decomposable, then* $\dim M_x = p$ *and if $x = w_1 \wedge \cdots \wedge w_p$, then* $\text{span}[w_1 \cdots w_p] = M_x$.

Proof. If $x = w_1 \wedge \cdots \wedge w_p \neq 0$, then $w_1, \ldots, w_p$ are linearly independent elements of M_x and hence, $\dim M_x = p$.

Corollary 8.18 *If* $\dim M_x = p$*, then x is decomposable.*

Proof. Use induction on p. For $p = 1$, there is nothing to prove. Let $v, w_2, \ldots, w_p$ be a basis of M_x with $[v] \oplus W = V$. Then $x = v \wedge y$ with $y \in \Lambda^{p-1}W$ by (iv). But $0 = x \wedge w_i = (y \wedge w_i) \wedge v$ for $i = 2, \ldots, p$ and $y \wedge w_i \in \Lambda^p W$. Since $y \in \Lambda^{p-1}W$, $\dim M_y \leq p-1$. Since $w_2, \ldots, w_p \in M_y$, $\dim M_y = p-1$ and y is decomposable (by induction). Thus $y = y_1 \wedge \cdots \wedge y_{p-1}$ and $x = v \wedge y_1 \wedge \cdots \wedge y_{p-1}$ is decomposable.

Corollary 8.19 *Let $x \in \Lambda^p V$ and let $\varphi_x \colon V \to \Lambda^{p+1}V$ be given by $\varphi_x(v) = x \wedge v$. Then φ_x is a linear map and x is decomposable if and only if the rank of φ_x is $n - p$.*

Proof. φ_x is obviously linear and $\operatorname{Ker} \varphi_x = M_x$.

Corollary 8.20 *Let $w_1, \ldots, w_p \in V$ and $x = w_1 \wedge \cdots \wedge w_p$. Then $\dim \operatorname{span}[w_1 \cdots w_p] = p$ if and only if φ_x has rank $n - p$.*

Now let Y be a p-dimensional subspace of V and let $y_1, \ldots, y_p$ and $y_1', \ldots, y_p'$ be bases of Y. Then $y_1 \wedge \cdots \wedge y_p = x \neq 0$ and $y_1' \wedge \cdots \wedge y_p' = x' \neq 0$. But

$$y_i' = \sum_{j=1}^{p} g_i^j y_j \tag{8.21}$$

where $g = (g_i^j)$ is a nonsingular $p \times p$ matrix (i.e., an element of $\mathrm{GL}(k;p)$). We have

$$y_1' \wedge \cdots \wedge y_p' = (\det g) y_1 \wedge \cdots \wedge y_p. \tag{8.22}$$

Hence, given the subspace Y, there is a unique (decomposable) element of $\mathbb{P}(\Lambda^p V) = \mathbb{P}^N$ where $N = \binom{n}{p} - 1$ which corresponds to Y. Conversely, if $z_1 \wedge \cdots \wedge z_p \neq 0$ and $Z = \operatorname{span}[z_1 \cdots z_p]$, then Z is a p-dimensional subspace of V. Moreover, if $z_1 \wedge \cdots \wedge z_p = c z_1' \wedge \cdots \wedge z_p'$ with $c \neq 0$, then $Z = \operatorname{span}[z_1' \cdots z_p']$. For, if $z' = z_1' \wedge \cdots \wedge z_p'$, then $z_i \wedge z' = 0$, $i = 1, \ldots, p$ and $z_i \in M_{z'}$, $i = 1, \ldots, p$. But $M_{z'} = \operatorname{span}[z_1' \cdots z_p']$ and $\dim M_{z'} = p$, so that $Z = \operatorname{span}[z_1' \cdots z_p']$. In other words, we can represent the p-dimensional subspaces of V as points of the projective space $\mathbb{P}(\Lambda^p V)$. If we fix a basis $v_1, \ldots, v_n$ of V, then the nonzero decomposable p-vectors can be represented in the form

$$\sum_{\alpha} \det[A \mid \alpha] \hat{v}_{\alpha} \tag{8.23}$$

where A is an $n \times p$ matrix of rank p and so p-dimensional subspaces correspond to points in $\mathbb{P}^N$, $N = \binom{n}{p} - 1$, with homogeneous coordinates

$$\xi_{\alpha} = \det\left[A_{1 \cdots p}^{i_1 \cdots i_p}\right] \tag{8.24}$$

with $\alpha = (i_1, \ldots, i_p)$, $1 \leq i_1 < \cdots < i_p \leq n$.

Example 8.25 Let $\dim V = 4$ and let ϵ_1, ϵ_2, ϵ_3, ϵ_4 be the standard basis of V. Let $p = 2$ and let W be a two-dimensional subspace of V with basis w_1, w_2. Then $\Lambda^2 V$ has basis $\epsilon_1 \wedge \epsilon_2$, $\epsilon_1 \wedge \epsilon_3$, $\epsilon_1 \wedge \epsilon_4$, $\epsilon_2 \wedge \epsilon_3$, $\epsilon_2 \wedge \epsilon_4$, $\epsilon_3 \wedge \epsilon_4$ and $\mathbb{P}(\Lambda^2 V)$ has dimension 5. We let $(Y_{12}, Y_{13}, Y_{14}, Y_{23}, Y_{24}, Y_{34})$ denote homogeneous coordinates on $\mathbb{P}(\Lambda^2 V)$. If $x = w_1 \wedge w_2$ with $w_i = \sum a_i^j \epsilon_j$, then $w_1 \wedge w_2 = \sum \det[A \mid \alpha]\hat{\epsilon}_\alpha$ where $A = (a_i^j)$. In other words, if

$$A = \begin{bmatrix} a_1^1 & a_2^1 \\ a_1^2 & a_2^2 \\ a_1^3 & a_2^3 \\ a_1^4 & a_2^4 \end{bmatrix}$$

then $w_1 \wedge w_2 = \det[A|_{12}^{12}]\epsilon_1 \wedge \epsilon_2 + \det[A|_{12}^{13}]\epsilon_1 \wedge \epsilon_3 + \det[A|_{12}^{14}]\epsilon_1 \wedge \epsilon_4 + \det[A|_{12}^{23}]\epsilon_2 \wedge \epsilon_3 + \det[A|_{12}^{24}]\epsilon_2 \wedge \epsilon_4 + \det[A|_{12}^{34}]\epsilon_3 \wedge \epsilon_4$ and the homogeneous coordinates of W are $y_{ij} = \det[A|_{12}^{ij}]$. We observe that

$$\det[A|_{12}^{12}] \det[A|_{12}^{34}] - \det[A|_{12}^{13}] \det[A|_{12}^{24}] + \det[A|_{12}^{14}] \det[A|_{12}^{23}] = 0$$

so that the non-zero decomposable 2-vectors lie on the variety $\mathrm{Gr}(2,4)$ given by

$$Y_{12}Y_{34} - Y_{13}Y_{24} + Y_{14}Y_{23} = 0 \tag{8.26}$$

in $\mathbb{P}(\Lambda^2 V) = \mathbb{P}^5$. This is clearly a proper subvariety since (say) no point of the form $(1, 0, 0, \xi, \eta, 1)$ lies on it. Suppose that $y_{12} = \det[A|_{12}^{12}] \neq 0$. Then $g = A_{12}^{12}$ is a nonsingular 2×2 matrix and, letting $B = Ag^{-1}$, we have

$$B = \begin{bmatrix} 1 & 0 \\ 0 & 1 \\ b_1^3 & b_2^3 \\ b_1^4 & b_2^4 \end{bmatrix}$$

and $w_1' \wedge w_2' = \sum \det[B \mid \alpha]\hat{\epsilon}_\alpha = (1/\det g) w_1 \wedge w_2$ $(w_i' = \sum g_i^j w_j)$ so that $y_{ij}' = \det[B \mid \alpha]$ are also homogeneous coordinates of W. The y_{ij}' are all the minors of B as $y_{12}' = 1$, $y_{13}' = b_2^3$, $y_{14}' = b_2^4$, $y_{23}' = -b_1^3$, $y_{24}' = -b_1^4$, $y_{34}' = b_1^3 b_2^4 - b_2^3 b_1^4$ and we also note that $1 \cdot (b_1^3 b_2^4 - b_2^3 b_1^4) - b_2^3(-b_1^4) + b_2^4(-b_1^3) = y_{12}'y_{34}' - y_{13}'y_{24}' + y_{14}'y_{23}' = 0$. If we let $x' = w_1' \wedge w_2'$, then $x' = \epsilon_1 \wedge \epsilon_2 + b_2^3 \epsilon_1 \wedge \epsilon_3 + b_2^4 \epsilon_1 \wedge \epsilon_4 - b_1^3 \epsilon_2 \wedge \epsilon_3 - b_1^4 \epsilon_2 \wedge \epsilon_4 + (b_1^3 b_2^4 - b_2^3 b_1^4)\epsilon_3 \wedge \epsilon_4$. Then, for example, $\varphi_{x'}(\epsilon_1) = -b_1^3 \epsilon_1 \wedge \epsilon_2 \wedge \epsilon_3 - b_1^4 \epsilon_1 \wedge \epsilon_2 \wedge \epsilon_4 + (b_1^3 b_2^4 - b_2^3 b_1^4)\epsilon_1 \wedge \epsilon_3 \wedge \epsilon_4$, and $\varphi_{x'}(\epsilon_2) = -b_2^3 \epsilon_1 \wedge \epsilon_2 \wedge \epsilon_3 - b_2^4 \epsilon_1 \wedge \epsilon_2 \wedge \epsilon_4 + (b_1^3 b_2^4 - b_2^3 b_1^4)\epsilon_2 \wedge \epsilon_3 \wedge \epsilon_4$. Let V^* be the dual space of V and let ϕ^1, ϕ^2, ϕ^3, ϕ^4 be the dual basis i.e., $\phi^i(\epsilon_j) = \delta_j^i$. The map $\epsilon_i \to \phi^i$ is an isomorphism of V and V^* which generates an isomorphism of $\Lambda^2 V$ with $\Lambda^2(V^*)$. We also have $(\Lambda^2 V)^* = \Lambda^2(V^*)$ in a natural way. More precisely, from $\epsilon_i \to \phi^i$ we have $\hat{\epsilon}_\alpha \to \hat{\phi}^\alpha$ i.e., $\epsilon_1 \wedge \epsilon_2 \to \phi^1 \wedge \phi^2$, $\epsilon_1 \wedge \epsilon_3 \to \phi^1 \wedge \phi^3$, etc. Consider $\Lambda^4 V$ which is a one-dimensional space with basis $\omega = \epsilon_1 \wedge \epsilon_2 \wedge \epsilon_3 \wedge \epsilon_4$. The dual basis is $\phi^1 \wedge \phi^2 \wedge \phi^3 \wedge \phi^4 = \omega^*$ and $\omega^*(\omega) = (\phi^1 \wedge \cdots \wedge \phi^4)(\epsilon_1 \wedge \cdots \wedge \epsilon_4) = \det(\phi^j(\epsilon_i)) = 1$. If x is a 2-vector, then we can define a map $\tilde{\varphi}_x \colon \Lambda^{4-2} V \to k$ by setting

$$\tilde{\varphi}_x(y) = \omega^*(x \wedge y). \tag{8.27}$$

Clearly, $\tilde{\varphi}_x$ is an element of $(\Lambda^{4-2} V)^*$ and the map $x \to \tilde{\varphi}_x$ is an isomorphism between $\Lambda^2 V$ and $(\Lambda^{4-2} V)^* = \Lambda^{4-2}(V^*)$. We shall write $\tilde{x}$ in place of $\tilde{\varphi}_x$ and $\tilde{x}$ is an element of $\Lambda^2 V^*$. There is a map $\psi_x \colon V^* \to \Lambda^3 V^*$ given by

$$\psi_x(V^*) = v^* \wedge \tilde{x}. \tag{8.28}$$

Then $x = w_1 \wedge w_2$ (i.e., x is decomposable) if and only if rank $\psi_x = 2$. If $x = w_1 \wedge w_2$, then Ann(Ker ψ_x) (the annihilator of the kernel of ψ_x) is M_x. Now, φ_x (Corollary 8.19) and ψ_x are linear maps and so have natural "transposes". Thus, $\varphi'_x \colon \Lambda^3 V^* \to V^*$ and $\psi'_x \colon \Lambda^3 V \to V$ (i.e., $\Lambda^3 V^*)^* \to (V^*)^*$) are given by

$$\begin{aligned} \varphi'_x(\mathbf{z})(v) &= \mathbf{z}(\varphi_x(v)) \\ \psi'_x(\mathbf{u})(v^*) &= \mathbf{u}(\psi_x(v^*)) \end{aligned} \tag{8.29}$$

where $\mathbf{z} \in \Lambda^3 V^*$, $v \in V$, $\mathbf{u} \in (\Lambda^3 V^*)^* = \Lambda^3 V$ and $v^* \in V^*$. Now $\varphi'_x(\mathbf{z})$ is an element of V^* and $\psi'_x(\mathbf{u})$ is an element of V so that

$$\varphi'_x(\mathbf{z})(\psi'_x(\mathbf{u})) = \langle \varphi'_x(\mathbf{z}), \psi'_x(\mathbf{u}) \rangle \tag{8.30}$$

makes sense. Moreover, by simple linear algebra, x is decomposable if and only if

$$\langle \varphi'_x(\mathbf{z}), \psi'_x(\mathbf{u}) \rangle = 0. \tag{8.31}$$

To illustrate, suppose that $x = \epsilon_1 \wedge \epsilon_2$. If $v = \sum a^j \epsilon_j$, then $\varphi_x(v) = a^3 \epsilon_1 \wedge \epsilon_2 \wedge \epsilon_3 + a^4 \epsilon_1 \wedge \epsilon_2 \wedge \epsilon_4$. If $y = \sum b^{ij} \epsilon_i \wedge \epsilon_j$, then $\tilde{\varphi}_x(y) = \omega^*(\epsilon_1 \wedge \epsilon_2 \wedge y) = \omega^*(\epsilon_1 \wedge \epsilon_2 \wedge (b^{34} \epsilon_3 \wedge \epsilon_4) = b^{34}$ so that $\tilde{x}(y) = (\phi^3 \wedge \phi^4)(y)$ and $\tilde{x} = \phi^3 \wedge \phi^4$. If $v^* = \sum a_i \phi^i$, then $\psi_x(v^*) = v^* \wedge \tilde{x} = v^* \wedge \phi^3 \wedge \phi^4 = a_1 \phi^2 \wedge \phi^3 \wedge \phi^4 + a_2 \phi^2 \wedge \phi^3 \wedge \phi^4$. Suppose that $\mathbf{z} = z_1 \phi^1 \wedge \phi^2 \wedge \phi^3 + z_2 \phi^1 \wedge \phi^2 \wedge \phi^4 + z_3\, \phi^1 \wedge \phi^3 \wedge \phi^4 + z_4 \phi^2 \wedge \phi^3 \wedge \phi^4$, then $\mathbf{z}(\varphi_x(v)) = z_1 a^3 + z_2 a^4$ and so, $\varphi'_x(\mathbf{z}) = z_1 \phi^3 + z_2 \phi^4$. Similarly, if $\mathbf{u} = u^1 \epsilon_1 \wedge \epsilon_2 \wedge \epsilon_3 + u^2 \epsilon_1 \wedge \epsilon_2 \wedge \epsilon_4 + u^3 \epsilon_1 \wedge \epsilon_3 \wedge \epsilon_4 + u^4 \epsilon_2 \wedge \epsilon_3 \wedge \epsilon_4$ then $\mathbf{u}(\psi_x(v^*)) = u^3 a_1 + u^4 a_2$ and $\psi'_x(\mathbf{u}) = u^3 \epsilon_1 + u^4 \epsilon_2$. Clearly $(z_1 \phi^3 + z_2 \phi^4)(u^3 \epsilon_1 + u^4 \epsilon_2) = 0$.

Now let us return to the general situation. V is a vector space of dimension n and V^* is its dual. Consider the vector space $\oplus \sum_{p=0}^{n} \Lambda^p V$ and define a product, called the *exterior* or *wedge product*, as follows: $\Lambda\colon (\Lambda^p V) \times (\Lambda^q V) \to \Lambda^{p+q} V$ by setting

$$\Lambda(v_1 \wedge \cdots \wedge v_p, w_1 \wedge \cdots \wedge w_q) = v_1 \wedge \cdots \wedge v_1 \wedge w_1 \wedge \cdots \wedge w_q \tag{8.32}$$

and extending by linearity. The exterior product is: (i) distributive i.e., $x \wedge (y_1 + y_2) = x \wedge y_1 + x \wedge y_2$; (ii) associative i.e., $x \wedge (y \wedge z) = (x \wedge y) \wedge z$; and, (iii) $x \wedge y = (-1)^{pq} y \wedge x$. Thus, $\oplus \sum_{p=0}^{n} \Lambda^p V$ becomes an algebra called the *exterior algebra of* V. If $T\colon V \to X$ is a linear map, then $\Lambda^p T\colon \Lambda^p V \to \Lambda^p X$ given by $(\Lambda^p T)(v_1 \wedge \cdots \wedge v_p) = (Tv_1) \wedge \cdots \wedge (Tv_p)$ and extended by linearity is also a linear map. Since $\Lambda^{p+q} T(x \wedge y) = (\Lambda^p T)(x) \wedge (\Lambda^q T)(y)$, we can define a natural homomorphism ΛT of exterior algebras.

Example 8.33 Let $V = \text{span}[v_1 \cdots v_n]$ and $X = \text{span}[w_1 \cdots w_m]$ and let $Tv_i = \sum a_{ij} w_j$ so that $A = (a_{ij})$ is the $n \times m$ matrix of T. Then

$$\begin{aligned}(\Lambda^p T)(v_{i_1} \wedge \cdots \wedge v_{i_p}) &= (Tv_i) \wedge \cdots \wedge (Tv_{i_p}) \\ &= \sum a_{i_1 k_1} \cdots a_{i_p k_p} w_{k_1} \wedge \cdots \wedge w_{k_p}\end{aligned} \tag{8.34}$$

and the matrix $\Lambda^p A$ of $\Lambda^p T$ is the so-called pth *compound of* A i.e., $\Lambda^p A = (\det A_{k_1 \cdots k_p}^{i_1 \cdots i_p}) = (p \times p \text{ minors of } A)$. This can be related to the Laplace expansion of the determinant of the Cauchy-Binet Formula ([R-2]).

Let $v_1, \ldots, v_n$ be a basis of V and let $v^{1*}, \ldots, v^{n*}$ be the dual basis so that

$$(v^{j*})(v_i) = \delta_i^j \tag{8.35}$$

and there is a natural isomorphism $v_i \to v^{i*}$ of V onto V^*. This extends to a natural isomorphism $v_{i_1} \wedge \cdots \wedge v_{i_p} \to v^{i_1 *} \wedge \cdots \wedge v^{i_p *}$ of $\Lambda^p V$ onto $\Lambda^p V^*$. Moreover, setting $(v^{i_1 *} \wedge \cdots \wedge v^{i_p *})(v_{j_1} \wedge \cdots \wedge v_{j_p}) = \det(v^{i_r *}(v_{j_t}))$ (and extending by linearity) we see that $\Lambda^p V^* \simeq (\Lambda^p V)^*$. Let $\omega^* = v^{1*} \wedge \cdots \wedge v^{n*} \in \Lambda^n V^*$. If x is a given p-vector, then

$$\tilde{\varphi}_x(y) = \omega^*(x \wedge y) \tag{8.36}$$

is a well-defined element of k for any $y \in \Lambda^{n-p} V$. Clearly $\tilde{\varphi}_x$ is an element of $(\Lambda^{n-p} V)^*$ and the map $x \to \tilde{\varphi}_x$ is an isomorphism of $\Lambda^p V$ with $\Lambda^{n-p} V^*$. We write $\tilde{x}$ in place of $\tilde{\varphi}_x$. Note that the isomorphism depends on the determinant of a basis change and so is defined naturally up to scalars. We recall the map $\varphi_x\colon V \to \Lambda^{p+1} V$ given by

$$\varphi_x(v) = x \wedge v \tag{8.37}$$

and we define an analogous map $\psi_x\colon V^* \to \Lambda^{n-p+1} V^*$ by setting

$$\psi_x(v^*) = v^* \wedge \tilde{x} \tag{8.38}$$

where $x \in \Lambda^p V$ and $\tilde{x}$ is the corresponding element of $\Lambda^{n-p}V^*$. Then, Corollary 8.19, $\mathrm{Ann}(\mathrm{Ker}\,\psi_x) = M_x$ and so x is decomposable if and only if rank $\psi_x = p$. Now, just as in Example 8.25, since φ_x and ψ_x are linear maps, there are natural "transposes" φ'_x and ψ'_x. In other words, $\varphi'_x \colon \Lambda^{p+1}V^* \to V^*$ and $\psi'_x \colon \Lambda^{n-p+1}V \to V$ (i.e., $(\Lambda^{n-p+1}V^*)^* \to (V^*)^*$) are given by

$$\begin{aligned} \varphi'_x(\mathbf{z})(v) &= \mathbf{z}(\varphi_x(v)) \\ \psi'_x(\mathbf{u})(v^*) &= \mathbf{u}(\psi_x(v^*)) \end{aligned} \tag{8.39}$$

where $\mathbf{z} \in \Lambda^{p+1}V^*$, $v \in V$, $\mathbf{u} \in \Lambda^{n-p+1}V$ and $v^* \in V^*$. Since $\varphi'_x(\mathbf{z})$ is an element of V^* and $\psi'_x(\mathbf{u})$ is an element of V,

$$\varphi'_x(\mathbf{z})(\psi'_x(\mathbf{u})) = \langle \varphi'_x(\mathbf{z}), \psi'_x(\mathbf{u}) \rangle \tag{8.40}$$

is a well-defined element of k. We also observe that x is decomposable if and only if

$$\langle \varphi'_x(\mathbf{z}), \psi'_x(\mathbf{u}) \rangle = 0 \tag{8.41}$$

for all $\mathbf{z}$, $\mathbf{u}$. The verification of this observation (P.8.4) is the type of tedious calculation best done quietly by the reader. However, we shall illustrate a part of it. Let $x = x_1 \wedge \cdots \wedge x_p$ be decomposable and let $X = \mathrm{span}[x_1, \ldots, x_p]$. Let $W = \mathrm{span}[w_{p+1}, \ldots, w_n]$ where $X \oplus W = V$. We let $x_1^*, \ldots, x_p^*$, $w_{p+1}^*, \ldots, w_n^*$ be the dual basis of V^* so that $\omega^* = x_1^* \wedge \cdots \wedge x_p^* \wedge w_{p+1}^* \wedge \cdots \wedge w_n^*$. If $v \in V$ and $v^* \in V^*$, then

$$\begin{aligned} v &= \sum_{i=1}^{p} s^i x_i + \sum_{j=1}^{n-p} t^j w_{p+j} \\ v^* &= \sum_{i=1}^{p} \alpha^i x_i^* + \sum_{j=1}^{n-p} \beta^j w_{p+j}^* \end{aligned} \tag{8.42}$$

and so,

$$\begin{aligned} \varphi_x(v) &= \sum_{j=1}^{n-p} t^j x_1 \wedge \cdots \wedge x_j \wedge w_{p+j} \\ \tilde{\varphi}_x(y) &= \omega^*(x \wedge y) = \omega^*(x \wedge \lambda(w_{p+1} \wedge \cdots \wedge w_n)) \\ &= (w_{p+1}^* \wedge \cdots \wedge w_n^*)(y) = \tilde{x}(y) \end{aligned} \tag{8.43}$$

where $y = y_1 + \lambda w_{p+1} \wedge \cdots \wedge w_n$ and all components of y_1 contain an x_i. Also,

$$\psi_x(v^*) = \sum_{i=1}^{p} \alpha^i x_i^* \wedge w_{p+1}^* \wedge \cdots \wedge w_n^*. \tag{8.44}$$

It follows that, writing

$$\begin{aligned} \mathbf{z} &= \sum_{j=1}^{n-p} z^j x_1^* \wedge \cdots \wedge x_p^* \wedge w_{p+j}^* + z_1 \\ \mathbf{u} &= \sum_{i=1}^{p} u^i x_i \wedge w_{p+1} \wedge \cdots \wedge w_n + u_1 \end{aligned} \tag{8.45}$$

we have

$$\begin{aligned}\varphi'_x(\mathbf{z}) &= \sum_{j=1}^{n-p} z^j w^*_{p+j} \\ \psi'_x(\mathbf{u}) &= \sum_{i=1}^{p} u^i x_i.\end{aligned} \tag{8.46}$$

Since $w^*_{p+j}(x_i) = 0$ for $i = 1, \ldots, p$, $j = 1, \ldots, n-p$, $\langle \varphi'_x(\mathbf{z}), \psi'_x(\mathbf{u})\rangle = 0$ for all $\mathbf{z}$, $\mathbf{u}$. While this intrinsic formulation is quite illuminating, we need a coordinate interpretation for computation.

Let $\epsilon_1, \epsilon_2, \ldots, \epsilon_n$ be the standard basis of V and let W be a p-dimensional subspace with basis $w_1, \ldots, w_p$. Suppose that

$$w_t = \sum_{j=1}^{n} \xi_t^j \epsilon_j, \quad t = 1, \ldots, p \tag{8.47}$$

and

$$\pi(\mathbf{i}) = \pi(i_1, \ldots, i_p) = \det(\xi_t^{i_j}) \tag{8.48}$$

for $\mathbf{i} = (i_1, \ldots, i_p)$ with $1 \le i_t \le n$. Then π is *alternating* in the sense that $\pi(\sigma\mathbf{i}) = (\text{sign }\sigma)\pi(\mathbf{i})$ where $\sigma\mathbf{i} = (i_{\sigma(1)}, \ldots, i_{\sigma(p)})$ and σ is a permutation of $(1, \ldots, p)$. Thus, the $\pi(\mathbf{i})$ are uniquely determined by the $\pi(i_1, \ldots, i_p)$ with $1 \le i_1 < \cdots < i_p \le n$. Moreover, if w'_t, $t = 1, \ldots, p$ is another basis of W with

$$w'_t = \sum_{s=1}^{p} g_t^s w_s \tag{8.49}$$

then

$$\xi_t'^j = \sum_{s=1}^{p} g_t^s \xi_s^j \tag{8.50}$$

with $\det(g_t^s) \neq 0$ and $\pi'(\mathbf{i}) = (\det g)\pi(\mathbf{i})$ for all $\mathbf{i}$. Hence, the point $(\pi(\mathbf{i}))$ in $\mathbb{P}^N$, $N = \binom{n}{p} - 1$, depends on W alone and not on the choice of basis. Thus, we have a well-defined map P of the set of all p-dimensional subspaces of V into $\mathbb{P}^N$ given by

$$P(W) = (\pi_W(\mathbf{i})) \tag{8.51}$$

where $\mathbf{i} = (i_1, \ldots, i_p)$, $1 \le i_1 < \cdots < i_p \le n$. The $\pi_W(\mathbf{i})$ are called the *Plücker coordinates of* W. We claim that P is injective. An element $v = \sum_{j=1}^{n} \eta^j \epsilon_j$ in V is in W if and only if the $n \times (p+1)$ matrix

$$\begin{pmatrix} \xi_1^1 \cdots \xi_p^1 & \eta^1 \\ \vdots & \vdots \\ \xi_1^n \cdots \xi_p^n & \eta^n \end{pmatrix} \tag{8.52}$$

has rank p. Since the w_t are independent, the rank is at least p and so v must be a linear combination of the w_t. In other words, v is in W if and only if every $p+1 \times p+1$ minor

$$\det \begin{pmatrix} \xi_1^{i_1} \cdots \xi_p^{i_1} & \eta^{i_1} \\ \vdots & \\ \xi_1^{i_{p+1}} \cdots \xi_p^{i_{p+1}} & \eta^{i_{p+1}} \end{pmatrix}$$

is zero. In other words, if and only if

$$\sum_{r=1}^{p+1} (-1)^{p+r-1} \eta^{i_r} \pi_W(i_1, \ldots, i_{r-1}, i_{r+1}, \ldots, i_{p+1}) = 0 \tag{8.53}$$

for all $\mathbf{i} = (i_1, \ldots, i_{p+1})$. Again, since π_W is alternating, we need only consider the equations with $1 \leq i_1 < \cdots < i_{p+1} \leq n$. Thus, for these $\mathbf{i}$, there are $\binom{n}{p+1}$ equations. Note that (8.53) implies the injectivity of P. P is clearly not surjective. Let $\pi_W(\mathbf{i})$, $1 \leq i_1 < \cdots < i_p \leq n$ be determined by W. Consider fixed $i_1, \ldots, i_{p-1}$ and let

$$v = \sum_{j=1}^{n} \pi_W(i_{1_j} \cdots i_{p-1}, j)\epsilon_j. \tag{8.54}$$

Then v will be in W if and only if

$$\pi_W(i_1, \ldots, i_{p-1}, j) = \sum_{t=1}^{p} \alpha^t \xi_t^j \tag{8.55}$$

for some α^t. Since

$$\pi_W(i_1, \ldots, i_{p-1}, j) = \det \begin{pmatrix} \xi_1^{i_1} & \cdots & \xi_p^{i_1} \\ \xi_1^{i_{p-1}} & \cdots & \xi_p^{i_{p-1}} \\ \xi_1^j & \cdots & \xi_p^j \end{pmatrix} \tag{8.56}$$

the assertion follows by expanding in term of the last row. Applying (8.53), we get

$$\begin{aligned} &\sum_{r=1}^{p+1} (-1)^{p+r-1} \pi_W(j_1, \ldots, j_{r-1}, i_r, j_{r+1}, \ldots, j_p) \\ &\qquad \pi_W(i_1, \ldots, i_{r-1}, i_{r+1}, \ldots, i_{p+1}) = 0 \end{aligned} \tag{8.57}$$

(and the alternating property). The equations (8.57) can be viewed as defining a variety (as yet only an algebraic set), $\mathrm{Gr}(p, n)$, in $\mathbb{P}^N$ and we shall show that every point in $\mathrm{Gr}(p,n)$ corresponds to a (unique) p-dimensional subspace W of V (i.e., P is surjective to $\mathrm{Gr}(p,n)$).

Definition 8.58 Let $\mathbf{S}(p,n) = \{\mathbf{i}\colon \mathbf{i} = (i_1, \ldots, i_p),\ 1 \le i_t \le n\}$ for $1 \le p < n$. A map $\pi\colon \mathbf{S}(p,n) \to k$ is *alternating* if $\pi(\sigma\mathbf{i}) = (\mathrm{sign}\ \sigma)\pi(\mathbf{i})$ where $\pi(\sigma\mathbf{i}) = \pi(i_{\sigma(1)}, \ldots, i_{\sigma(p)})$ for $\sigma \in S_p$ (the symmetric group). An alternating map π satisfies the *Plücker relations* if

$$\sum_{k=1}^{p+1} (-1)^{k-1}\pi(\mathbf{j}(\hat{i},k))\pi(\mathbf{i}(\hat{k})) = 0 \tag{8.59}$$

for all $\mathbf{j} \in \mathbf{S}(p,n)$, $\mathbf{i} \in \mathbf{S}(p+1,n)$ where

$$\mathbf{i}(\hat{k}) = (i_1, \ldots, i_{k-1}, i_{k+1}, \ldots, i_{p+1}) \in \mathbf{S}(p,n)$$
$$\mathbf{j}(\hat{i},k) = (j_1, \ldots, j_{i-1}, i_k, j_{i+1}, \ldots, j_p) \in \mathbf{S}(p,n)$$

(cf. 8.57).

The equations (8.59) can be written

$$\sum_{k=1}^{p+1} (-1)^{k-1}\pi(j_1, \ldots, j_{i-1}, i_k, j_{i+1}, \ldots, j_p)\pi(i_1, \ldots, i_{k-1}, i_{k+1}, \ldots, i_{p+1}) = 0 \tag{8.60}$$

and, since π is alternating, we need only consider the equations with $j_1 < \cdots < j_p$ and $i_1 < \cdots < i_{p+1}$. We note that if A is $n \times p$ and $\pi_A(\mathbf{i}) = \det[A|_{1\cdots p}^{i_1\cdots i_p}]$, then $\pi_A(\mathbf{i})$ satisfies the Plücker relations.

Proposition 8.61 *Suppose that $\pi(\cdot) \neq 0$ satisfies the Plücker relations. Then $x = \sum \pi(\mathbf{i})\hat{\epsilon}_{\mathbf{i}}$ is decomposable.*

Proof. Let $v_r = \sum_{t=1}^{n} \pi(i_1, \ldots, i_{r-1}, t, i_{r+1}, \ldots, i_p)\epsilon_t$ for $r = 1, \ldots, p$ ($v_r = v_r(\mathbf{i})$). Then $v_r \wedge x = \sum b^{\mathbf{j}}\hat{\epsilon}_{\mathbf{j}}$, $\mathbf{j} = (j_1, \ldots, j_{p+1})$ where

$$\begin{aligned} b^{\mathbf{j}} &= \sum_{k=1}^{p+1} (-1)^{k-1}\pi(i_1, \ldots, i_{r-1}, k, i_{r+1}, \ldots, i_p)\pi(j_1, \ldots, j_{k-1}, j_{k+1}, \ldots, j_{p+1}) \\ &= 0 \end{aligned}$$

by the Plücker relations. Since $\pi(\mathbf{i}) \neq 0$, the coefficient of ϵ_{i_r} in v_r is not zero. In other words, $v_r = \alpha_r\epsilon_{i_r} + w_r$ where $w_r \in \mathrm{span}[\epsilon_t\colon t \notin \mathbf{i}]$. Thus $v_1, \ldots, v_p$ are independent elements of M_x and so x is decomposable.

Corollary 8.62 *V_h (Plücker relations) $=$ set of p-dimensional subspaces of $V = \mathrm{Gr}(p,n)$.*

Definition 8.63 $\mathrm{Gr}(p,n)$ is called the *Grassmann variety* of p-dimensional subspaces of n space.

Since $\mathbb{P}(V)$ is an $n-1$ dimensional projective space and a p-dimensional subspace W of V corresponds to a $p-1$ dimensional linear subspace of $\mathbb{P}(V)$, we can also view $\mathrm{Gr}(p,n)$ as the variety of $p-1$ dimensional linear subspaces of $\mathbb{P}^{n-1}$. When we do so, we write $\mathbb{G}r(p-1,n-1)$ in place of $\mathrm{Gr}(p,n)$ i.e., $\mathbb{G}\mathrm{r}(p-1,n-1) = \{L_{p-1}\colon L_{p-1}$ is a linear subspace of $\mathbb{P}^{n-1}\}$. For example if $p=2$, $n=4$, then $\mathbb{G}\mathrm{r}(1,3)$ is the variety of lines in $\mathbb{P}^3$.

Corollary 8.62 states that $\mathrm{Gr}(p,n)$ is cutout by the quadratic Plücker relations.

Example 8.64 Consider $\mathrm{Gr}(2,4) \subset \mathbb{P}^5$ and let $\mathbf{Y} = (Y_{12}, Y_{13}, Y_{14}, Y_{23}, Y_{24}, Y_{34})$ be the coordinates on $\mathbb{P}^5$. Then the Plücker relations become

$$F(\mathbf{Y}) = Y_{12}Y_{34} + Y_{14}Y_{23} - Y_{13}Y_{34} = 0 \tag{8.65}$$

and $\mathrm{Gr}(2,4) = V_h(F(\mathbf{Y})) = V_h(Y_{12}Y_{34} + Y_{14}Y_{23} - Y_{13}Y_{24})$ is a hypersuface in $\mathbb{P}^5$ (so that $\dim \mathrm{Gr}(2,4) = 4$). Since the form $F(\mathbf{Y}) = Y_{12}Y_{34} + Y_{14}Y_{23} - Y_{13}Y_{24}$ is irreducible, $\mathrm{Gr}(2,4)$ is irreducible as well. Let w_1, w_2 be vectors in k^4 and associate with them the 4×2 matrix

$$Y = [w_1 \; w_2].$$

If $\mathbf{w} = w_1 \wedge w_2$, then the coordinate $\mathbf{w}_\alpha$ of $\mathbf{w}$ is the minor $\det[Y|_{12}^{\alpha}] = y_\alpha$ where $\alpha = (ij)$, $1 \le i < j \le 4$. If $\mathrm{span}[w_1 \; w_2] = W = \mathrm{span}[w_1' \; w_2']$, then $Y' = Yg$ with $g \in \mathrm{GL}(2;k)$ and $Y_\alpha' = Y_\alpha g$ for all α. Let $M_{*2} \subset k^8 = \mathbb{A}_k^8$ be the set of Y of rank 2. If $Y \in M_{*2}$, then there is an α with $\det[Y|_{12}^{\alpha}] \neq 0$. We let $(\mathbb{A}_k^8)_\alpha$ be the open set in $\mathbb{A}_k^8$ where $\det[Y|_{12}^{\alpha}] \neq 0$. Then $M_{*2} = \bigcup_\alpha (\mathbb{A}_k^8)_\alpha$ is open in $\mathbb{A}_k^8$ and so is irreducible. Note that there are $6 = \binom{4}{2}$ principal open sets $(\mathbb{A}_k^8)_\alpha$. The map $Y \to (y_\alpha)$ is clearly a morphism into $\mathbb{P}^5$ and generates a morphism $\varphi\colon M_{*2} \to \mathrm{Gr}(2,4)$ which is surjective. If $Y \in (\mathbb{A}_k^8)_\alpha$, then $\tilde{Y} = YY_\alpha^{-1}$ (where $Y_\alpha = Y_{12}^{\alpha}$) has $\tilde{Y}_\alpha = I_2$ and $\varphi(\tilde{Y}) = \varphi(Y)$. Let $(\mathbb{A}_k^8)_{\tilde{\alpha}}$ be the set of $\tilde{Y} \in (\mathbb{A}_k^8)_\alpha$ with $\tilde{Y}_\alpha = I_2$. Then it is clear that $(\mathbb{A}_k^8)_{\tilde{\alpha}} \simeq k^{2(4-2)} \simeq \mathbb{A}_k^4$. For example, if $\alpha = (12)$, then the elements look like

$$\begin{bmatrix} 1 & 0 \\ 0 & 1 \\ a_1^3 & a_2^3 \\ a_1^4 & a_2^4 \end{bmatrix}$$

with the a_j^i arbitrary. The restriction of φ to $(\mathbb{A}_k^8)_{\tilde{\alpha}}$ gives an isomorphism between $\mathbb{A}_k^4$ and an affine open set U_α of $\mathrm{Gr}(2,4)$. If $V_\alpha = (\mathbb{P}_k^5)_\alpha$ (e.g., if $\alpha = (12)$, then $Y_{12} \neq 0$), then $U_{\boldsymbol{\alpha}} = \mathrm{Gr}(2,4) \cap V_{\boldsymbol{\alpha}}$ and $\varphi^{-1}(U_{\boldsymbol{\alpha}}) = (\mathbb{A}_k^8)_{\boldsymbol{\alpha}}$. On the other hand, if $(\xi_{12}, \xi_{13}, \xi_{14}, \xi_{23}, \xi_{24}, \xi_{34}) \in U_{\boldsymbol{\alpha}}$, then the morphism $\psi_{\boldsymbol{\alpha}}\colon U_{\boldsymbol{\alpha}} \to (\mathbb{A}_k^8)_{\boldsymbol{\alpha}}$ (given in the obvious way) is such that $\psi_{\boldsymbol{\alpha}}(\varphi) =$ identity. For example, if $\boldsymbol{\alpha} = (12)$ so that $\xi_{12} \neq 0$, then $\psi_{\boldsymbol{\alpha}}(1, \xi_{13}/\xi_{12}, \ldots, \xi_{34}/\xi_{12})$ is given by

$$\begin{bmatrix} 1 & 0 \\ 0 & 1 \\ -\xi_{23}/\xi_{12} & \xi_{13}/\xi_{12} \\ -\xi_{24}/\xi_{12} & \xi_{14}/\xi_{12} \end{bmatrix}.$$

Note that by the Plücker relations, $\xi_{34} = (\xi_{13}\xi_{24} - \xi_{14}\xi_{23})/\xi_{12}$ so that $\xi_{34}/\xi_{12} = (\xi_{13}\xi_{24} - \xi_{14}\xi_{23})/\xi_{12}^2 = \det[\psi_{\boldsymbol{\alpha}}(1, \ldots, \xi_{34}/\xi_{12})\big|_{12}^{34}]$.

Of course, the analysis of the example applies in general. Thus the map $\varphi\colon M_*(p,n) \to \mathrm{Gr}(p,n)$ given by $\varphi(\pi_A(\mathbf{i})) = (\det[A\big|_{1\cdots p}^{i_1\cdots i_p}])$ is a surjective morphism. If $\boldsymbol{\alpha} = (i_1, \ldots, i_p)$, $1 \leq i_1 < \cdots < i_p \leq n$, and we let $M_*(p,n)_{\boldsymbol{\alpha}} = (\mathbb{A}_k^{pn})_{\boldsymbol{\alpha}}$ be the part of $\mathbb{A}_k^{pn}$ where $\pi_A(\boldsymbol{\alpha}) \neq 0$, then

$$M_*(p,n) = \bigcup_{\boldsymbol{\alpha}} M_*(p,n)_{\boldsymbol{\alpha}} = \bigcup_{\boldsymbol{\alpha}} (\mathbb{A}_k^{pn})_{\boldsymbol{\alpha}}$$

is irreducible. If $A \in (\mathbb{A}_k^{pn})_{\boldsymbol{\alpha}}$, then $\tilde{A} = AA_{\boldsymbol{\alpha}}^{-1}$ (where $A_{\boldsymbol{\alpha}} = (A_{1\cdots p}^{i_1\cdots i_p})$) has $\tilde{A}_{\boldsymbol{\alpha}} = I_p$ and $\varphi(\tilde{A}) = \varphi(A)$. Let $(\mathbb{A}_k^{pn})_{\tilde{\alpha}} = \{\tilde{A} \in (A_k^{pn})_{\boldsymbol{\alpha}}\colon \tilde{A}_{\boldsymbol{\alpha}} = I_p\}$. Then clearly $(\mathbb{A}_k^{pn})_{\tilde{\alpha}} \simeq \mathbb{A}_k^{p(n-p)}$ and the restriction of φ to $(\mathbb{A}_k^{pn})_{\tilde{\alpha}}$ gives an isomorphism of $\mathbb{A}_k^{p(n-p)}$ and an affine open set $U_{\boldsymbol{\alpha}}$ of $\mathrm{Gr}(p,n)$. If $V_{\boldsymbol{\alpha}} = (\mathbb{P}_k^n)_{\boldsymbol{\alpha}}$, then $U_{\boldsymbol{\alpha}} = \mathrm{Gr}(p,n) \cap V_{\boldsymbol{\alpha}}$ and $\varphi^{-1}(U_{\boldsymbol{\alpha}}) = (\mathbb{A}_k^{pn})_{\boldsymbol{\alpha}}$. If $\boldsymbol{\alpha} = (i_1, \ldots, i_p)$, $\boldsymbol{\beta} = (j_1, \ldots, j_p)$ and $\boldsymbol{\alpha} \cap \boldsymbol{\beta} = \{\ell_1, \ldots, \ell_r\}$, $\boldsymbol{\alpha} - \boldsymbol{\alpha} \cap \boldsymbol{\beta} = \{m_1, \ldots, m_{p-r}\}$, $\boldsymbol{\beta} - \boldsymbol{\alpha} \cap \boldsymbol{\beta} = \{n_1, \ldots, n_{p-r}\}$, then the matrix

$$E(\boldsymbol{\alpha}, \boldsymbol{\beta}) = \left[\epsilon_{\ell_1} \cdots \epsilon_{\ell_r} \epsilon_{m_1+n_1} \cdots \epsilon_{m_{p-r}+n_{p-r}}\right]$$

is an element of $(\mathbb{A}_k^{pn})_{\tilde{\boldsymbol{\alpha}}} \cap (\mathbb{A}_k^{pn})_{\tilde{\boldsymbol{\beta}}}$ and $\varphi(E(\boldsymbol{\alpha}, \boldsymbol{\beta}))$ is in $U_{\boldsymbol{\alpha}} \cap U_{\boldsymbol{\beta}}$. Thus, we have:

Proposition 8.66 $\mathrm{Gr}(p,n)$ *is an irreducible projective variety of dimension* $p(n-p)$ *which is cutout by the quadratic Plücker relations.*

Proposition 8.67 *There is no nontrivial linear form vanishing on* $\mathrm{Gr}(p,n)$.

Proof. If $\sum a_{\boldsymbol{\alpha}} \pi_x(\boldsymbol{\alpha}) = 0$ for all decomposable x, then $\sum a_{\boldsymbol{\alpha}} \det[A\big|_{1\cdots p}^{i_1\cdots i_p}] = 0$ for all $n \times p$ A of rank p. Let $A_{\boldsymbol{\alpha}}$ be $n \times p$ with $A_{1\cdots p}^{i_1\cdots i_p} = I_p$ and 0 elsewhere. Then $\det[A_{\boldsymbol{\alpha}}\big|_{1\cdots p}^{i_1\cdots i_p}] = 1$ and $\det[A_{\boldsymbol{\beta}}\big|_{1\cdots p}^{j_1\cdots j_p}] = 0$ if $\boldsymbol{\beta} = (j_1, \ldots, j_p) \neq \boldsymbol{\alpha}$. Thus all $a_{\boldsymbol{\alpha}} = 0$.

It is in fact true that $I_h(\mathrm{Gr}(p,n))$ is generated by the quadratic Plücker relations but we shall not prove that here ([H-2], [H-9]).

Now let $W \subset \mathbb{A}_k^n$ with $W \in \mathrm{Gr}(p,n)$ and suppose that $W = \mathrm{span}[w_1 \cdots w_p]$. If $\omega \in (\mathbb{A}^n)^*$ with $\omega(W) = 0$, then $\omega \in \mathrm{Ann}(W) = W^\perp$. Since $\dim W^\perp = n-p$, $W^\perp = \mathrm{span}[\omega_1, \dots, \omega_{n-p}] \in \mathrm{Gr}(n-p,n)$ and the map $W \to W^\perp$ is an isomorphism between $\mathrm{Gr}(p,n)$ and $\mathrm{Gr}(n-p,n)$. If we view the ω_i as defining hyperplanes in $\mathbb{A}^n$, then $W = \ker[\omega_1 \cdots \omega_{n-p}]$ and if $\tilde{A} = (\omega_j)$ an $n-p \times n$ matrix, then $W = \{w \mid \tilde{A}w = 0\} = \mathrm{Ker}\,\tilde{A}$. Let $\mathbf{j} = (j^1, \dots, j^{n-p})$ with $1 \le j^1 < \cdots < j^{n-p} \le n$ and we let

$$p^{\mathbf{j}} = \det\big[\tilde{A}\big|_{j^1 \dots j^{n-p}}^{1 \cdots n-p}\big]. \tag{8.68}$$

Similarly (as earlier), let $A = (w_j)$ an $n \times p$ matrix with $W = \{w \mid w = Ax\} = \mathcal{R}(A)$ and let $\mathbf{i} = (i_1, \dots, i_p)$ with $1 \le i_1 < \cdots < i_p \le n$ and

$$p_{\mathbf{i}} = \det\big[A\big|_{1 \cdots p}^{i_1 \cdots i_p}\big]. \tag{8.69}$$

Let $\sigma(1, \dots, n) = (i_1, \dots, i_n)$ with $\sigma \in S_n$. Let $(p_{\mathbf{i}}) = W = (p^{\mathbf{j}})$ with $\mathbf{j} = (i^{p+1}, \dots, i^n) = \mathbf{n} - \mathbf{i}$ be coordinates and *dual coordinates* of W. If we *define* quantities $q_{\mathbf{i}} = q_{i_1 \cdots i_p}$ by

$$q_{\mathbf{i}} = q_{i_1 \cdots i_p} = \epsilon_{i_1 \cdots i_p i_{p+1} \cdots i_n} p^{i_{p+1} \cdots i_n} \tag{8.70}$$

$\epsilon_{\mathbf{ij}} = \mathrm{sign}\,\sigma$. Then there is a $\lambda \neq 0$ such that $q_{\mathbf{i}} = \lambda p_{\mathbf{i}}$ and so $W = (q_{\mathbf{i}}) = (\epsilon_{\mathbf{ij}} p^{\mathbf{j}})$ (where $\mathbf{j} = (i_{p+1}, \dots, i_n)$. We verify this as follows ([H-9]): let $g \in \mathrm{Gr}(n,k)$ and let $\tilde{x} = gx$ so that $(\tilde{A}g^{-1})\tilde{x} = 0$ and

$$\begin{aligned}
\tilde{p}^{j^1 \dots j^{n-p}} &= \det\big[\tilde{A}g^{-1}\big|_{j^1 \dots j^{n-p}}^{1 \cdots n-p}\big] \\
&= \sum_{i_1 \cdots i_{n-p}} \det\big[\tilde{A}\big|_{i_1 \cdots i_{n-p}}^{1 \cdots n-p}\big] \det\big[g^{-1}\big|_{j^1 \dots j^{n-p}}^{i_1 \cdots i_{n-p}}\big] \\
&= \sum_{i_1 \cdots i_{n-p}} p^{i_1 \cdots i_{n-p}} \det\big[g^{-1}\big|_{j^1 \dots j^{n-p}}^{i_1 \cdots i_{n-p}}\big].
\end{aligned}$$

It follows that if $\tilde{g}_{i_1 \cdots i_p} = (\mathrm{sign}\,\sigma)\tilde{p}^{i_{p+1} \cdots i_n}$ $(\sigma(1, \dots, n) = (i_1, \dots, i_n))$, then

$$\begin{aligned}
\tilde{q}_{i_1 \cdots i_p} &= (\mathrm{sign}\,\sigma) \sum_{j^{p+1} \dots j^n} p^{j^{p+1} \dots j^n} \det\big[g^{-1}\big|_{i_{p+1} \cdots i_n}^{j^{p+1} \dots j^n}\big] \\
&= \sum \delta_{j^1 \dots j^n}^{i_1 \cdots i_n} \det\big[g^{-1}\big|_{i_{p+1} \cdots i_n}^{j^{p+1} \dots j^n}\big]
\end{aligned}$$

where the sum is over all distinct $(j^1 \cdots j^p)(j^{p+1} \cdots j^n)$ and using (8.70). However, by the usual characterization of inverses in terms of cofactors, we obtain

$$\tilde{q}_{i_1 \cdots i_p} = (\det g^{-1}) \sum_{j_1 \cdots j_p} \det\big[g\big|_{j_1 \cdots j_p}^{i_1 \cdots i_p}\big] q_{j_1 \cdots j_p}. \tag{8.71}$$

In other words, g transforms the $q_{\mathbf{i}}$ in the same way as the $p_{\mathbf{i}}$ except for the scalar factor, $\det g^{-1}$. If we show that, for a judicious choice of coordinates, $q_{\mathbf{i}} = p_{\mathbf{i}}$, then our assertion holds (for $\lambda = 1$). If we choose coordinates with $W = \operatorname{span}[\epsilon_1 \cdots \epsilon_p]$, then

$$A = \begin{bmatrix} I_p \\ O_{n-p,p} \end{bmatrix}.$$

$p_{1\cdots p} = 1$ and $p_{\mathbf{i}} = 0$ unless $\mathbf{i} = \sigma(1 \cdots p)$, $\sigma \in S_p$ in which case $p_{\mathbf{i}} = (\operatorname{sign} \sigma)$. Then $\omega_j(\cdot) = V(X_j)$, $j = p+1, \ldots, n$, and

$$\tilde{A} = (O_{n-p,p}, I_{n-p})$$

so that $q_{1\cdots p} = p^{p+1\cdots n} = 1$, $q_{i_1\cdots i_p} = \pm p^{i_{p+1}\cdots i_n} = 0$ unless $\mathbf{i} = (i_1 \cdots i_p) = \sigma(1 \cdots p)$, $\sigma \in S_p$ in which case $q_{\mathbf{i}} = (\operatorname{sign} \sigma)$. In other words, we are free to use dual coordinates for W. Dual coordinates play a role in one approach to output feedback.

Exercises

(1) Verify equation (8.3).

(2) Verify equation (8.22).

(3) Show that the map ψ_x of (8.28) is linear and that x is decomposable if and only if rank $\psi_x = 2$. Show that if x is decomposable then $\operatorname{Ann}(\operatorname{Ker} \psi_x) = M_x$. Show that x is decomposable if and only if (8.31) holds.

(4) Show that $x \in \Lambda^p V$ is decomposable if and only if (8.41) holds for all $\mathbf{z}$, $\mathbf{u}$.

(5) Show that if A is an $n \times p$ matrix and $\pi_A(\mathbf{i}) = \det[A|_{1\cdots p}^{i_1\cdots i_p}]$, then the $\pi_A(\mathbf{i})$ satisfy the Plücker relations ([H-9]).

(6) Show that the map $W \to W^\perp$ ($W^\perp \subset V^*$) of $\operatorname{Gr}(p, n) \to \operatorname{Gr}(n-p, n)$ is an isomorphism of projective varieties.

(7) Develop dual coordinates in $\operatorname{Gr}(2, 4)$.

9

The Laurent Isomorphism Theorem: I

Let $\mathbf{F}(z)$ be an element of $\mathbf{Rat}(n, m, p)$. We may assume that $p \leq m$ (otherwise deal with $\mathbf{F}'(z)$). Then, Chapter 4, $\mathbf{F}(z) = \mathbf{P}(z)\mathbf{Q}^{-1}(z)$ with $\mathbf{P}$, $\mathbf{Q}$ coprime, $\mathbf{Q}$ column proper, $\det \mathbf{Q}(z)$ monic, and $\partial_1 \geq \cdots \geq \partial_m$ where $\partial_i = \partial_i(\mathbf{Q})$. Let

$$M_{\mathbf{F}}(z) = \begin{bmatrix} \mathbf{P}(z) \\ \mathbf{Q}(z) \end{bmatrix} \tag{9.1}$$

be the $p+m \times m$ matrix corresponding to $\mathbf{F}$. We call $M_{\mathbf{F}}(z)$ the *system matrix*. Then $M_{\mathbf{F}}(z)$ has rank m for every $z \in k$ and so corresponds to an element $W_{\mathbf{F}}(z)$ of $\mathrm{Gr}(m, p+m) \subset \mathbb{P}^N$, $N = \binom{m+p}{p} - 1 = \binom{m+p}{m} - 1$, for every $z \in k$. Let $\boldsymbol{\alpha} = (\alpha_1, \ldots, \alpha_m)$ with $1 \leq \alpha_1 < \cdots < \alpha_m \leq m+p$ and let

$$\psi_{F,\boldsymbol{\alpha}}(z) = \det\left[M_{\mathbf{F}}(z)\Big|_{1\cdots m}^{\alpha_1\cdots\alpha_m}\right]. \tag{9.2}$$

The $\psi_{F,\boldsymbol{\alpha}}(z)$ are the Plücker coordinates of $W_{\mathbf{F}}(z)$. Since $\mathbf{F}(z) = \mathbf{P}(z)\mathbf{Q}^{-1}(z) = \mathbf{P}(z)\mathrm{Adj}\ \mathbf{Q}(z)/\det \mathbf{Q}(z)$, we have

$$\mathbf{F}_j^i(z) = \mathbf{P}^i(\mathrm{Adj}\ \mathbf{Q})_j/\det \mathbf{Q} \tag{9.3}$$

for $i = 1, \ldots, p$, $j = 1, \ldots, m$. We can recover the $\mathbf{F}_j^i(z)$ and more, from the $\psi_{F,\boldsymbol{\alpha}}(z)$. Suppose that

$$\boldsymbol{\alpha} = (i_1, \ldots, i_t, j_1 + p, \ldots, j_s + p) \tag{9.4}$$

where $1 \leq i_1 < \cdots < i_t \leq p$, $1 \leq j_1 < \cdots < j_s \leq m$, and

$$t + s = m. \tag{9.5}$$

P. Falb, *Methods of Algebraic Geometry in Control Theory: Part II*,
Modern Birkhäuser Classics, https://doi.org/10.1007/978-3-319-96574-1_9

Then

$$\psi_{F,\boldsymbol{\alpha}}(z) = \det \begin{bmatrix} \mathbf{P}^{i_1}(z) \\ \vdots \\ \mathbf{P}^{i_t}(z) \\ \mathbf{Q}^{j_1}(z) \\ \vdots \\ \mathbf{Q}^{j_s}(z) \end{bmatrix} = \det \begin{bmatrix} \mathbf{P}^{i_1}(z) \\ \vdots \\ \mathbf{P}^{i_t}(z) \\ \mathbf{Q}^{j_1}(z) \\ \vdots \\ \mathbf{Q}^{j_s}(z) \end{bmatrix} \mathbf{Q}^{-1}(z)\mathbf{Q}(z) \tag{9.6}$$

$$= \det \begin{bmatrix} \mathbf{P}^{i_1}(\text{Adj } \mathbf{Q})_1/\det \mathbf{Q} & \cdots & \mathbf{P}^{i_1}(\text{Adj } \mathbf{Q})_m/\det \mathbf{Q} \\ \vdots & & \vdots \\ \mathbf{P}^{i_1}(\text{Adj } \mathbf{Q})_1/\det \mathbf{Q} & \cdots & \mathbf{P}^{i_t}(\text{Adj } \mathbf{Q})_m/\det \mathbf{Q} \\ & \epsilon^{j_1} & \\ & \epsilon^{j_2} & \\ & \vdots & \\ & \epsilon^{j_s} & \end{bmatrix} \cdot \det \mathbf{Q}.$$

Let $\mathbf{m} = (1, \dots, m)$ and $\mathbf{j}_{\boldsymbol{\alpha}} = \mathbf{m} - \{j_1, \dots, j_s\} = \{\ell_1, \dots, \ell_t\}$. If

$$\pi_{\boldsymbol{\alpha}}(\mathbf{F}) = \det\left[\mathbf{F}(z)\Big|_{\ell_1 \cdots \ell_t}^{i_1 \cdots i_t}\right] \tag{9.7}$$

then

$$\psi_{F,\boldsymbol{\alpha}}(z) = (-1)^{\sigma_{\mathbf{j}\boldsymbol{\alpha}}} \pi_{\boldsymbol{\alpha}}(\mathbf{F}) \det \mathbf{Q} = (-1)^{\sigma_{\mathbf{j}\boldsymbol{\alpha}}} \pi_{\boldsymbol{\alpha}}(\mathbf{F}) \psi_{F,(p+1\cdots p+m)}(z) \tag{9.8}$$

and so, for $t = 1$,

$$\mathbf{F}_{\mathbf{j},\boldsymbol{\alpha}}^{i_1}(z) = (-1)^{\sigma_{\mathbf{j}\boldsymbol{\alpha}}} \psi_{F,\boldsymbol{\alpha}}(z)/\psi_{F,(p+1\cdots p+m)}(z) \tag{9.9}$$

(which recovers $\mathbf{F}$ from the $\psi_{F,\boldsymbol{\alpha}}(z)$).

Example 9.10 Let $p = 2$, $m = 3$. Consider $\boldsymbol{\alpha} = (1\ 3\ 4) = (1\ 1+2\ 2+2)$ so that $t = 1$, $s = 2$, $i_1 = 1$, $j_1 = 1$, $j_2 = 2$. Then $\mathbf{j}_{\boldsymbol{\alpha}} = (\ell_1) = \mathbf{m} - (1, 2) = (1\ 2\ 3) - (1\ 2) = (3)$ and

$$\psi_{F,\boldsymbol{\alpha}}(z) = \det \begin{bmatrix} \mathbf{P}^1 \\ \mathbf{Q}^1 \\ \mathbf{Q}^2 \end{bmatrix}$$

$$= \det \begin{bmatrix} \mathbf{P}^1(\text{Adj } \mathbf{Q})_1/\det \mathbf{Q} & \mathbf{P}^2(\text{Adj } \mathbf{Q})_2/\det \mathbf{Q} & \mathbf{P}^3(\text{Adj } \mathbf{Q})_3/\det \mathbf{Q} \\ 1 & 0 & 0 \\ 0 & 1 & 0 \end{bmatrix}$$

$$= \mathbf{F}_3^1(z) \cdot \det \mathbf{Q}(z)$$

so that $\mathbf{F}_3^1(z) = \psi_{F,(1\ 3\ 4)}(z)/\psi_{F,(3\ 4\ 5)}(z)$.

Let s be the number of rows of $\mathbf{Q}(z)$ in $\boldsymbol{\alpha}$ (so that $\boldsymbol{\alpha}$ is of the form (9.4)). Since $\mathbf{Q}$ is column proper and $\mathbf{F}$ is strictly proper, $\partial_i(\mathbf{P}(z)) \leq \partial_i - 1$ for $i = 1, \ldots, m$. It follows that

$$\deg \psi_{F,\boldsymbol{\alpha}}(z) \leq n + s - m \tag{9.11}$$

where $s = m - p, m - p + 1, \ldots, m$. For simplicity, we set

$$\psi_{F,0}(z) = \psi_{F,(p+1 \cdots p+m)}(z) = \det \mathbf{Q}(z). \tag{9.12}$$

Then the minors of $\mathbf{F}(z)$ are of the form

$$\pi_{\boldsymbol{\alpha}}(\mathbf{F}) = (-1)^{\sigma_{J\boldsymbol{\alpha}}} \psi_{F,\boldsymbol{\alpha}}(z)/\psi_{F,0}(z) \tag{9.13}$$

and the coordinates in $\mathbb{P}^N$ of $W_{\mathbf{F}}(z)$ are the $\psi_{F,\boldsymbol{\alpha}}$ or, equivalently, the minors (all) of $\mathbf{F}(z)$. We note that the number of these minors is given by

$$\sum_{r=0}^{p} \binom{p}{r} \binom{m}{m-r} = \binom{m+p}{m} = \binom{m+p}{p}. \tag{9.14}$$

This can also be established by noting that

$$M_{\mathbf{F}}(z) = M_{\mathbf{F}}(z)\mathbf{Q}^{-1}(z)\mathbf{Q}(z) = \begin{bmatrix} \mathbf{F}(z) \\ I_m \end{bmatrix} \mathbf{Q}(z)$$

and hence that

$$\psi_{F,\boldsymbol{\alpha}}(z) = \det \left[\begin{matrix} \mathbf{F}(z) \\ I_m \end{matrix} \middle| \begin{matrix} \alpha_1 \cdots \alpha_m \\ 1 \cdots m \end{matrix} \right] \det \mathbf{Q}(z).$$

Letting $z = x_1/x_0$ and homogenizing, we see that $\psi_{\mathbf{F}}(x_0, x_1) = (\psi_{F,\boldsymbol{\alpha}}(x_0, x_1))$ consists of forms of degree n which are relatively prime for all $(x_0, x_1) \in \mathbb{P}^1$. We also have

$$\psi_{\mathbf{F}}(\xi_\infty) = \psi_{\mathbf{F}}(0, 1) = (0, \ldots, 0, 1) \tag{9.15}$$

and

$$\#\text{coeff } \psi_{F,\boldsymbol{\alpha}}(x_0, x_1) \leq n + 1 + s - m \tag{9.16}$$

where $s = m - p, m - p + 1, \ldots, m$ in view of the degree conditions (9.11). Note that in (9.16) s is the number of rows of $\mathbf{Q}$ in $\boldsymbol{\alpha}$ and $t = m - s$ is the number of rows of $\mathbf{P}$ in $\boldsymbol{\alpha}$ so that (9.16) can also be written

$$\#\text{coeff } \psi_{F,\boldsymbol{\alpha}}(z) \leq n + 1 - t \tag{9.17}$$

where $t = 0, 1, \dots, p$. Let $\mu = \sum_{\alpha} \#\,\text{coeff}\ \psi_{F,\alpha}$. Then $\mu \leq \gamma(n, m, p)$ where

$$\begin{aligned}\gamma(n,m,p) &= \sum_{s=m-p}^{m} (n+1+s-m)\binom{p}{m-s}\binom{m}{s} \\ &= \sum_{t=0}^{p}(n+1-t)\binom{p}{t}\binom{m}{m-t} \\ &= \sum_{t=0}^{p}(n+1-t)\binom{p}{t}\binom{m}{t}.\end{aligned} \tag{9.18}$$

In view of (9.14), an element $(\psi_0(x_0, x_1), \dots, \psi_N(x_0, x_1))$ of $\mathbb{P}^N$, $N = \binom{m+p}{p} - 1$, with the ψ_i forms of degree n, has (at most)

$$\lambda(n,m,p) = (n+1)\sum_{t=0}^{p}\binom{p}{t}\binom{m}{m-t} \tag{9.19}$$

coefficients. Thus, the degree conditions determine a linear space of dimension $\gamma(n, m, p)$ which is defined by $\lambda(n, m, p) - \gamma(n, m, p)$ equations. But, by direct computation,

$$\lambda(n,m,p) - \gamma(n,m,p) = \sum_{p=0}^{t} t\binom{p}{t}\binom{m}{t} = \sum_{t=1}^{p} t\binom{p}{t}\binom{m}{t} \tag{9.20}$$

which is *independent* of n.

Now let us recall that if $W_{\boldsymbol{\alpha}} \subset \mathbb{A}_k^{p+m}$ is the subspace spanned by $e_{\alpha_1}, \dots, e_{\alpha_m}$ where $\boldsymbol{\alpha} = (\alpha_1, \dots, \alpha_m)$ and $1 \leq \alpha_1 < \alpha_2 < \cdots < \alpha_m \leq m+p$, then $\text{Gr}(m, p+m)_{\boldsymbol{\alpha}}$ is the affine open subset of $\text{Gr}(m, p+m)$ where the $m \times m$ minor $|_{1\cdots m}^{\alpha_1\cdots\alpha_m}$ is not zero and any element of $\text{Gr}(m, p+m)_{\boldsymbol{\alpha}}$ is represented by the column space of a (unique) $p+m \times m$ matrix of the form

$$\begin{bmatrix} F_m^{1\cdots\alpha_1-1} \\ e_{\alpha_1} \\ F_m^{\alpha_1+1,\dots,\alpha_2-1} \\ e_{\alpha_2} \\ \vdots \\ e_{\alpha_m} \\ F_m^{\alpha_m+1,\dots,m+p} \end{bmatrix}. \tag{9.21}$$

Note that $F_m^{\alpha_j+1,\ldots,\alpha_{j+1}-1}$ is an $(\alpha_{j+1}-\alpha_j)-1\times m$ matrix. If $\mathbf{F}_m^p$ is the $p\times m$ matrix given by

$$\mathbf{F}_m^p = \begin{bmatrix} F_m^{1\cdots\alpha_1-1} \\ F_m^{\alpha_1+1,\ldots,\alpha_2-1} \\ \vdots \\ F_m^{\alpha_m+1,\ldots,m+p} \end{bmatrix} \tag{9.22}$$

then the affine coordinates on $\mathrm{Gr}(m,p+m)_{\boldsymbol{\alpha}}$ are just the $m\times m$ minors of (9.21) i.e., *all* the minors of $\mathbf{F}_m^p$. The expansion of these determinants (via rows or columns) gives the Plücker relations and so determines $\mathrm{Gr}(m,p+m)_{\boldsymbol{\alpha}}$.

Consider now the set of all regular maps $\psi\colon \mathbb{P}_k^1 \to \mathbb{P}_k^N$, $N = \binom{m+p}{m}-1$, with $\psi = (\psi_{\boldsymbol{\alpha}}(x_0,x_1))$, $\boldsymbol{\alpha}=(\alpha_1,\ldots,\alpha_m)$, $1\le\alpha_1<\cdots<\alpha_m\le p+m$, where each $\psi_{\boldsymbol{\alpha}}(x_0,x_1)$ is a form of degree n. We may view ψ as an element of $\mathbb{P}_k^{\binom{m+p}{m}(n+1)-1}$ (in terms of the coefficients). Since the condition that the forms $\psi_{\boldsymbol{\alpha}}(x_0,x_1)$ be relatively prime is algebraic on the coefficients we are really working on an open set in $\mathbb{P}_k^{\binom{m+p}{m}(n+1)-1}$. This is implicitly assumed in the sequel. The degree conditions define a linear space $L_{\gamma(n,m,p)}$ in this projective space. If $\psi\in\mathrm{Gr}(m,p+m)$ satisfies the quadratic Plücker relations, then these relations determine a variety $\mathbf{W}(n,m,p)$ in $\mathbb{P}^{\nu(n,m,p)}$ where $\nu(n,m,p) = \binom{m+p}{m}(n+1)-1$. Then $\mathbf{W}(n,m,p)\cap L_{\gamma(n,m,p)}$ is an algebraic set in $\mathbb{P}^{\nu(n,m,p)}$. If we further suppose that $\psi(0,1)=(0,\ldots,0,1)$, then ψ is an element of the open set

$$\mathbf{V}(n,m,p) = \left[\mathbf{W}(n,m,p)\cap L_{\gamma(n,m,p)}\right]_{Y_{\nu(n,m,p)}} \tag{9.23}$$

contained in $\mathbb{P}^{\nu(n,m,p)}$. Clearly, if $\mathbf{F}(z)\in\mathbf{Rat}(n,m,p)$, then $\psi_{\mathbf{F}}\in\mathbf{V}(n,m,p)$ and the map $\mathbf{F}\to\psi_{\mathbf{F}}$ is injective in view of (9.3). We claim that it is surjective as well. Let $\psi = (\psi_{\boldsymbol{\alpha}}(x_0,x_1))$ be an element of $\mathbf{V}(n,m,p)$ and let $\psi(z) = (\psi_{\boldsymbol{\alpha}}(1,z))$. Suppose that $1\le i\le p$ and $1\le j\le m$. Then set

$$F_{\psi j}^{\,i}(z) = (-1)^{\sigma_{ij}}\psi_{\boldsymbol{\alpha}}(z)/\psi_{(p+1,\ldots,p+m)}(z) \tag{9.24}$$

where $\boldsymbol{\alpha}=(i,p+1,\ldots,p+j-1,p+j+1,\ldots,p+m)$. Then $F_\psi(z) = (F_{\psi j}^{\,i}(z))$ is a $p\times m$ rational transfer matrix and the minors of F_ψ, $\pi_{\boldsymbol{\beta}}(F_\psi)$, are given by

$$\pi_{\boldsymbol{\beta}}(F_\psi) = (-1)^{\sigma_{1\boldsymbol{\beta}}}\psi_{\boldsymbol{\beta}}(z)/\psi_{(p+1,\ldots,p+m)}(z). \tag{9.25}$$

Since F_ψ is a $p\times m$ rational transfer matrix, perhaps of order n_1, there are $\varphi_{\boldsymbol{\beta}}(z)$ with

$$\pi_{\boldsymbol{\beta}}(F_\psi) = (-1)^{\sigma_{1\boldsymbol{\beta}}}\varphi_{\boldsymbol{\beta}}(z)/\varphi_{(p+1,\ldots,p+m)}(z). \tag{9.26}$$

But the $\psi_\beta(z)$ and the $\varphi_\beta(z)$ are relatively prime so that

$$\begin{aligned} 1 &= \sum_\beta f_\beta(z)\psi_\beta(z) \\ 1 &= \sum_\beta g_\beta(z)\varphi_\beta(z). \end{aligned} \tag{9.27}$$

It follows that

$$\begin{aligned} \varphi_{(p+1,\dots,p+m)}(z) &= \sum f_\beta(z)\psi_\beta(z)\varphi_{(p+1,\dots,p+m)}(z) \\ &= \left(\sum f_\beta(z)\varphi_\beta(z)\right)\psi_{(p+1,\dots,p+m)}(z) \end{aligned}$$

(by (9.25) and (9.26)) and, similarly, that $\psi_{(p+1,\dots,p+m)}(z) = (\sum g_\beta(z)\psi_\beta(z)) \cdot \varphi_{(p+1,\dots,p+m)}(z)$. In other words, $\varphi_{(p+1,\dots,p+m)}(z)$ and $\psi_{(p+1,\dots,p+m)}(z)$ are associates and hence have the same degree n. Thus $F_\psi(z) \in \mathbf{Rat}(n,m,p)$. We have thus shown that $\mathbf{Rat}(n,m,p)$ can be identified with $\mathbf{V}(n,m,p)$ which gives $\mathbf{Rat}(n,m,p)$ a projective algebraic structure.

Now let us recall (Definition 3.69) that, as subsets of $\mathbb{A}^{n(pm+1)}$,

$$\begin{aligned} \mathrm{Hank}(n,m,p) &= \{(\mathbf{H}_1,\dots,\mathbf{H}_n,\xi_1,\dots,\xi_n) : \rho(\mathcal{H}(\mathbf{H}_j)) = n, \\ &\qquad \chi_1 = -\xi_n,\dots,\chi_n = -\xi_1\} \\ \mathrm{Rat}(n,m,p) &= \{(\mathbf{P}_0,\dots,\mathbf{P}_{n-1},a_0,\dots,a_{n-1}) : \rho(\mathcal{H}(\mathbf{P})) = n, \\ &\qquad \chi_1 = -a_{n-1},\dots,\chi_n = -a_0\} \end{aligned} \tag{9.28}$$

and that the Laurent map L: $\mathrm{Rat}(n,m,p) \to \mathrm{Hank}(n,m,p)$ is the natural restriction of the isomorphism $\psi_L \colon \mathbb{A}_k^{n(pm+1)} \to \mathbb{A}_k^{n(pm+1)}$ given by

$$\psi_L(\mathbf{X}_1,\dots,\mathbf{X}_n,x_{n+1},\dots,x_{2n}) = (\mathbf{Y}_1,\dots,\mathbf{Y}_n,y_{n+1},\dots,y_{2n})$$

where

$$\begin{aligned} y_{n+j} &= x_{n+j}, \quad j = 1,\dots,n \\ \mathbf{Y}_1 &= \mathbf{X}_n \\ \mathbf{Y}_s &= \mathbf{X}_{n+1-s} - \sum_{j=1}^{s-1} \mathbf{Y}_j x_{n+s-j} \end{aligned}$$

for $s = 2,\dots,n$ (cf. (3.86)). In other words, we have

$$\begin{aligned} \xi_j &= a_{j-1}, \quad j = 1,\dots,n \\ \mathbf{H}_1 &= \mathbf{P}_{n-1} \\ \mathbf{H}_s &= \mathbf{P}_{n-s} - \sum_{j=1}^{s-1} \mathbf{P}_j a_{s-j-1} \end{aligned} \tag{9.29}$$

for $s = 2, \ldots, n$, as the equations defining the Laurent map L. If $\mathbf{F}(z) \in \text{Rat}(n, m, p)$, then

$$\mathbf{F}(z) = \sum_{t=0}^{n-1} \mathbf{P}_t z^t / a_0 + \cdots + a_{n-1} z^{n-1} + z^n \tag{9.30}$$

and consequently

$$\mathbf{F}_j^i(z) = \sum_{t=0}^{n-1} p_{t,j}^i z^t / \psi(z) \tag{9.31}$$

where $\mathbf{P}_t = (p_{t,j}^i)$ and $\psi(z) = a_0 + \cdots + a_{n-1} z^{n-1} + z^n$. It follows that

$$\begin{aligned} \psi(z) &= \psi_{F,(p+1,\ldots,p+m)}(z) \\ (-1)^{\sigma_{ij}} \psi_{F,\alpha}(z) &= \sum_{t=0}^{n-1} p_{t,j}^i z^t \end{aligned} \tag{9.32}$$

for $\alpha = (i, p+1, \ldots, p+j-1, p+j+1, \ldots, p+m)$. Hence, the degree conditions and the Grassmann equations define a variety in $\mathbb{A}_k^{n(pm+1)}$ and $\text{Rat}(n, m, p)$ is an open subset of this variety. Since L is a morphism and ψ_L is an isomorphism, the Laurent Isomorphism Theorem holds.

Example 9.33 Let $p = 2$, $m = 2$ so that we work in $\text{Gr}(2, 4)$ and $\mathbb{A}^{n(4+1)}$. Viewing $\text{Gr}(2, 4)$ as a subvariety of $\mathbb{P}^5$ with coordinates $(Y_0, \ldots, Y_5)$, the equation for $\text{Gr}(2, 4)$ is

$$Y_0 Y_5 - Y_1 Y_4 + Y_2 Y_3 = 0.$$

Let $\psi_{12}(x_0, x_1)$, $\psi_{13}(x_0, x_1)$, $\psi_{14}(x_0, x_1)$, $\psi_{23}(x_0, x_1)$, $\psi_{24}(x_0, x_1)$, $\psi_{34}(x_0, x_1)$ be relatively prime forms of degree n which satisfy the (degree) conditions

$$\begin{aligned} \#\,\text{coeff}\ \psi_{12}(x_0, x_1) &\le n - 1 \\ \#\,\text{coeff}\ \psi_{13}(x_0, x_1) &\le n \\ \#\,\text{coeff}\ \psi_{14}(x_0, x_1) &\le n \\ \#\,\text{coeff}\ \psi_{23}(x_0, x_1) &\le n \\ \#\,\text{coeff}\ \psi_{24}(x_0, x_1) &\le n \\ \#\,\text{coeff}\ \psi_{34}(x_0, x_1) &\le n + 1. \end{aligned}$$

Then the Grassmann equation for $\text{Rat}(n, 2, 2)$ becomes

$$\psi_{12}(x_0, x_1)\psi_{34}(x_0, x_1) - \psi_{13}(x_0, x_1)\psi_{24}(x_0, x_1) + \psi_{14}(x_0, x_1)\psi_{23}(x_0, x_1) = 0.$$

Let $\psi_{34}(z) = \psi_{34}(1, z) = \psi_0(z)$ and let $\psi_\alpha(z) = \psi_\alpha(1, z)$. If the $\psi_\alpha(z)$ are relatively prime, then

$$\begin{bmatrix} \psi_{14}(z)/\psi_0(z) & -\psi_{13}(z)/\psi_0(z) \\ \psi_{24}(z)/\psi_0(z) & -\psi_{23}(z)/\psi_0(z) \end{bmatrix}$$

determines an element of $\mathrm{Rat}(n,2,2)$. Moreover, $\psi_{12}(z)\psi_0(z) - \psi_{13}(z)\psi_{24}(z) + \psi_{14}(z)\psi_{23}(z) = 0$ determines $\psi_{12}(z)$. If $\psi_0(\boldsymbol{\xi}) \neq 0$, then the 4×2 matrix representation is

$$\begin{bmatrix} a_1^1 & a_2^1 \\ a_1^2 & a_2^2 \\ 1 & 0 \\ 0 & 1 \end{bmatrix}$$

and $\psi_{12}(\xi) = a_1^1 a_2^2 - a_2^1 a_1^2$, $\psi_{13}(\xi) = -a_2^1$, $\psi_{14}(\xi) = a_1^1$, $\psi_{23}(\xi) = -a_2^2$, $\psi_{24}(\xi) = a_1^2$, and $\psi_0(\xi) = 1$. If $\psi_0(\xi) = 0$, then some $\psi_{\boldsymbol{\alpha}}(\xi) \neq 0$. Say, for instance, $\psi_{14}(\xi) \neq 0$. Then the 4×2 matrix representation is

$$\begin{bmatrix} 1 & 0 \\ a_1^2 & a_2^2 \\ a_1^3 & a_2^3 \\ 0 & 1 \end{bmatrix}$$

and $\psi_{12}(\xi) = a_2^2$, $\psi_{13}(\xi) = a_2^3$, $\psi_{14}(\xi) = 1$, $\psi_{23}(\xi) = a_1^2 a_2^3 - a_2^2 a_1^3$, $\psi_{24}(\xi) = a_1^2$, $\psi_{34}(\xi) = a_1^3 = 0$. Observe that on $\mathbf{V}_{Y_5}$ with coordinates $(Y_0, \ldots, Y_4, 1)$, the equation is $Y_0 - Y_1Y_4 + Y_2Y_3 = 0$; and, that on $\mathbf{V}_{Y_2}$ with coordinates $(x_0, x_1, 1, x_3, x_4, x_5)$ the equation is $x_3 - x_1x_4 + x_0x_5 = 0$. Thus on $\mathbf{V}_{Y_2} \cap \mathbf{V}_{Y_5}$ with coordinates $(z_0, z_1, 1, z_3, z_4, 1)$ the equation is $z_0 - z_1z_4 + z_3 = 0$. Now suppose that $n = 2$. Then $\lambda(n,m,p) = 18$ and we work in $\mathbb{P}^{17}$. We let $\psi_{ij}(z) = a_{ij,0} + a_{ij,1}z + a_{ij,2}z^2$ and we let the $a_{ij,\ell}$ be the coordinates in $\mathbb{P}^{17}$ ($ij = (12)$, (13), (14), (23), (24), (34)). Then the degree conditions are given by

$$\begin{aligned} a_{12,1} &= 0, \quad & a_{12,2} &= 0 \\ a_{13,2} &= 0, \quad & a_{14,2} &= 0 \\ a_{23,2} &= 0, \quad & a_{24,2} &= 0 \end{aligned}$$

and define a $\mathbb{P}^{11} \subset \mathbb{P}^{17}$. We let $(W_0, \ldots, W_{11})$ be the coordinates on $\mathbb{P}^{11}$ so that

$$\begin{aligned} &\psi_{12}(z) = W_0, \psi_{13}(z) = W_1 + W_2z, \psi_{14}(z) = W_3 + W_4z \\ &\psi_{23}(z) = W_5 + W_6z, \psi_{24}(z) = W_7 + W_8z, \psi_{34}(z) = W_9 + W_{10}z + W_{11}z^2. \end{aligned}$$

Then, the variety $\mathbf{W}(2,2,2) \cap \mathbb{P}^{11}$ is defined by the equations

$$\begin{aligned} &W_0W_9 - W_1W_7 + W_3W_5 = 0 \\ &W_0W_{11} - W_2W_8 + W_6W_4 = 0 \\ &W_0W_{10} - W_1W_8 - W_2W_7 + W_3W_6 + W_4W_5 = 0 \end{aligned} \tag{9.34}$$

and $\mathbf{V}(2,2,2) = [\mathbf{W}(2,2,2) \cap \mathbb{P}^{11}]_{W_{11}}$ has dimension 8. $\mathbf{V}(2,2,2)$ represents Rat(2, 2, 2). If $(W_0, \ldots, W_{10}, 1) \in \mathbf{V}(2,2,2)$, then the corresponding element of Rat(2, 2, 2) is

$$\mathbf{P}_0 = \begin{bmatrix} W_3 & -W_1 \\ W_7 & -W_5 \end{bmatrix}, \quad \mathbf{P}_1 = \begin{bmatrix} W_4 & -W_2 \\ W_8 & W_6 \end{bmatrix}, W_9, W_{10}$$

and the image of this element under the Laurent map is $(\mathbf{H}_1, \mathbf{H}_2, W_9, W_{10})$ with

$$\mathbf{H}_1 = \begin{bmatrix} W_4 & W_2 \\ W_8 & -W_6 \end{bmatrix}, \quad \mathbf{H}_2 = \begin{bmatrix} W_3 - W_{10}W_4 & -W_1 + W_{10}W_2 \\ W_7 - W_{10}W_8 & -W_5 + W_{10}W_6 \end{bmatrix}.$$

The Grassmann equations (9.34) become

$$\begin{aligned} W_0 &= W_2W_8 - W_4W_6 = \det \mathbf{H}_1 \\ W_0W_9 &= W_1W_7 - W_3W_5 = \det \mathbf{P}_0 = \det[\mathbf{H}_2 + W_{10}\mathbf{H}_1] \\ W_0W_{10} &= -\det[\mathbf{H}_{1,1}\ \mathbf{H}_{2,2}] + \det[\mathbf{H}_{1,2}\ \mathbf{H}_{2,1}]. \end{aligned}$$

As we saw in Example 2.43, the rank of the Hankel matrix is 2. This gives an explicit representation of the Laurent isomorphism. We shall revisit this example again and we note that we have not yet given an appropriate algebraic structure to Hank(2, 2, 2).

Example 9.35 Let $p = 2$, $m = 3$ so that we work in $\mathbb{A}_k^{n(6+1)}$. Viewing Gr(3, 5) as a subvariety of $\mathbb{P}^9$ with coordinates $(Y_0, Y_1, \ldots, Y_9)$, then, on $\mathrm{Gr}(3,5)_{Y_9}$ ($Y_9 \neq 0$), the equations are

$$\begin{aligned} Y_0 + Y_4Y_6 - Y_3Y_7 &= 0 \\ Y_1 + Y_5Y_6 - Y_3Y_8 &= 0 \\ Y_2 + Y_5Y_7 - Y_4Y_8 &= 0. \end{aligned}$$

Let $\psi_{123}, \psi_{124}, \psi_{125}, \psi_{134}, \psi_{135}, \psi_{145}, \psi_{234}, \psi_{235}, \psi_{245}, \psi_{345}$ be relatively prime forms of degree n which satisfy the (degree) conditions

$$\begin{array}{lll} \#\,\mathrm{coeff}\ \psi_{123} \leq n-1, & \#\,\mathrm{coeff}\ \psi_{124} \leq n-1, & \#\,\mathrm{coeff}\ \psi_{125} \leq n-1 \\ \#\,\mathrm{coeff}\ \psi_{134} \leq n, & \#\,\mathrm{coeff}\ \psi_{135} \leq n, & \#\,\mathrm{coeff}\ \psi_{145} \leq n \\ \#\,\mathrm{coeff}\ \psi_{234} \leq n, & \#\,\mathrm{coeff}\ \psi_{235} \leq n, & \#\,\mathrm{coeff}\ \psi_{245} \leq n \\ \#\,\mathrm{coeff}\ \psi_{345} \leq n+1. & & \end{array}$$

Let $\psi_{i_1i_2i_3}(z) = \psi_{i_1i_2i_3}(1, z)$ and set $\psi_0(z) = \psi_{345}(z)$. Since the $\psi_\alpha(z)$ are reltively prime,

$$\begin{bmatrix} \psi_{145}(z)/\psi_0(z) & -\psi_{135}(z)/\psi_0(z) & \psi_{134}(z)/\psi_0(z) \\ \psi_{245}(z)/\psi_0(z) & -\psi_{235}(z)/\psi_0(z) & \psi_{234}(z)/\psi_0(z) \end{bmatrix}$$

determines an element of $\mathrm{Rat}(n,3,2)$. Moreover, the Grassmann equations give

$$\psi_{123}(z) = [-\psi_{135}(z)\psi_{234}(z) + \psi_{134}(z)\psi_{235}(z)]/\psi_0(z)$$
$$\psi_{124}(z) = [-\psi_{145}(z)\psi_{234}(z) + \psi_{134}(z)\psi_{245}(z)]/\psi_0(z)$$
$$\psi_{125}(z) = [-\psi_{145}(z)\psi_{235}(z) + \psi_{135}(z)\psi_{245}(z)]/\psi_0(z).$$

These determine ψ_{123}, ψ_{124} and ψ_{125}. If $\psi_0(\xi) \neq 0$, then the 5×3 matrix representation is

$$\begin{bmatrix} a_1^1 & a_2^1 & a_3^1 \\ a_1^2 & a_2^2 & a_3^2 \\ 1 & 0 & 0 \\ 0 & 1 & 0 \\ 0 & 0 & 1 \end{bmatrix}$$

and $\psi_{123}(\xi) = a_2^1 a_3^2 - a_3^1 a_2^2$, $\psi_{124}(\xi) = -(a_1^1 a_3^2 - a_3^1 a_1^2)$, $\psi_{125}(\xi) = a_1^1 a_2^2 - a_2^1 a_1^2$, $\psi_{134}(\xi) = a_3^1$, $\psi_{135}(\xi) = -a_2^1$, $\psi_{145}(\xi) = a_1^1$, $\psi_{234}(\xi) = a_3^2$, $\psi_{235}(\xi) = -a_2^2$, $\psi_{245}(\xi) = a_1^2$ and $\psi_{345}(\xi) = \psi_0(\xi) = 1$. If $\psi_0(\xi) = 0$, then some $\psi_{\boldsymbol{\alpha}}(\xi) \neq 0$ and we can work on a different piece (cf. Example 2.61).

Example 9.36 Let $n = 2$, $m = 2$, $p = 2$ and consider $\mathrm{Hank}(2,2,2)$. Let $(\mathbf{H}_1, \mathbf{H}_2, \xi_1, \xi_2)$ be a typical element of $\mathrm{Hank}(2,2,2)$. Let $\boldsymbol{\alpha}_0 = (1\ 2)$ and let $\lambda_0 = \lambda_{\boldsymbol{\alpha}_{0,0}} \in \mathbb{A}^1$. Consider the variety in $\mathbb{A}^1 \times \mathbb{A}^8 \times \mathbb{A}^2$ (with coordinates $(\lambda, \mathbf{H}_1, \mathbf{H}_2, \xi_1, \xi_2)$) defined by the equation(s)

$$\det[\mathbf{H}_1 z + \xi_1 \mathbf{H}_1 + \mathbf{H}_2] = \lambda_0(\xi_1 + \xi_2 z + z^2)$$

or, by expanding the determinant, the equations

$$\lambda_0 = \det \mathbf{H}_1$$
$$\lambda_0 \xi_1 = \det[\xi_1 \mathbf{H}_1 + \mathbf{H}_2]$$
$$\lambda_0 \xi_2 = \det[(\xi_1 \mathbf{H}_1 + \mathbf{H}_2)_1\ (\mathbf{H}_1)_2] - \det[(\xi_1 \mathbf{H}_1 + \mathbf{H}_2)_2\ (\mathbf{H}_1)_1].$$

Then $\mathrm{Hank}(2,2,2)$ is an open subset of this variety. This gives an algebraic structure to $\mathrm{Hank}(2,2,2)$.

Now, we have defined an algebraic structure $\mathbf{V}(n,m,p)$ for $\mathbf{Rat}(n,m,p)$ and $\mathrm{Rat}(n,m,p)$. In order to do the same for $\mathrm{Hank}(n,m,p)$, we shall need the notion of a product for projective varieties. We shall do this in the next chapter and shall then return to the Laurent Isomorphism Theorem.

Exercises

(1) Let $p = 2$, $m = 3$. Determine all $\mathbf{F}^i_j(z)$ in terms of the $\psi_{F,\alpha}(z)$.

(2) Prove (9.14).

(3) Verify (9.20) for $p = 2$, $m = 3$ and any n.

(4) Show that $\det[\mathbf{H}_1 z + \xi_1 \mathbf{H}_1 + \mathbf{H}_2] = (\det \mathbf{H}_1)z^2 + \{\det[(\xi_1 \mathbf{H}_1 + \mathbf{H}_2)_1\ (\mathbf{H}_1)_2] - \det[(\xi_1 \mathbf{H}_1 + \mathbf{H}_2)_2\ (\mathbf{H}_1)]\}z + \det[\xi_1 \mathbf{H}_1 + \mathbf{H}_2]$.

10

Projective Algebraic Geometry III: Products, Graphs, Projections

We wish to define the product $V \times W$ where $V \subset \mathbb{P}^N$ and $W \subset \mathbb{P}^M$ are quasi projective varieties. Consider the set $V \times W = \{(v, w)\colon v \in V,\ w \in W\}$. To make this a *product*, we require the following:

(a) $V \times W$ should be *quasi-projective* i.e., there is an isomorphism $\psi\colon V \times W \to \mathbb{P}^K$ with $\psi(V \times W)$ a quasi-projective variety in $\mathbb{P}^K$;

(b) the notion should be *local* i.e., if $\xi \in V$, $\eta \in W$ then there are open affine neighborhoods U_ξ, U_η of ξ, η respectively such that $\psi(U_\xi \times U_\eta)$ is open in $\psi(U \times V)$ and $\psi(U_\xi \times U_\eta) \simeq U_\xi \times U_\eta$ (as a product of quasi-affine varieties); and,

(c) the notion is *categorical* i.e., the projections $\pi_V\colon V \times W \to V$, $\pi_W\colon V \times W \to W$ are regular, and, given Z with regular maps $\varphi_1\colon Z \to V$, $\varphi_2\colon Z \to W$, then there is a unique morphism $\varphi_1 \times \varphi_2\colon Z \to V \times W$ with $\pi_V \circ (\varphi_1 \times \varphi_2) = \varphi_1$, $\pi_W \circ (\varphi_1 \times \varphi_2) = \varphi_2$.

We shall develop the product in several ways.

Method 1 *Multi-degree* Let $F(Y_0, \ldots, Y_N; Z_0, \ldots, Z_M)$ be an element of $S = k[Y_0, \ldots, Y_N, Z_0, \ldots, Z_M]$. F is a *bi-form* of *bi-degree* (q, r) if F is homogeneous of degree q in $Y_0, \ldots, Y_N$ and homogeneous of degree r in $Z_0, \ldots, Z_M$. In other words, F is a bi-form of bi-degree (q, r) if and only if

$$F = \sum_{\substack{i_0 + \cdots i_N = q \\ j_0 + \cdots + j_M = r}} a_{\mathbf{ij}} Y_0^{i_0} \cdots Y_N^{i_N}\, Z_0^{j_0} \cdots Z_M^{j_M}. \tag{10.1}$$

P. Falb, *Methods of Algebraic Geometry in Control Theory: Part II*,
Modern Birkhäuser Classics, https://doi.org/10.1007/978-3-319-96574-1_10

We say that a set $Z \subset \mathbb{P}^N \times \mathbb{P}^M$ is *closed* if $Z = V(F_1, \ldots, F_\nu)$ where the F_i are bi-forms. This defines the Zariski topology on $\mathbb{P}^N \times \mathbb{P}^M$ with local affine pieces $(\mathbb{P}^N_{Y_i}) \times (\mathbb{P}^M_{Z_j}) \simeq \mathbb{A}^N \times \mathbb{A}^M$. The topology has a base of principal open sets $(\mathbb{P}^N \times \mathbb{P}^M)_F$ where F is a bi-form of bi-degree (q, q). This is so because if G is of bi-degree (q, r) with (say) $q \geq r$, then

$$\{(\boldsymbol{\xi}, \boldsymbol{\eta}) \colon G(\boldsymbol{\xi}, \boldsymbol{\eta}) \neq 0\} = \bigcup_{j=0}^{M} \{(\boldsymbol{\xi}, \boldsymbol{\eta}) \colon (Z_j^{q-r} G)(\boldsymbol{\xi}, \boldsymbol{\eta}) \neq 0\}.$$

We note also that if F is of bi-degree (q, q), then $F = G(Y_0 Z_0, \ldots, Y_i Z_j, \ldots)$. If $S_{q,r} = \{F \colon F \text{ a bi-form of bi-degree } (q, r) \text{ or } 0\}$, then $S = \oplus \sum S_{q,r}$ is bi-graded. If $V = V(G_1, \ldots, G_\mu)$ and $W = V(H_1, \ldots, H_\nu)$ where we may assume the G_i and H_j are forms of degree q, then $V \times W$ is $V(Z_\alpha^q G_i,\ Y_\beta^q H_j,\ G_i H_j)$ $(\alpha = 0, \ldots, M,\ \beta = 0, \ldots, N,\ i = 1, \ldots, \mu,\ j = 1, \ldots, \nu)$ and is a closed set. It is clear that this notion of a product is categorical as well.

Example 10.2 Consider $\mathbb{P}^1 \times \mathbb{P}^1$ and let $\Delta = V(Y_0 Z_1 - Y_1 Z_0)$. Then Δ is closed ($Y_0 Z_1 - Y_1 Z_0$ is a bi-form of bi-degree $(1, 1)$) and $\Delta = \{(\boldsymbol{\xi}, \boldsymbol{\eta}) \colon \boldsymbol{\xi} = \boldsymbol{\eta}\}$ is the diagonal. For, if $\boldsymbol{\xi} = \boldsymbol{\eta}$, then $\xi_0 = \lambda \eta_0$, $\xi_1 = \lambda \eta_1$, $\lambda \neq 0$ and so, $\xi_0 \lambda \eta_1 - \xi_1 \lambda \eta_0 = \lambda^2 \eta_0 \eta_1 - \lambda^2 \eta_1 \eta_0 = 0$; conversely, if $\xi_0 \eta_1 = \xi_1 \eta_0$, then (say) $\xi_0 \neq 0$ and $\xi_1 / \xi_0 = \eta_1 / \eta_0$ (if $\eta_0 = 0$, then $\eta_1 = 0$ a contradiction) so that $\boldsymbol{\xi} = \boldsymbol{\eta}$. In $\mathbb{P}^N \times \mathbb{P}^N$, $\Delta = V(Y_i Z_j - Y_j Z_i)$ is also closed.

Proposition 10.3 *Let $\varphi \colon V \to W$ be a morphism and let $\operatorname{gr}(\varphi) = \{(\boldsymbol{\xi}, \boldsymbol{\eta}) \in V \times W \colon \boldsymbol{\eta} = \varphi(\boldsymbol{\xi})\}$ be the graph of φ. Then $\operatorname{gr}(\varphi)$ is closed (cf. I.12.15).*

Proof. Consider the morphism $\psi \colon V \times W \to W \times W$ given by $\psi(\boldsymbol{\xi}, \boldsymbol{\eta}) = (\varphi(\boldsymbol{\xi}), \boldsymbol{\eta})$. Then $\operatorname{gr}(\varphi) = \psi^{-1}(\Delta(W))$ is closed.

Method 2 *The Segre Map* Let $Y_0, \ldots, Y_N$ be coordinates on $\mathbb{P}^N$, $Z_0, \ldots, Z_M$ be coordinates on $\mathbb{P}^M$ and let U_{ij}, $i = 0, 1, \ldots, N$, $j = 0, 1, \ldots, M$ be coordinates on $\mathbb{P}^{(N+1)(M+1)-1} = \mathbb{P}^{NM+M+N}$. We define a map $\psi \colon \mathbb{P}^N \times \mathbb{P}^M \to \mathbb{P}^{NM+M+N}$, called the *Segre map*, by setting

$$\psi(Y_0, \ldots, Y_N, Z_0, \ldots, Z_M) = (Y_0 Z_0, \ldots, Y_i Z_j, \ldots, Y_N Z_M) \tag{10.4}$$

i.e., $U_{ij} = Y_i Z_j$. For example, if $N = M = 1$, then

$$\psi(Y_0, Y_1, Z_0, Z_1) = (Y_0 Z_0, Y_0 Z_1, Y_1 Z_0, Y_1 Z_1).$$

We have:

Proposition 10.5 (1) *ψ is well-defined;* (2) *ψ is injective;* (3) *$\psi(\mathbb{P}^N \times \mathbb{P}^M)$ is closed in $\mathbb{P}^{NM+M+N}$ and $\psi(\mathbb{P}^N \times \mathbb{P}^M) = X = V(U_{ij} U_{k\ell} - U_{kj} U_{i\ell})$ where $i, k =$*

$0, 1, \ldots, N$, $j, \ell = 0, 1, \ldots, M$; *and* (4) ψ *is an isomorphism of* $(\mathbb{P}^N_{Y_i}) \times (\mathbb{P}^M_{Z_j})$ *onto* $X \cap (\mathbb{P}^{NM+N+M}_{U_{ij}})$ *as affine varieties.*

Proof. (i) $\psi(\lambda\boldsymbol{\xi}, \mu\boldsymbol{\eta}) = (\lambda\xi_i\mu\eta_j) = \lambda\mu(\xi_i, \eta_j)$. (2) Suppose that $\xi_i\eta_j = \lambda\xi_i'\eta_j'$ for all i, j with $\lambda \neq 0$. Then we may assume (say) $\xi_0, \eta_0 \neq 0$. Since $0 \neq \xi_0\eta_0 = \lambda\xi_0'\eta_0'$, $\xi_0', \eta_0' \neq 0$ and so $\xi_1/\xi_0 = \xi_i\eta_0/\xi_0\eta_0 = \lambda\xi_i'\eta_0'/\lambda\xi_0'\eta_0' = \xi_i'/\xi_0'$. Similarly $\eta_j/\eta_0 = \eta_j'/\eta_0'$ and so $\boldsymbol{\xi} = \boldsymbol{\xi}'$, $\boldsymbol{\eta} = \boldsymbol{\eta}'$. (3) Let X be the variety in $\mathbb{P}^{NM+N+M}$ defined by the equations

$$U_{ij}U_{k\ell} - U_{kj}U_{i\ell} = 0 \tag{10.6}$$

for $i, k = 0, 1, \ldots, N$, $j, \ell = 0, 1, \ldots, M$. We claim that $\psi(\mathbb{P}^N \times \mathbb{P}^M) = X$. Since $\xi_i\eta_j\xi_k\eta_\ell = \xi_k\eta_j\xi_i\eta_\ell$, $\psi(\mathbb{P}^N \times \mathbb{P}^M) \subset X$. If $\mathbf{x} = (x_{ij})$ is a point in X with (say) $x_{i_0j_0} \neq 0$, then set $\xi_i = x_{ij_0}/x_{i_0j_0}$ and $\eta_j = x_{i_0j}/x_{i_0j_0}$. It follows that $\psi(\boldsymbol{\xi}, \boldsymbol{\eta}) = (\xi_i\eta_j)$ where

$$\xi_i\eta_j = \frac{x_{ij_0}x_{i_0j}}{x^2_{i_0j_0}} = \frac{x_{ij}x_{i_0j_0}}{x^2_{i_0j_0}} = \lambda x_{ij}$$

with $\lambda = 1/x_{i_0j_0} \neq 0$. In other words, $\psi(\boldsymbol{\xi}, \boldsymbol{\eta}) = \mathbf{x}$ and ψ maps onto X. (4) We may assume that $i = 0$, $j = 0$. We let $y_1, \ldots, y_N$, $y_i = Y_i/Y_0$ be affine coordinates on $\mathbb{P}^N_{Y_0}$ and we let $z_1, \ldots, z_M$, $z_j = Z_j/Z_0$, be affine coordinates on $\mathbb{P}^M_{Z_0}$, and we let $u_{ij} = U_{ij}/U_{00}$, $i \geq 1$ or $j \geq 1$, be affine coordinates on $\mathbb{P}^{NM+M+N}_{U_{00}}$. Then $\psi((\xi_1, \ldots, \xi_N, \eta_1, \ldots, \eta_M) = (\xi_i, \eta_j, \xi_i\eta_j)$ i.e., $u_{ij} = \xi_i\eta_j$, $i \geq 1$, $j \geq 1$ and $u_{i0} = \xi_i$, $u_{0j} = \eta_j$. The image $\psi(\mathbb{P}^N_{Y_0}) \times \mathbb{P}^M_{Z_0}) = X \cap (\mathbb{P}^{NM+N+M}_{U_{00}}) = V(u_{ij} - u_{i0}u_{j0})$. But then the affine coordinate ring

$$k[\psi(\mathbb{P}^N_{Y_0} \times \mathbb{P}^M_{Z_0})] = k[U_{ij}]/(U_{ij} - U_{i0}U_{j0}) \simeq k[U_{i0}, U_{0j}] \quad (i, j \geq 1)$$

which is a polynomial ring. But ψ^* maps this isomorphically onto the polynomial ring $k[y_1, \ldots, y_N, z_1, \ldots, z_M]$. But $k[y_1, \ldots, y_N, z_1, \ldots, z_M]$ is $k[(\mathbb{P}^N_{Y_0}) \times (\mathbb{P}^M_{Z_0})]$ and the result follows.

Corollary 10.7 $\operatorname{Dim} \mathbb{P}^N \times \mathbb{P}^M = N + M$.

Corollary 10.8 $\mathbb{P}^N \times \mathbb{P}^M$ *is irreducible.*

Corollary 10.9 *If* V, W *are irreducible, then* $V \times W$ *is irreducible.*

Proof. Apply I.12.6 proof 2.

Corollary 10.10 *If* V, W *are irreducible, then*

$$\mathfrak{o}_{(\boldsymbol{\xi},\boldsymbol{\eta}),V\times W} = \left(\mathfrak{o}_{\boldsymbol{\xi},V} \otimes_k \mathfrak{o}_{\boldsymbol{\eta},W}\right)_{(\mathfrak{m}_{\boldsymbol{\xi}}\mathfrak{o}_{\boldsymbol{\eta},W} + \mathfrak{m}_{\boldsymbol{\eta}}\mathfrak{o}_{\boldsymbol{\xi},V})} \tag{10.11}$$

Proof. Use the fact that the local rings are defined using affine neighborhoods and apply I.12.7.

Example 10.12 Let $N = 1$, $M = 1$. Then $\psi(Y_0, Y_1, Z_0, Z_1) = (Y_0Z_0,\ Y_0Z_1,\ Y_1Z_0,\ Y_1Z_1)$ so that $U_{00} = Y_0Z_0$, $U_{01} = Y_0Z_1$, $U_{10} = Y_1Z_0$, $U_{11} = Y_1Z_1$ and $X = V(U_{00}U_{11} - U_{01}U_{10}) = \psi(\mathbb{P}^1 \times \mathbb{P}^1)$ is a quadratic in $\mathbb{P}^3$. $X \cap \mathbb{P}^3_{U_{00}} = V(u_{11} - u_{01}u_{10})$ and $k[X \cap \mathbb{P}^3_{u_{00}}] = k[u_{11}, u_{01}, u_{10}]/(u_{11} - u_{01}u_{10}) \simeq k[u_{01}, u_{10}] \simeq k[y_1, z_1] = k[\mathbb{P}^1_{Y_0} \times \mathbb{P}^1_{Z_0}]$. Thus $X \cap \mathbb{P}^3_{U_{00}} \simeq \mathbb{A}^1 \times \mathbb{A}^1 \simeq \mathbb{P}^1_{Y_0} \times \mathbb{P}^1_{Z_0}$. If we view $(U_{00}, U_{01}, U_{10}, U_{11})$ as entries in a 2×2 matrix, then the equation of X shows that the rank of the matrix is 1.

We now turn our attention to a critical result which is often called the Main Theorem of Elimination Theory.

Definition 10.13 A variety V is *complete* if, for every variety W, the projection $\pi_W : V \times W \to W$ is a closed map.

Example 10.14 $\mathbb{A}^1$ is not complete for if $Z = V(XY - 1)$ in $\mathbb{A}^1 \times \mathbb{A}^1$, then $\pi_2(Z) = \mathbb{A}^1 - \{0\}$ is not closed.

Proposition 10.15 (1) *If V is complete and V_1 is closed in V, then V_1 is complete;* (2) *if V, W are complete, then $V \times W$ is complete; and,* (3) *if V is complete and $\psi: V \to Z$ is a morphism, then $\psi(V)$ is closed in Z and $\psi(V)$ is complete.*

Proof. (1) If V_1 is closed in V, then $V_1 \times W$ is closed in $V_1 \times W$ and in $V \times W$ for any W. If Z_1 is closed in $V_1 \times W$, then Z_1 is closed in $V \times W$ and $\pi_W(Z_1)$ is closed as V is complete. (2) Since $(V \times W) \times Z = V \times (W \times Z)$ and $\pi_Z(X)$ (in $(V \times W) \times Z) = \pi_Z(\tilde{\pi}_{W \times Z}(X))$ (in $W \times Z$) where $\tilde{\pi}_{W \times Z}$ is the projection in $V \times (W \times Z)$, the assertion holds. (3) Since $\mathrm{gr}(\psi)$ is closed in $V \times Z$, $\pi_Z(\mathrm{gr}(\psi)) = \psi(V)$ is closed in Z. The rest is an easy exercise.

We now have:

Theorem 10.16 (*Main Theorem of Elimination Theory*) *Every projective variety is complete.*

Proof. By (1) of the proposition, it is enough to show that $\mathbb{P}^N$ is complete. Since closure is a local property, it is enough to show that the projection $\pi_{\mathbb{A}^M} : \mathbb{P}^N \times \mathbb{A}^M \to \mathbb{A}^M$ is closed. Let $X = V(F_1, \ldots, F_\nu)$ be closed in $\mathbb{P}^N \times \mathbb{A}^M$. The F_i are homogeneous in $Y_0, \ldots, Y_N$ of degree d_i. If $\boldsymbol{\xi} \in \mathbb{A}^M$, then $\boldsymbol{\xi} \notin \pi_{\mathbb{A}^M}(X)$ if and only if the forms $F_i(Y_0, \ldots, Y_N, \xi_1, \ldots, \xi_M)$ have no common zeros in $\mathbb{P}^N$ which holds if and only if

$$\sqrt{(F_1(Y_0, \ldots, Y_N, \boldsymbol{\xi}), \ldots, F_\nu(Y_0, \ldots, Y_N, \boldsymbol{\xi})} = \sqrt{\mathfrak{a}_{\boldsymbol{\xi}}} = (Y_0, \ldots, Y_N) \quad (10.17)$$

i.e., $(Y_0, \ldots, Y_N)^q \subset \mathfrak{a}_{\boldsymbol{\xi}}$ for some q. Thus it will suffice to show that $\{\boldsymbol{\xi}\colon \sqrt{\mathfrak{a}_{\boldsymbol{\xi}}} = (Y_0, \ldots, Y_N)\}$ is open in $\mathbb{A}^M$. Let $S_q \subset S = k[Y_0, \ldots, Y_N]$ be the k-space of homogeneous polynomials of degree q in $Y_0, \ldots, Y_N$ and consider the map $\psi_q(\boldsymbol{\xi})\colon S_{q-d_1} \oplus \cdots \oplus S_{q-d_\nu} \to S_q$ given by

$$\psi_q(\boldsymbol{\xi})(G_1, \ldots, G_\nu) = \sum G_i F_i(Y_0, \ldots, Y_N, \xi_1, \ldots, \xi_M). \quad (10.18)$$

If we fix (say) monomial bases of these spaces, then $\psi_q(\boldsymbol{\xi})$ is a linear map for each $\boldsymbol{\xi}$ and the matrix $\psi_q(\boldsymbol{\xi})$ of this map is polynomial in the ξ_i. Then $(Y_0, \ldots, Y_N)^q \subset \mathfrak{a}_{\boldsymbol{\xi}}$ if and only if $\psi_q(\boldsymbol{\xi})$ is surjective. But $\psi_q(\boldsymbol{\xi})$ will be surjective if and only if some minor of $\psi_q(\boldsymbol{\xi})$ is nonzero. It follows that $\{\boldsymbol{\xi}\colon \sqrt{\mathfrak{a}_{\boldsymbol{\xi}}} = (Y_0, \ldots, Y_N)\}$ is open. (This proof follows [M-5].)

Corollary 10.19 *If $\psi\colon V \to W$ is a morphism of projective varieties, then $\psi(V)$ is closed in W.*

Proof. Since $\mathrm{gr}(\psi)$ is closed, $\pi_W(\mathrm{gr}(\psi)) = \psi(V)$ is closed.

Corollary 10.20 *If $\pi_{\boldsymbol{\xi}}\colon \mathbb{P}^N - \{\boldsymbol{\xi}\} \to \mathbb{P}^{N-1}$ is the projection with center $\{\boldsymbol{\xi}\}$, then $\pi_{\boldsymbol{\xi}}(V)$ is closed if V is closed in $\mathbb{P}^N - \{\boldsymbol{\xi}\}$.*

Proof. Simply note that $\pi_{\boldsymbol{\xi}}$ is a morphism.

We can also deduce the theorem from Corollary 10.20 which we prove independently. More precisely, we have (following [H-2]):

Proof 2. We proceed by induction on N. For $N = 1$, consider Z closed in $\mathbb{P}^1 \times \mathbb{A}^M$ with $Z = V(F_1, \ldots, F_\nu)$ where $F_i(Y_0, Y_1, X_1, \ldots, X_M)$ is homogeneous in Y_0, Y_1 of degree q. If we consider the resultant $\rho(\sum s_i F_i(Y_0, 1; X_1, \ldots, X_M), \sum t_j F_j(Y_0, 1; X_1, \ldots, X_M)) = \sum R_{\alpha\beta}(X_1, \ldots, X_M) s_i^\alpha t_j^\beta$, then $\pi(Z) = \cap V(R_{\alpha\beta})$ is closed in $\mathbb{A}^M$. So assume the result for $N - 1$. Let $\boldsymbol{\xi}_\infty$ be a point in $\mathbb{P}^N$ and let $\pi_\infty\colon \mathbb{P}^N - \{\boldsymbol{\xi}_\infty\} \to \mathbb{P}^{N-1}$ be the projection. Then, *assuming* the Lemma 10.22 which follows, we consider

$$\mathbb{P}^N \times W \xrightarrow{\pi_\infty} \mathbb{P}^{N-1} \times W \xrightarrow{\pi_W^{N-1}} W. \quad (10.21)$$

By induction, the map π_W^{N-1} is closed. Suppose that Z is closed in $\mathbb{P}^N \times W$ and let $Z_\infty = \{w\colon (\boldsymbol{\xi}_\infty, w) \in Z\} = Z \cap \{\boldsymbol{\xi}_\infty\} \times W$. Then Z_∞ is closed. For, if

$G_\alpha(Y_0, \ldots, Y_N, Z_1, \ldots, Z_M)$ define Z, then $G_\alpha(\boldsymbol{\xi}_\infty, Z_1, \ldots, Z_M)$ define Z_∞ i.e., if $w \in W$ and $(\boldsymbol{\xi}_\infty, w) \in Z$, then $G_\alpha(\boldsymbol{\xi}_\infty, w) = 0$ and conversely. If $G_\alpha(\boldsymbol{\xi}_\infty, w) = 0$, then $(\boldsymbol{\xi}_\infty, w) \in Z$ and $w \in Z_\infty$. Then $W - Z_\infty$ is open in W. But $(\pi_W^{N-1} \circ \pi_\infty)(Z) \cap W - Z_\infty$ is closed in $W - Z_\infty$ and $Z_\infty \subset (\pi_W^{N-1} \circ \pi_\infty)(Z)$ so that $(\pi_W^{N-1} \circ \pi_\infty)(Z) = (\pi_W^{N-1} \circ \pi_\infty)(Z) \cap (W - Z_\infty) \cup (\pi_W^{N-1} \circ \pi_\infty)(Z) \cap Z_\infty$ is closed.

Lemma 10.22 *Let $\boldsymbol{\xi}_\infty = (0, \ldots, 0, 1)$ and let $\pi_\infty \colon \mathbb{P}^N - \{\boldsymbol{\xi}_\infty\} \to \mathbb{P}^{N-1}$ be the projection so that $\pi_\infty(\xi_0, \ldots, \xi_N) = (\xi_0, \ldots, \xi_{N-1})$. If $V \subset \mathbb{P}^N$ is an algebraic set with $\boldsymbol{\xi}_\infty \notin V$, then $\pi_\infty(V) = \tilde{V}$ is an algebraic set i.e., is closed.*

Proof. Let $R = k[Y_0, \ldots, Y_{N-1}]$. Then R is a UFD. If F_1, F_2 are forms in $S = k[Y_0, \ldots, Y_N] = R[Y_N]$, then the resultant, $\mathrm{Res}(F_1, F_2)$, of F_1, F_2 as elements of $R[Y_N]$ is a well-defined form in R (I.3.9ff) and $\mathrm{Res}(F_1, F_2)(\xi_0, \ldots, \xi_{N-1}) = 0$ if and only if either $F_1(\xi_0, \ldots, \xi_{N-1}, Y_N)$ and $F_2(\xi_0, \ldots, \xi_{N-1}, Y_N)$ have a common zero or the leading coefficients of both F_1 and F_2 vanish. Let $\mathfrak{a} = I_h(V)$ and let $\rho(\mathfrak{a}) = \{\mathrm{Res}(F_1, F_2)\colon F_1, F_2 \in \mathfrak{a}\}$. Then $V(\rho(\mathfrak{a}))$ is an algebraic set in $\mathbb{P}^{N-1}$ and we claim that $\tilde{V} = V(\rho(\mathfrak{a}))$. Let $\boldsymbol{\eta} = (\eta_0, \ldots, \eta_{N-1}, 0) \in \mathbb{P}^{N-1}$ and consider the line $\ell = J(\boldsymbol{\eta}, \boldsymbol{\xi}_\infty) = \{(\lambda\eta_0, \ldots, \lambda\eta_{N-1}, \mu)\colon \lambda, \mu \text{ not both } 0\}$. If ℓ meets V at $\boldsymbol{\xi} = (\lambda\boldsymbol{\eta}, \mu)$, then any forms F_1, F_2 in $\mathfrak{a} = I_h(V)$ vanish at $\boldsymbol{\xi}$ and hence $\mathrm{Res}(F_1, F_2)(\lambda\boldsymbol{\eta}) = 0$ so that $\boldsymbol{\eta} \in \tilde{V}$. If ℓ does not meet V, then there is an $F_1 \in \mathfrak{a}$ such that $\ell \cap V(F_1)$ is a finite set of points. Since none of these points is in V, there is an $F_2 \in \mathfrak{a}$ which does not vanish at any of these points. For example, say $\ell \cap V(F_1) = \{\boldsymbol{\xi}_1, \boldsymbol{\xi}_2\}$ and let $F^1 \in \mathfrak{a}$ with $F^1(\boldsymbol{\xi}_1) \neq 0$. If $F^1(\boldsymbol{\xi}_2) \neq 0$, then done so assume $F^1(\boldsymbol{\xi}_2) = 0$. Let $F^2 \in \mathfrak{a}$ with $F^2(\boldsymbol{\xi}_2) \neq 0$ and (as before) may assume $F^2(\boldsymbol{\xi}_1) = 0$. But then $F^1 + F^2 \in \mathfrak{a}$ and $(F^1 + F^2)(\boldsymbol{\xi}_i) \neq 0$. In other words, $\boldsymbol{\eta} \notin V(\rho(\mathfrak{a}))$.

Corollary 10.23 *Let $\Lambda_\infty \simeq \mathbb{P}^r$ and let $\pi_\infty \colon \mathbb{P}^N - \Lambda_\infty \to \mathbb{P}^{N-r-1}$ be the projection. If $V \subset \mathbb{P}^N - \Lambda_\infty$ is an algebraic set, then $\pi_\infty(V) = \tilde{V}$ is also an algebraic set.*

Most of the basic properties of morphisms remain true for projective (and quasi-projective) varieties. However, the fact that globally regular functions are constant means that rational functions play a more important role than regular functions in the projective case. We gather now some simple propositions which mimic the affine situation.

Proposition 10.24 *Let $\psi \colon V \to W$ be a morphism. Then $\mathrm{gr}(\psi) \simeq V$.*

Proof. Let $\tilde{\psi} \colon V \to \mathrm{gr}(\psi)$ be given by $\tilde{\psi}(\boldsymbol{\xi}) = (\boldsymbol{\xi}, \psi(\boldsymbol{\xi}))$. Then $\tilde{\psi}$ is bijective and if $\tilde{\pi} = \pi_V \mid \mathrm{gr}(\psi)$, then $\tilde{\pi}$ is bijective and $(\tilde{\pi} \circ \tilde{\psi})(\boldsymbol{\xi}) = \boldsymbol{\xi}$ and $(\tilde{\psi} \circ \tilde{\pi})(\boldsymbol{\xi}, \psi(\boldsymbol{\xi})) = (\boldsymbol{\xi}, \psi(\boldsymbol{\xi}))$.

Proposition 10.25 *Let $\varphi\colon V \to W$, $\psi\colon V \to W$ be morphisms. Then the set $\{\boldsymbol{\xi}\colon \varphi(\boldsymbol{\xi}) = \psi(\boldsymbol{\xi})\}$ is closed.*

Proof. Since closure is a local property, we may apply Proposition I.12.13.

Corollary 10.26 *If $\varphi = \psi$ on a dense set, then $\varphi = \psi$.*

Corollary 10.27 *If $\psi\colon V \to W$ is a morphism, then there is an injective morphism i and a surjective morphism π with $\varphi = \pi \circ i$.*

Proposition 10.28 *Let V be a quasi-projective variety and let $\psi\colon V \to \mathbb{P}^M$ be a morphism. If U is a constructible subset of V (I.18.18), then $\psi(U)$ is constructible.*

Proof. We shall reduce to where we can apply Theorem I.18.17. First we *claim* that $\psi(U)$ contains an open set $\mathcal{O}$ of $\overline{\psi(U)}$. If this is so, then $U_1 = U \cap (V - \psi^{-1}(\mathcal{O}))$ is constructible and $\psi(U) = \psi(U_1) \cup \psi(U \cap \psi^{-1}(\mathcal{O}))$. If $\psi(U_1)$ is constructible, then $\psi(U \cap \psi^{-1}(\mathcal{O})) = \mathcal{O}$ so that $\psi(U)$ would be constructible. Continuing we get a chain of closed sets $U_1 > \cdots > U_t$ which eventually stops and the result would follow. We may replace U by an open subset which can be assumed affine and we may also assume that the range is affine contained in $\mathbb{A}^M$. Since $V \simeq \mathrm{gr}(\psi)$, we may suppose that ψ is the restriction of a projection. Finally, since $\mathbb{A}^M - \overline{\psi(U)}$ is closed, we may assume ψ is dominant.

Example 10.29 Let $M_0, \ldots, M_N$ be the monomials of degree q in $Y_0, \ldots, Y_m$ so that $N = \binom{m+q}{q} - 1$ and consider the space $\mathbb{P}^N$ with coordinates $p_{i_0 \cdots i_m}$ with $i_0 + \cdots + i_m = q$. This space represents all forms of degree q. We claim that the set of irreducible polynomials is open, or, equivalently, the set of reducible forms is closed. For example, if $q = 2$ and $m = 2$, then a plane conic $\sum_{i,j=0}^{2} a_{ij} Y_i Y_j$ is reducible if and only if $\det(a_{ij}) = 0$. To establish the claim in general, let $V = \{\boldsymbol{\xi} \in \mathbb{P}^N\colon F_{\boldsymbol{\xi}} \text{ is reducible}\}$ and let $V_j \subset V$ be the set $\{\boldsymbol{\xi} \in \mathbb{P}^N\colon F_{\boldsymbol{\xi}} = G_{j,\boldsymbol{\xi}} G_{q-j,\boldsymbol{\xi}},$ $G_{j,\boldsymbol{\xi}}$ of degree j, $G_{q-j,\boldsymbol{\xi}}$ of degree $q-j\}$ for $j = 1, \ldots, q-1$. Since $V = \cup V_j$, we need only show that each V_j is closed. Let $N_j = \binom{m+j}{j} - 1$, $N_{q-j} = \binom{m+q-j}{q-j} - 1$. Then the map $\psi\colon \mathbb{P}^{N_j} \times \mathbb{P}^{N_{q-j}} \to \mathbb{P}^N$ given by $\psi(G_j, G_{q-j}) = G_j G_{q-j}$ is a morphism and so by Theorem 10.16, $\psi(\mathbb{P}^{N_j} \times \mathbb{P}^{N_{q-j}}) = V_j$ is closed.

We shall give some further illustrations of the uses of Theorem 10.16 in Chapter 12.

Exercises

(1) Complete the proof of (3), Proposition 10.15.

(2) Show that the closed sets in $\mathbb{P}^N \times \mathbb{A}^M$ are of the form $V(F_1, \ldots, F_\nu)$ where $F_i(Y_0, \ldots, Y_N, X_1, \ldots, X_M)$ is homogeneous in $Y_0, \ldots, Y_N$.

(3) Prove Proposition 10.28 using Theorem I.18.19.

11

The Laurent Isomorphism Theorem: II

Now we wish to give an appropriate algebraic structure to $\text{Hank}(n,m,p)$. One approach would be to consider the image of $\mathbf{V}(n,m,p)$ under the Laurent map, which by Theorem 10.16, would be a quasi-projective variety and then to show the image is bijective to $\text{Hank}(n,m,p)$. We shall use a different approach here.

Let

$$\begin{aligned}\tau(n,m,p) &= \sum_{j=2}^{p}(n+1-j)\binom{p}{j}\binom{m}{m-j} \\ &= \gamma(n,m,p) - n(pm+1) - 1 \end{aligned} \tag{11.1}$$

and let

$$X = \mathbb{A}^{\tau(n,m,p)} \times \mathbb{P}^n \times M((n+1)p,(n+1)m) \tag{11.2}$$

with coordinates

$$\mathbf{x} = (\lambda_{\boldsymbol{\alpha}_s,0},\ldots,\lambda_{\boldsymbol{\alpha}_s,n+s-m},\xi_0,\ldots,\xi_n,\mathbf{M}) \tag{11.3}$$

where $\mathbf{M} = (\mathbf{M}_j^i)$, $i,j = 1,\ldots,n+1$ with each $\mathbf{M}_j^i$ a $p\times m$ block, and

$$\boldsymbol{\alpha}_s = (i_1,\ldots,i_t,p+j_1,\ldots,p+j_s) \tag{11.4}$$

$s = m-p,\ldots,m-2$, $s+t=m$, $1\le i_1<\cdots<i_t\le p$, $1\le j_1<\cdots<j_s\le m$. The $\lambda_{\boldsymbol{\alpha}_s,j}$ represent (ultimately) the coefficients of minors which are 2×2, 3×3, up to $p\times p$. We consider the following sets of equations:

(a) *Block Symmetry*

$$\mathbf{M}_j^i - \mathbf{M}_i^j = 0, \quad i,j = 1,\ldots,n+1 \tag{11.5}$$

P. Falb, *Methods of Algebraic Geometry in Control Theory: Part II*,
Modern Birkhäuser Classics, https://doi.org/10.1007/978-3-319-96574-1_11

or, equivalently,

$$(\mathbf{M}_j^i)_r^q - (\mathbf{M}_i^j)_r^q = 0, \quad i,j = 1,\dots,n+1, q = 1,\dots,p, r = 1,\dots,m. \tag{11.6}$$

There are $n(n+1)pm/2$ such equations.

(b) *Hankel Structure*

$$\mathbf{M}_{j+1}^i - \mathbf{M}_j^{i+1} = 0, \quad i = 1,\dots,j-1,\ j = 2,\dots,n \tag{11.7}$$

or, equivalently,

$$(\mathbf{M}_{j+1}^i)_r^q - (\mathbf{M}_j^{i+1})_r^q = 0, \quad i = 1,\dots,j-1,\ j = 2,\dots,n,\ q = 1,\dots,p,\ r = 1,\dots,m. \tag{11.8}$$

There are $n(n-1)pm/2$ such equations.

(c) *Dependence*

$$\xi_n \mathbf{M}_{n+1}^i + \sum_{\ell=0}^{n-1} \xi_\ell \mathbf{M}_{\ell+1}^i = 0, \quad i = 1,\dots,n+1 \tag{11.9}$$

or, equivalently,

$$\xi_n (\mathbf{M}_{n+1}^i)_r^q + \sum_{\ell=0}^{n-1} \xi_\ell (\mathbf{M}_{\ell+1}^i)_r^q = 0,\ i = 1,\dots,n+1,\ q = 1,\dots,p,\ r = 1,\dots,m. \tag{11.10}$$

There are $(n+1)pm$ such equations.

The variety $\mathcal{L}(n+1;m,p)$, defined by these equations in X has dimension $\gamma(n,m,p)-1$ since $\tau(n,m,p)+n+(n+1)^2pm - \left[\frac{n(n+1)}{2} + \frac{n(n-1)}{2} + n+1\right]pm = \tau(n,m,p)+n(pm+1)$.

Example 11.11 Let $C_s^r = \binom{r}{s} = r!/s!(r-s)!$. If $m = p = 1$, then $\gamma(n,1,1) = (n+1)+n = 2n+1$ and $\dim \mathcal{L}(n+1,1,1) = 2n$. If $m = 2$, $p = 2$, then $\gamma(n,2,2) = (n+1)C_0^2C_2^2 + nC_1^2C_1^2 + (n-1)C_2^2C_0^2 = (n+1)+4n-1 = 6n$ and $\dim \mathcal{L}(n+1,2,2) = 6n-1$ (cf. 2.47). If $p = 2$, $m = 3$ then $\gamma(n,3,2) = (n+1)C_0^2C_3^3 + nC_1^2C_2^3 + (n-1)C_2^2C_1^3 = (n+1)+6n+3(n-1) = 10n-2$ and $\dim \mathcal{L}(n+1,3,2) = 10n-3$ (cf. 2.61). When $m = 2$, $p = 2$, $n(m+p) = 4n = \dim \mathcal{L}(n+1,2,2) - (2n-1)$. When $m = 3$, $p = 2$, $n(m+p) = 5n = \dim \mathcal{L}(n+1,3,2) - (5n-3)$.

Let $t+s=m$ and let

$$\boldsymbol{\alpha}_s(\mathbf{ij}) = (i_1, \dots, i_t, p+j_1, \dots, p+j_s) \tag{11.12}$$

$\mathbf{i}=(i_1,\dots,i_t)$, $\mathbf{j}=(j_1,\dots,j_s)$, $1 \le i_1 < \cdots < i_t \le p$, $1 \le j_1 < \cdots < j_s \le m$. Let

$$\begin{aligned} \psi_0(x_0,x_1) &= \xi_0 x_0^n + \cdots + \xi_n x_1^n \\ \psi_{\boldsymbol{\alpha}_s(\mathbf{i},\mathbf{j})}(x_0,x_1) &= \lambda_{\boldsymbol{\alpha}_s(\mathbf{i},\mathbf{j}),0} x_0^n \\ &\quad + \cdots + \lambda_{\boldsymbol{\alpha}_s(\mathbf{i},\mathbf{j}),n+s-m} x_0^{m-s} x_1^{n+s-m} \end{aligned} \tag{11.13}$$

for $s=m-p,\dots,m-2$, or, equivalently,

$$\begin{aligned} \psi_0(x_0,x_1) &= \xi_0 x_0^n + \cdots + \xi_n x_1^n \\ \psi_{\boldsymbol{\beta}_t(\mathbf{i},\mathbf{j})}(x_0,x_1) &= \lambda_{\boldsymbol{\alpha}_{m-t}(\mathbf{i},\mathbf{j}),0} x_0^n \\ &\quad + \cdots + \lambda_{\boldsymbol{\alpha}_{m-t}(\mathbf{i},\mathbf{j}),n-t} x_0^t x_1^{n-t} \end{aligned} \tag{11.14}$$

for $t=2,\dots,p$ (and $\boldsymbol{\beta}_t(\mathbf{i},\mathbf{j}) = \boldsymbol{\alpha}_{m-t}(\mathbf{i},\mathbf{j})$). If $z = x_1/x_0$, ($x_0 \neq 0$), then we have (abuse of notation)

$$\begin{aligned} \psi_0(z) &= \xi_0 + \xi_1 z + \cdots + \xi_n z^n \\ \psi_{\boldsymbol{\alpha}_s(\mathbf{i},\mathbf{j})}(z) &= \lambda_{\boldsymbol{\alpha}_s(\mathbf{i},\mathbf{j}),0} + \lambda_{\boldsymbol{\alpha}_s(\mathbf{i},\mathbf{j}),1} z \\ &\quad + \cdots + \lambda_{\boldsymbol{\alpha}_s(\mathbf{i},\mathbf{j}),n+s-m} z^{n+s-m}. \end{aligned} \tag{11.15}$$

Let

$$\Lambda(\mathbf{M})(z) = \sum_{r=0}^{n-1} \left[\sum_{\ell=r+1}^{n} \xi_\ell \mathbf{M}_{\ell-r}^1 \right] z^r. \tag{11.16}$$

If $(\ell_1,\dots,\ell_t) = \mathbf{m} - \{j_1,\dots,j_s\}$ ($\mathbf{m}=(1,\dots,m)$), then we set

$$\Lambda_{\boldsymbol{\alpha}_s(\mathbf{i},\mathbf{j})}(\mathbf{M})(z) = \det\left[\Lambda(\mathbf{M})(z)\Big|_{\ell_1\cdots\ell_t}^{i_1\cdots i_t}\right] \tag{11.17}$$

and we have:

(d) *Grassmann*

$$\Lambda_{\boldsymbol{\alpha}_s(\mathbf{i},\mathbf{j})}(\mathbf{M})(z) = \psi_0^{m-s-1}(z)\psi_{\boldsymbol{\alpha}_s(\mathbf{i},\mathbf{j})}(z) \tag{11.18}$$

for $s=m-p,\dots,m-2$, or, for $t=m-s$,

$$\Lambda_{\boldsymbol{\alpha}_t(\mathbf{i},\mathbf{j})}(\mathbf{M})(z) = \psi_0^{t-1}(z)\psi_{\boldsymbol{\alpha}_t(\mathbf{i},\mathbf{j})}(z) \tag{11.19}$$

for $t=2,\dots,p$.

The variety defined by the equations (d) in X is denoted by $V_{\text{Gr}}(n,m,p)$. We observe that if $(\boldsymbol{\lambda}, \boldsymbol{\xi}, \mathbf{M})$ is an element of $\mathcal{L}(n+1,m,p) \cap V_{\text{Gr}}(n,m,p)$, then the rank of $\mathbf{M}$, $\rho(\mathbf{M}) \leq n$ (e.g. by the Lemma of Risannen 3.42).

Definition 11.20 The open set $\mathcal{H}(n,m,p)$ in X given by

$$\mathcal{H}(n,m,p) = \{(\boldsymbol{\lambda}, \boldsymbol{\xi}, \mathbf{M}) \in \mathcal{L}(n+1,m,p) \cap V_{\text{Gr}}(n,m,p) \colon \rho(\mathbf{M}) = n\}$$

is called the *Hankel variety.*

Suppose that $\mathbf{x} = (\mathbf{H}_1, \dots, \mathbf{H}_n, \xi_0, \dots, \xi_{n-1})$ is an element of $\mathbb{A}^{n(pm+1)}$ (not necessarily in $\text{Hank}(n,m,p)$) and define a morphism $\psi \colon \mathbb{A}^{n(pm+1)} \to X$ by setting

$$\psi(\mathbf{x}) = (\boldsymbol{\lambda}(\mathbf{x}), \boldsymbol{\xi}(\mathbf{x}), \mathbf{M}(\mathbf{x})) \tag{11.21}$$

where

$$\begin{aligned} \xi_{i+1}(\mathbf{x}) &= \xi_i, & i &= 0,1,\dots,n-1 \\ \mathbf{M}_j^i(\mathbf{x}) &= (\mathbf{H}_{i+j-1}), & i,j &= 1,\dots,n+1 \end{aligned} \tag{11.22}$$

with $\mathbf{H}_{n+\sigma} = -\xi_0 \mathbf{H}_\sigma - \cdots - \xi_{n-1}\mathbf{H}_{n+\sigma-1}$ for $\sigma = 1, \dots,$ and

$$\lambda_{\boldsymbol{\alpha}_s(\mathbf{i},\mathbf{j})}(\mathbf{x}) + \cdots + \lambda_{\boldsymbol{\alpha}_s(\mathbf{i},\mathbf{j})}(\mathbf{x}) z^{n+s-m} = \det\left[\Lambda(\mathbf{x})(z)\Big|_{\ell_1 \cdots \ell_t}^{i_1 \cdots i_t}\right] \tag{11.23}$$

where

$$\Lambda(\mathbf{x})(z) = \sum_{r=0}^{n-1} \left[\sum_{\ell=r+1}^{n} \xi_\ell \mathbf{H}_{\ell-r} \right] z^r \tag{11.24}$$

and the $\boldsymbol{\alpha}_s(\mathbf{i},\mathbf{j})$ are as in (11.12). We note immediately that $\psi(\mathbb{A}^{n(pm+1)}) \subset \mathcal{L}(n+1,m,p)$. If $\psi(\mathbf{x}) = \psi(\tilde{\mathbf{x}})$, then $\xi_i = \xi_{i+1}(\mathbf{x}) = \xi_{i+1}(\tilde{\mathbf{x}}) = \tilde{\xi}_i$ and $\mathbf{H}_j = \mathbf{M}_j^1(\mathbf{x}) = \mathbf{M}_j^1(\tilde{\mathbf{x}}) = \tilde{\mathbf{H}}_j$ so that ψ is injective. Now let us define a morphism $\varphi \colon \mathcal{H}(n,m,p) \to \mathbb{A}^{n(pm+1)}$ by setting

$$\varphi(\mathbf{y}) = (\mathbf{M}_1^1, \dots, \mathbf{M}_n^1, \xi_0, \dots, \xi_{n-1}) \tag{11.25}$$

where $\mathbf{y} = (\boldsymbol{\lambda}, \boldsymbol{\xi}, \mathbf{M}) \in \mathcal{H}(n,m,p)$ and $\xi_i(\mathbf{y}) = \xi_{i+1}$ for $i = 0,1,\dots,n-1$. Clearly $\varphi(\mathbf{y})$ is an element of $\text{Hank}(n,m,p)$ so that $\varphi(\mathcal{H}(n,m,p)) \subset \text{Hank}(n,m,p)$. Since $\psi(\varphi(\mathbf{y})) = \mathbf{y}$, the map $\psi \circ \varphi \colon \mathcal{H}(n,m,p) \to \mathcal{H}(n,m,p)$ is the identity. Now if $\mathbf{x} = (\mathbf{H}_1, \dots, \mathbf{H}_n,\ \xi_0, \dots, \xi_{n-1})$ is an element of $\text{Hank}(n,m,p)$, then

$$\psi_L^{-1}(\mathbf{x}) = (P_0(\mathbf{x}), \dots, P_{n-1}(\mathbf{x}), \xi_0, \dots, \xi_{n-1}) \tag{11.26}$$

where

$$P_r(\mathbf{x}) = \sum_{\ell=r+1}^{n} \xi_\ell \mathbf{H}_{\ell-r}, \quad r = 0,1,\dots,n-1 \tag{11.27}$$

(with $\xi_n = 1$) is an element of $\text{Rat}(n,m,p)$. But then $\psi_L^{-1}(\mathbf{x})$ satisfies the Grassmann equations (d) for

$$\Lambda(\psi_L^{-1}(\mathbf{x}))(z) = \sum_{r=0}^{n-1} \mathbf{P}_r(\mathbf{x}) z^r = \sum_{r=0}^{n} \left[\sum_{\ell=r+1}^{n} \xi_\ell \mathbf{H}_{\ell-r} \right] z^r$$
$$= \Lambda(\mathbf{x})(z). \tag{11.28}$$

In other words, if $\mathbf{x} \in \text{Hank}(n,m,p)$, then the Grassmann equations (d) are satisfied by $\Lambda(\mathbf{x})(z)$. Thus, if $\mathbf{x} \in \text{Hank}(n,m,p)$, then $\psi(\mathbf{x}) \in \mathcal{H}(n,m,p)$ i.e., $\psi(\text{Hank}(n,m,p)) \subset \mathcal{H}(n,m,p)$ and, in fact, ψ is surjective. As φ is a morphism, $\varphi(\mathcal{H}(n,m,p)) = \text{Hank}(n,m,p)$ is quasi-projective. In other words, we have proven:

Theorem 11.29 $\text{Hank}(n,m,p)$ *and* $\mathcal{H}(n,m,p)$ *are isomorphic.*

This gives $\text{Hank}(n,m,p)$ an algebraic structure and we have:

Theorem 11.30 (Laurent Isomorphism) $\text{Rat}(n,m,p)$ *and* $\text{Hank}(n,m,p)$ *are isomorphic.*

Proof. Let $\mathbf{y} = (\mathbf{P}_0, \dots, \mathbf{P}_{n-1}, \xi_0, \dots, \xi_{n-1})$ be an element of $\text{Rat}(n,m,p)$ and let

$$\Lambda(\mathbf{y})(z) = \sum_{r=0}^{n-1} \mathbf{P}_r z^r$$
$$\psi_{\mathbf{y},\boldsymbol{\alpha}_s(\mathbf{i},\mathbf{j})}(z) = \det\left[\Lambda(\mathbf{y})(z)\Big|_{\ell_1\cdots\ell_t}^{i_1\cdots i_t}\right]$$

where $\boldsymbol{\alpha}_s(\mathbf{i},\mathbf{j}) = (i_1,\dots,i_t, p+j_1,\dots,p+j_s)$, $(\ell_1,\dots,\ell_t) = \mathbf{m} - \{j_1,\dots,j_s\}$, etc. Since $\mathbf{y} \in \text{Rat}(n,m,p)$, the $\psi_{\mathbf{y},\boldsymbol{\alpha}_s(\mathbf{i},\mathbf{j})}(z)$ satisfy the degree conditions, the relatively prime conditions, and the Grassmann equations. We define the Laurent map as the restriction to $\text{Rat}(n,m,p)$ of the isomorphism $\psi_L \colon \mathbb{A}^{n(mp+1)} \to \mathbb{A}^{n(mp+1)}$ so that $\psi_L(\mathbf{y}) = (\mathbf{H}_1(\mathbf{y}), \dots, \mathbf{H}_n(\mathbf{y}), \xi_0, \dots, \xi_{n-1})$ where

$$\mathbf{H}_1(\mathbf{y}) = \mathbf{P}_{n-1}, \ \mathbf{H}_t(\mathbf{y}) = \mathbf{P}_{n-t} - \sum_{j=2}^{t} \xi_{n+1-j} \mathbf{H}_{t+1-j}(\mathbf{y})$$

$t = 2, \dots, n$ and $\xi_n = 1$, and

$$\mathbf{H}_{n+j}(\mathbf{y}) + \sum_{\ell=0}^{n-1} \xi_\ell \mathbf{H}_{\ell+j}(\mathbf{y}) = 0, \quad j = 1, \dots .$$

If $\mathbf{x} = \psi_1(\mathbf{y})$, then $\mathbf{x} \in \text{Hank}(n, m, p)$ and $\Lambda(\mathbf{x})(z) = \Lambda(\mathbf{y})(z)$ so that all the relevant conditions are satisfied. Thus, $\psi_L(\text{Rat}(n, m, p)) \subset \text{Hank}(n, m, p)$. Similarly, $\psi_L^{-1}(\text{Hank}(n, m, p)) \subset \text{Rat}(n, m, p)$ and the theorem is established.

Example 11.31 (Example 9.33 revisited) Let $p = 2$, $m = 2$, $n = 2$. First, $\text{Rat}(2,2,2) = \mathbf{V}(2,2,2)$. We have $\nu(2,2,2) = \binom{4}{2} \cdot 3 - 1 = 17$ and the $\boldsymbol{\alpha}$ are (12), (13), (14), (23), (24), (35). The coordinates on $\mathbb{P}^{17}$ are $(x_{12,0}, x_{12,1}, x_{12,2}, x_{13,0}, \dots, x_{34,2})$ and $L_{\gamma(2,2,2)} = L_{12}$ is defined by the six equations

$$\begin{aligned} x_{12,1} = 0, &\quad x_{12,2} = 0 \\ x_{13,2} = 0, &\quad x_{23,2} = 0 \\ x_{14,2} = 0, &\quad x_{24,2} = 0. \end{aligned}$$

The variety $\mathbf{W}(2,2,2)$ is given by the five equations (writing x_{120} in place of $x_{12,0}$ etc.)

$$\begin{aligned} &x_{120}x_{340} - x_{130}x_{240} + x_{140}x_{230} = 0 \\ &x_{120}x_{341} + x_{121}x_{340} - x_{130}x_{241} - x_{131}x_{240} + x_{140}x_{231} + x_{141}x_{230} = 0 \\ &x_{120}x_{342} + x_{121}x_{341} + x_{122}x_{340} - x_{130}x_{242} - x_{131}x_{241} - x_{132}x_{240} \\ &\qquad + x_{140}x_{232} + x_{141}x_{231} + x_{142}x_{230} = 0 \\ &x_{121}x_{342} + x_{122}x_{341} - x_{131}x_{242} - x_{132}x_{242} + x_{141}x_{232} + x_{142}x_{231} = 0 \\ &x_{122}x_{342} - x_{132}x_{242} + x_{142}x_{232} = 0 \end{aligned}$$

and $\mathbf{W}(2,2,2) \cap L_{12}$ is given by the three equations

$$\begin{aligned} &x_{120}x_{340} - x_{130}x_{240} + x_{140}x_{230} = 0 \\ &x_{120}x_{341} - x_{130}x_{241} - x_{131}x_{240} + x_{140}x_{231} + x_{141}x_{230} = 0 \\ &x_{120}x_{342} - x_{131}x_{241} + x_{141}x_{231} = 0. \end{aligned}$$

The polynomials $\psi_{12}(z) = x_{120}$, $\psi_{13}(z) = x_{130} + x_{131}z$, $\psi_{14}(z) = x_{140} + x_{141}z$, $\psi_{23}(z) = x_{230} + x_{231}z$, $\psi_{24}(z) = x_{240} + x_{241}z$, $\psi_{34}(z) = x_{340} + x_{341}z + x_{342}z^2$ will be relatively prime if and only if $\text{Res}(s_0\psi_{12} + \cdots + s_5\psi_{34},\ t_0\psi_{12} + \cdots + t_5\psi_{34}) \equiv 0$ as a polynomial in $s_0, \dots, s_5$, $t_0, \dots, t_5$. If $\text{Res}(s_0\psi_{12} + \cdots + s_r\psi_{34},\ t_0\psi_{12} + \cdots + t_5\psi_{34}) = \sum R_{\alpha\beta}(x_{120}, \dots, x_{342}) s_{\rm i}^{\alpha} t_{\rm j}^{\beta}$ and if $\mathbf{X}(2,2,2) = \cap V(R_{\alpha\beta})$, then $\mathbf{V}(2,2,2) = \mathbf{W}(2,2,2) \cap L_{12} \cap (\mathbb{P}^{17} - \mathbf{X}(2,2,2))$. More, conveniently, if we identify L_{12} with $\mathbb{P}^{11}$, then $\text{Rat}(2,2,2) \simeq \mathbf{V}(2,2,2) = \mathbf{W}(2,2,2) \cap (\mathbb{P}^{11} - \mathbf{X}(2,2,2))$. For simplicity, we shall suppose $x_{120} \neq 0$ (this amounts to having $\det \mathbf{P}(z) \neq 0$) and we shall work on the open subset $\mathbf{W}(2,2,2) \cap \mathbb{P}^{11}_{x_{120}} \cap \mathbb{P}^{11}_{x_{312}}$ of $\mathbf{V}(2,2,2)$ and we shall take $x_{342} = 1$. Thus we consider the points $(x, x_{130}, x_{131}, x_{140}, x_{141}, x_{230}, x_{231}, x_{240}, x_{241}, \xi_0, \xi_1)$ in $\mathbb{A}^{11}_x$ $(x \neq 0)$ with

$$\left.\begin{aligned} &x\xi_0 - x_{130}x_{240} + x_{140}x_{230} = 0 \\ &x\xi_1 - x_{130}x_{241} - x_{131}x_{240} + x_{140}x_{231} + x_{141}x_{230} = 0 \\ &x - x_{131}x_{241} + x_{141}x_{231} = 0. \end{aligned}\right\} \tag{11.32}$$

We shall, by abuse of language, refer to this as Rat(2, 2, 2). Let $X = \mathbb{A}^1 \times \mathbb{P}^2 \times M(6,6)$ with coordinates $(\lambda, \xi_0, \xi_1, \xi_2, \mathbf{M})$, $\mathbf{M} = (\mathbf{M}_j^i)$, $i, j = 1, 2, 3$ and $\mathbf{M}_j^i$ are 2×2 blocks. As $\xi_2 \neq 0$, we shall work on $X_{\xi_2} \simeq \mathbb{A}^{39}$ with coordinates, $(\lambda, \xi_0, \xi_1, \mathbf{M})$. $\mathcal{H}(2,2,2) = \{(\lambda, \xi_0, \xi_1, \mathbf{M}) \in \mathcal{L}(2,2,2) \cap V_{\mathrm{Gr}}(2,2,2)\colon \rho(\mathbf{M}) = 2\}$. We define a map $\psi\colon \mathrm{Rat}(2,2,2) \to X$ as follows: let $\mathbf{x} = (x, x_{130}, \ldots, x_{241}, \xi_0, \xi_1)$ and set $\psi(\mathbf{x}) = (x, \xi_0, \xi_1, \mathbf{M}(\mathbf{x}))$ where $\mathbf{M}_j^i(\mathbf{x}) = \mathbf{H}_{i+j-1}(\mathbf{x})$ and

$$\mathbf{H}_1(\mathbf{x}) = \begin{bmatrix} -x_{131} & x_{141} \\ x_{231} & -x_{241} \end{bmatrix}, \quad \mathbf{H}_2(\mathbf{x}) = \begin{bmatrix} -x_{130} + \xi_1 x_{131} & x_{140} - \xi_1 x_{141} \\ x_{230} - \xi_1 x_{231} & -x_{240} + \xi_1 x_{241} \end{bmatrix}$$

and $\mathbf{H}_{2+j}(\mathbf{x}) = \xi_1 \mathbf{H}_{2+j-1}(\mathbf{x}) + \xi_0 \mathbf{H}_{2+j-2}(\mathbf{x})$, $j = 1, 2, 3$. Clearly ψ is an injective morphism. We *claim* that $\psi(\mathbf{x}) \in \mathcal{H}(2,2,2)$. Clearly (a), (b) and (c) hold. What about (d)? Since $\Lambda(\mathbf{M})(z) = [\xi_1 \mathbf{H}_1 + \mathbf{H}_2] + \mathbf{H}_1 z$, we must show that

$$\begin{aligned} \xi_0 x + \xi_1 x z + x z^2 &= \det[\xi_1 \mathbf{H}_1 + \mathbf{H}_2 + \mathbf{H}_1 z] = \det[\xi_1 \mathbf{H}_1 + \mathbf{H}_2] + \det[\mathbf{H}_1] z^2 \\ &\quad + \{\det[(\xi_1 \mathbf{H}_1 + \mathbf{H}_2)_1\ (\mathbf{H}_1)_2] - \det[(\xi_1 \mathbf{H}_1 + \mathbf{H}_2)_2\ (\mathbf{H}_1)_1]\} z. \end{aligned}$$

So, (1) is $x = \det \mathbf{H}_1$? But $x = x_{131} x_{241} - x_{141} x_{231}$ by (11.32) and done for (1). Now, (2) is $\xi_0 x = \det[\xi_1 \mathbf{H}_1 + \mathbf{H}_2]$? But

$$\xi_1 \mathbf{H}_1 + \mathbf{H}_2 = \begin{bmatrix} -x_{130} & x_{140} \\ x_{230} & -x_{240} \end{bmatrix}$$

and $\det[\xi_1 \mathbf{H}_1 + \mathbf{H}_2] = x_{130} x_{240} - x_{140} x_{230} = x \xi_0$ by (11.32). Now, (3) is $x \xi_1 = \det[(\xi_1 \mathbf{H}_1 + \mathbf{H}_2)_1\ (\mathbf{H}_1)_2] - \det[(\xi_1 \mathbf{H}_1 + \mathbf{H}_2)_2\ (\mathbf{H}_1)_1]$? Computation gives $x_{130} x_{241} - x_{141} x_{230} - x_{140} x_{231} + x_{131} x_{240}$ for the sum of the determinants which is $x \xi_1$ by (11.32). Since $x \neq 0$ (by assumption) $\psi(\mathbf{x}) \in \mathcal{H}(2,2,2)$. Conversely, let $\mathbf{y} = (\lambda_0, \xi_0, \xi_1, \mathbf{M})$ be an element of $\mathcal{H}(2,2,2)$ with $\lambda_0 \neq 0$. Let us set $\mathbf{M}_1^1 = (m_j^{1,i})$, $\mathbf{M}_1^2 = (m_j^{2,i})$, $i, j = 1, 2$. Then $\psi_0(z) = \xi_0 + \xi_1 z + z^2$, $\psi_{\alpha_2}(z) = \lambda_0$, and $\Lambda(\mathbf{M})(z) = \xi_1 \mathbf{M}_1^1 + \mathbf{M}_2^1 + \mathbf{M}_1^1 z$ so that the Grassmann equations become

$$\begin{aligned} \lambda_0 &= \det \mathbf{M}_1^1 \\ \lambda_0 \xi_0 &= \det[\xi_1 \mathbf{M}_1^1 + \mathbf{M}_2^1] \\ \lambda_0 \xi_1 &= \det[(\xi_1 \mathbf{M}_1^1 + \mathbf{M}_2^1)_1\ (\mathbf{M}_1^1)_2] - \det[(\xi_1 \mathbf{M}_1^1 + \mathbf{M}_2^1)_2\ (\mathbf{M}_1^1)_1]. \end{aligned} \tag{11.33}$$

We define a map $\varphi\colon \mathcal{H}(2,2,2) \to \mathbb{A}_x^{11}$ by setting $\varphi(\mathbf{y}) = (x, x_{130}, \ldots, x_{241}, \xi_0, \xi_1)$ where

$$\begin{aligned} x &= \lambda_0, \quad \xi_0 = \xi_0, \quad \xi_1 = \xi_1 \\ x_{130} &= -\xi_1 m_1^{1,1} - m_1^{2,1}, \quad x_{131} = -m_1^{1,1} \\ x_{140} &= \xi_1 m_2^{1,1} + m_2^{2,1}, \quad x_{141} = m_2^{1,1} \\ x_{230} &= \xi_1 m_1^{1,2} + m_1^{2,2}, \quad x_{231} = m_1^{1,2} \\ x_{240} &= -\xi_1 m_2^{1,2} - m_2^{2,2}, \quad x_{241} = -m_1^{1,2}. \end{aligned}$$

It is clear that φ is an injective morphism. We *claim* that $\varphi(\mathbf{y}) \in \mathrm{Rat}(2,2,2)$ and that $(\psi \circ \varphi)(\mathbf{y}) = \mathbf{y}$ (clear). To show this, we must show that (11.32) holds (since $x = \lambda_0 \neq 0$, the relatively prime condition holds). By the Grassmann equations (11.33) in $\mathcal{H}(2,2,2)$, we have

$$\lambda_0 = m_1^{1,1} m_2^{1,2} - m_2^{1,1} m_1^{1,2} = (-x_{131})(-x_{241}) - x_{141}x_{231}$$

so that $x = x_{131}x_{241} - x_{141}x_{231}$. Next,

$$\begin{aligned}\lambda_0\xi_0 &= \det\begin{bmatrix}\xi_1 m_1^{1,1} + m_1^{2,1} & \xi_1 m_2^{1,1} + m_2^{2,1} \\ \xi_1 m_1^{1,2} + m_1^{2,2} & \xi_1 m_2^{1,2} + m_2^{2,2}\end{bmatrix} \\ &= \det\begin{bmatrix}-x_{130} & x_{140} \\ x_{230} & -x_{240}\end{bmatrix} = x_{130}x_{240} - x_{140}x_{230}\end{aligned}$$

so that $x\xi_0 = x_{130}x_{240} - x_{140}x_{230}$. Finally,

$$\lambda_0\xi_1 = \det\begin{bmatrix}-x_{130} & x_{141} \\ x_{230} & -x_{241}\end{bmatrix} - \det\begin{bmatrix}x_{140} & -x_{131} \\ -x_{240} & x_{231}\end{bmatrix}$$

so that $x\xi_1 = x_{130}x_{241} - x_{141}x_{230} - x_{140}x_{231} + x_{131}x_{240}$. In other words, $\varphi(\mathbf{y}) \in \mathrm{Rat}(2,2,2)$ and so $\mathrm{Rat}(2,2,2)$ and $\mathcal{H}(2,2,2)$ are isomorphic.

We now give some examples relating to the various conditions (a), (b), (c) and (d).

Example 11.34 Let

$$\mathbf{H}_1 = \begin{bmatrix}0 & 0 \\ 1 & 0\end{bmatrix}, \quad \mathbf{H}_2 = 0, \quad \mathbf{H}_3 = \begin{bmatrix}1 & 0 \\ 0 & 0\end{bmatrix}.$$

Then the matrix

$$\mathbf{M} = \begin{bmatrix}\mathbf{H}_1 & \mathbf{H}_2 \\ \mathbf{H}_2 & \mathbf{H}_3\end{bmatrix} = \begin{bmatrix}0 & 0 & 0 & 0 \\ 1 & 0 & 0 & 0 \\ 0 & 0 & 1 & 0 \\ 0 & 0 & 0 & 0\end{bmatrix}$$

satisfies the block symmetry and Hankel structure equations and has rank 2. But the dependence equations are *not* satisfied and $\mathbf{M}$ does not come from an element of $S_{2,2}^2$.

Example 11.35 Let

$$\mathbf{H}_1 = \begin{bmatrix}1 & 0 \\ 0 & 1\end{bmatrix}, \quad \mathbf{H}_2 = \begin{bmatrix}1 & 0 \\ 0 & 0\end{bmatrix}, \quad \mathbf{H}_3 = \begin{bmatrix}0 & 0 \\ 0 & 1\end{bmatrix}$$

so that $\mathbf{H}_3 = \mathbf{H}_1 - \mathbf{H}_2 = -\xi_0\mathbf{H}_1 - \xi_1\mathbf{H}_2$ with $\xi_0 = -1$, $\xi_1 = 1$. Then the matrix

$$\mathbf{M} = \begin{bmatrix} \mathbf{H}_1 & \mathbf{H}_2 \\ \mathbf{H}_2 & \mathbf{H}_3 \end{bmatrix} = \begin{bmatrix} 1 & 0 & 1 & 0 \\ 0 & 1 & 0 & 0 \\ 1 & 0 & 0 & 0 \\ 0 & 0 & 0 & 1 \end{bmatrix}$$

satisfies the block symmetry, Hankel structure, and dependence equations. However, the Grassmann equations are *not* satisfied and $\mathbf{M}$ (of rank 4) does not come from an element of $S_{2,2}^2$.

Example 11.36 Let $p = 2$, $m = 3$ and $n = 2$. Then $\gamma(2,3,2) = 18$ and $\tau(2,3,2) = 18 - 2(7) - 1 = 3$. Consider the matrix

$$\mathbf{H} = \begin{bmatrix} H_1 & H_2 \\ H_2 & -\xi_0 H_1 - \xi_1 H_2 \end{bmatrix}.$$

The Grassmann equations are

$$\begin{aligned} (\xi_0 + \xi_1 z + z^2)\lambda_{23} &= \det\left[(\xi_1 H_1 + H_2) + H_1 z\big|_{23}\right] \\ (\xi_0 + \xi_1 z + z^2)\lambda_{12} &= \det\left[(\xi_1 H_1 + H_2) + H_1 z\big|_{12}\right] \\ (\xi_0 + \xi_1 z + z^2)\lambda_{13} &= \det\left[(\xi_1 H_1 + H_2) + H_1 z\big|_{13}\right]. \end{aligned}$$

It follows that

$$\lambda_{12} = \det\left[H_1\big|_{12}\right], \quad \lambda_{13} = \det\left[H_1\big|_{13}\right], \quad \lambda_{23} = \det\left[H_1\big|_{23}\right]$$

and that

$$\begin{aligned} \xi_0\lambda_{12} &= \det\left[\xi_1 H_1 + H_2\big|_{12}\right] \\ \xi_0\lambda_{13} &= \det\left[\xi_1 H_1 + H_2\big|_{13}\right] \\ \xi_0\lambda_{23} &= \det\left[\xi_1 H_1 + H_2\big|_{23}\right]. \end{aligned}$$

If $\lambda_{12}, \lambda_{13}, \lambda_{23}$ are all non-zero, then we must have

$$\frac{\det[\xi_1 H_1 + H_2|_{12}]}{\det[H_1|_{12}]} = \frac{\det[\xi_1 H_1 + H_2|_{13}]}{\det[H_1|_{13}]} = \frac{\det[\xi_1 H_1 + H_2|_{23}]}{\det[H_1|_{23}]}. \tag{11.37}$$

If $x = (A, B, C)$ with $CB = H_1$, $CAB = H_2$, then (11.37) becomes

$$\frac{\det[\xi_1 CB + CAB|_{12}]}{\det[CB|_{12}]} = \frac{\det[\xi_1 CB + CAB|_{13}]}{\det[CB|_{13}]} = \frac{\det[\xi_1 CB + CAB|_{23}]}{\det[CB|_{23}]}.$$

Since C is 2×2 and (say) $\det\left[CB\big|_{12}\right] \neq 0$, $\det C \neq 0$ and we have

$$\frac{\det[(\xi_1 I + A)B|_{12}]}{\det[B|_{12}]} = \frac{\det[(\xi_1 I + A)B|_{13}]}{\det[B|_{13}]} = \frac{\det[(\xi_1 I + A)B|_{23}]}{\det[B|_{23}]}$$

which is an identity (for $\det\left[B_i\big|_{ij}\right] \neq 0$). Now let

$$H_1 = \begin{bmatrix} 0 & 0 & 0 \\ 0 & 0 & 0 \end{bmatrix}, \quad H_2 = \begin{bmatrix} 1 & 0 & 0 \\ 1 & 0 & 0 \end{bmatrix}$$
$$H_3 = \begin{bmatrix} 2 & 0 & 0 \\ 2 & 0 & 0 \end{bmatrix}, \quad H_4 = \begin{bmatrix} 3 & 0 & 0 \\ 3 & 0 & 0 \end{bmatrix}$$

and $\xi_0 = 1$, $\xi_1 = -2$. Let

$$\mathbf{M} = \begin{bmatrix} H_1 & H_2 & H_3 \\ H_2 & H_3 & H_4 \\ H_3 & H_4 & -\xi_0 H_3 - \xi_1 H_4 \end{bmatrix}.$$

Then $(0, 0, 0, \xi_0, \xi_1, \mathbf{M}) \in \mathcal{H}(2, 3, 2)$ since all the equations (a), (b), (c), (d) are satisfied and $\mathbf{M}$ has rank 2. The corresponding transfer matrix is

$$\mathbf{F}(z) = \begin{bmatrix} 1(z-1)^2 & 0 & 0 \\ 1/(z-1)^2 & 0 & 0 \end{bmatrix}$$

and it is clear what the significance of the vanishing of λ_{12}, λ_{13}, λ_{23} is.

The Grassmann equations (d) impose significant structural constraints. This is why we used the approach we did.

Exercises

(1) Show that $n(pm+1) \leq \gamma(n, m, p) - 1 = \dim \mathcal{L}(n, m, p)$.

(2) Verify all the details in Example 11.31.

(3) Let $n = 2$, $m = 3$, $p = 2$ and consider (A, B, C) with A 2×2, B 2×3 and C 2×2. Let $\mathbf{M}_j^i = CA^{i+j-2}B$ and let $\lambda_{12} = \det\left[\mathbf{M}_1^1\big|_{12}\right]$, $\lambda_{13} = \det\left[\mathbf{M}_1^1\big|_{13}\right]$, $\lambda_{23} = \det\left[\mathbf{M}_1^1\big|_{23}\right]$. Let $\det[zI - A] = z^2 + \xi_1 z + \xi_0$. Show that $(\lambda_{12}, \lambda_{13}, \lambda_{23}, \xi_0, \xi_1, \mathbf{M}) \in \mathcal{H}(2, 3, 2)$.

12

Projective Algebraic Geometry IV: Families, Projections, Degree

We shall use the Main Theorem of Elimination Theory (10.16) to develop some families of varieties.

Consider $\mathbb{P}^N \times \mathbb{P}^{N^*} = \{(\boldsymbol{\xi}, \mathbf{u})\colon H = V(\sum u_i Y_i) \text{ is a hyperplane in } \mathbb{P}^N\}$. If $Y_0, \ldots, Y_N$ are coordinates on $\mathbb{P}^N$ and $U_0, \ldots, U_N$ are coordinates on $\mathbb{P}^{N^*}$, then $\Theta = V\left(\sum_{i=0}^{N} u_i Y_i\right)$ is closed in $\mathbb{P}^N \times \mathbb{P}^{N^*}$. A point $(\boldsymbol{\xi}, \mathbf{u})$ is in Θ if and only if $\boldsymbol{\xi}$ and the hyperplane $H = V(\sum u_i Y_i)$ are incident (i.e., $\boldsymbol{\xi} \in H$ and H passes through $\boldsymbol{\xi}$). If $\pi\colon \Theta \to \mathbb{P}^N$ is the restriction of the projection $\mathbb{P}^N \times \mathbb{P}^{N^*} \to \mathbb{P}^N$ and if V is a variety in $\mathbb{P}^N$, then $\pi^{-1}(V) = \{(\boldsymbol{\xi}, \mathbf{u})\colon \boldsymbol{\xi} \in V \cap H, H = V(\sum u_i Y_i)\}$ is a subvariety of Θ called the *family of hyperplane sections of* V.

Example 12.1 If $N = 2$, then $\Theta = \{(\xi_0, \xi_1, \xi_2, u_0, u_1, u_2)\colon u_0\xi_0 + u_1\xi_1 + u_2\xi_2 = 0\}$. If $V = \{Y_0^2 - Y_1Y_2 = 0\}$, then $\pi^{-1}(V) = \{(\xi_0, \xi_1, \xi_2, u_0, u_1, u_2)\colon u_0\xi_0 + u_1\xi_1 + u_2\xi_2 = 0 \text{ and } \xi_0^2 - \xi_1\xi_2 = 0\}$. The point $(0, 1, 1, 1, 1, -1)$ is in Θ but not in $\pi^{-1}(V)$ so that this point is not a hyperplane section of V. On the other hand, the points $\left(\frac{1+\sqrt{5}}{2}, 1, \frac{3+\sqrt{5}}{2}, 1, 1, -1\right), \left(\frac{1-\sqrt{5}}{2}, 1, \frac{3-\sqrt{5}}{2}, 1, 1, -1\right)$ are a hyperplane section of V. Note since $\dim V = 1$, $\dim(V \cap H) = 0$ for hyperplanes H (unless $V \subset H$).

Example 12.2 Let $\mathbb{P}^{N-1} = V(Y_N) \subset \mathbb{P}^N$ and let $\boldsymbol{\xi}_\infty = (0, \ldots, 0, 1)$. Let $V \subset \mathbb{P}^{N-1}$ with $V = V(F_1, \ldots, F_r)$ F_i, $i = 1, \ldots, r$ homogeneous in $Y_0, \ldots, Y_{N-1}$ and let $W = \{\boldsymbol{\xi} \in \mathbb{P}^N\colon F_i(\boldsymbol{\xi}) = 0, i = 1, \ldots, r\} = C(V, \boldsymbol{\xi}_\infty)$ be the cone through V with vertex $\boldsymbol{\xi}_\infty$. Let $\pi^{-1}(W) = \{(\boldsymbol{\xi}, \mathbf{u})\colon \boldsymbol{\xi} \in H \cap W, H = V(\sum u_i Y_i)\}$ be the family of hyperplane sections of W. We *claim* that there is an α in $\mathrm{PGL}(N, k)$ such that $\alpha(H \cap W) = V$ for almost all $H \in \mathbb{P}^{N^*}$. If $H = (u_0, \ldots, u_N)$ with

P. Falb, *Methods of Algebraic Geometry in Control Theory: Part II*,
Modern Birkhäuser Classics, https://doi.org/10.1007/978-3-319-96574-1_12

$u_N \neq 0$, then the affine coordinates of H are $(h_0, \dots, h_{N-1}, 1)$ (i.e., $H \in \mathbb{P}^{N^*}_{u_N}$ open in $\mathbb{P}^{N^*}$). Let $\alpha\colon \mathbb{P}^N \to \mathbb{P}^N$ be the element of $\mathrm{PGL}(N,k)$ given by

$$\alpha(\xi_i) = \xi_i, \quad i = 0, 1, \dots, N-1$$
$$\alpha(\xi_N) = \xi_N + \sum_{j=0}^{N-1} h_j \xi_j.$$

Then $\alpha(H \cap W) = V$. If $\boldsymbol{\xi} = (\xi_0, \dots, \xi_N) \in H \cap W$, then $\alpha(\xi_i) = \xi_i$, $i = 0, 1, \dots, N-1$ and $F_j(\alpha(\boldsymbol{\xi})) = 0$ for $j = 1, \dots, r$ and $\alpha(\xi_N) = 0$.

Now consider in $\mathbb{P}^N$ the set of hypersurfaces of degree q. In other words, let $S = k[Y_0, \dots, Y_N]$ and $S_q = \{F \in S\colon F \text{ a form of degree } q\}$. Then $\dim_k S_q = \binom{N+q}{q} = M + 1$ and S_q is a projective space $\mathbb{P}^M$. In $\mathbb{P}^N \times \mathbb{P}^M$, we let

$$\Theta(q) = \{(\boldsymbol{\xi}, F)\colon \boldsymbol{\xi} \in V(F), F \in S_q\}. \tag{12.3}$$

This is a closed set in $\mathbb{P}^N \times \mathbb{P}^M$. If $V \subset \mathbb{P}^N$ and $\pi\colon \Theta(q) \to \mathbb{P}^N$ is the restriction of the projection $\pi\colon \mathbb{P}^N \times \mathbb{P}^M \to \mathbb{P}^N$, then $\pi^{-1}(V) = \{(\boldsymbol{\xi}, F)\colon \boldsymbol{\xi} \in V \cap V(F), F \in S_q\}$ is the *family of hypersurface sections of V of degree q*.

Example 12.4 Let $N = 2$, $q = 2$. Then $F \in S_2$ means that $F = u_{00}Y_0^2 + u_{01}Y_0Y_1 + u_{02}Y_0Y_2 + u_{11}Y_1^2 + u_{12}Y_1Y_2 + u_{22}Y_2^2$ and the coordinates of F are $(u_{00}, \dots, u_{22}) \in \mathbb{P}^5$. $\Theta(2) = \{(\xi_0, \xi_1, \xi_2)\colon F(\xi_0, \xi_1, \xi_2) = 0, F \in S_2\} = V(U_{00}Y_0^2 + \cdots + U_{22}Y_2^2)$ where U_{00}, U_{01}, U_{02}, U_{11}, U_{12}, U_{22} are coordinates in $\mathbb{P}^5$. If $V \subset \mathbb{P}^2$, then $\pi^{-1}(V) = \{(\boldsymbol{\xi}, F)\colon F(\boldsymbol{\xi}) = 0, \boldsymbol{\xi} \in V, F \in S_2\}$. If $\dim V = 1$, then $\dim V \cap V(F) = 0$ and, in general, a hypersurface section of degree 2 will be two points.

Proposition 12.5 *If V is irreducible, then the family $\Theta_V(1)$ of hyperplane sections of V is irreducible.*

Proof. ([H-2]) $\Theta_V(1) = \pi^{-1}(V)$ and we let $\Theta_{V,\boldsymbol{\xi}} = \pi^{-1}(\boldsymbol{\xi})$ be the fiber over $\boldsymbol{\xi}$. If $\Theta_V(1) = W_1 \cup \cdots \cup W_r$ is the decomposition into irreducible components, then we let $\tilde{W}_i = \{\boldsymbol{\xi} \in V\colon \Theta_{V,\boldsymbol{\xi}} \subset W_i\}$. We *claim* that $\tilde{W}_i$ is closed in V. Since $V = \cup \tilde{W}_i$ and $\Theta_{V,\boldsymbol{\xi}} = \pi^{-1}(\boldsymbol{\xi})$ is irreducible and isomorphic to $\mathbb{P}^{N-1}$, the closure of $\tilde{W}_i$ for all i would imply that $V = \tilde{W}_{i_0}$ (for some i_0) and hence, that $\Theta_V(1) = W_{i_0}$. Since closure is a local property we need only show that $\tilde{W}_i \cap V_{Y_j}$ are all closed. Consider (say) V_{Y_0}. Then $\tilde{W}_i \subset V_{Y_0} = \cap \pi(W_i \cap X_u)$ where $X_u \subset V_{Y_0} \times \mathbb{P}^{N^*}$ is given by: $U_0 = -u_1Y_1 - \cdots - u_NY_N$, $U_i = u_iY_0$, $i = 1, \dots, N$ (U_i coordinates on $\mathbb{P}^{N^*}$), $u = (u_1, \dots, u_N)$. Then X_u is closed in $V_{Y_0} \times \mathbb{P}^{N^*}$ and $X_u \cap \Theta_{V,\xi}$ is a single point. Moreover, $\pi^{-1}(V_{Y_0}) = \cup X_u$. Thus, $\tilde{W}_i \cap V_{Y_0} = \cap\{\boldsymbol{\xi} \in V_{Y_0}\colon X_u \cap \Theta_{V,\xi} \in W_i\} = \cap \pi(W_i \cap X_u)$. But $\pi(W_i \cap X_u)$ is closed by Theorem 10.16 and hence so is $\tilde{W}_i \cap V_{Y_0}$.

We note that the proposition is also a consequence of a more general result (Theorem 14.10) since $\pi \colon \theta_V(1) \to V$ is surjective, V is irreducible, and the fibers $\Theta_{V,\boldsymbol{\xi}}$ are irreducible of dimension $N-1$.

Example 12.6 Let $N = 3$ and let H_1, H_2 be hyperplanes in $\mathbb{P}^3$. If H_1, H_2 are viewed as elements of $(\mathbb{P}^3)^*$ with coordinates $(u_0^1, u_1^1, u_2^1, u_3^1)$ and $(u_0^2, u_1^2, u_1^2, u_3^2)$ respectively, then H_1 and H_2 are independent if and only if $\operatorname{rank}(u_j^i)_{j=0,1,2,3}^{i=1,2}$ is 2. In other words, the set $U = \{(H_1, H_2) \colon H_i \in (\mathbb{P}^3)^*,\ H_1, H_2 \text{ independent }\}$ is open in $(\mathbb{P}^3)^* \times (\mathbb{P}^3)^*$. The set $\{(\boldsymbol{\xi}, H_1, H_2) \colon \boldsymbol{\xi} \in H_1 \cap H_2\} \subset \mathbb{P}^3 \times U$ is closed and is a family of lines.

We now have some further consequences of the Main Theorem of Elimination Theory, particularly relating to projections.

Theorem 12.7 (Noether Normalization-Geometric form) *Let V be an r-dimensional variety in $\mathbb{P}^N$. Then there is a linear space $\mathbf{L} = L_{N-r-1}$ of dimension $N-r-1$ such that $\mathbf{L} \cap V = \emptyset$. The projection $\pi_{\mathbf{L}} \colon \mathbb{P}^N - \mathbf{L} \to \mathbb{P}^r$ restricted to V is a surjective closed map with finite fibers. Moreover, $S(V) = k[Y_0, \ldots, Y_N]/I_h(V)$ is a finite module over $k[Z_0, \ldots, Z_r]$ (and hence is integral over $k[Z_0, \ldots, Z_r]$).*

Proof. That the projection is a closed map is Theorem 10.16. If $N - r = 0$, then $V = \mathbb{P}^N$ and the result is obvious. So we use induction on $N - r$ and we suppose $N - r > 0$. Let $N - r = 1$ (i.e., $r = N - 1$) and let $\mathbf{L} = L_0 = \boldsymbol{\xi} \notin V$. Then $\pi_{\boldsymbol{\xi}} \colon V \to \mathbb{P}^{N-1}$ with $\pi_{\boldsymbol{\xi}}(V) = \tilde{V}$ a variety in $\mathbb{P}^{N-1}$. If $\boldsymbol{\eta} \in \tilde{V}$, then $\pi_{\boldsymbol{\xi}}^{-1}(\boldsymbol{\eta})$ is the set $J(\boldsymbol{\xi}, \boldsymbol{\eta}) - \{\boldsymbol{\xi}\}$ (where $J(\boldsymbol{\xi}, \boldsymbol{\eta})$, the join of $\boldsymbol{\xi}, \boldsymbol{\eta}$, is a line through $\boldsymbol{\xi}$). If $J(\boldsymbol{\xi}, \boldsymbol{\eta}) \subset V$, then $\boldsymbol{\xi}$ would be in V and so, $J(\boldsymbol{\xi}, \boldsymbol{\eta}) \cap V = \pi_{\boldsymbol{\xi}}^{-1}(\boldsymbol{\eta})$ is a finite set. We may suppose that $\boldsymbol{\xi} = \boldsymbol{\xi}_\infty = (0, \ldots, 0, 1)$ and that $\pi_{\boldsymbol{\xi}_\infty} \colon \mathbb{P}^N - \{\boldsymbol{\xi}_\infty\} \to \mathbb{P}^{N-1}$ is given by $(Y_0, \ldots, Y_N) \to (Y_0, \ldots, Y_{N-1})$. Then $k[Y_0, \ldots, Y_N]/I_h(V) \supset k[Y_0, \ldots, Y_{N-1}]/I_h(\tilde{V})$ and since $\boldsymbol{\xi}_\infty \notin V$, there is an $F \in I_h(V)$ with $F = Y_N^t + f_1(Y_0, \ldots, Y_{N-1})Y_N^{t-1} + \cdots + f_t(Y_0, \ldots, Y_{N-1})$. Thus, $S(V) = k[Y_0, \ldots, Y_N]/I_h(V)$ is generated by $1, \ldots, Y_N^{t-1}$ as an $S(\tilde{V}) = k[Y_0, \ldots, Y_{N-1}]/I_h(\tilde{V})$ module. Hence $\operatorname{tr} \deg_k S(V) = \dim_k V + 1 = \operatorname{tr} \deg_k S(\tilde{V}) = \dim_k \tilde{V} + 1$. In other words, $\dim_k \tilde{V} = \dim_k V = N - 1$ and so $\tilde{V} = \mathbb{P}^{N-1}$. Consider now the general case. Let $\boldsymbol{\xi} \in \mathbb{P}^N - V$ and let $\pi_{\boldsymbol{\xi}}(V) = \tilde{V} \subset \mathbb{P}^{N-1}$. By the induction hypothesis, there is an L_{N-r-2} with $L_{N-r-2} \cap \tilde{V} = \emptyset$ and hence, $L_{N-r-1} = J(\boldsymbol{\xi}, L_{N-r-2})$ is an $N - r - 1$ dimensional linear space with $L_{N-r-1} \cap V = \emptyset$. Since $\pi_{L_{N-r-1}} = \pi_{L_{N-r-2}} \circ \pi_{\boldsymbol{\xi}}$, the result follows. [Note that the result holds for any L_{N-r-1} which does not meet V.]

Corollary 12.8 $\dim V = \min\{s\colon$ *there is an* L_{N-s-1} *with* $L_{N-s-1} \cap V = \emptyset\}$ *(cf. Proposition 6.59).*

Proof. Let $s^* = \min\{s\colon$ there is an L_{N-s-1} with $L_{N-s-1} \cap V = \emptyset\}$. By the theorem, $r \geq s^*$. If $r > s^*$, then $r - s^* \geq 1$ and $N - s^* - 1 + r \geq N$. So $V \cap L_{N-s^*-1} \neq \emptyset$.

Example 12.9 Let $\mathbb{G}\mathrm{r}(p, N)$ $(= \mathrm{Gr}(p+1, N+1))$ be the Grassmann of p-planes L_p in $\mathbb{P}^N$ (see Chapter 8). Let $\Theta(p, N) = \{(\boldsymbol{\xi}, L_p)\colon \boldsymbol{\xi} \in L_p\} \subset \mathbb{P}^N \times \mathbb{G}\mathrm{r}(p, N)$. Then $\Theta(p, N) = \{(\boldsymbol{\xi}, v_1 \wedge \cdots \wedge v_{p+1})\colon \boldsymbol{\xi} \wedge v_1 \wedge \cdots \wedge v_{p+1} = 0\}$ is a projective variety (or by using the Plücker relations and projecting). If W is a subvariety of $\mathbb{G}\mathrm{r}(p, N)$, then $\bigcup_{L_p \in W} L_p \subset \mathbb{P}^N$ is also a variety since

$$\bigcup_{L_p \in W} L_p = \pi_1(\pi_2^{-1}(W)) \tag{12.10}$$

where $\pi_1\colon \mathbb{P}^N \times \mathbb{G}\mathrm{r}(p, N) \to \mathbb{P}^N$ and $\pi_2\colon \mathbb{P}^N \times \mathbb{G}\mathrm{r}(p, N) \to \mathbb{G}\mathrm{r}(p, N)$ are the projections restricted to $\Theta(p, N)$. If V is a variety in $\mathbb{P}^N$ and

$$\mathcal{L}_p(V) = \{L_p\colon V \cap L_p \neq \emptyset\} \tag{12.11}$$

then $\mathcal{L}_p(V) = \pi_2(\pi_1^{-1}(V))$ is a variety in $\mathbb{G}\mathrm{r}(p, N)$ and hence, its complement is open. This, if $\dim V = r$, then $\mathcal{L}_{N-r-1}(V)$ is closed and $\mathcal{L}_{N-r-1}(V) < \mathbb{G}\mathrm{r}(N-r-1, N)$. In other words, almost all L_{N-r-1} do not meet V. If V is quasi-projective, then $\overline{\mathcal{L}_p(V)} = \mathcal{L}_p(\overline{V})$ since $\pi_2(\pi_1^{-1}(\overline{V})) = \overline{\pi_2(\pi_1^{-1}(V))}$. Since $\dim V = \dim \overline{V}$, almost all L_{N-r-1} again do not meet V. We often say that "a general L_{N-r-1} does not meet V". In fact, if $r < p+1$, then $\mathcal{L}_{N-p-1}(V) < \mathbb{G}\mathrm{r}(N-p-1, N)$ and a general L_{N-p-1} $(r < p+1)$ does not meet V. Note that here $N - p - 1 + r < N$. If $\dim V = N - 1$, then any L_p for $p \geq 1$ will meet V. If $r + 2 \leq N$ then $\mathcal{L}_1(V) < \mathbb{G}\mathrm{r}(1, N)$ and a general line does not meet V. For instance, if $r = N - 2$, then $L_1 \cap V = \emptyset$ for almost all lines L_1. This type of analysis often allows us to choose projections freely.

Example 12.12 Let $V = V(F)$ be a hypersurface in $\mathbb{P}^N$ and let $\mathcal{F}(p, N) = \{L_p\colon L_p \subset V\}$. Suppose F is homogeneous of degree q (i.e., $F \in S_q$ where $S = k[Y_0, \ldots, Y_N]$). Let $L_p \in \mathbb{G}\mathrm{r}(p, N)$ and let $I_h(L_p)$ be the ideal of L_p in S. Then $I_h(L_p)_q \subset S_q$ and is a (k-)subspace of dimension $\binom{N+q}{q} - \binom{p+q}{q}$. In other words, if

$$\mu = \begin{pmatrix} N+q \\ q \end{pmatrix} - \begin{pmatrix} p+q \\ q \end{pmatrix}, \quad \nu = \begin{pmatrix} N+q \\ q \end{pmatrix}$$

then we have a regular map $\psi\colon \mathbb{G}\mathrm{r}(p, N) \to \mathbb{G}\mathrm{r}(\mu, \nu)$ given by

$$\psi(L_p) = I_h(L_p)_q. \tag{12.13}$$

But $L_p \subset V$ if and only if $F \in I_h(L_p)_q$. However, the set of L_μ in $\mathrm{Gr}(\mu,\nu)$ with $F \in L_\mu$ is a subvariety $\mathbf{X}$ of $\mathrm{Gr}(\mu,\nu)$ and $\psi^{-1}(\mathbf{X}) = \mathcal{F}(p,N)(V)$. Thus $\mathcal{F}(p,N)(V)$ is a subvariety of $\mathbb{G}\mathrm{r}(p,N)$. If $V \subset \mathbb{P}^N$ is an r-dimensional variety, then

$$\mathcal{F}(p,N)(V) = \{L_p \colon L_p \subset V\} = \bigcap_H \mathcal{F}(p,N)(H),$$

$H = V(F)$ with $V \subseteq H$, so that $\mathcal{F}(p,N)(V)$ is always a subvariety of $\mathbb{G}\mathrm{r}(p,N)$. It is called the *Fano variety* of *p-planes contained in* V.

Example 12.14 Suppose that V is an r-dimensional variety in $\mathbb{P}^N$ and that V is not a linear space. Let Δ be the diagonal in $\mathbb{P}^N \times \mathbb{P}^N$ and consider the map $\psi_s \colon V \times V - \Delta \cap V \times V \to \mathbb{G}\mathrm{r}(1,N)$ given by $\psi_s(\boldsymbol{\xi},\boldsymbol{\xi}') = J(\boldsymbol{\xi},\boldsymbol{\xi}') = \ell(\boldsymbol{\xi},\boldsymbol{\xi}')$, the line joining $\boldsymbol{\xi}$ and $\boldsymbol{\xi}'$. Since ψ_s is regular, its image is a quasi-projective variety. If $\ell(\boldsymbol{\xi},\boldsymbol{\xi}') \not\subset V$ (i.e., $\ell(\boldsymbol{\xi},\boldsymbol{\xi}') \notin \mathcal{F}(1,N)(V)$), then $\psi_s^{-1}(\ell(\boldsymbol{\xi},\boldsymbol{\xi}'))$ is a finite set. Let $\mathcal{S}_1(V) = \{\ell(\boldsymbol{\xi},\boldsymbol{\xi}')\}$ = image of ψ_s. Then $\mathcal{S}_1(V)$ is the *secant line variety of* V. We note that $\mathcal{F}(1,N)(V) \subset \mathcal{S}_1(V) \subset \mathcal{L}_1(V)$ in general.

Example 12.15 Consider $\mathbb{G}\mathrm{r}(1,3)$ ($= \mathrm{Gr}(2,4)$) the space of lines in $\mathbb{P}^3$. We view it as the quadratic hypersurface $Y_0Y_5 - Y_2Y_3 + Y_1Y_4 = 0$ in $\mathbb{P}^5$. An element L_1 of $\mathbb{G}\mathrm{r}(1,3)$ corresponds to a 2×4 matrix Λ_1 of rank 2 with (say)

$$\Lambda_1 = \begin{bmatrix} \lambda_{11} & \lambda_{12} & \lambda_{13} & \lambda_{14} \\ \lambda_{21} & \lambda_{22} & \lambda_{23} & \lambda_{24} \end{bmatrix}$$

and $Y_0(L_1) = \det[\Lambda_1|_{12}] = (\lambda_{11}\lambda_{22} - \lambda_{12}\lambda_{21}), \ldots, Y_5(L_1) = \det[\Lambda_1|_{34}] = (\lambda_{13}\lambda_{24} - \lambda_{14}\lambda_{23})$. If $H_1 = V(\lambda_{11}Y_0 + \lambda_{12}Y_1 + \lambda_{13}Y_2 + \lambda_{14}Y_3)$ and $H_2 = V(\lambda_{21}Y_0 + \lambda_{22}Y_1 + \lambda_{23}Y_2 + \lambda_{24}Y_3)$, then $L_1 = H_1 \cap H_2$ and $\boldsymbol{\xi} = (\xi_0,\xi_1,\xi_2,\xi_3) \in L_1$ if and only if $\Lambda_1\boldsymbol{\xi} = 0$. In other words, $\Theta(1,3) = \{(\boldsymbol{\xi},L_1) \colon \boldsymbol{\xi} \in L_1\} = \{(\boldsymbol{\xi},L_1) \colon \Lambda_1\boldsymbol{\xi} = 0\}$. For instance, if $Y_0(L_1) \neq 0$, then we may take $Y_0(L_1) = 1$ and Λ_1 in the form

$$\Lambda_1 = \begin{bmatrix} 1 & 0 & \alpha & \beta \\ 0 & 1 & \gamma & \delta \end{bmatrix}$$

and so, $\Theta(1,3) \cap (\mathbb{P}^3 \times \mathbb{G}\mathrm{r}(1,3)_{Y_0}) = \{(\boldsymbol{\xi},L_1) \colon \boldsymbol{\xi} = (\xi_0,\xi_1,\xi_2,\xi_3),\ L_1 = (1,\gamma,\delta,-\alpha,-\beta,\alpha\delta-\beta\gamma),\ \xi_0 + \alpha\xi_2 + \beta\xi_3 = 0,\ \xi_1 + \gamma\xi_2 + \delta\xi_3 = 0\}$. If, for example, $V = V(Z_0Z_1 - Z_2Z_3) \subset \mathbb{P}^3$, then the elements of $\pi_1^{-1}(V)$ are of the form

$$(-\alpha\xi_2 - \beta\xi_3, -\gamma\xi_2 - \delta\xi_3, \xi_2, \xi_3, 1, \gamma, \delta, -\alpha, -\beta, \alpha\delta - \beta\gamma)$$

where $\alpha\gamma\xi_2^2 + [(\alpha\delta+\beta\gamma) - 1]\xi_2\xi_3 + \beta\delta\xi_3^2 = 0$. This defines (via π_2) $\mathcal{L}_1(V)$. Note that $\dim V = 2$ and that $\mathcal{L}_1(V) = \mathcal{S}_1(V)$ here. If $L_1 \subset V$, then $(\alpha\gamma)Y_2^2 + [(\alpha\delta + \beta\gamma) - 1]Y_2Y_3 + (\beta\delta)Y_3^2 \equiv 0$ so that $\alpha\gamma = 0$, $\alpha\delta + \beta\gamma = 1$ and $\beta\delta = 0$. Since either

α or $\gamma \neq 0$ (but not both) we have L_α with coordinates $(1, 0, \alpha^{-1}, -\alpha, 0, 1)$ (or $\xi_0 = -\alpha\xi_2$, $\xi_1 = -\alpha^{-1}\xi_3$) and L_γ with coordinates $(1, \gamma, 0, 0, -\gamma^{-1}, -1)$ (or $\gamma_0 = -\gamma^{-1}\xi_3$, $\xi_1 = -\gamma\xi_2$) as families of lines lying on V. In other words, $\mathcal{F}(1,3)(V)_{Y_0} = \{L_\alpha, L_\gamma\}$. If, for example, $\alpha = 1$, $\beta = 1$, $\gamma = 1$ and $\delta = 0$ so that $\xi_0 = -\xi_2 - \xi_3$, $\xi_1 = -\xi_3$, then the line meets V where $\xi_2^2 = 0$ i.e., at the point $(\xi_0, \xi_0, 0, -\xi_0) = (1, 1, 0, -1)$ only. Thus $\mathcal{F}(1,3)(V) < \mathcal{L}_1(V)$. Now let $W \subset \mathbb{P}^3$ be the (plane) conic given by $Y_0 = 0$, $Y_3^2 + Y_1Y_2 = 0$. Clearly $\mathcal{L}_1(W) < \mathbb{G}\mathrm{r}(1,3)$ (for instance, the line $Y_1 = 0$, $Y_2 = 0$ does not meet W). Let $L_1 \in \mathbb{G}\mathrm{r}(1,3)$ be the line $(1, 0, -1, -1, -1, -1)$ so that

$$\Lambda_1 = \begin{bmatrix} 1 & 0 & 1 & 1 \\ 0 & 1 & 0 & -1 \end{bmatrix}$$

and

$$\Lambda_1 \cdot \begin{bmatrix} 0 \\ \xi_1 \\ \xi_2 \\ \xi_3 \end{bmatrix} = \begin{bmatrix} \xi_2 + \xi_3 \\ \xi_1 - \xi_3 \end{bmatrix}.$$

Then L_1 meets W in the single point $(0\ 1\ {-1}\ 1)$ and L_1 is in $\mathcal{L}_1(W)$ but not in $\mathcal{S}_1(W)$. Let L_2 be the line $(1, 1, 0, 0, 0, 0)$ so that

$$\Lambda_2 = \begin{bmatrix} 1 & 0 & 0 & 0 & 0 \\ 0 & 1 & 1 & 0 & 0 \end{bmatrix}$$

and

$$\Lambda_2 \begin{bmatrix} 0 \\ \xi_1 \\ \xi_2 \\ \xi_3 \end{bmatrix} = \begin{bmatrix} 0 \\ \xi_1 + \xi_2 \end{bmatrix}.$$

Then L_2 meets W in the two distinct points $(0\ {-1}\ 1\ 1)$, $(0\ {-1}\ 1\ {-1})$ (if char $k \neq 2$) but $L_2 \not\subset W$. In other words, $L_2 \in \mathcal{S}_1(W)$ but $L_2 \notin \mathcal{F}(1,3)(W)$. The elements of $\pi_1^{-1}(W)$ are of the form $(0, -\gamma\xi_2 - \delta\xi_3, \xi_2, \xi_3, 1, \gamma, \delta, -\alpha, -\beta, \alpha\delta - \beta\gamma)$ where

$$\begin{aligned} &\alpha\xi_2 + \beta\xi_3 = 0 \\ &\xi_3^2 + (-\gamma\xi_2 - \delta\xi_3)\xi_2 = \xi_3^2 - \delta\xi_2\xi_3 - \gamma\xi_2^2 = 0. \end{aligned} \tag{12.16}$$

This defines (via π_2) $\mathcal{L}_1(W)$. Since 12.16 is not 0 as a polynomial, $\mathcal{F}(1,3)(W)$ is empty.

Example 12.17 Let Z be a fixed element of $\mathrm{Gr}(m, m+p)$ and let $\sigma_k(Z) = \{W \in \mathrm{Gr}(p, m+p)\colon \dim(W \cap Z) \geq k\}$. Let $\Psi = \{(\Gamma, W)\colon \Gamma \subset W\} \subset \mathrm{Gr}(k, Z) \times$

$\mathrm{Gr}(p, m+p)$ (where $\mathrm{Gr}(k, Z)$ is the variety of k-dimensional subspaces of Z). Then Ψ is a variety (e.g., $(\Gamma, W) \in \Psi$ if and only if $(z_1 \wedge \cdots \wedge z_k) \wedge (w_1 \wedge \cdots \wedge w_p) = 0$). Since $\Psi = \pi_1^{-1}(\mathrm{Gr}(k, Z))$ Ψ is irreducible. Since $\pi_1^{-1}(\Gamma) = \{(\Gamma, W)\colon W \in \mathrm{Gr}(p, m+p), \Gamma \subset W\} \simeq \mathrm{Gr}(p-k, m+p-k)$, we have $\dim \Psi = \dim \mathrm{Gr}(k, Z) + \dim \mathrm{Gr}(p-k, m+p-k) = k(m-k) + (p-k)m = mp - k^2$. In particular, if $k = 1$, $\dim \sigma_1(Z) = mp - 1$ and $\sigma_1(Z)$ is a hypersurface in $\mathrm{Gr}(p, m+p)$ (called a *Schubert hypersurface*). This will play a role in studying pole placement via output feedback (Chapter 20).

Now we recall (Example 6.72) that if V is a hypersurface, then a general line L_1 meets V in degree $V = d$ points. Suppose that $V \subset \mathbb{P}^N$ is an r-dimensional variety with $r \leq N-2$. Then every L_{N-r} meets V and almost all L_{N-r-1} do not meet V. If $\dim(V \cap L_{N-r}) \geq 1$, then V contains a line. Since a general L_{N-r} is the join of a general L_{N-r-1} and a general L_1, it follows that $\dim(V \cap L_{N-r}) = 0$ for almost all L_{N-r}. In other words, $V \cap L_{N-r}$ is a finite set of points for almost all L_{N-r}.

Theorem 12.18 *If $V \subset \mathbb{P}_k^N$ is an r-dimensional variety (irreducible), then the number of points of intersection of a general L_{N-r} with V is a fixed number, $\delta(V)$.*

Proof. Let us consider a general L_{N-r} with $\dim(V \cap L_{N-r}) = 0$. We may suppose that such an L_{N-r} does not meet V at infinity i.e., (say) on $H_0 = V(Y_0)$. In other words, we work on $\mathbb{P}_{Y_0}^N = \mathbb{A}^N$. Let $X_1, \ldots, X_N$ be coordinates on $\mathbb{A}^N$ and let $U = (U_{ij})$, $i = 1, \ldots, r$, $j = 1, \ldots, N$ and let $Y_u = U \cdot X$ so that

$$(Y_u)_i = \sum_{j=1}^{N} U_{ij} X_j \tag{12.19}$$

for $i = 1, \ldots, r$. Since k is a perfect field, letting $\alpha = (\alpha_{ij})$ and $y_\alpha = \alpha x$ where $k(V) = k(x_1, \ldots, x_N)$ with $x_i = \overline{X}_i = X_i \bmod I(V)$, we have $y_\alpha = (y_{\alpha_1}, \ldots, y_{\alpha_r})$ as a separating transcendence basis of $k(V)$ over k for almost all α. (Cf. I.Appendix D.) Let $Z_\beta = \sum \beta_i X_i$ and let $z_\beta = \sum \beta_i x_i$. For almost all β, z_β is a primitive element of $k(V)$ over $k(y_\alpha)$ i.e., $k(V) = k(y_{\alpha_1}, \ldots, y_{\alpha_r}, z_\beta)$ and z_β is separable algebraic over $k(y_\alpha)$ [by the Theorem of Primitive Element I.D.5]. Consider the affine projection $\pi_{\alpha,\beta}\colon \mathbb{A}^N \to \mathbb{A}^{r+1}$ given by

$$\pi_{\alpha,\beta}(X_1, \ldots, X_N) = \left(\sum \alpha_{1j} X_j, \ldots, \sum \alpha_{rj} X_j, \sum \beta_i X_i\right). \tag{12.20}$$

Then $\pi_{\alpha,\beta}(V)$ is a hypersurface H_V in $\mathbb{A}^{r+1}$ with $k(V) \simeq k(H_V)$ i.e., V and H_V are "birational". In fact (12.20) is the affine projection from an $N-r-1$ dimensional subspace M_{N-r-1} defined by $y_\alpha = 0$, $z_\beta = 0$. Thus, when viewed as a map from $\mathbb{P}^N \to \mathbb{P}^{r+1}$, (12.20) is a projection from a general L_{N-r-2} (e.g., assuming $V \cap H_0 = \emptyset$, from the L_{N-r-2} generated by (12.20) and $Y_0 = 0$). So we have shown that the projection from almost all L_{N-r-2} is "birational" to a hypersurface. If $M_{N-r} = L_{N-r} \cap \mathbb{P}^N_{Y_0}$ and if $\boldsymbol{\xi} \in M_{N-r} \cap V$, then $\pi_{\alpha,\beta}(\boldsymbol{\xi}) = \pi_{\alpha,\beta}(V) \cap \pi_{\alpha,\beta}(M_{N-r})$. Conversely, if $\boldsymbol{\eta} \in \mathbb{A}^{r+1}$, then $\pi^{-1}_{\alpha,\beta}(\boldsymbol{\eta}) \cap V$ consists of only one point for almost all $\boldsymbol{\eta}$. In other words, the number of points in $M_{N-r} \cap V$ (a fortiori in $L_{N-r} \cap V$) is the degree of the hypersurface H_V since $\pi_{\alpha,\beta}(M_{N-r})$ is a line. But the degree of H_V, $\deg H_V = [k(V) \colon k(y_\alpha)] = [k(x_1, \ldots, x_N) \colon k(y_{\alpha_1}, \ldots, y_{\alpha_r})] = [k(y_{\alpha_1}, \ldots, y_{\alpha_r}, z_\beta) \colon k(y_{\alpha_1}, \ldots, y_{\alpha_r})]$ is a fixed number (for almost all α, β) (see e.g. [Z-3]), $\delta(V)$.

Corollary 12.21 *If $V \subset \mathbb{P}^N_k$ is an r-dimensional (irreducible) variety, then there is a hypersurface $H_V \subset \mathbb{P}^{r+1}$ with $k(V) = k(H_V)$.*

Theorem 12.22 *Let $V \subset \mathbb{P}^N_k$ be an r-dimensional variety and let d_V = degree V so that the Hilbert polynomial $h_V(t) = \frac{d_V t^2}{r!} + \cdots$. Then $d_V = \delta(V)$ = the number of points of intersection of a general L_{N-r} with V (see 6.66).*

Proof. ([M-5]) Let $L = L_{N-r-2}$ with $L \cap V = \emptyset$ and $\pi_L(V) = H_V$ a hypersurface in $\mathbb{P}^{r+1}$ and $k(V) = k(H_V)$. Then $\delta(V) = \deg H_V$ and, by 6.67, the Hilbert polynomial h_{H_V} of H_V is of the form $(\deg H_V)\frac{t^r}{r!} + \cdots$. Thus it will be enough to show that h_V and h_{H_V} differ only by terms of degree less than r. We may assume that

$$\begin{aligned} L_{N-r-2} &= V(Y_0, \ldots, Y_{r+1}) \\ S(V) &= k[Y_0, \ldots, Y_N]/I_h(V) \\ S(H_V) &= k[Y_0, \ldots, Y_{r+1}]/(F) \end{aligned} \tag{12.23}$$

where F is a form and $H_V = V(F)$. Since π_L is birational, $S(V)$ and $S(H_V)$ have the same quotient field. Moreover, by Theorem 12.17, $S(V)$ is a finite $S(H_V)$-module. But then (Exercise 3), there is an element s of $S(H_V)$ with $sS(V) \subset S(H_V)$. Since $S(V)$, $S(H_V)$ are graded and $s = \sum s_q$, $s_q S(V) \subset S(H_V)$ as well. Thus, we may suppose that s is homogeneous of degree q_0. Then

$$s_{q_0} S(V)_{q-q_0} \subset S(H_V)_q \subset S(V)_q \tag{12.24}$$

and, hence, for the Hilbert polynomials,

$$h_V(q - q_0) \leq h_{H_V}(q) \leq h_V(q) \tag{12.25}$$

and the result follows.

Example 12.26 Let $\mathbb{P}^M$, $\mathbb{P}^{N-M-1}$ be complementary linear spaces in $\mathbb{P}^N$ and let $V \subset \mathbb{P}^M$, $W \subset \mathbb{P}^{N-M-1}$ be disjoint varieties. Let $J(V, W)$ be the *join* of V and W i.e.,

$$J(V, W) = \bigcup_{v \in V, w \in W} J(v, w)$$

is the union of the lines joining points of V and W. If $\mathcal{L}(V,W) = \{L_1 \colon L_1 \text{ is a line joining } V \text{ and } W\}$, then $\mathcal{L}(V,W) = \mathcal{L}_1(V) \cap \mathcal{L}_1(W)$ is a subvariety of $\mathbb{G}\mathrm{r}(1,N)$ and, by (12.10), $J(V,W)$ is a variety. Then $S(J(V,W)) = S(V) \otimes_k S(W)$ as a graded ring and, letting $J = J(V,W)$,

$$S(J)_q = \oplus \sum S(V)_\lambda \otimes_k S(W)_{q-\lambda}. \tag{12.27}$$

It follows that

$$h_J(q) = \sum_{\lambda=0}^{q} h_V(\lambda) h_W(q-\lambda) \tag{12.28}$$

where h_J, h_V, h_W are Hilbert polynomials. But

$$\begin{aligned}
\sum_{\lambda=0}^{q} h_V(\lambda) h_W(q-\lambda) &= \sum_{\lambda=0}^{q} \left(\deg V \cdot \binom{\lambda+r}{r} + \cdots\right) \\
&\quad \cdot \left(\deg W \cdot \binom{q-\lambda+s}{0} + \cdots\right) \\
&= \deg V \cdot \deg W \sum_{\lambda=0}^{q} \binom{\lambda+r}{r}\binom{q-\lambda+s}{s} + \mathrm{order}(q^{s+r}) \\
&= \deg V \cdot \deg W \frac{q^{r+s+1}}{(r+s+1)!} + \mathrm{order}(q^{s+r})
\end{aligned}$$

where $r = \dim V$, $s = \dim W$. Since $\dim J = \dim V + \dim W + 1 = r + s + 1$, we have, for the degree of the join,

$$\deg J(V,W) = \deg V \cdot \deg W. \tag{12.29}$$

If W is a linear space L, then $\deg J(V,L) = \deg V$. We shall use this example later in dealing with Bezout's Theorem.

Example 12.30 Let $V \subset \mathbb{P}^N$ be an r-dimensional variety and let $H = V(F)$ be a hypersurface with F a form of degree m such that $V \not\subset H$. We note that every irreducible component of $V \cap H$ has dimension $r-1$ (i.e., $V \cap H$ is of *pure dimension* $r-1$). Since $\dim(V \cap H) = r-1 = r+(N-1)-N = \dim V + \dim H - N$, we say that V and H *intersect properly*. Since $I_h(V \cap H) = I_h(V) + I_h(H)$ and letting $M = S/I_h(V \cap H) = S/I_h(V) + I_h(H)$ where $S = k[Y_0, \ldots, Y_N]$, we have

$$h_{V\cap H}(q) = \deg(V \cap H)\frac{q^{r-1}}{(r-1)!} + \cdots = h_M(q) = \dim_k M_q.$$

If we consider the exact sequence of graded S-modules

$$0 \longrightarrow \frac{S}{I_h(V)}(-m) \xrightarrow{F} \frac{S}{I_h(V)} \longrightarrow M \longrightarrow 0$$

then we have

$$\begin{aligned} h_{V\cap H}(q) &= h_V(q) - h_V(q-m) \\ &= \deg V \frac{q^r}{r!} + \cdots - \left[\deg V \frac{(q-m)^r}{r!} + \cdots\right] \\ &= m \deg V \frac{q^{r-1}}{(r-1)!} + \cdots \\ &= \deg H \cdot \deg V \frac{q^{r-1}}{(r-1)!} + \cdots . \end{aligned}$$

In other words,

$$\deg(V \cap H) = m \deg V = \deg H \cdot \deg V \tag{12.31}$$

where H is a hypersurface of degree m. In particular, if H is a hyperplane, then $\deg(V \cap H) = \deg V$.

Exercises

(1) Generalize Example 12.6.

(2) Verify the computations in Example 12.15.

(3) Show there is an s in $S(H_V)$ with $sS(V) \subset S(H_V)$. [Theorem 12.22].

13

The State Space: Realizations, Controllability, Observability, Equivalence

We have already introduced "realizations" in dealing with the transfer and Hankel matrices (see Chapter 3). In this chapter, we extend the theory developed in Part I (e.g., Chapters 10 and 11).

Definition 13.1 The triple $x = (A, B, C)$ is a *(state space) realization of* $\mathbf{F}(z)$ if $\mathbf{F}(z) = C(zI - A)^{-1}B$, or, equivalently, if $\mathbf{F}(z) = \sum_{j=1}^{\infty} CA^{j-1}Bz^{-j}$. Similarly, $x = (A, B, C)$ is a *(state space) realization* of the Hankel matrix $\mathcal{H} = (H_{i+j-1})$ if $H_j = CA^{j-1}B$ for all j. (Here $A \in M(n, n)$, $B \in M(n, m)$, $C \in M(p, n)$.)

We view $x = (A, B, C)$ as a point in $\mathbb{A}_k^{n^2+n(m+p)}$ and we use the polynomial ring $k[X, Y, Z] = k[(X_j^i), (Y_r^i), (Z_j^s)]$. We define the various realization maps.

First, let $\mathbf{P}_0, \ldots, \mathbf{P}_{n-1}$, $\xi_0, \ldots, \xi_{n-1}$ denote the coordinate functions on $\mathbb{A}^{n(mp+1)}$ and we define a morphism $\mathcal{R}_f \colon \mathbb{A}^{n^2+n(m+p)} \to \mathbb{A}^{n(mp+1)}$ by

$$\begin{aligned}\mathcal{R}_f(X, Y, Z) = (&\mathbf{P}_0(X, Y, Z), \ldots, \mathbf{P}_{n-1}(X, Y, Z), \\ &\xi_0(X, Y, Z), \ldots, \xi_{n-1}(X, Y, Z))\end{aligned} \tag{13.2}$$

where

$$\begin{aligned}F(z, X, Y, Z) &= Z\,\mathrm{adj}(zI - X)Y / \det(zI - X) \\ &= \frac{\mathbf{P}_0(X, Y, Z) + \cdots + \mathbf{P}_{n-1}(X, Y, Z)z^{n-1}}{\xi_0(X, Y, Z) + \cdots + \xi_{n-1}(X, Y, Z)z^{n-1} + z^n}\end{aligned} \tag{13.3}$$

and

$$\xi_j(X, Y, Z) = -\chi_{n-j}(X) \tag{13.4}$$

P. Falb, *Methods of Algebraic Geometry in Control Theory: Part II*,
Modern Birkhäuser Classics, https://doi.org/10.1007/978-3-319-96574-1_13

for $j = 0, \dots, n-1$ and $\det(zI - X) = z^n - \chi_1(X)z^{n-1} - \cdots - \chi_n(X)$. We call $\mathcal{R}_f$ the *transfer matrix realization map.*

Let $\mathbf{H}_1, \dots, \mathbf{H}_{2n}$ denote the coordinate functions on $\mathbb{A}^{2nmp}$ and define a morphism $\mathcal{R}_h \colon \mathbb{A}^{n^2+n(m+p)} \to \mathbb{A}^{2nmp}$ by

$$\mathcal{R}_h(X, Y, Z) = (\mathbf{H}_1(X, Y, Z), \dots, \mathbf{H}_{2n}(X, Y, Z)) \tag{13.5}$$

where

$$\mathbf{H}_j(X, Y, Z) = (ZX^{j-1}Y) \tag{13.6}$$

for $j = 1, \dots, 2n$. We let $\mathbf{H}(X, Y, Z) = (ZX^{i+j-2}Y)_{i,j=1}^{\infty}$ as well. We call $\mathcal{R}_h$ the *Hankel matrix realization* map. Note that $n(mp + 1) = 2nmp$ if and only if $m = 1$, $p = 1$.

Finally, let $\mathbf{U}_1, \dots, \mathbf{U}_n, u_{n+1}, \dots, u_{2n}$ denote the coordinate functions on $\mathbb{A}^{n(mp+1)}$ and define a morphism $\mathcal{R}_\chi \colon \mathbb{A}^{n^2+n(m+p)} \to \mathbb{A}^{n(mp+1)}$ by

$$\begin{aligned}\mathcal{R}_\chi(X, Y, Z) = (\mathbf{U}(X, Y, Z), \dots, \mathbf{U}_n(X, Y, Z),\\ u_{n+1}(X, Y, Z), \dots, u_{2n}(X, Y, Z))\end{aligned} \tag{13.7}$$

where

$$\begin{aligned}\mathbf{U}_j(X, Y, Z) &= ZX^{j-1}Y, & j &= 1, \dots, n\\ u_{n+t}(X, Y, Z) &= -\chi_{n+1-t}(X), & t &= 1, \dots, n\end{aligned} \tag{13.8}$$

(and $\det(zI - X) = z^n - \chi_1(X)z^{n-1} - \cdots - \chi_n(X)$). We call $\mathcal{R}_\chi$ the *characteristic function realization map.*

None of the realization maps are injective and none are surjective in general. Now let us introduce some intertwining morphisms. We let $\psi_\chi \colon \mathbb{A}^{n(mp+1)} \to \mathbb{A}^{n(mp+1)}$ be given by

$$\psi_\chi(\mathbf{U}_j, u_{n+t}) = (\mathbf{P}_r(\mathbf{U}, u), \xi_j(\mathbf{U}, u)) \tag{13.9}$$

where

$$\begin{aligned}\xi_j(\mathbf{U}, u) &= u_{n+j+1}, & j &= 0, 1, \dots, n-1\\ \mathbf{P}_r(\mathbf{U}, u) &= \sum_{\ell=r+1}^{n} u_{n+1+\ell}\mathbf{U}_{\ell-r}, & r &= 0, 1, \dots, n-1\end{aligned} \tag{13.10}$$

(with $u_{2n+1} = 1$ for convenience). Note that $\psi_\chi = \psi_L^{-1}$ is the inverse of the Laurent map.

Proposition 13.11 *ψ_χ is an isomorphism and $\psi_\chi \circ \mathcal{R}_\chi = \mathcal{R}_f$.*

Proof. Let $\varphi\colon \mathbb{A}^{n(mp+1)} \to \mathbb{A}^{n(mp+1)}$ be the morphism given by $\varphi(\mathbf{P}_r, \xi_j) = (\mathbf{U}_t(\mathbf{P},\xi), u_{n+s}(\mathbf{P},\xi))$ where

$$\begin{aligned} u_{n+s}(\mathbf{P},\xi) &= \xi_{s-1}, \quad s = 1,\dots,n \\ \mathbf{U}_1(\mathbf{P},\xi) &= \mathbf{P}_{n-1} \\ \mathbf{U}_t(\mathbf{P},\xi) &= \mathbf{P}_{n-t} - \sum_{j=2}^{t} \xi_{n+1-j}\mathbf{U}_{t+1-j}(\mathbf{P},\xi) \end{aligned}$$

for $t = 2,\dots,n$. Then $\varphi \circ \psi_\chi$ is the identity for: $(\varphi \circ \psi_\chi)(\mathbf{U},u)_{n+s} = \xi_{s-1}(\mathbf{U},u) = u_{n+s-1+1} = u_{n+s}$ for $s = 1,\dots,n$ and $(\varphi \circ \psi_\chi)(\mathbf{U},u)_1 = \mathbf{P}_{n-1}(\mathbf{U},u) = \mathbf{U}_1$ and, for $t = 2,\dots,n$,

$$\begin{aligned} (\varphi \circ \psi_\chi)(\mathbf{U},u)_t &= \mathbf{P}_{n-t}(\mathbf{U},u) - \sum_{j=2}^{t} \xi_{n+1-j}\mathbf{U}_{t+1-j}(\mathbf{P}(\mathbf{U},u), \xi(\mathbf{U},u)) \\ &= \sum_{\ell=n-t+1}^{n} \xi_\ell \mathbf{U}_{t+\ell-n} - \sum_{j=2}^{t} \xi_{n+1-j}\mathbf{U}_{t+1-j} \\ &= \mathbf{U}_t + \sum_{j=2}^{t} \xi_{n+1-j}\mathbf{U}_{t+1-j} - \sum_{j=2}^{t} \xi_{n+1-j}\mathbf{U}_{t+1-j} \\ &= \mathbf{U}_t. \end{aligned} \tag{13.12}$$

Thus ψ_χ is an isomorphism and, clearly, $\psi_\chi \circ \mathcal{R}_\chi = \mathcal{R}_f$.

Example 13.13 (To illustrate (13.12)) Let $t = 2$. Then $(\varphi \circ \psi_\chi)(\mathbf{U},u)_2 = \mathbf{P}_{n-2}(\mathbf{U},u) - \xi_{n-1}\mathbf{U}_1 = \mathbf{U}_2 + \xi_{n-1}\mathbf{U}_1 - \xi_{n-1}\mathbf{U}_1 = \mathbf{U}_2$. Let $t = 3$. Then $(\varphi \circ \psi_\chi)(\mathbf{U},u)_3 = \mathbf{P}_{n-3}(\mathbf{U},u) - \sum_{j=2}^{3} \xi_{n+1-j}\mathbf{U}_{4-j}(\mathbf{P},\xi) = \mathbf{U}_3 + \xi_{n-1}\mathbf{U}_2 + \xi_{n-2}\mathbf{U}_1 - \xi_{n-1}\mathbf{U}_2(\mathbf{P},\xi) - \xi_{n-2}\mathbf{U}_1(\mathbf{P},\xi) = \mathbf{U}_3 + \xi_{n-1}\mathbf{U}_2 + \xi_{n-1}\mathbf{U}_1 - \xi_{n-1}(\mathbf{P}_{n-2}(\mathbf{U},u) - \xi_{n-1}\mathbf{U}_1) - \xi_{n-2}\mathbf{P}_{n-1}(\mathbf{U},u) = \mathbf{U}_3 + \xi_{n-1}\mathbf{U}_2 + \xi_{n-2}\mathbf{U}_1 - \xi_{n-1}\mathbf{U}_2 - \xi_{n-2}\mathbf{U}_1 = \mathbf{U}_3$ since $\mathbf{P}_{n-2} = \mathbf{U}_2 + \xi_{n-1}\mathbf{U}_1$ and $\mathbf{P}_{n-1}(\mathbf{U},u) = \mathbf{U}_1$.

Now let $\psi_h\colon \mathbb{A}^{n(mp+1)} \to \mathbb{A}^{2nmp}$ be the morphism given by

$$\psi_h(\mathbf{H}_1,\dots,\mathbf{H}_n,\xi_0,\dots,\xi_{n-1}) = (\mathbf{H}_1,\dots,\mathbf{H}_{2n}) \tag{13.14}$$

where

$$\mathbf{H}_{n+j} = -\xi_0\mathbf{H}_j - \cdots - \xi_{n-1}\mathbf{H}_{n+j-1} \tag{13.15}$$

for $j = 1,\dots,n$. Then $\psi_h \circ \mathcal{R}_\chi = \mathcal{R}_h$ and also $\psi_h \circ \psi_\chi^{-1} \circ \mathcal{R}_f = \mathcal{R}_h$.

Definition 13.16 Let $\mathrm{Hankel}(n,m,p) \subset \mathbb{A}^{2nmp}$ be the set of $(\mathbf{H}_1,\dots,\mathbf{H}_{2n})$ such that there is an $\mathcal{H}_\infty^\infty$ in $\mathbf{Hank}(n,m,p)$ with $(\mathcal{H}_\infty^\infty)_j^1 = \mathbf{H}_j$, $j = 1,\dots,2n$.

We claim that $\mathbf{Hank}(n,m,p)$ can be identified with Hankel(n,m,p). Define a map $\psi\colon \mathbf{Hank}(n,m,p) \to \text{Hankel}(n,m,p)$ by

$$\psi(\mathcal{H}) = (\mathcal{H}_1^1, \mathcal{H}_2^1, \ldots, \mathcal{H}_{2n}^1). \tag{13.17}$$

By definition ψ is surjective and well-defined.

Proposition 13.18 *ψ is injective.*

Proof. Suppose that $\psi(\mathcal{H}) = \psi(\mathcal{H}_1)$ and that $x = (A,B,C)$, $x_1 = (A_1,B_1,C_1)$ are realizations of $\mathcal{H}$, $\mathcal{H}_1$, respectively, of order n. Then

$$CA^{t-1}B = C_1A_1^{t-1}B_1 \tag{13.19}$$

for $t = 1,\ldots,2n$. Let

$$\mathbf{Z} = \begin{bmatrix} C \\ CA \\ \vdots \\ CA^{n-1} \end{bmatrix}, \quad \mathbf{Y} = [B\ AB \cdots A^{n-1}B]$$

$$\mathbf{Z}_1 = \begin{bmatrix} C_1 \\ C_1A_1 \\ \vdots \\ C_1A_1^{n-1} \end{bmatrix}, \quad \mathbf{Y}_1 = [B_1\ A_1B_1 \cdots A_1^{n-1}B_1]$$

so that

$$\mathbf{Z}\mathbf{Y} = \mathbf{Z}_1\mathbf{Y}_1 \tag{13.20}$$

by (13.19) and $\rho(\mathbf{Z},\mathbf{Y}) = n = \rho(\mathbf{Z}_1,\mathbf{Y}_1)$ (by the lemma of Risannen). Thus, $\mathbf{Z}$, $\mathbf{Z}_1$, $\mathbf{Y}$, $\mathbf{Y}_1$ are all of full rank n. Let $\mathbf{Z}^\alpha$ be the first n independent rows of $\mathbf{Z}$ and let $\mathbf{Y}_\beta$ be the first n independent columns of $\mathbf{Y}$. Then $\mathbf{Z}^\alpha\mathbf{Y}_\beta = \mathbf{Z}_1^\alpha\mathbf{Y}_{1\beta}$ and $\mathbf{Z}^\alpha$, $\mathbf{Z}_1^\alpha$, $\mathbf{Y}_\beta$, $\mathbf{Y}_{1\beta}$ are all non-singular (i.e., are elements of $G = \text{GL}(n;k)$). In view of (13.19), we have $\mathbf{Z}^\alpha B = \mathbf{Z}_1^\alpha B_1$, $C\mathbf{Y}_\beta = C_1\mathbf{Y}_{1\beta}$, and $\mathbf{Z}^\alpha A\mathbf{Y}_\beta = \mathbf{Z}_1^\alpha A_1\mathbf{Y}_{1\beta}$ (since $CA^{2n-1}B = C_1A_1^{2n-1}B_1$). It follows that if $g = (\mathbf{Z}_1^\alpha)^{-1}\mathbf{Z}_\alpha = \mathbf{Y}_{1\beta}\mathbf{Y}_\beta^{-1}$, then $g \cdot x = x_1$ and, hence, $\mathcal{H} = \mathcal{H}_1$ as $CA^{t-1}B = ((g^{-1})(gA^{t-1}g^{-1})(gB) = C_1A_1^{t-1}B_1$ for *all* t.

Corollary 13.21 *If $x = (A,B,C)$ and $x_1 = (A_1,B_1,C_1)$ are order n realizations of $\mathcal{H} \in \mathbf{Hank}(n,m,p)$, then there is a $g \in \text{GL}(n;k)$ with $g \cdot x = x_1$.*

Example 13.22 Let $m = p = 2$ and $n = 2$. Then (I_2, O_2, O_2, I_2) is *not* an element of Hankel(2, 2, 2) even though $\rho(\mathbf{H}) = 2$ where

$$\mathbf{H} = \begin{bmatrix} I_2 & O_2 \\ O_2 & O_2 \end{bmatrix}.$$

For if $x = (A, B, C)$ were a realization of order 2, then $CB = I_2$, $CAB = O_2$, $CA^2B = O_2$, $CA^3B = I_2$ but $CA^3B = \chi_1(A)CA^2B + \chi_2(A)CAB = O_2$ a contradiction. Thus, Hankel$(n, m, p) < \{(\mathbf{H}_1, \ldots, \mathbf{H}_{2n})\colon \rho(\mathcal{H}(\mathbf{H}_j)) = n\}$.

Proposition 13.23 *ψ_h: Hank$(n, m, p) \to$ Hankel(n, m, p) is a bijective morphism.*

Proof. Let $(\mathbf{H}_1, \ldots, \mathbf{H}_n, \xi_0, \ldots, \xi_{n-1})$ be an element of Hank(n, m, p) and let $\mathbf{H}_{n+j} = -\xi_0\mathbf{H}_j - \cdots - \xi_{n-1}\mathbf{H}_{n+j-1}$ for $j = 1, \ldots$. Then $\mathcal{H}(\mathbf{H}_j) \in$ **Hank**(n, m, p) and (by (13.17)) $\psi(\mathcal{H}(\mathbf{H}_j)) = \psi_h(\mathbf{H}_1, \ldots, \mathbf{H}_n, \xi_0, \ldots, \xi_{n-1})$. In other words, ψ_h(Hank(n, m, p)) $\subset$ Hankel(n, m, p). If $x = (A, B, C)$ is an order n realization of $(\mathbf{H}_1, \ldots, \mathbf{H}_{2n}) \in$ Hankel(n, m, p), then $\psi_h(CB, \ldots, CA^{n-1}B, -\chi_1(A), \ldots, -\chi_n(A)) = (\mathbf{H}_1, \ldots, \mathbf{H}_{2n})$ so that ψ_h is surjective. If

$$\psi_h(\mathbf{H}_1, \ldots, \mathbf{H}_n, \ \xi_0, \ldots, \xi_{n-1}) = \psi_h(\mathbf{H}'_1, \ldots, \mathbf{H}'_n, \xi'_0, \ldots, \xi'_{n-1}),$$

then $\mathbf{H}_i = \mathbf{H}'_i$ and $\mathbf{H}_{n+i} = \mathbf{H}'_{n+i}$ for $i = 1, \ldots, n$. Since ψ (given by (13.17)) is bijective, there are realizations x, x' of order n with $g \cdot x = x'$. Thus $\chi_i(x) = \chi_i(x')$ and $\xi_j = \xi'_j$ and ψ_h is injective.

Example 13.24 Let $m = p = 2$ and $n = 2$. Let $\mathbf{H} = (H_1, H_2, H_3, H_4)$ and consider the equations

$$\begin{aligned} x_0H_1 + x_1H_2 &= H_3 \\ x_0H_2 + x_1H_3 &= H_4. \end{aligned}$$

Suppose that $\rho(\mathcal{H}) = 2$ where $\mathcal{H} = (\mathbf{H}_{i+j-1})_{i,j=1}^2$. Take, for example, $H_1 = I_2$, $H_2 = O_2$, $H_3 = O_2$, $H_4 = O_2$. Then $x_0 = 0$, x_1 *any* is a solution. Thus, simply inverting the appropriate set of subequations does not show that ψ_h^{-1} is a morphism. Of course, here (O_2, I_2, I_2) is a realization of order 2 and the Grassmann equations $\lambda_0 = \det H_1$, $\lambda_0 x_0 = \det[x_1H_1 + H_2]$, $\lambda_0 x_1 = \det[(x_1H_1 + H_2)_1 \ (H_1)_2] - \det[(x_1H_1 + H_2)_2 \ (H_1)_1] = \det[(H_2)_2 \ (H_1)_1] - \det[(H_2)_1 \ (H_1)_2]$ have a unique solution for x_0, x_1.

To show that ψ_h^{-1}: Hankel$(n, m, p) \to$ Hank(n, m, p) is a morphism will require an appropriate local coordinatization ([C-1]). Let $\mathbf{e} = (e^1, \ldots, e^n)$, $1 \le e^1 < \cdots < e^n$ be a (row) index and call $\mathbf{e}$ *admissible* if $e^i + p \in \mathbf{e}$ implies $e^i \in \mathbf{e}$. Let $\mathbf{x} = (\mathbf{H}_1, \ldots, \mathbf{H}_{2n})$ be an element of Hankel(n, m, p) and let

$\mathcal{H}_\infty^\infty(\mathbf{x}) = \mathcal{H}(\mathbf{H}_t)$ be the element of $\mathbf{Hank}(n, m, p)$ with $\psi(\mathcal{H}_\infty^\infty(\mathbf{x})) = \mathbf{x}$. Then there is an admissible index $\mathbf{e} = \mathbf{e}(\mathbf{x})$ and (unique) matrices $C(\mathbf{x}) \in M(p, n)$ and $A(\mathbf{x}) \in M(n, n)$ such that

$$\begin{aligned} C(\mathbf{x})\mathcal{H}_\infty^{\mathbf{e}}(\mathbf{x}) &= \mathcal{H}_\infty^{\mathbf{p}}(\mathbf{x}) \\ A(\mathbf{x})\mathcal{H}_\infty^{\mathbf{e}}(\mathbf{x}) &= \mathcal{H}_\infty^{\mathbf{e}+p}(\mathbf{x}) \end{aligned} \tag{13.25}$$

where $\mathbf{p} = (1, \ldots, p)$ and $\mathbf{e} + p = (e^1 + p, \ldots, e^n + p)$. Set $B(\mathbf{x}) = \mathcal{H}_{(1\cdots m)}^{\mathbf{e}}(\mathbf{x})$. Then $(A(\mathbf{x}), B(\mathbf{x}), C(\mathbf{x}))$ is an order n realization of $\mathcal{H}_\infty^\infty(\mathbf{x})$ or, equivalently, of $\mathbf{x}$. We *claim* that

$$\mathcal{H}_\infty^{\mathbf{e}}(\mathbf{x}) \subset \begin{bmatrix} \mathcal{H}_\infty^{\mathbf{p}}(\mathbf{x}) \\ \mathcal{H}_\infty^{\mathbf{e}+p}(\mathbf{x}) \end{bmatrix} \tag{13.26}$$

(i.e., rows are subrows). If $e^1 < \cdots < e^n$ and, say, $e^1 \notin (1, \ldots, p)$, then $e^1 = \tilde{e}^1 + p$ and $\mathbf{e}$ admissible, together imply that $\tilde{e}^1 \in \mathbf{e}$ with $\tilde{e}^1 < e^1$ (a contradiction). So let $e^1, \ldots, e^s \in \mathbf{p}$ and $e^{s+1} \notin \mathbf{p}$, then $e^{s+1} = \tilde{e}^{s+1} + p$ and $\mathbf{e}$ admissible together imply that $\tilde{e}^{s+1} \in \mathbf{e}$ and, hence, that $e^{s+1} \in \mathbf{e} + p$. We now let $\mathbf{L} = \mathbf{L}(\mathbf{x})$ be the element of $M(p + n, n)$ given by

$$\mathbf{L} = \begin{bmatrix} C(\mathbf{x}) \\ A(\mathbf{x}) \end{bmatrix}. \tag{13.27}$$

Then

$$\mathbf{L}\mathcal{H}_\infty^{\mathbf{e}}(\mathbf{x}) = \begin{bmatrix} \mathcal{H}_\infty^{\mathbf{p}}(\mathbf{x}) \\ \mathcal{H}_\infty^{\mathbf{e}+p}(\mathbf{x}) \end{bmatrix} \tag{13.28}$$

and, by virtue of (13.26), $\mathbf{L}^{\mathbf{e}} = I_n$. Let $\mathbf{Q} = \mathbf{Q}(\mathbf{x}) = \mathbf{L} - \mathbf{L}^{\mathbf{e}}$ (abuse of notation) be the element of $M(p, n)$ formed by deleting the $\mathbf{e}$ rows of $\mathbf{L}$. Let $U^{\mathbf{e}} = \{\mathbf{y} \in \text{Hankel}(n, m, p)\colon \rho(\mathcal{H}_\infty^{\mathbf{e}}(\mathbf{y})) = n\}$. Note that $\mathbf{x} \in U^{\mathbf{e}}$ and that $U^{\mathbf{e}}$ is open (e.g. via the Lemma of Risannen). Define a map $\phi^{\mathbf{e}} = \phi\colon U^{\mathbf{e}} \to \mathbb{A}_k^{n(m+p)}$ by

$$\phi(\mathbf{y}) = (\mathbf{Q}(\mathbf{y}), B(\mathbf{y})). \tag{13.29}$$

If $\phi(\mathbf{y}) = \phi(\mathbf{y}')$, then $A(\mathbf{y}) = A(\mathbf{y}')$, $B(\mathbf{y}) = B(\mathbf{y}')$ and $C(\mathbf{y}) = C(\mathbf{y}')$ so that $\mathbf{y} = \mathbf{y}'$ and ϕ is injective. Let $(\mathbb{A}_k^{n(m+p)})_*$ be the open set where $\mathbf{Z}(\mathbf{Q}, B)$ and $\mathbf{Y}(\mathbf{Q}, B)$ have rank n. The matrices $\mathbf{Z}(\mathbf{Q}, B)$, $\mathbf{Y}(\mathbf{Q}, B)$ are given by

$$\begin{aligned} \mathbf{Z}(\mathbf{Q}, B) &= \mathbf{Z}(\mathbf{L}^{p+1\cdots p+n}, B, \mathbf{L}^{1\cdots p}) \\ \mathbf{Y}(\mathbf{Q}, B) &= \mathbf{Y}(\mathbf{L}^{p+1\cdots p+n}, B, \mathbf{L}^{1\cdots p}) \end{aligned} \tag{13.30}$$

where $\mathbf{L} = \mathbf{Q} + I_n^{\mathbf{e}}$ (abuse of notation). Then $\phi(\mathbf{U}^{\mathbf{e}}) = (\mathbb{A}_k^{n(m+p)})_*$. Since $\rho(\mathcal{H}_\infty^{\mathbf{e}}(\mathbf{x})) = n$, there is a (column) index $\mathbf{f} = (f_1, \ldots, f_n)$ with $\det \mathcal{H}_{\mathbf{f}}^{\mathbf{e}}(\mathbf{x}) \neq 0$ and also, $\det \mathcal{H}_{\mathbf{f}}^{\mathbf{e}}(\mathbf{y}) \neq 0$ on a neighborhood $U_{\mathbf{f}}^{\mathbf{e}}$ of $\mathbf{x}$. By Cramer's rule, $A(\mathbf{y})$, $B(\mathbf{y})$, $C(\mathbf{y})$ are regular on $U_{\mathbf{f}}^{\mathbf{e}}$. Thus ϕ is a bijective morphism. If $(\mathbf{Q}, B) \in$

$(\mathbb{A}_k^{n(m+p)})_*$, then $\phi^{-1}(\mathbf{Q}, B) = \mathbf{L}^{1\cdots p}(\mathbf{L}^{p+1\cdots p+n})^{j-1}B$ for $j = 1, \ldots, 2n$ where $\mathbf{L} = \mathbf{Q} + I_n^{\mathbf{e}}$ (abuse of notation). In other words, ϕ^{-1} is a morphism and so the $U^{\mathbf{e}}$, $\phi^{\mathbf{e}}$ give a local coordinatization. This means that locally Hankel(n, m, p) looks like $(\mathbb{A}_k^{n(m+p)})_*$, an open set in $\mathbb{A}_k^{n(m+p)}$. Thus,

Proposition 13.31 Hankel(n, m, p) *has dimension* $n(m+p)$ *and is normal.*

Proposition 13.32 Hank(n, m, p) *and* Hankel(n, m, p) *are isomorphic.*

Proof. If $\mathbf{x} = (\mathbf{H}_1, \ldots, \mathbf{H}_{2n}) \in U^{\mathbf{e}}$, then

$$\phi_h^{-1}(\mathbf{y}) = (C(\mathbf{y})A(\mathbf{y})^{j-1}B(\mathbf{y}), -\chi_i(A(\mathbf{y})))$$

is regular on $U_{\mathbf{f}}^{\mathbf{e}}$ and so, is a morphism.

Corollary 13.33 Hank(n, m, p) *has dimension* $n(m+p)$ *and is normal.*

Now let $\mathbf{Y}\colon \mathbb{A}_k^{n^2+n(m+p)} \to M(n, (n+1)m)$ be given by

$$\mathbf{Y}(X, Y, Z) = \begin{bmatrix} Y & XY & \cdots & X^n Y \end{bmatrix}.$$

The morphism Y is called the *controllability map*. Similarly, let $\mathbf{Z}$: $\mathbb{A}_k^{n^2+n(m+p)} \to M((n+1)p, n)$ be the map given by

$$\mathbf{Z}(X, Y, Z) = \begin{bmatrix} Z \\ ZX \\ \vdots \\ ZX^n \end{bmatrix}.$$

The morphism $\mathbf{Z}$ is called the *observability map*. Finally, we let $\mathbf{H}$: $\mathbb{A}_k^{n^2+n(m+p)} \to M((n+1)p, (n+1)m)$ be given by

$$\mathbf{H}(X, Y, Z) = \mathbf{Z}(X, Y, Z)\mathbf{Y}(X, Y, Z)$$

and we call the morphism $\mathbf{H}$ the *Hankel matrix map*. (Frequently we consider only the first nm columns of $\mathbf{Y}$ or the first np rows of $\mathbf{Z}$ and still refer to these as "the" controllability and observability maps. This abuse of language should not be unduly burdensome.) We often use an alternative form of these maps, namely:

$$\tilde{\mathbf{Y}}(X,Y,Z) = [Y_1 \; XY_1 \cdots X^n Y_m]$$

$$\tilde{\mathbf{Z}}(X,Y,Z) = \begin{bmatrix} Z^1 \\ Z^1 X \\ \vdots \\ Z^p X^n \end{bmatrix}$$

$$\tilde{\mathbf{H}}(X,Y,Z) = \tilde{\mathbf{Z}}(X,Y,Z)\tilde{\mathbf{Y}}(X,Y,Z)$$

where the Y_j are the columns of Y and the Z^i are the rows of Z. This form is called the *Hermite form.* It is clear that the ranks are the same and that

$$\rho(\mathbf{H}(A,B,C)) \leq \min[\rho(\mathbf{Y}(A,B,C)), \rho(\mathbf{Z}(A,B,C))] \leq n$$

for all (A,B,C) in $\mathbb{A}_k^{n^2+n(m+p)}$.

Definition 13.34 The linear system (A,B,C) is *controllable* if $\rho(\mathbf{Y}(A,B,C)) = n$ and *observable* if $\rho(\mathbf{Z}(A,B,C)) = n$. If $\rho(\mathbf{H}(A,B,C)) = n$, then (A,B,C) is *minimal.*

A system is minimal if and only if it is both controllable and observable. We have:

Definition 13.35 $S_{m,p}^n = \{x = (A,B,C)\colon x \text{ is minimal}\}$ is the set of *linear systems of degree n with m inputs and p outputs.*

Let $\mathbf{e} = (e^1, \ldots, e^p)$ be a p-partition of n and let $\mathbf{f} = (f_1, \ldots, f_m)$ be an m-partition of n (i.e., $0 \leq e^i \leq n$, $0 \leq f_j \leq n$ and $\sum_{i=1}^{p} e^i = n$, $\sum_{j=1}^{m} f_j = n$). Let $\alpha_{\mathbf{e}} = (1 \cdots e^1 \; n+1 \cdots n+e^2 \cdots n(p-1)+e^p)$ and let $\beta_{\mathbf{f}} = (1 \cdots f_1 \; n+1 \cdots n+f_2 \cdots n(m-1)+f_m)$. The $\alpha_{\mathbf{e}}$ is an indexing of n of the rows of $\tilde{\mathbf{Z}}$ and $\beta_{\mathbf{f}}$ is an indexing of n of the columns of $\tilde{\mathbf{Y}}$. Let

$$\begin{aligned} \omega_{\mathbf{e}}(X,Y,Z) &= \det[\tilde{\mathbf{Z}}(X,Y,Z)|^{\alpha_{\mathbf{e}}}] \\ \gamma_{\mathbf{f}}(X,Y,Z) &= \det[\tilde{\mathbf{Y}}(X,Y,Z)|_{\beta_{\mathbf{f}}}] \end{aligned} \tag{13.36}$$

Then $x = (A,B,C)$ is controllable if and only if $\gamma_{\mathbf{f}}(A,B,C) \neq 0$ for some $\mathbf{f}$ and $x = (A,B,C)$ is observable if and only if $\omega_{\mathbf{e}}(X,Y,Z) \neq 0$ for some $\mathbf{e}$. Obviously, if $\gamma_{\mathbf{f}}(A,B,C) \neq 0$, then x is controllable. To demonstrate the converse, let $V_j = \text{span}[b_j \; Ab_j \cdots A^{n-1}b_j]$ ($= \text{span}[b_j \cdots A^n b_j]$ by Cayley-Hamilton), $W_j = V_1 + \cdots + V_j$, and $f_j = \dim W_j - \dim W_{j-1}$ for $j = 1, \ldots, m$ (where $W_0 = (0)$). Then $0 \leq f_j \leq n$ and $\sum_{j=1}^{m} f_j = \dim W_m = \dim[V_1 + \cdots + V_m] = \dim[\text{span}\, \mathbf{Y}(A,B,C)]$.

Observe that $AV_j \subset V_j$ and $AW_j \subset W_j$ for $j = 1, \ldots, m$. Moreover, $\dim W_t = \sum_{j=1}^{t} f_j$ for $t = 1, \ldots, m$. Set $d_t = \dim W_t$ and let $w_1, \ldots, w_{d_{t-1}}$ be a basis of W_{t-1}. We *claim* that $w_1, \ldots, w_{d_{t-1}}, b_t, \ldots, A^{f_t-1}b_t$ is a basis of W_t. If b_t were in W_{t-1}, then $Ab_t \in AW_{t-1} \subset W_{t-1}$ and so on. In other words, f_t would be 0. So suppose that $w_1, \ldots, w_{d_{t-1}}, b_t, \ldots, A^s b_t$ are independent. Then $s \leq f_t - 1$. If $s < f_t - 1$ and $w_1, \ldots, w_{d_{t-1}}, b_t, \ldots, A^s b_t, A^{s+1}b_t$ are dependent so that $A^{s+1}b_t \in \text{span}[w_1, \ldots, w_{d_{t-1}}, b_t, \ldots, A^s b_t]$, then $A^{s+2}b_t \in \text{span}[w_1, \ldots, w_{d_{t-1}}, b_t, \ldots, A^s b_t]$, etc. Thus, we have $s = f_t - 1$. If $x = (A, B, C)$ is controllable, then $n = \dim[\text{span}\, \mathbf{Y}(A, B, C)] = \dim[\text{span}\, \tilde{\mathbf{Y}}(A, B, C)] = \dim W_m = \sum_{j=1}^{m} f_j$ and $\gamma_{\mathbf{f}}(A, B, C) \neq 0$.

Let $\mathbf{I}_0 = \{\omega_{\mathbf{e}}\text{: all } \mathbf{e}\}$ and let $\mathbf{I}_c = \{\gamma_{\mathbf{f}}\text{: all } \mathbf{f}\}$. Then $V_0 = V(\mathbf{I}_0)$ and $V_c = V(\mathbf{I}_c)$ are algebraic sets in $\mathbb{A}_k^{n^2+n(m+p)}$ and x is observable (controllable) if and only if $x \notin V_0$ ($x \notin V_c$).

Proposition 13.37 $x = (A, B, C) \in S_{m,p}^n$ *if and only if* $x \notin V_0 \cup V_c$ *if and only if* $\det \tilde{\mathbf{H}}_{\beta_{\mathbf{f}}}^{\alpha_{\mathbf{e}}} \neq 0$ *for some* $\mathbf{e}$, $\mathbf{f}$.

Proof. The last assertion follows since $\det \tilde{\mathbf{H}}_{\beta_{\mathbf{f}}}^{\alpha_{\mathbf{e}}} = \det \tilde{\mathbf{Z}}^{\alpha_{\mathbf{e}}} \det \tilde{\mathbf{Y}}_{\beta_{\mathbf{f}}}$.

Proposition 13.38 (1) *If* $x \in S_{m,p}^n$, *then* $\mathcal{R}_\chi(x) \in \text{Hank}(n, m, p)$ *and conversely, if* $\mathbf{x} = (\mathbf{H}_1, \ldots, \mathbf{H}_n, \xi_0, \ldots, \xi_{n-1}) \in \text{Hank}(n, m, p)$, *then there is an* x *in* $S_{m,p}^n$ *with* $\mathcal{R}_\chi(x) = \mathbf{x}$. *In other words,* $\mathcal{R}_\chi\colon S_{m,p}^n \to \text{Hank}(n, m, p)$ *is a surjective morphism;* (2) *If* $x \in S_{m,p}^n$, *then* $\mathcal{R}_f(x) \in \text{Rat}(n, m, p)$ *and conversely, if* $\mathbf{F}(z) \in \text{Rat}(n, m, p)$, *then there is an* $x_{\mathbf{F}} \in S_{m,p}^n$ *with* $\mathcal{R}_f(x_{\mathbf{F}}) = \mathbf{F}(z)$. *In other words,* $\mathcal{R}_f\colon S_{m,p}^n \to \text{Rat}(n, m, p)$ *is a surjective morphism; and,* (3) *If* $x \in S_{m,p}^n$, *then* $\mathcal{R}_h(x) \in \text{Hankel}(n, m, p)$ *and conversely, if* $\mathbf{x} = (\mathbf{H}_1, \ldots, \mathbf{H}_{2n}) \in \text{Hankel}(n, m, p)$, *then there is an* $x \in S_{m,p}^n$ *with* $\mathcal{R}_h(x) = \mathbf{x}$. *In other words,* $\mathcal{R}_h\colon S_{m,p}^n \to \text{Hankel}(n, m, p)$ *is a surjective morphism.*

Proof. (2) will follow from (1) via the Laurent Isomorphism and (3) will follow from (1) since the map ψ_h is an isomorphism (Propositions 13.23 and 13.32). If $x \in S_{p,m}^n$, then x gives an order n realization of $\mathcal{R}_\chi(x)$. Conversely, if $\mathbf{x} = (\mathbf{H}_1, \ldots, \mathbf{H}_n, \xi_0, \ldots, \xi_{n-1}) \in \text{Hank}(n, m, p)$ and x is an order n realization, then $\rho(\mathcal{H}(\mathbf{x})) = \rho(\mathcal{H}(x)) = n$ so that $x \in S_{m,p}^n$.

In view of Proposition 13.37, $S_{m,p}^n$ is open in $\mathbb{A}_k^{n^2+n(m+p)}$ and is, therefore, irreducible. It follows that $\mathcal{R}_\chi(S_{m,p}^n) = \text{Hank}(n, m, p)$ (or $\mathcal{R}_f(S_{m,p}^n) = \text{Rat}(n, m, p)$ or $\mathcal{R}_h(S_{m,p}^n) = \text{Hankel}(n, m, p)$) is also irreducible. Moreover, in view of Corollaries 13.31 and 13.33 and Proposition 13.32 and the Laurent Isomorphism Theorem, all the varieties $\text{Hank}(n, m, p)$, $\text{Rat}(n, m, p)$, and $\text{Hankel}(n, m, p)$ are all normal (in fact, non-singular) of dimension $n(m + p)$.

Let $G = \mathrm{GL}(n, k)$. Then G acts on $\mathbb{A}_k^{n^2+n(m+p)}$ via

$$g \cdot (X, Y, Z) = (gXg^{-1}, g^Y, Zg^{-1}) \tag{13.38}$$

where $g \in G$. We have

Proposition 13.39 (1) $\mathbf{Y}(g \cdot x) = g\mathbf{Y}(x)$ *and* $\mathbf{Z}(g \cdot x) = \mathbf{Z}(x)g^{-1}$; (2) $\mathcal{R}_\chi(g \cdot x) = \mathcal{R}_\chi(x)$, $\mathcal{R}_f(g \cdot x) = \mathcal{R}_f(x)$ *and* $\mathcal{R}_h(g \cdot x) = \mathcal{R}_h(x)$ *so that* $\mathcal{R}_\chi$, $\mathcal{R}_f$, $\mathcal{R}_h$ *are invariant under the action of* G.

Proof. Observe that

$$\mathbf{Y}(gXg^{-1}, gY, Zg^{-1}) = [gY\ gXg^{-1}Y \cdots] = g[Y\ XY \cdots X^nY]$$

as a polynomial identity and similarly for $\mathbf{Z}$. (2) is obvious.

Corollary 13.40 $\mathbb{A}^{n^2+n(m+p)} - V_0$, $\mathbb{A}^{n^2+n(m+p)} - V_c$ *and* $S_{m,p}^n$ *are invariant under* G. *In other words, if* $x = (A, B, C)$ *is observable, controllable or minimal, then so is* $g \cdot x$ *for all* $g \in G$.

Proof. Say x is controllable and $\gamma_{\beta_f}(x) \neq 0$. Then $\mathbf{Y}_{\beta_f}(x) \in G$ and $\mathbf{Y}_{\beta_f}(g \cdot x) = g \cdot \mathbf{Y}_{\beta_f}(x)$ is also nonsingular.

Corollary 13.41 *If* x *is controllable (or observable) and* $g \cdot x = x$, *then* $g = I$ *so that* $S_G(x) = \{I\}$. *Consequently, if* x *is controllable (or observable) and* $g \cdot x = \tilde{g} \cdot x$, *then* $g = \tilde{g}$. [Cf. I.11.15 and I.11.18.]

In view of Corollaries 13.40, 13.41, we have established:

Theorem 13.42 (State Space Isomorphism Theorem) *If* x, x_1 *are minimal realizations of* $(\mathbf{H}_1, \ldots, \mathbf{H}_n,\ \xi_0, \ldots, \xi_{n-1}) \in \mathrm{Hank}(n, m, p)$ *(or* $\mathbf{F}(z) \in \mathrm{Rat}(n, m, p)$ *or* $(\mathbf{H}_1, \ldots, \mathbf{H}_{2n}) \in \mathrm{Hankel}(n, m, p)$*), then there is a unique* $g \in G$ *with* $g \cdot x = x_1$.

We shall eventually show that $(\mathrm{Hank}(n, m, p), \mathcal{R}_\chi)$ gives a geometric quotient of $S_{p,m}^n$ modulo G (see Part I, Chapter 14 and in particular, Definition I.14.19). Here, we first note that if $x = (A, B, C) \in S_{p,m}^n$, then $S_G(x) = \{g\colon g \cdot x = x\} = \{I\}$ implies that $\dim O_G(x) = n^2$ and hence that the *orbits* $O_G(x)$ *are closed in* $S_{p,m}^n$ (in view of Lemma I.15.9). Also, if $x, x_1 \in S_{p,m}^n$ with $\mathcal{R}_\chi(x) = \mathcal{R}_\chi(x_1)$, then, by the State Space Isomorphism Theorem, $x \sim_G x_1$ and $O_G(x) = O_G(x_1)$. In other words, $\mathcal{R}_\chi$ is *injective on orbits.* Moreover, if $\mathbf{x} = (\mathbf{H}_1, \ldots, \mathbf{H}_n, \xi_0, \ldots, \xi_{n-1}) \in \mathrm{Hank}(n, m, p)$, then there is an $x \in S_{p,m}^n$ with $\mathcal{R}_\chi(x) = \mathbf{x}$. It follows that $\mathcal{R}_\chi^{-1}(\mathbf{x}) = O_G(x)$ is a closed orbit.

We wish now to consider serveral representations of $S_{m,p}^n$ which will be useful in dealing with geometric quotients. Let $\mathcal{C}_m^n = \{y = (A, B)\colon y \text{ is controllable}\}$ and let $\mathcal{O}_p^n = \{z = (A, C)\colon z \text{ is observable}\}$. We view $\mathcal{C}_m^n \subset \mathbb{A}_k^{n^2+nm}$ and $\mathcal{O}_p^n \subset \mathbb{A}_k^{n^2+np}$. G acts on $\mathcal{C}_m^n$ and $\mathcal{O}_p^n$ via

$$g \cdot (A, B) = (gAg^{-1}, gB), \quad g \cdot (A, C) = (gAg^{-1}, Cg^{-1}) \tag{13.43}$$

and both are G-invariant. The maps $\mathbf{Y}\colon \mathcal{C}_m^n \to M_*(n, (n+1)m)$ and $\mathbf{Z}\colon \mathcal{O}_p^n \to M_*((n+1)p, n)$ given by

$$\mathbf{Y}(A, B) = [B\ AB \cdots A^n B]$$

$$\mathbf{Z}(A, C) = \begin{bmatrix} C \\ CA \\ \vdots \\ CA^n \end{bmatrix}$$

are injective G-morphisms. Let Δ_A be the A-diagonal in $\mathcal{O}_p^n \times \mathcal{C}_m^n$ i.e., $\Delta_A = \{(A_1, C) \times (A_2, B)\colon A_1 = A_2\}$. Clearly, $S_{m,p}^n \simeq_G \Delta_A$ and so Δ_A is irreducible of dimension $n^2 + n(m+p)$. The map $\psi\colon \Delta_A \to M_*((n+1)p, n) \times M_*(n, (n+1)m) \times \mathbb{A}^{n^2}$ given by

$$\psi((A, C) \times (A, B)) = (\mathbf{Z}(A, C), \mathbf{Y}(A, B), A) \tag{13.44}$$

is an injective G-morphism (if G acts on $\mathbb{A}^{n^2}$ via gAg^{-1}). Let $\mathbf{Z}^1, \ldots, \mathbf{Z}^{n+1}(p \times n)$, $\mathbf{Y}_1, \ldots, \mathbf{Y}_{n+1}(n \times m)$ and $\mathbf{X}(n \times n)$ be (block) coordinates on $M_*((n+1)p, n) \times M_*(n, (n+1)m) \times \mathbb{A}^{n^2}$. The equations

$$\mathbf{Z}^i = \mathbf{Z}^1 \mathbf{X}^{i-1}, \quad \mathbf{Y}_j = \mathbf{X}^{j-1} \mathbf{Y}_1 \tag{13.45}$$

$i, j = 1, \ldots, n+1$ define a variety $\mathbf{V}_A(n, m, p)$ in $M_*((n+1)p, n) \times M_*(n, (n+1)m) \times \mathbb{A}^{n^2}$.

Proposition 13.46 *ψ is a G-isomorphism between Δ_A and $\mathbf{V}_A(n, m, p)$. Hence, $S_{m,p}^n$ and $\mathbf{V}_A(n, m, p)$ are G-isomorphic.*

Proof. We have $\psi(\Delta_A) \subset \mathbf{V}_A(n, m, p)$ and we have noted that ψ is an injective G-morphism. If $v \in \mathbf{V}_A(n, m, p)$ with $v = (\mathbf{Z}_v, \mathbf{Y}_v, \mathbf{X}_v)$, then $(\mathbf{X}_v, \mathbf{Z}_v^1) \in \mathcal{O}_p^n$ (as $\mathbf{Z}_v \in M_*((n+1)p, n)$ i.e., rank n) and $(\mathbf{X}_v, \mathbf{Y}_{v1}) \in \mathcal{C}_m^n$ (as $\mathbf{Y}_v \in M_*(n, (n+1)m)$ i.e., rank n) and $\psi((\mathbf{X}_v, \mathbf{Z}_v^1) \times (\mathbf{X}_v, \mathbf{Y}_{v1})) = v$.

Since $S_{m,p}^n \simeq_G \mathbf{V}_A(n, m, p)$, a geometric quotient of $\mathbf{V}_A(n, m, p)$ will also be a geometric quotient of $S_{m,p}^n$. Let $\psi_A\colon \mathbf{V}_A(n, m, p) \to \mathrm{Hank}(n, m, p)$ be given by

$$\begin{aligned}\psi_A(\mathbf{Z},\mathbf{Y},\mathbf{X}) &= (\mathbf{Z}^1\mathbf{Y}_1,\ldots,\mathbf{Z}^1\mathbf{Y}_n,-\chi_i(\mathbf{X}))\\ &= (\mathbf{Z}^1\mathbf{Y}_1,\mathbf{Z}^1\mathbf{X}\mathbf{Y}_1,\ldots,\mathbf{Z}^1\mathbf{X}^{n-1}\mathbf{Y}_1,-\chi_i(\mathbf{X})). \qquad (13.47)\end{aligned}$$

Clearly, ψ_A is a G-morphism of $\mathbf{V}_A(n,m,p)$ into Hank(n,m,p). Since any element of Hank(n,m,p) has a (minimal) realization, ψ_A is surjective. Moreover, by the State Space Isomorphism Theorem, ψ_A is injective on G-orbits. We shall show in the sequel that (Hank$(n,m,p),\psi_A$) is a geometric quotient of $\mathbf{V}_A(n,m,p)$ modulo G.

We next develop another representation of $S^n_{m,p}$ based on the Hermite form. This complex representation indicates some of the deeper structure of $S^n_{m,p}$. As usual, we let $M_*((n+1)p,n) = \{\mathbf{Z} \in M(n(p+1),n)\colon \rho(\mathbf{Z}) = n\}$ and $M_*(n,(n+1)m) = \{\mathbf{Y} \in M(n,(n+1)m)\colon \rho(\mathbf{Y}) = n\}$. If $\mathbf{Z} \in M((n+1)p,n)$ and $\mathbf{Y} \in M(n,(n+1)m)$, then $\mathbf{Z} = [\mathbf{Z}^i]$, $i = 1,\ldots,p$, $\mathbf{Z}^i \in M((n+1),n)$ and $\mathbf{Y} = [\mathbf{Y}_j]$, $j = 1,\ldots,m$, $\mathbf{Y}_j \in M(n,n+1)$. Let

$$\begin{aligned}V^i(\mathbf{Z}) &= \mathrm{span}[\mathbf{Z}^i], \quad W^i(\mathbf{Z}) = \textstyle\sum_{r=1}^i V^r(\mathbf{Z})\\ V_j(\mathbf{Y}) &= \mathrm{span}[\mathbf{Y}_j], \quad W_j(\mathbf{Y}) = \textstyle\sum_{s=1}^j W_j(\mathbf{Z})\end{aligned} \qquad (13.48)$$

for $i = 1,\ldots,p$, $j = 1,\ldots,m$. Let

$$\begin{aligned}e^i(\mathbf{Z}) &= \dim W^i(\mathbf{Z}) - \dim W^{i-1}(\mathbf{Z})\\ f_j(\mathbf{Y}) &= \dim W_j(\mathbf{Y}) - \dim W_{j-1}(\mathbf{Y})\end{aligned} \qquad (13.49)$$

where $W^0(\mathbf{Z}) = (0)$, $W_0(\mathbf{Y}) = (0)$, $i = 1,\ldots,p$, $j = 1,\ldots,m$. Then $0 \le e^i(\mathbf{Z}) \le n$, $0 \le f_j(\mathbf{Y}) \le n$ and $\sum_{i=1}^{p} e^i(\mathbf{Z}) \le n$ with equality if and only if $\mathbf{Z} \in M_*((n+1)p,n)$ and $\sum_{j=1}^{m} f_j(\mathbf{Y}) \le n$ with equality if and only if $\mathbf{Y} \in M_*(n,(n+1)m)$. Let $G = \mathrm{GL}(n,k)$ so that G acts on $M((n+1)p,n)$ and $M(n,(n+1)m)$ via

$$\mathbf{Z} \to \mathbf{Z}g^{-1}, \quad \mathbf{Y} \to g\mathbf{Y} \qquad (13.50)$$

for $g \in G$. Note that

$$\begin{aligned}&(\mathbf{Z}g^{-1})^i = \mathbf{Z}^i g^{-1},\ V^i(\mathbf{Z}g^{-1}) = V^i(\mathbf{Z})g^{-1},\ W^i(\mathbf{Z}g^{-1}) = W^i(\mathbf{Z})g^{-1}\\ &(g\mathbf{Y})_j = g\mathbf{Y}_j,\ V_j(g\mathbf{Y}) = gV_j(\mathbf{Y}),\ W_j(g\mathbf{Y}) = gW_j(\mathbf{Y})\end{aligned} \qquad (13.51)$$

for all $g \in G$, $i = 1,\ldots,p$, $j = 1,\ldots,m$. It follows that $\mathbf{e}(\mathbf{Z}) = (e^1(\mathbf{Z}),\ldots,e^p(\mathbf{Z}))$ and $\mathbf{f}(\mathbf{Y}) = (f_1(\mathbf{Y}),\ldots,f_m(\mathbf{Y}))$ are invariant for the action of G. These are called the *Hermite indices.* Label the rows of $\mathbf{Z}$ as $(11\cdots(n+1)1\ 12\cdots(n+1)2\cdots(n+1)p)$ and the columns of $\mathbf{Y}$ as $(11\cdots 1n+1\ 21\cdots 2n+1\cdots m(n+1))$ and consider multi-indices $\alpha = (\alpha^1,\ldots,\alpha^n)$ where $\alpha^i = \ell^i s^i$,

$\ell^i \in (1,\dots,n+1)$, $s^i \in (1,\dots,p)$ and $\beta = (\beta_1,\dots,\beta_n)$ where $\beta_j = r_j\ell_j$, $r_j \in (1,\dots,m)$, $\ell_j \in (1,\dots,n+1)$. Let $U^\alpha \subset M_*((n+1)p,n)$ and $U_\beta \subset M_*(n,(n+1)p)$ be given by

$$\begin{aligned} U^\alpha &= \{\mathbf{Z}\colon \det[\mathbf{Z}|^{\alpha^1,\dots,\alpha^n}] \neq 0\} \\ U_\beta &= \{\mathbf{Y}\colon \det[\mathbf{Y}|_{\beta_1,\dots,\beta_n}] \neq 0\}. \end{aligned} \tag{13.52}$$

We observe that the U^α, U_β are G-invariant and that the U^α cover $M_*((n+1)p,n)$ and the U_β cover $M_*(n,(n+1)m)$.

Definition 13.53 Let $\mathbf{e} = (e^1,\dots,e^p)$ where $e^i \in (0,1,\dots,n)$ and let $\mathbf{f} = (f_1,\dots,f_m)$ where $f_j \in (0,1,\dots,n)$. If $\sum_{i=1}^{p} e^i = n$, $\mathbf{e}$ is called a *p-partition of n* and if $\sum_{j=1}^{m} f_j = n$, $\mathbf{f}$ is called an *m-partition of n*. Let $\mathbf{e}$ be a p-partition of n and call $\alpha = (\alpha^1,\dots,\alpha^n)$ *nice relative to* $\mathbf{e}$ if $\alpha = (11\cdots e^1 1\ 12\cdots e^2 2\cdots e^p p)$. Similarly, if $\mathbf{f}$ is an m-partition of n, then $\beta = (\beta_1,\dots,\beta_m)$ is *nice relative to* $\mathbf{f}$ if $\beta = (11\cdots 1f_1\ 21\cdots 2f_2\cdots mf_m)$. [The terms $\cdot i$ are omitted if $e^i = 0$ and the terms $j\cdot$ are omitted if $f_j = 0$.]

Note that $\alpha = (11\cdots e^1 1\cdots e^p p)$ and $\beta = (11\cdots 1f_1\cdots mf_m)$ are the *unique* nice indices relative to $\mathbf{e}$ and $\mathbf{f}$, respectively. Let $\lambda_{n,m}$ (or λ_n^p) be the number of distinct m-partitions (p-partitions) of n. Then

Proposition 13.54 $\lambda_{n,m} = \binom{n+m-1}{n} = (n+m-1)!/(m-1)!n!$.

Proof. Clearly $\lambda_{1,m} = m$, $\lambda_{n,1} = 1$. But $\lambda_{n,m} = \lambda_{n,m-1} + \lambda_{n-1,m-1} + \cdots + \lambda_{1,m-1} + 1 = \lambda_{n,m-1} + \lambda_{n-1,m}$. Then use induction.

Consider the morphisms

$$\tilde{\mathbf{Z}}\colon \mathbb{A}_k^{n^2+n(m+p)} \to M((n+1)p,n)$$

and

$$\tilde{\mathbf{Y}}\colon \mathbb{A}_k^{n^2+n(m+p)} \to M(n,(n+1)m)$$

given by

$$\tilde{\mathbf{Z}}(A,B,C) = \begin{bmatrix} \mathbf{Z}^1(A,B,C) \\ \vdots \\ \mathbf{Z}^p(A,B,C) \end{bmatrix}, \quad \tilde{\mathbf{Y}}(A,B,C) = [\mathbf{Y}_1(A,B,C)\cdots\mathbf{Y}_m(A,B,C)]$$

where

$$\mathbf{Z}^i(A,B,C) = \begin{bmatrix} c^i \\ c^iA \\ \vdots \\ c^iA^n \end{bmatrix}, \quad \mathbf{Y}_j(A,B,C) = \begin{bmatrix} b_j \; Ab_j \cdots A^n b_j \end{bmatrix}$$

$i = 1,\dots,p$, $j = 1,\dots,m$ and c^i are the rows of C and b_j are the columns of B. Since $\tilde{\mathbf{Z}}(gAg^{-1}, gB, Cg^{-1}) = \tilde{\mathbf{Z}}(A,B,C)g^{-1}$ and also $\tilde{\mathbf{Y}}(gAg^{-1}, gB, Cg^{-1}) = g\mathbf{Y}(A,B,C)$, the maps $\tilde{\mathbf{Z}}$, $\tilde{\mathbf{Y}}$ naturally intertwine the action of G on $\mathbb{A}_k^{n^2+n(m+p)}$ and on $M((n+1)p,n)$ or $M(n,(n+1)m)$. Thus the diagrams

$$\begin{array}{ccccccc} \mathbb{A}^{n^2+n(m+p)} & \xrightarrow{\tilde{\mathbf{Z}}} & M(n(p+1),n) & & \mathbb{A}^{n^2+n(m+p)} & \xrightarrow{\tilde{\mathbf{Y}}} & M(n(n+1),m) \\ g\downarrow & & \downarrow g & & g\downarrow & & \downarrow g \\ \mathbb{A}^{n^2+n(m+p)} & \xrightarrow{\tilde{\mathbf{Z}}} & M(n(p+1),n) & & \mathbb{A}^{n^2+n(m+p)} & \xrightarrow{\tilde{\mathbf{Y}}} & M(n(n+1),m) \end{array}$$

commute and $\tilde{\mathbf{Z}}, \tilde{\mathbf{Y}}$ are G-morphisms.

Proposition 13.55 *$x = (A,B,C)$ is controllable if and only if* $\det \tilde{\mathbf{Y}}(A,B,C)_\beta \neq 0$ *for β nice relative to $\mathbf{f} = \mathbf{f}(x) = \mathbf{f}(\tilde{\mathbf{Y}}(A,B,C))$. Similarly, $x = (A,B,C)$ is observable if and only if* $\det \tilde{\mathbf{Z}}(A,B,C)^\alpha \neq 0$ *for α nice relative to $\mathbf{e} = \mathbf{e}(x) = \mathbf{e}(\tilde{\mathbf{Z}}(A,B,C))$.*

Corollary 13.56 *$\tilde{\mathbf{Z}}(S^n_{m,p}) \subset \cup U^{\alpha_e}$, α_e nice relative to $\mathbf{e}$; and, $\tilde{\mathbf{Y}}(S^n_{m,p}) \subset \cup U_{\beta_f}$, β_f nice relative to $\mathbf{f}$.*

Proposition 13.57 *Let $x = (A,B,C)$, $x_1 = (A_1,B_1,C_1)$ be elements of $S^n_{m,p}$. If $\tilde{\mathbf{Z}}(x) = \tilde{\mathbf{Z}}(x_1)$, then $A = A_1$, $C = C_1$. If $\tilde{\mathbf{Y}}(x) = \tilde{\mathbf{Y}}(x_1)$, then $A = A_1$, $B = B_1$. Consequently, if $\tilde{\mathbf{Z}}(x) = \tilde{\mathbf{Z}}(x_1)$ and $\tilde{\mathbf{Y}}(x) = \tilde{\mathbf{Y}}(x_1)$, then $x = x_1$.*

Proof. We give the proof for $\tilde{\mathbf{Y}}$. If $\tilde{\mathbf{Y}}(x) = \tilde{\mathbf{Y}}(x_1)$, then $A^tB = A_1^tB_1$ for $t = 0,1,\dots,n$. In particular, $B = B_1$. Let $A^{t_j}b_j = A_1^{t_j}b_{1j} = v_j$, $j = 1,\dots,n$ be a basis for $\operatorname{span}[\tilde{\mathbf{Y}}(x)] = \operatorname{span}[\tilde{\mathbf{Y}}(x_1)] = k^n$ (note $t_j \in (0,1,\dots,n-1)$ by Cayley-Hamilton). Then $Av_1 = A_1v_1,\dots,Av_n = A_1v_n$ so that $A = A_1$.

The proposition means that the G-morphisms $\tilde{\mathbf{Z}}\colon \mathbb{A}^{n^2+n(m+p)} - V_0 \to M_*((n+1)p,n)$, $\tilde{\mathbf{Y}}\colon \mathbb{A}^{n^2+n(m+p)} - V_c \to M_*(n,(n+1)m)$, and $(\tilde{\mathbf{Z}},\tilde{\mathbf{Y}})\colon S^n_{m,p} \to M_*((n+1)p,n) \times M_*(n,(n+1)m)$ are injective. Let $\mathcal{C}^n_{m,p} = \{x = (A,B,C)\colon x \text{ is controllable}\}$, $\mathcal{O}^n_{m,p} = \{x = (A,B,C)\colon x \text{ is observable}\}$. Then $\mathcal{C}^n_{m,p}$, $\mathcal{O}^n_{m,p}$, $S^n_{m,p}$ are irreducible (being open in $\mathbb{A}^{n^2+n(m+p)}$) and so are $\tilde{\mathbf{Z}}(\mathcal{O}^n_{m,p})$, $\tilde{\mathbf{Y}}(\mathcal{C}^1_{m,p})$ and $(\tilde{\mathbf{Z}},\tilde{\mathbf{Y}})(S^n_{m,p})$.

Now let us consider $M_*((n+1)p, n) \times M_*(n, (n+1)m) \times \mathbb{A}^n$ with coordinates $(\tilde{\mathbf{Z}}, \tilde{\mathbf{Y}}, \xi_0, \ldots, \xi_{n-1})$. We suppose that $\tilde{\mathbf{Z}} = [\tilde{\mathbf{Z}}^i]$, $\tilde{\mathbf{Y}} = [\tilde{\mathbf{Y}}_j]$ where $\tilde{\mathbf{Z}}^i \in M((n+1), n)$, $\tilde{\mathbf{Y}}_j \in M(n, n+1)$, $i = 1, \ldots, p$, $j = 1, \ldots, m$. The following sets of equations define a variety:

(a) *Block Symmetry*

$$(\tilde{\mathbf{Z}}^i\tilde{\mathbf{Y}}_j)_s^r = (\tilde{\mathbf{Z}}^i\tilde{\mathbf{Y}}_j)_r^s \tag{13.58}$$

for $s = 1, \ldots, n+1$, $r = 1, \ldots, n+1$;

(b) *Hankel Structure*

$$(\tilde{\mathbf{Z}}^i\tilde{\mathbf{Y}}_j)_{s+1}^r = (\tilde{\mathbf{Z}}^i\tilde{\mathbf{Y}}_j)_r^{s+1} \tag{13.59}$$

for $r = 1, \ldots, s-1$, $s = 2, \ldots, n$;

(c) *Dependence*

$$(\tilde{\mathbf{Z}}^i\tilde{\mathbf{Y}}_j)_{n+1}^r = -\sum_{\ell=0}^{n-1} \xi_\ell (\tilde{\mathbf{Z}}^i\tilde{\mathbf{Y}}_j)_{\ell+1}^r \tag{13.60}$$

for $r = 1, \ldots, n+1$; and, for $i = 1, \ldots, p$, $j = 1, \ldots, m$.

Let

$$v^{*\ell_i i} = (\tilde{\mathbf{Z}}^i)^{\ell_i}, \quad w_{j\lambda_j} = (\tilde{\mathbf{Y}}_j)_{\lambda_j} \tag{13.61}$$

so that

$$v^{*\ell_i i} w_{j\lambda_j} = (\tilde{\mathbf{Z}}^i\tilde{\mathbf{Y}}_j)_{\lambda_j}^{\ell_i} \tag{13.62}$$

for $\ell_i = 1, \ldots, n+1$, $\lambda_j = 1, \ldots, n+1$, $i = 1, \ldots, p$, $j = 1, \ldots, m$. Then the equations take the form:

(a) *Block Symmetry*

$$v^{*\ell_i i} w_{j\lambda_j} = v^{*\lambda_j i} w_{j\ell_i} \tag{13.63}$$

for $\ell_i = 1, \ldots, n+1$, $\lambda_j = 1, \ldots, n+1$;

(b) *Hankel Structure*

$$v^{*\ell_i i} w_{j\lambda_{j+1}} = v^{*\ell_{i+1} i} w_{j\lambda_j} \tag{13.64}$$

for $\ell_i = 1, \ldots, \lambda_{j-1}$, $\lambda_j = 2, \ldots, n$;

(c) *Dependence*

$$\begin{aligned} v^{*\ell_i i} w_{jn+1} &= -\sum_{\ell=0}^{n-1} \xi_\ell v^{*\ell_i i} w_{j\ell+1} \\ v^{*n+1 i} w_{j\lambda_j} &= -\sum_{\ell=0}^{n-1} \xi_\ell v^{*\ell+1 i} w_{j\lambda_j} \end{aligned} \tag{13.65}$$

for $\ell_i = 1, \ldots, n+1$, $\lambda_j = 1, \ldots, n+1$.

We then have another form of the Lemma of Risannen.

Lemma 13.66 (Risannen) *Suppose that the equations (a), (b), (c) are satisfied and that $\rho(\tilde{\mathbf{Y}}) = n$. If*

$$v^{*ri} = \sum_{t=1}^{r-1} \alpha_t v^{*ti} + \sum_{s<i} \beta_{s\ell_s} v^{*\ell_s s} = \phi_{ir}$$

then

$$v^{*r+1i} = \sum_{t=1}^{r-1} \alpha_t v^{*t+1i} + \sum_{s<i} \beta_{s\ell_s} v^{*\ell_s+1s} = \phi_{ir+1}.$$

In other words, if $v^{*ri} \in \mathrm{span}[v^{*1i} \cdots v^{*r-1i},\ v^{*\ell_s s},\ s < i]$ then so is v^{*r+1i}. [And similarly for the $w_{j\lambda_j}$ if $\rho(\tilde{\mathbf{Z}}) = n$.]

Proof. First let $\lambda_j = r+1, \ldots, n$. Then, for $r = 1, \ldots, \lambda_j - 1$,

$$\begin{aligned} v^{*r+1i} w_{j\lambda_j} &= v^{*ri} w_{j\lambda_j+1} \\ &= \sum_{t=1}^{r-1} \alpha_t v^{*ti} w_{j\lambda_j+1} + \sum_{s<i} \beta_{s\ell_s} v^{*\ell_s s} w_{j\lambda_j+1} \\ &= \sum_{t=1}^{r-1} \alpha_t v^{*t+1i} w_{j\lambda_j} + \sum_{s<i} \beta_{s\ell_s} v^{*\ell_s+1s} w_{j\lambda_j} \\ &= \phi_{ir+1} w_{j\lambda_j}. \end{aligned}$$

Now, let $\lambda_j = 1, \ldots, r$. Then

$$\begin{aligned} v^{*r+1i} w_{j\lambda_j} &= v^{*\lambda_j i} w_{jr+1} = v^{*\lambda_j+1i} w_{jr} = v^{*ri} w_{j\lambda_j+1} \\ &= \sum_{t=1}^{r-1} \alpha_t v^{*ti} w_{j\lambda_j+1} + \sum_{s<i} \beta_{s\ell_s} v^{*\ell_s s} w_{j\lambda_j+1} \\ &= \phi_{ir+1} w_{j\lambda_j}. \end{aligned}$$

Thus $v^{*r+1i} = \phi_{ir+1}$ (as $w_{j\lambda_j}$ can be taken from a basis).

Corollary 13.67 *Suppose that the equations (a), (b), (c) are satisfied and that $\rho(\tilde{\mathbf{Z}}) = n$, $\rho(\tilde{\mathbf{Y}}) = n$. Then there is a (unique) p-partition $\mathbf{e} = (e^1, \ldots, e^p)$ of n and a (unique) m-partition $\mathbf{f} = (f_1, \ldots, f_m)$ of n such that $v^{*11}, \ldots, v^{*e^1 1}, \ldots, v^{*e^p p}$ is a basis of* $\mathrm{span}[\tilde{\mathbf{Z}}]$ *and $w_{11}, \ldots, w_{1f_1}, w_{21}, \ldots, w_{mf_m}$ is a basis of* $\mathrm{span}[\tilde{\mathbf{Y}}]$.

The power of the equations (a), (b), (c) is clearly demonstrated in the lemma. The point of the next example is to show that the equations (a), (b), (c) are *not* enough to define the variety of linear systems. We need a version of the "Grassmann equations".

Example 13.68 Let

$$\tilde{\mathbf{Y}} = \begin{bmatrix} 1 & 0 & a & a(1-a) & 0 & 0 & 0 & 0 \\ 0 & 1 & 1-a & a+(1-a)^2 & 0 & 0 & 0 & 0 \\ 0 & 0 & 0 & 0 & 1 & 1 & 1 & 1 \end{bmatrix}$$

so that $m = 2$, $n = 3$. Let

$$A = \begin{bmatrix} 0 & a & 0 \\ 1 & 1-a & 0 \\ 0 & 0 & 1 \end{bmatrix}, \quad B = \begin{bmatrix} 1 & 0 \\ 0 & 0 \\ 0 & 1 \end{bmatrix}.$$

Then $\det[zI - A] = z^3 + (a-2)z^2 + (1-2a)z + a$ so that $\xi_0(A) = a$, $\xi_1(A) = 1-2a$, $\xi_2(A) = a - 2$. Moreover,

$$\tilde{\mathbf{Y}}(A, B) = \begin{bmatrix} b_1 & Ab_1 & A^2b_1 & A^3b_1 & b_2 & Ab_2 & A^2b_2 & A^3b_2 \end{bmatrix} = \tilde{\mathbf{Y}}.$$

We note that

$$-\xi_0(A)b_1 - \xi_1(A)Ab_1 - \xi_2(A)A^2b_1 = \begin{bmatrix} a(1-a) \\ a+(1-a)^2 \\ 0 \end{bmatrix} = A^3b_1$$

$$-\xi_0(A)b_2 - \xi_1(A)Ab_2 - \xi_2(A)A^2b_2 = \begin{bmatrix} 0 \\ 0 \\ 1 \end{bmatrix} = A^3b_2.$$

Let $\alpha_0 = 0$, $\alpha_1 = -a$, $\alpha_2 = a - 1$. Then, also,

$$-\alpha_0 b_1 - \alpha_1 Ab_1 - \alpha_2 A^2b_1 = \begin{bmatrix} a(1-a) \\ a+(1-a)^2 \\ 0 \end{bmatrix} = A^3b_1$$

$$-\alpha_0 b_2 - \alpha_1 Ab_2 - \alpha_2 A^2b_2 = \begin{bmatrix} 0 \\ 0 \\ 1 \end{bmatrix} = A^3b_2.$$

Note that $\xi_0(A) + \xi_1(A) + \xi_2(A) = \alpha_0 + \alpha_1 + \alpha_2$, $\alpha_0 - \xi_0(A) = (\xi_2(A) - \alpha_2)a$, $(\alpha_1 - \xi_1(A)) = (1-a)(\xi_2(A) - \alpha_2)$ and that $(\xi_0(A), \xi_1(A), \xi_2(A)) \neq (\alpha_0, \alpha_1, \alpha_2)$. Now let

$$C = \begin{bmatrix} 1 & 0 & 0 \\ 0 & 0 & 1 \end{bmatrix}$$

and let $\tilde{\mathbf{Z}} = \tilde{\mathbf{Z}}(A, B, C)$ so that

$$\tilde{\mathbf{Z}} = \begin{bmatrix} 1 & 0 & 0 \\ 0 & a & 0 \\ a & a(1-a) & 0 \\ a(1-a) & a^2 + a(1-a)^2 & 0 \\ 0 & 0 & 1 \\ 0 & 0 & 1 \\ 0 & 0 & 1 \\ 0 & 0 & 1 \end{bmatrix}.$$

If $a \neq 0$, then $\rho(\tilde{\mathbf{Z}}) = 3$. Thus $x = (A, B, C) \in S_{2,2}^3$, $\mathbf{e} = (2, 1)$, $\mathbf{f} = (2, 1)$. We *claim* that the equations (a), (b), (c) are satisfied by both $(\tilde{\mathbf{Z}}, \tilde{\mathbf{Y}}, \xi_0(A), \xi_1(A), \xi_2(A))$ *and* $(\tilde{\mathbf{Z}}, \tilde{\mathbf{Y}}, \alpha_0, \alpha_1, \alpha_2)$. This follows since the product $\tilde{\mathbf{Z}}\tilde{\mathbf{Y}}$ is given by

$$\begin{bmatrix} 1 & 0 & a & a(1-a) & 0\,0\,0\,0 \\ 0 & a & a(1-a) & a(1-a+a^2) & 0\,0\,0\,0 \\ a & a(1-a) & a(1-a+a^2) & a(1-a+a^2-a^3) & 0\,0\,0\,0 \\ a(1-a) & a(1-a+a^2) & a(1-a+a^2-a^3) & a(1-a+a^2-a^3+a^4) & 0\,0\,0\,0 \\ 0 & 0 & 0 & 0 & 1\,1\,1\,1 \\ 0 & 0 & 0 & 0 & 1\,1\,1\,1 \\ 0 & 0 & 0 & 0 & 1\,1\,1\,1 \\ 0 & 0 & 0 & 0 & 1\,1\,1\,1 \end{bmatrix}.$$

If we consider $(\tilde{\mathbf{Z}}, \tilde{\mathbf{Y}}, \alpha_0, \alpha_1, \alpha_2)$, then there is *no* $x_1 = (A_1, B_1, C_1) \in S_{2,2}^3$ with $\tilde{\mathbf{Z}} = \tilde{\mathbf{Z}}(x_1)$, $\tilde{\mathbf{Y}} = \tilde{\mathbf{Y}}(x_1)$ *and* $\det[zI - A_1] = z^3 + \alpha_2 z^2 + \alpha_1 z + \alpha_0$ (for by Proposition 13.57 $A_1 = A$ but the polynomials are different). Thus, there is an element of the variety defined by (a), (b), (c) which does not correspond to an element of $S_{2,2}^3$.

So now let us consider $M_*((n+1)p, n) \times M_*(n, (n+1)m) \times \mathbb{A}^n \times \mathbb{A}^{\tau(n,m,p)}$ with coordinates $(\mathbf{Z}, \mathbf{Y}, \boldsymbol{\xi}, \boldsymbol{\lambda})$. Letting

$$\mathbf{M}_s^r = \begin{bmatrix} \mathbf{Z}^{(r-1)p+1} \\ \vdots \\ \mathbf{Z}^{rp} \end{bmatrix} \left[\mathbf{Y}_{(s-1)m+1} \cdots \mathbf{Y}_{sm}\right] \tag{13.69}$$

for $r = 1, \ldots, n+1$, $s = 1, \ldots, n+1$, we then have the equations

(a) *Block Symmetry*

$$\mathbf{M}_j^i - \mathbf{M}_i^j = O \tag{13.70}$$

(b) *Hankel Structure*

$$\mathbf{M}_{j+1}^i - \mathbf{M}_j^{i+1} = O \tag{13.71}$$

for $i = 1, \ldots, j-1$, $j = 2, \ldots, n$;

(c) *Dependence*

$$\mathbf{M}_{n+1}^{i} + \sum_{\ell=0}^{n-1} \xi_{\ell} \mathbf{M}_{\ell+1}^{i} = O \tag{13.72}$$

for $i = 1, \dots, n+1$;

(d) *Grassmann*

$$\det\left[\Lambda(\mathbf{M})(z)\Big|_{\ell_1 \cdots \ell_t}^{i_1 \cdots i_t}\right] = \psi_0^{t-1}(z)\psi_{\boldsymbol{\alpha}_t}(z) \tag{13.73}$$

where $\psi_{\boldsymbol{\alpha}_t}(z) = \lambda_{\boldsymbol{\alpha}_{m-t,0}} + \cdots + \lambda_{\boldsymbol{\alpha}_{m-t,n-t}} z^{n-t}$, $\mathbf{M} = (\mathbf{M}_j^i)$, and

$$\Lambda(\mathbf{M})(z) = \sum_{r=0}^{n-1} \left[\sum_{\ell=r+1}^{n} \xi_{\ell} \mathbf{M}_{\ell-r}^{1} \right] z^r$$

[Cf. the corresponding equations in Chapter 11 viz. (11.5–11.10), (11.15–11.19), etc.] Let $\Sigma(n,m,p)$ be the variety defined by these equations and let G act on $\Sigma(n,m,p)$ via

$$g \cdot (\mathbf{Z}, \mathbf{Y}, \boldsymbol{\xi}, \boldsymbol{\lambda}) = (\mathbf{Z}g^{-1}, g\mathbf{Y}, \boldsymbol{\xi}, \boldsymbol{\lambda}) \tag{13.74}$$

(note that $G \cdot \Sigma(n,m,p) \subset \Sigma(n,m,p)$). We observe that there is a natural map $\psi\colon \Sigma(n,m,p) \to \mathcal{H}(n,m,p)$ given by

$$\psi(\mathbf{Z}, \mathbf{Y}, \boldsymbol{\xi}, \boldsymbol{\lambda}) = (\mathbf{ZY}, \boldsymbol{\xi}, \boldsymbol{\lambda}) \tag{13.75}$$

(see Definition 11.20). If G acts trivially on $\mathcal{H}(n,m,p)$, then ψ is a G-morphism. Let $\varphi\colon S_{m,p}^n \to \Sigma(n,m,p)$ be given by

$$\varphi(A,B,C) = (\mathbf{Z}(A,B,C), \mathbf{Y}(A,B,C), -\chi_i(A), \boldsymbol{\lambda}(A,B,C)) \tag{13.76}$$

where the $\boldsymbol{\lambda}(A,B,C)$ are the determinantal coefficients from

$$\Lambda(A,B,C)(z) = \sum_{r=0}^{n-1} \left[\sum_{\ell=r+1}^{n} (CA^{\ell-r-1}B) \right] z^r.$$

Clearly $\varphi(S_{m,p}^n) \subset \Sigma(n,m,p)$ and φ is a G-morphism. In view of Proposition 13.57, φ is injective. If $(\mathbf{Z}, \mathbf{Y}, \boldsymbol{\xi}, \boldsymbol{\lambda}) = \mathbf{v} \in \Sigma(n,m,p)$, then $\psi(\mathbf{v}) \in \mathcal{H}(n,m,p)$ and there is an $x = (A,B,C) \in S_{m,p}^n$ with $(\mathbf{H}(x), \boldsymbol{\xi}(x), \boldsymbol{\lambda}(x)) = \psi(\mathbf{v})$ (e.g., by Proposition 13.38). But $\mathbf{Z}(x)\mathbf{Y}(x) = \mathbf{ZY}$ and all of rank n together imply there is a $g \in G$ with $\mathbf{Z} = \mathbf{Z}(x)g^{-1}$, $\mathbf{Y} = g\mathbf{Y}(x)$. Let $x_1 = g \cdot x$. Then $\varphi(x_1) = \mathbf{v}$ and φ is surjective. In other words, φ is a bijective G-morphism between $S_{m,p}^n$ and $\Sigma(n,m,p)$.

Proposition 13.77 *φ^{-1} is a morphism and hence, $S_{m,p}^n$ and $\Sigma(n,m,p)$ are G-isomorphic.*

Proof. Let $\mathbf{v} = (\mathbf{Z}, \mathbf{Y}, \boldsymbol{\xi}, \boldsymbol{\lambda}) \in \Sigma(n,m,p)$ and let $x = (A,B,C) \in S^n_{m,p}$ with $\varphi(x) = \mathbf{v}$. Let $\tilde{\mathbf{Z}} = \tilde{\mathbf{Z}}(x)$, $\tilde{\mathbf{Y}} = \tilde{\mathbf{Y}}(x)$ and note that the entries in $\tilde{\mathbf{Z}}$, $\tilde{\mathbf{Y}}$ are simply those in $\mathbf{Z}$, $\mathbf{Y}$ rearranged. Let $(\tilde{\mathbf{Y}}_1)_1, \ldots, (\tilde{\mathbf{Y}}_m)_{f_m}$ be a basis of span$[\tilde{\mathbf{Y}}]$. Since $B_j = (\tilde{\mathbf{Y}}_j)_1$, $C^i = (\tilde{\mathbf{Z}}^i)^1$, we see that $B(\varphi^{-1}(\mathbf{v}))$, $C(\varphi^{-1}(\mathbf{v}))$ are morphisms (more precisely, the maps $\mathbf{v} \to B(\varphi^{-1}(\mathbf{v}))$, $\mathbf{v} \to C(\varphi^{-1}(\mathbf{v}))$). Let $g = [(\tilde{\mathbf{Y}}_1)_1 \cdots (\tilde{\mathbf{Y}}_m)_{f_m}]$ so that $g \in G$. Noting that

$$(\tilde{\mathbf{Y}}_j)_{f_j+1} = \sum_{s_1=1}^{f_1} \alpha^1_{js_1} (\tilde{\mathbf{Y}}_1)_{s_1} + \cdots + \sum_{s_j=1}^{f_j} \alpha^j_{js_j} (\tilde{\mathbf{Y}}_j)_{s_j} \tag{13.78}$$

for $j = 1, \ldots, m$, we have, by the Lemma of Risannen (13.65),

$$(\tilde{\mathbf{Y}}_j)_{f_j+t_j} = \sum_{s_1=1}^{f_1} \alpha^1_{js_1} (\tilde{\mathbf{Y}}_1)_{s_i+t_j-1} + \cdots + \sum_{s_j=1}^{f_j} \alpha^j_{js_j} (\tilde{\mathbf{Y}}_j)_{s_j+t_j-1} \tag{13.79}$$

with the *same* coefficients. Since $(\tilde{\mathbf{Y}}_1)_1, \ldots, (\tilde{\mathbf{Y}}_m)_{f_m}$ are a basis, the $\alpha^t_{js_t}$ depend via Cramer's rule morphically on the $(\tilde{\mathbf{Y}}_j)_{\lambda_j}$. Let

$$A_g = \begin{bmatrix} L_{f_1,f_1-1} & & O_{f_1,f_2-1} & & \vdots \\ & x_{f_1} & L_{f_2,f_2-1} & x_{f_2} & \vdots \\ O_{n-f_1,f_1-1} & & O_{n-(f_1+f_2),f_2-1} & & \vdots \end{bmatrix} \tag{13.80}$$

where

$$L_{f_j,f_j-1} = \begin{bmatrix} O_{1,f_j-1} \\ I_{f_j-1,f_j-1} \end{bmatrix} = \begin{bmatrix} 0 & 0 & \cdots & 0 \\ 1 & 0 & & \\ & 1 & & \vdots \\ \vdots & \vdots & & 0 \\ 0 & \vdots & & 1 \end{bmatrix} \tag{13.81}$$

and

$$\mathbf{x}_{f_j} = \begin{bmatrix} \alpha_{j1}^1 \\ \vdots \\ \alpha_{js_1}^1 \\ \alpha_{j1}^2 \\ \vdots \\ \alpha_{jf_j}^j \\ \vdots \\ 0 \end{bmatrix}$$

We *claim* that $A(\varphi^{-1}(\mathbf{v})) = gA_g g^{-1}$ and hence that $\mathbf{v} \to A(\varphi^{-1}(\mathbf{v}))$ is a morphism. Let $\tilde{\mathbf{Y}}_g = g^{-1}\tilde{\mathbf{Y}}$. Then $B = gB_g$ where $B_g = [\epsilon_1\ \epsilon_{f_1+1} \cdots \epsilon_{f_{m+1}}]$ (assuming all $f_j \neq 0$; see Chapter 4 for the case where some $f_j = 0$). If $\tilde{\mathbf{Y}}_g = [y_{j\lambda_j}]$, then $y_{j\lambda_j}$ for $\lambda_j = 1, \ldots, f_j$ are unit basis vectors and

$$y_{jf_j+t_j} = \sum_{s_1=1}^{f_1} \alpha_{js_1}^1 y_{1s_1+t_j-1} + \cdots + \sum_{s_j=1}^{f_j} \alpha_{js_j}^j y_{js_j+t_j-1}.$$

It follows that $y_{j\lambda_j} = A_g^{\lambda_j-1}(B_g)_j$ and that $gy_{j\lambda_j} = (\tilde{\mathbf{Y}}_j)_{\lambda_j} = gA_g^{\lambda_j-1}g^{-1}g(B_g)_j$. Thus $A(\varphi^{-1}(\mathbf{v}))$ depends morphically on $\mathbf{Y}$ and the result follows.

Thus, $\Sigma(n, m, p)$ is still another representation of the space of linear systems of degree n with m inputs and p outputs. Moreover, $S_{m,p}^n/G \simeq \Sigma(n, m, p)/G$ and we shall show in the sequel that $(\mathcal{H}(n, m, p), \psi)$ is a geometric quotient of $\Sigma(n, m, p)$ modulo G.

Exercises

(1) Treat the case $t = 3$ in (13.12).

(2) Work out in detail the local coordinatization of Hankel(2, 2, 2).

(3) Carry out the induction in the proof of Proposition 13.53.

(4) Give the Grassmann equations (13.73) explicitly for $n = 2$, $m = 2$, $p = 2$.

(5) Calculate the determinantal form of the $\alpha_{js_t}^t$ in (13.78). Exhibit B_g when (say) $f_1 = 0$.

14

Projective Algebraic Geometry V: Fibers of Morphisms

Our goal here is to extend and amplify the results of Part I, Chapter 18 for the projective situation. The term "variety" means either a projective or quasi-projective variety.

Definition 14.1 Let $\psi\colon X \to Y$ be a morphism of varieties. ψ is *dominant* if $\overline{\psi(X)} = Y$ i.e., $\psi(X)$ is dense in Y. If Z is a component of $\psi^{-1}(W)$, then Z *dominates* W if $\psi(Z)$ is dense in W. A morphism is *finite* if every $y \in Y$ has an affine neighborhood U with $V = \psi^{-1}(U)$ affine and the (restriction of) map $\psi\colon V \to U$ a finite morphism i.e., $k[V]$ is integral over $\psi^*(k[U])$. If $y \in Y$, the closed set $\psi^{-1}(y) = \{x \in X\colon \psi(x) = y\}$ is the *fiber of* ψ *over* y; and, if W is a closed irreducible subset of Y, then the closed set $\psi^{-1}(W) = \{x \in X\colon \psi(x) \in W\}$ is the *fiber of* ψ *over* W.

Theorem 14.2 *Let* $\psi\colon X \to Y$ *be a dominant morphism and let* W *be a closed irreducible subset of* Y. *If* Z *is a component of* $\psi^{-1}(W)$ *which dominates* W, *then* $\dim Z \geq \dim W + \dim X - \dim Y$.

Proof. If U is an open affine in Y with $U \cap W \neq \emptyset$, then $U \cap W$ is dense in W and we may assume that Y is affine. If $t = \operatorname{codim}_Y W$, then W is a component of $V(g_1, \ldots, g_t)$, $g_i \in k[Y]$. Let $f_i = \psi^*(g_i)$. Then $Z \subset \psi^{-1}(W) \subset V(f_1, \ldots, f_t)$ and argue as in Theorem I.18.5.

Corollary 14.3 (cf. I.18.6) $\dim \psi^{-1}(y) \geq \dim X - \dim Y$.

Proposition 14.4 (cf. I.18.12) *Let* $\psi\colon X \to Y$ *be a dominant finite morphism. Then* (i) ψ *is closed (and therefore surjective); and,* (ii) *if* W *is a closed*

P. Falb, *Methods of Algebraic Geometry in Control Theory: Part II*,
Modern Birkhäuser Classics, https://doi.org/10.1007/978-3-319-96574-1_14

irreducible subset of Y *and* Z *is a component of* $\psi^{-1}(W)$*, then* $\psi(Z)$ *is closed,* $\dim \psi(Z) = \dim Z$*, and for some* Z*,* $\psi(Z) = W$*.*

Proof. (i) Since closure is a local property, we may assume things are affine and apply I.18.12. (ii) By (i), $\psi(Z)$ is closed. Since ψ is finite and dominant, it follows from I.18.12(b), that $\dim \psi(Z) = \dim Z$. Replacing W by an open affine U with $U \cap W \neq \emptyset$ and considering the map $\psi\colon \psi^{-1}(U) \to U$, we have, for some Z with $Z \cap \psi^{-1}(U) \neq \emptyset$, $\psi(Z) \supset U$. But $U \cap W$ is dense in W and $\psi(Z)$ is closed so that $\psi(Z) = W$.

We remark that if Y is normal i.e., if all $\mathfrak{o}_{\xi,Y}$ are integrally closed, then $\psi(Z) = W$ for all components Z (replacing W by an open affine U with $U \cap W \neq \emptyset$ and using I.18.12(c)).

Corollary 14.5 *If* $\psi\colon X \to Y$ *is a finite surjective morphism,* Y *is normal, and* W *is a closed irreducible subset of* Y*, then* $\dim Z = \dim W + \dim X - \dim Y$ *for every component* Z *of* $\psi^{-1}(W)$*.*

Theorem 14.6 *If* $\psi\colon X \to Y$ *is a dominant morphism, then there is an open* $U \subset Y$ *such that* (a) $U \subset \psi(X)$*; and,* (b) *if* W *is a closed irreducible subset of* Y *with* $W \cap U \neq \emptyset$ *and if* Z *is a component of* $\psi^{-1}(W)$ *with* $Z \cap \psi^{-1}(U) \neq \emptyset$*, then* $\dim Z = \dim W + \dim X - \dim Y$*.*

Proof. We may assume Y is affine by the argument in Theorem 14.2. We may also suppose that X is affine. For if there are $U_i \subset Y$ which work for ψ restricted to X_i affine in X with $\cup X_i = X$, then $U = \cap U_i$ works. The result then follows from I.18.17.

Corollary 14.7 *There is a* U *open in* Y *such that every component* Z *of* $\psi^{-1}(y)$*,* $y \in U$*, has* $\dim Z = \dim X - \dim Y$*. Hence,* $\dim \psi^{-1}(y) = \dim X - \dim Y$ *for all* $y \in U$*.*

Proof. Take U as in the theorem. Since $y \in U$, Z a component of $\psi^{-1}(y)$ implies $Z \subset \psi^{-1}(y) \subset \psi^{-1}(U)$. Hence $Z \cap \psi^{-1}(U) \neq \emptyset$ and $\dim Z = \dim X - \dim Y$.

Corollary 14.8 *If* $\psi\colon X \to Y$ *is bijective (so that* $\dim X = \dim Y$*), then there is an affine open* $V \subset X$ *and* $U \subset Y$ *with* $\psi(V) \subset U$ *and* $\psi \mid V$ *finite.* (Cf. Proof of I.18.17.)

Corollary 14.9 *Let* $\psi\colon X \to Y$ *be a morphism and let* $e_t(\psi) = \{x \in X\colon \dim \psi^{-1}(\psi(x)) \geq t\}$*. Then* $e_t(\psi)$ *is closed.*

Proof. ([M-2]) Use induction on $\dim Y$. If ψ is not dominant, then $\overline{\psi(X)} < Y$ and $\dim Y > \dim \overline{\psi(X)}$ so done. If ψ is dominant, let U be as in Corollary

14.7. Then (i) $e_t(\psi) = X$ if $t \leq \dim X - \dim Y$ and (ii) $e_t(\psi) \subset \psi^{-1}(Y - U)$ if $t > \dim X - \dim Y$. Let $Y - U = Y_1 \cup \cdots \cup Y_\ell$ be the decomposition into irreducible components and let $\psi^{-1}(Y_i) = Z_{i,1} \cup \cdots \cup Z_{i,\ell_i}$ be the decomposition into irreducible components. Let $\psi_{ij}\colon \psi \mid Z_{ij} \to Y_i$. Then $e_t(\psi) = \bigcup_{i,j} e_t(\psi_{ij})$ for $t > \dim X - \dim Y$. Since U is open in Y, $\overline{U} = Y$ and $\dim Y_i < \dim Y$. Thus the $e_t(\psi_{ij})$ are closed by induction.

Theorem 14.10 *Suppose that X is an algebraic set and that $\psi\colon X \to Y$ is a surjective morphism. If Y is irreducible, if $\psi^{-1}(y)$ is irreducible and $\dim \psi^{-1}(y) = n$ for all y, then X is irreducible.*

Proof. ([S-2]) Suppose that $X = \bigcup_{i=1}^{t} X_i$ is the decomposition into irreducible components. Then the $\psi(X_i)$ are closed and irreducible. Say, $\psi(X_j) = Y$ for $j = 1, \ldots, s$. Let $Y' = Y - \bigcup_{\ell=s+1}^{t} \psi(X_\ell)$ and $\psi^{-1}(Y') = X'$. Then $X' = \cup X_j'$, $X_j' = \psi^{-1}(Y') \cap X_j$. Consider $\psi_j\colon X_j' \to Y'$. Let $n_j = \min \dim_{y \in Y'} \psi_j^{-1}(y)$ so that n_j is obtained from $y \in U$ open in Y'. But $\cup \psi_j^{-1}(y) = \psi^{-1}(y)$ is irreducible of dimension n and so, $\max n_j = n$. Say, $\max n_j = n_1 = n$. Then $\dim \psi_1^{-1}(y) = n$ for $y \in U$ and hence for all $y \in Y$. But $\psi^{-1}(y) = \cup \psi_j^{-1}(y)$ for all y then leads to $\psi_1^{-1}(y) = \psi^{-1}(y)$ (both dimension n). In other words, $X_1' = X$ which means that $X_1 = X$.

Theorem 14.11 (cf. I.18.19) *A morphism maps constructible sets into constructible sets.*

Proof. Same as I.18.19.

Corollary 14.12 (cf. I.18.20) *If $\psi\colon X \to Y$ is a dominant morphism and if, for every closed irreducible subset W of Y and each component Z of $\psi^{-1}(W)$; $\dim Z = \dim W + \dim X - \dim Y$, then ψ is an open map.*

Corollary 14.13 *If $\psi\colon X \to Y$ is a dominant finite morphism and Y is normal, then ψ is an open map.*

Corollary 14.14 (cf. I.18.22) *Let $\psi\colon X \to Y$ be a dominant morphism and let $r = \dim X - \dim Y$. If Y is normal and all components of $\psi^{-1}(y)$, $y \in Y$, are r-dimensional, then ψ is an open map.*

Proof. Since $\psi(X)$ contains an open affine U and since U_1 open in X contains an open affine U_0, we may suppose that X, Y are affine and then apply I.18.22.

Corollary 14.15 *If $\psi: X \to Y$ is a dominant morphism, then there is an open U_X in X such that $\psi \mid U_X$ is an open map.*

Proof. By Theorem 14.6, there is a U_y open in Y with $U_y \subset \psi(X)$ such that if W is a closed irreducible subset in Y with $W \cap U_y \neq \emptyset$ and if Z is a component of $\psi^{-1}(W)$ with $Z \cap \psi^{-1}(U_y) \neq \emptyset$, then $\dim Z = \dim W + \dim X - \dim Y$. Let $U_X = \psi^{-1}(U_y)$ and let $X' = U_X$, $Y' = U_y$, $\psi' = \psi \mid X'$. Since $U_y \subset \psi(X)$, $\psi'(X') = Y'$ and ψ' is dominant. If W' is a closed, irreducible subset of Y', then W' (or better $\overline{W}'$) meets U_y and any component Z' of ${\psi'}^{-1}(W')$ (or better $\overline{Z}'$ of $\psi^{-1}(\overline{W}'))$ meets U_X. Hence $\dim Z' = \dim W + \dim X' - \dim Y'$. It follows from Corollary 14.12 that ψ' is an open map.

Definition 14.16 Let $\psi: X \to Y$ be a dominant morphism so that $k(X)$ may be viewed as an extension field of $k(Y)$. If $k(X) = k(Y)$, then ψ is a *birational morphism* or, simply, *birational.*

Let $\psi: X \to Y$ be a dominant morphism and recall (Chapter 7) that $k(X)$ and $k(Y)$ are the fields of rational functions on X and Y, respectively. If $\mathbf{h} \in k(Y)$, then $\mathbf{h} = (h, U)$, U open affine in Y and $h \in \mathfrak{o}_Y(U) = k[U]$. We may suppose that $V = \psi^{-1}(U)$ is open affine in X. Then $\psi^*(\mathbf{h}) = (h \circ \psi, V) \in k(X)$ since $\psi^*(h) = h \circ \psi \in \mathfrak{o}_X(V) = k[V]$. In other words, $\psi^*: k(Y) \to k(X)$. Now, by Theorem 14.6, we may assume that $U \subset \psi(X)$ so that $\psi(V) = \psi(\psi^{-1}(U)) = U$. Then $\psi \mid V$ is a dominant morphism of affine varieties. Therefore, $\psi^*: k(Y) = k(U) = k(\psi(V)) \to k(V) = k(X)$ is injective. It is in this sense that we view $k(X)$ as an extension of $k(Y)$.

Proposition 14.17 *ψ is birational if and only if there are non-empty open sets $U_x \subset X$ and $U_y \subset Y$ such that $\psi: U_x \to U_y$ is an isomorphism.*

Proof. Since $k(U_x) = k(X)$ and $k(U_y) = k(Y)$, it is clear that if $\psi: U_x \to U_y$ is an isomorphism then ψ is birational. Conversely, if ψ is birational, then there is an open affine $X' \subset X$ and an open affine $Y' \subset Y$ such that $k(X') = k(X)$ and $k(Y') = k(Y)$. In other words, we may assume that X and Y are affine. Then $k[X] = k[Y][h_1, \ldots, h_m]$ with $h_i \in k(Y)$. If $h_i = h_i'/h$ with $h_i', h \in k[Y]$, then ψ induces an isomorphism $\psi^*: k[Y]_h \to k[X]_h$ and so, since X, Y are affine, ψ is an isomorphism between X_h and Y_h (Corollary I.7.17).

We have shown that if $\psi: X \to Y$ is a dominant morphism of varieties, then there is a non-empty U_X open in X such that (i) $\psi \mid U_X$ is an open morphism, and (ii) if W is an irreducible closed subvariety of Y and Z is an irreducible component of $\psi^{-1}(W)$ with $Z \cap U_X \neq \emptyset$, then $\dim Z = \dim W + \dim X - \dim Y$. If $W = \{y\}$ is a point of Y, then any component of the fiber $\psi^{-1}(y)$ which meets U_X has dimension $\dim X - \dim Y$. We observe that

$$\dim X - \dim Y = \operatorname{tr}\deg_{k(Y)} k(X) \tag{14.18}$$

and hence, that $\dim X = \dim Y$ if and only if $k(X)$ is a finite algebraic extension of $k(Y)$. We let $[k(X) : k(Y)]$ be the degree of this extension. In this case, $\psi^{-1}(\psi(x))$ for $x \in U_X$ has dimension 0 and is, therefore, a finite set of points. We shall show that on (a perhaps smaller) open set, this number is constant. In fact, we have:

Proposition 14.19 *If $k(X)$ is a finite algebraic extension of $k(Y)$, then there is an open set $U_X \subset X$ such that if $x \in U_X$, then*

$$\#\psi^{-1}(\psi(x)) = [k(X) : k(Y)]_s \tag{14.20}$$

*i.e., the number of points in the fiber $\psi^{-1}(\psi(x))$ is the separable degree of the extension $k(X)/k(Y)$).**

Proof. By factoring, we may assume that $k(X)$ is a simple extension of $k(Y)$, i.e., $k(X) = k(Y)(\theta)$. First, suppose that θ is purely inseparable over $k(Y)$ and assume, without loss of generality, that $\theta^p \in k(Y)$. Then, on an open set in X, we have $k[X] = k[Y][\theta]$ with $\theta^p = g(Y) \in k[Y]$ (by clearing denominators, etc.). We can view X as $\{(y, g(y)^{1/p})\colon y \in Y,\ g(y)^{1/p} \in \mathbb{A}^1_k\}$ and ψ as the projection $\pi_Y : Y \times \mathbb{A}^1 \to Y$. Since $T^p - g(Y)$ is irreducible over $k[Y]$, the only point in $\pi_Y^{-1}(\boldsymbol{\xi})$ is $(\boldsymbol{\xi}, g(\boldsymbol{\xi})^{1/p})$. In other words, $\#\pi_Y^{-1}(\pi_Y(x)) = 1 = [k(X) : k(Y)]_s$ in this case. Next assume that θ is separable algebraic over $k(Y)$. Let $F(z) = \sum_{i=1}^{n} f_i(Y) z^i$ be the minimal polynomial of θ over $k(Y)$ (note that $n = [k(X) : k(Y)]$) with $f_i(Y) \in k(Y)$. We suppose $f_i(Y) = g_i(Y)/g_0(Y)$ with $g_0, g_i \in k[Y]$. Let $U_1 = Y_{g_0}$ and $U_2 = Y_{\Delta(F)}$, $\Delta(F)$ being the discriminant of F (I.C.16). Let $U = U_1 \cap U_2$. Then $G(\boldsymbol{\xi}, z) = \sum g_i(\boldsymbol{\xi}) z^i$ has n distinct roots for $\boldsymbol{\xi} \in U$. Consider $V = V(G(Y, z)) = \{(\boldsymbol{\xi}, a) \in U \times \mathbb{A}^1_k\colon G(\boldsymbol{\xi}, a) = 0\}$ and let $\pi_1 : V \to Y$ be the projection on the first factor. π_1 is an open map (I.12.12). If $I(V) = (h_j(Y, z))$, then, in $k[Y][z]$, each h_j is a multiple of G i.e., there are α_j, β_j with $\alpha_j(Y) h_j(Y, z) \equiv \beta_j G \bmod I(Y)$. If $\prod \alpha_j(\boldsymbol{\xi}) \neq 0$, then every solution $(\boldsymbol{\xi}, a)$ is a point of V and so, the number of points in $\pi_1^{-1}(Y) \cap V$ is the same as the number of solutions of $G(\boldsymbol{\xi}, z) = 0$ which is $n = [k(X) : k(Y)]$ in this case.

Corollary 14.21 *If ψ is birational, then $\#\psi^{-1}(\psi(x)) = 1$ almost everywhere (i.e., ψ is generically one-to-one).*

The case of a purely inseparable extension shows that a morphism may be generically one-to-one, without being birational.

---

*If K is a field and $L \supset K$ is a finite algebraic extension (i.e., $[L : K] < \infty$), then $L_s = \{u \in L\colon u \text{ separable over } K\}$ is a field with $K \subset L_s \subset L$ and $[L_s : K] = [L : K]_s$ is the *separable degree of K over L*. The extension L/L_s is *purely inseparable* i.e., $u \in L$ implies $u^{p^e} \in L_s$. Note $[L : K] = [L : L_s] \cdot [L_s : L]$.

Proposition 14.22 *If $k(X) = k(Y)(z_1, \ldots, z_r)$ is a finitely generated transcendental extension where $\psi\colon X \to Y$ is a dominant morphism, then ψ is an open map. If W is a closed irreducible subvariety of Y, then $\psi^{-1}(W)$ is irreducible and* $\dim \psi^{-1}(W) = \dim W + r$.

Proof. We may assume that $k[X] = k[Y][Z]$ (by induction) with Z transcendental over $k(Y)$, and hence that $X = Y \times \mathbb{A}^1$ with ψ the projection π_Y. Then π_Y is open (I.12.12). If W is irreducible with $\mathfrak{p}_W = I(W)$ prime in $k[Y]$, then $\pi_Y^{-1}(W) = V(\mathfrak{p}_W k[X])$ for: $(\boldsymbol{\xi}, a) \in \pi_Y^{-1}(W)$ if and only if $\pi_Y(\boldsymbol{\xi}, a) = \boldsymbol{\xi} \in W$ and so, $(\boldsymbol{\xi}, a) \in \pi_Y^{-1}(W)$ implies $(\boldsymbol{\xi}, a) \in V(\mathfrak{p}_W k[X])$ and, conversely, if $(\boldsymbol{\xi}, a) \in V(\mathfrak{p}_W k[X])$, then $\boldsymbol{\xi} \in V(\mathfrak{p}_W) = W$. Moreover, $k[X]/\mathfrak{p}_W k[X] \simeq k[Y][Z]/\mathfrak{p}_W k[Y][Z] \simeq (k[Y]/\mathfrak{p}_W)[Z]$ so that $\mathfrak{p}_W k[X]$ is prime, $\pi_Y^{-1}(W)$ is irreducible, and $\dim \pi_Y^{-1}(W) = \dim W + 1$.

Consider now the variety $\mathbf{V}_A(n, m, p) \simeq S^n_{m,p}$ (Chapter 13, viz. (13.45)) and the G-morphism $\psi_A\colon \mathbf{V}_A(n, m, p) \to \text{Hank}(n, m, p)$ given by (13.47). In view of Corollary 13.33, $\text{Hank}(n, m, p)$ is normal and of dimension $n(m + p)$.

Proposition 14.23 *If $\boldsymbol{\xi} = (\mathbf{H}_1, \ldots, \mathbf{H}_n, \xi_0, \ldots, \xi_{n-1}) \in \text{Hank}(n, m, p)$, then the "fiber" $\psi_A^{-1}(\boldsymbol{\xi}) = \{v \in \mathbf{V}_A(n, m, p)\colon \psi_A(v) = \boldsymbol{\xi}\}$ is a closed irreducible variety and* $\dim_k \psi_A^{-1}(\boldsymbol{\xi}) = n^2$.

Proof. Since ψ_A is a G-morphism, $\psi_A^{-1}(\boldsymbol{\xi}) = O_G(v)$ for some $v \in \mathbf{V}_A(n, m, p)$. By the State Space Isomorphism Theorem 13.42, $S_G(v) = \{I\}$ and $O_G(v) \simeq G \cdot v \simeq G$. Since $O_G(v)$ is closed (Chapter 13), the result follows.

Theorem 14.24 *If U is an invariant open set in $\mathbf{V}_A(n, m, p)$, then there is an open set U_0 in* $\text{Hank}(n, m, p)$ *such that* $\psi_A^{-1}(U_0) = U$.

Proof. In view of Corollary 14.14 and Proposition 14.23, ψ_A is an open map. Let U be an invariant open set in $\mathbf{V}_A(n, m, p)$. Then $U_0 = \psi_A(U)$ is open in $\text{Hank}(n, m, p)$ and $\psi_A^{-1}(U_0)$ is an open invariant set in $\mathbf{V}_A(n, m, p)$ with $U \subset \psi_A^{-1}(U_0)$. We claim that $U = \psi_A^{-1}(U_0)$. Let $v_0 \in \psi_A^{-1}(U_0)$ so that $\psi_A(v_0) = \boldsymbol{\xi}_0 \in U_0 = \psi_A(U)$. Then there is a v in U with $\psi_A(v) = \boldsymbol{\xi}_0$. Therefore $v_0 \in O_G(v)$. But U is invariant so that $v \in U$ implies $O_G(v) \subset U$. Thus $v_0 \in U$.

Theorem 14.24 shows that the second requirment of a geometric quotient is satisfied by $(\text{Hank}(n, m, p), \psi_A)$.

We now develop some results required in the next chapter (particularly, Theorem 15.31 which includes Proposition I.23.52).

Definition 14.25 Let $\psi\colon X \to Y$ be a dominant morphism. Then ψ is *separable* if $k(X)$ is a *separable* (i.e., separably generated) extension of $k(Y)$ (I.D.15).

If k has characteristic 0, then all morphisms ψ are separable.

Example 14.26 Let k have characteristic p and let $X \subset \mathbb{A}_k^2$ be the curve $f(X_1, X_2) = X_2 - X_1^p = 0$. Let $Y = \mathbb{A}_k^1$. The projection $\pi\colon X \to Y$ with $\pi(\xi_1, \xi_2) = \xi_2$ is a surjective (hence dominant) morphism. We have $k(Y) = k(X_2)$ and $k(X) = k(X_2, \overline{X}_1)$ where $\overline{X}_1^p = X_2$. In other words, $k(X)$ is a purely inseparable extension of $k(Y)$ and, therefore, π is *not* a separable morphism.

We showed, Theorem I.D.16, that if K/L is a separably generated extension of fields, then tr. $\deg K/L = \dim_K \mathrm{Der}_L(K,K)$ where $\mathrm{Der}_L(K,K) = \{D\colon D$ a derivation of K into K with $D(L) = 0\}$. Letting $K = k(X)$, $L = k(Y)$ we have:

Theorem 14.27 *If ψ is a separable morphism, then*

$$r - s = \text{tr. } \deg k(X)/k(Y) = \mathrm{Dim}_{k(X)} \mathrm{Der}_{k(Y)}(k(X), k(X))$$

where $r = \dim X$ and $s = \dim Y$.

Proof. Apply I.D.16.

We wish to establish the converse of this theorem. We begin with a lemma.

Lemma 14.28 *Let K/L be a finitely generated extension of fields and let $\nu = \mathrm{Dim}_K \mathrm{Der}_L(K,K)$. Then ν is the least number of elements $z_1, \ldots, z_t$ such that $K/L(z_1, \ldots, z_t)$ is separable algebraic.*

Proof. If L has characteristic 0, then K/L is separable and $\nu = \text{tr } \deg K/L$. So let us suppose that L has characteristic $p \neq 0$. Then every element of $\mathrm{Der}_L(K,K)$ is trivial on $L(K^p)$ and K is purely inseparable over $L(K^p)$. In other words, $\mathrm{Der}_L(K,K) = \mathrm{Der}_{L(K^p)}(K,K)$. We have

Case 1. $K = L(z)$, $z^p = \xi \in L$, $z \notin L$ (i.e., K/L is purely inseparable and $[K:L] = p$). Then $K = L[Z]/(z^p - \xi)$. The derivation $\partial/\partial z$ of $L[Z]$ has $\partial(z^p - \xi)/\partial z = 0$ and so extends to a $D \in \mathrm{Der}_L(K,K)$ with $Dz = 1$.

Case 2. $[K:L] = p^e$ and K/L purely inseparable. Then $\mathrm{Dim}_K \mathrm{Der}_L(K,K) = e$ and there are $u_1, \ldots, u_e$, $D_1, \ldots, D_e$ with $K = L(u_1, \ldots, u_e)$, $u_j \notin L(u_1, \ldots, u_{j-1})$, $D_i(u_j) = \delta_{ij}$ and $[L(u_1, \ldots, u_j) : L(u_1, \ldots, u_{j-1})] = p$. The u_j are constructed by induction using Case 1 and extending by 0. In other words, by Case 1 (and induction), there is a $\tilde{D}_j$ with $\tilde{D}_j(L(u_1, \ldots, u_{j-1})) = 0$, $\tilde{D}_j(u_j) = 1$. Set $D_j = \tilde{D}_j$ on $L(u_1, \ldots, u_j)$ and $D_j(u_{j+1}) = \cdots = D_j(u_e) = 0$. Since $K/L(K^p)$ is purely inseparable with $[K : L(K^p)] = p^\nu$, $\mathrm{Dim}_K(\mathrm{Der}_{L(K^p)}(K,K)) = \nu$ and there are $u_1, \ldots, u_\nu$ with $K = L(K^p)(u_1, \ldots, u_\nu)$ and derivations $D_i \in \mathrm{Der}_{L(K^p)}(K,K) = \mathrm{Der}_L(K,K)$ with $D_i(u_j) = \delta_{ij}$. If $D \in \mathrm{Der}_L(K,K)$ and $D(u_i) = \alpha_i$, then $D - \sum \alpha_i D_i = 0$

on $L(u_1, \ldots, u_\nu)$ and hence, $D - \sum \alpha_i D_i = 0$ on $K = L(K^p)(u_1, \ldots, u_\nu)$. It follows that $\nu = \mathrm{Dim}_K \mathrm{Der}_L(K, K)$ and that $K/L(u_1, \ldots, u_\nu)$ is separable algebraic by I.D.13. If there were fewer elements than ν, then $\mathrm{Dim}_K \mathrm{Der}_L(K, K)$ would be less than ν (a contradiction).

Corollary 14.29 *If* $\mathrm{Dim}_K \mathrm{Der}_L(K, K) = \mathrm{tr.\ deg}\, K/L$ *then* K *is separable over* L.

Theorem 14.30 *If* $\psi\colon X \to Y$ *is a dominant morphism,* $\dim X = r$, $\dim Y = s$ *and if* $r - s = \mathrm{Dim}_{k(X)} \mathrm{Der}_{k(Y)}(k(X), k(X))$, *then* ψ *is a separable morphism.*

Proof. Apply Corollary 14.29 with $K = k(X)$, $L = k(Y)$.

15

Projective Algebraic Geometry VI: Tangents, Differentials, Simple Subvarieties

We recall (I.20) that if V_a is an affine variety and $\xi \in V_a$, then the (Zariski) *tangent space to* V_a *at* ξ, $T_{V_a,\xi}$, is given by any of the following:

(i) $V\left(\sum_{i=1}^{N} \frac{\partial f(\xi)}{\partial X_i}(X_i - \xi_i)\right)$ where f runs through $\mathfrak{p} = I(V_a)$;

(ii) $\mathrm{Der}_k(k[V_a], k_\xi)$;

(iii) $\mathrm{Der}_k(\mathfrak{o}_{\xi,V_a}, k_\xi)$; and,

(iv) $(\mathfrak{m}_{\xi,V_a}/\mathfrak{m}_{\xi,V_a}^2)^*$.

We observe that

$$\dim_k T_{V_a,\xi} = N - \mathrm{rank}[J(f_1, \dots, f_\alpha; X_1, \dots, X_N)(\xi)] \tag{15.1}$$

where $\mathfrak{p} = I(V_a) = (f_1, \cdots, f_\alpha)$ and $V_a \subset \mathbb{A}_k^N$. We called ξ a *simple point* of V_a if $\dim_k T_{V_a,\xi} = \dim_k V_a$. In other words, if $\dim V_a = r$, then ξ is simple if and only if $\dim_k(\mathfrak{m}_{\xi,V_a}/\mathfrak{m}_{\xi,V_a}^2)^* = r$, or equivalently, if and only if $\mathrm{rank}[J(f_1, \dots, f_\alpha; X_1, \dots, X_N)(\xi)] = N - r$.

Proposition 15.2 *ξ is a simple point of V_a if and only if $\mathfrak{o}_{\xi,V_a}$ is a regular local ring.*

Proof. (See Appendix A.) If ξ is simple, then $\dim_k \mathfrak{m}_{\xi,V_a}/\mathfrak{m}_{\xi,V_a}^2 = r = \dim V_a = K\,\mathrm{Dim}\, \mathfrak{o}_{\xi,V_a}$ (Krull dimension). Let $f_1, \dots, f_r$ be elements of $\mathfrak{m}_{\xi,V_a}$ such that $\overline{f}_1, \dots, \overline{f}_r$ are a basis of $\mathfrak{m}_{\xi,V_a}/\mathfrak{m}_{\xi,V_a}^2$. In view of Nakayama's Lemma 7.45,

P. Falb, *Methods of Algebraic Geometry in Control Theory: Part II*, Modern Birkhäuser Classics, https://doi.org/10.1007/978-3-319-96574-1_15

$\mathfrak{m}_{\xi,V_a} = (f_1,\dots,f_r)$ and $\mathfrak{o}_{\xi,V_a}$ is a regular local ring. Conversely, if $\mathfrak{o}_{\xi,V_a}$ is regular, then $\mathfrak{m}_{\xi,V_a} = (f_1,\dots,f_r)$ (i.e., there is a regular system of parameters) so that $K\,\mathrm{Dim}\,\mathfrak{o}_{\xi,V_a} = \dim_k \mathfrak{m}_{\xi,V_a}/\mathfrak{m}^2_{\xi,V_a} = \dim V_a = \dim_k T_{V_a,\xi}$ and ξ is simple.

Corollary 15.3 (Appendix A) *If ξ is a simple point of V_a, then $\mathfrak{o}_{\xi,V_a}$ is a UFD (unique factorization domain).*

We note that if $f \in \mathfrak{o}_{\xi,V_a}$, then f defines a linear map $d_\xi f\colon T_{V_a,\xi} \to k_\xi = k$ by

$$d_\xi f(D) = D(f), \quad D \in \mathrm{Der}_k(\mathfrak{o}_{\xi,V_a}, k_\xi) \tag{15.4}$$

or, equivalently, by

$$d_\xi f(D) = D(f), \quad D \in \mathrm{Der}_k(k[V_a], k_\xi) \tag{15.5}$$

for $f \in k[V_a]$. Since $d_\xi a = 0$ for $a \in k$, d_ξ may be viewed as a map of $\mathfrak{m}_{\xi,V_a} \to (T_{V_a,\xi})^*$ and is, in fact, an isomorphism between $\mathfrak{m}_{\xi,V_a}/\mathfrak{m}^2_{\xi,V_a}$ and $(T_{V_a,\xi})^*$. We call $d_\xi f$ the *differential of f at ξ*.

Corollary 15.6 *ξ is a simple point of V_a if and only if there are $f_1,\dots,f_r \in \mathfrak{m}_{\xi,V_a}$ such that $d_\xi f_1,\dots,d_\xi f_r$ are independent i.e., form a k-basis of $(T_{V_a,\xi})^* = \mathfrak{m}_{\xi,V_a}/\mathfrak{m}^2_{\xi,V_a}$. Such f_i form a regular system of parameters at ξ.*

Corollary 15.7 *If ξ is simple on V_a, then there is a regular system of parameters $g_1,\dots,g_r$ with the $g_i \in k[V_a]$. If $V_a^i = V(g_i) \cap V_a$, then ξ is simple on V_a^i and $\cap T_{V_a^i,\xi} = (0)$ (i.e., the V_a^i meet transversally at ξ).*

Proof. Let $f_i = g_i/g$, $g_i, g \in k[V_a]$, $g_i(\xi) = 0$, $g(\xi) \neq 0$ be regular parameters at ξ. But $d_\xi f_i = d_\xi(g_i/g) = d_\xi g_i/g(\xi)$ as $f_i(\xi) = 0 = g_i(\xi)$ and so the $d_\xi g_i$ are independent. Since $V_a^i \subset V_a$, $T_{V_a^i,\xi} \subset T_{V_a,\xi}$. But $T_{V_a^i,\xi} \subset \mathrm{Ker}\, d_\xi f_i$ so that $\dim T_{V_a^i,\xi} \leq r-1$. Since $\dim T_{V_a^i,\xi} \geq \dim V_a^i \geq r-1$, ξ is simple on V_a^i. If Z is a component of $\cap V_a^i$ with $\xi \in Z$ and $\dim Z > 0$, then $T_{Z,\xi} \subset \cap T_{V_a^i,\xi}$ and there is a $D \neq 0$ with $(d_\xi f_i)(D) = 0$, $i = 1,\dots,r$. This contradicts the independence of the $d_\xi f_i$.

We can use simple points to develop a method for obtaining affine varieties ([M-5]).

Lemma 15.8 *$k[X_1,\dots,X_N] \subset k[X_1,\dots,X_N]_{\mathfrak{m}_0} \subset k[[X_1,\dots,X_N]]$ where $\mathfrak{m}_0 = (X_1,\dots,X_N) = \mathfrak{m}_{0,\mathbf{A}^N}$.*

Proof. ([M-5]). If $g(X_1,\dots,X_N) \notin \mathfrak{m}_0$, then

$$g(X_1,\dots,X_N) = \alpha_0 - \sum_{i=1}^{N} \alpha_i(X_1,\dots,X_N)X_i$$

with $\alpha_0 \neq 0$ and $\alpha_i(X_1, \dots, X_N) \in k[X_1, \dots, X_N]$. Then $g(X)^{-1} = \alpha_0^{-1}(1 - \sum \alpha_0^{-1}\alpha_i(X)X_i)^{-1}$. But

$$\left(1 - \sum \alpha_0^{-1}\alpha_i(X)X_i\right)^{-1} = 1 + \sum_1^\infty \left(\sum_{t=1}^N \alpha_0^{-1}\alpha_i(X)X_i\right)^t$$

(in $k[[X_1, \dots, X_N]]$). This yields the inclusion $k[X]_{\mathfrak{m}_0} \subset k[[X]]$.

We let $R = k[X_1, \dots, X_N]$, $R_{\mathfrak{m}_0} = k[X_1, \dots, X_N]_{\mathfrak{m}_0}$, and $R^* = k[[X_1, \dots, X_N]]$.

Theorem 15.9 ([M-5]) *Let $f_1, \dots, f_{N-r} \in R$ with $f_i(0) = 0$ (i.e., $f_i \in \mathfrak{m}_0$) and independent linear terms. Let*

$$\mathfrak{p} = \left\{ g \in R \colon g = \sum_{i=1}^{N-r} h_i f_i / h, h_i, h \in R, h(0) \neq 0 \right\}. \tag{15.10}$$

Then $\mathfrak{p}$ is a prime ideal in R, $\dim V(\mathfrak{p}) = r$, and 0 is a simple point of $V(\mathfrak{p})$. Finally, $V(f_1, \dots, f_{N-r}) = V(\mathfrak{p}) \cup X$ with $0 \notin X$.

Proof. Let $\mathfrak{p}' = \mathfrak{p}R_{\mathfrak{m}_0} = (f_1, \dots, f_{N-r})$ in $R_{\mathfrak{m}_0}$ and let $\mathfrak{p}'' = \mathfrak{p}R^* = (f_1, \dots, f_{N-r})$ in R^*. Clearly $\mathfrak{p}' \cap R = \mathfrak{p}$. Now $\mathfrak{p}'' \cap R_{\mathfrak{m}_0} = \{g \in R_{\mathfrak{m}_0} \colon g = \sum \theta_i f_i, \theta_i \in R^*\} \subset \bigcap_{t=1}^{\infty} (\mathfrak{p}' + \mathfrak{m}_0^t R_{\mathfrak{m}_0}) = \mathfrak{p}'$ by Krull's Theorem (Corollary A.4). But $\mathfrak{p}' \subset \mathfrak{p}'' \cap R_{\mathfrak{m}_0}$ and so, $\mathfrak{p}' = \mathfrak{p}'' \cap R_{\mathfrak{m}_0}$. This it will be sufficient to show that $\mathfrak{p}''$ is a prime ideal. However, the Weierstrass Preparation Theorem (A.49) gives $R^*/\mathfrak{p}'' \simeq k[[X_{N-r+1}, \dots, X_N]]$ which is an integral domain. Thus $\mathfrak{p}''$ and a fortiori $\mathfrak{p}'$, $\mathfrak{p}$ are prime ideals. Since $R/\mathfrak{p} \subset R^*/\mathfrak{p}''$, tr. $\deg R/\mathfrak{p} = \dim V(\mathfrak{p}) \geq r$. But $T_{V(\mathfrak{p}),0}$ is determined by the independent linear terms of $f_1, \dots, f_{N-r}$ so that $\dim T_{V(\mathfrak{p}),0} = N - (N-r) = r \geq \dim V(\mathfrak{p})$. Hence $\dim V(\mathfrak{p}) = r$ and 0 is a simple point. As for the final point, if $\mathfrak{p} = (g_1, \dots, g_t)$, then there are β_j, $\beta_j(0) \neq 0$, with $\beta_j g_j \in (f_1, \dots, f_{N-r})$. If $\beta = \prod \beta_j$, then $\beta\mathfrak{p} \subset (f_1, \dots, f_{N-r})$ and $V(\mathfrak{p}) \subset V(f_1, \dots, f_{N-r}) \subset V(\beta) \cup V(\mathfrak{p})$. The result follows as $\beta \notin \mathfrak{m}_0$.

Corollary 5.11 *If $\dim V(\mathfrak{p}) = r$ and 0 is a simple point of $V(\mathfrak{p})$, then there are $f_1, \dots, f_{N-r} \in \mathfrak{p}$ such that*

$$\mathfrak{p} = \{g \in R \colon g = \sum h_i f_i / h, h_i, h \in R, h(0) \neq 0\} \tag{15.12}$$

(by translation this applies at any simple point).

Proof. Since $\dim T_{V(\mathfrak{p}),0} = r$, there are $f_1, \dots, f_{N-r} \in \mathfrak{p}$ with independent linear terms. Let $\tilde{\mathfrak{p}} = \{g \in R \colon g = \sum h_i f_i / h,\ h_i, h \in R,\ h(0) \neq 0\}$. Then $\tilde{\mathfrak{p}}$ is prime by the Theorem and $V(\tilde{\mathfrak{p}})$ has dimension r. But $\mathfrak{p} \supset \tilde{\mathfrak{p}}$ so that $V(\mathfrak{p}) \subset V(\tilde{\mathfrak{p}})$. Since $\dim V(\mathfrak{p}) = r = \dim V(\tilde{\mathfrak{p}})$, $V(\mathfrak{p}) = V(\tilde{\mathfrak{p}})$ and $\mathfrak{p} = \tilde{\mathfrak{p}}$.

If V is *any* (quasi-projective) variety and $\boldsymbol{\xi} \in V$, then there is an open affine $U \subset V$ with $\boldsymbol{\xi} \in U$ and we make the following:

Definition 15.13 The (Zariski) *tangent space to* V *at* $\boldsymbol{\xi}$, $T_{V,\boldsymbol{\xi}}$, is given by $T_{U,\boldsymbol{\xi}}$.

In view of the local character of the tangent space, $T_{V,\boldsymbol{\xi}}$ is well-defined e.g., via $\mathrm{Der}_k(\mathfrak{o}_{\boldsymbol{\xi},V}, k_{\boldsymbol{\xi}})$ or $(\mathfrak{m}_{\boldsymbol{\xi},V}/\mathfrak{m}^2_{\boldsymbol{\xi},V})^*$. Alternatively, if $U_i = \mathbb{P}^N - V(Y_i) \simeq \mathbb{A}^N$, then $T_{V,\boldsymbol{\xi}} = T_{V\cap U_i,\boldsymbol{\xi}}$ (for $\boldsymbol{\xi} \in V \cap U_i$) and if $\mathfrak{p}_i = I(V \cap U_i)$, then

$$\dim_k T_{V,\boldsymbol{\xi}} = N - \mathrm{rank}[J(f_1, \dots, f_\alpha; X_1, \dots, X_N)(\boldsymbol{\xi})] \tag{15.14}$$

where $\mathfrak{p}_i = (f_1, \dots, f_\alpha)$ and $X_1, \dots, X_N$ are affine coordinates on U_i. Also, if $\tilde{U}$ is another affine open with $\tilde{U} \subset V$ and $\boldsymbol{\xi} \in \tilde{U}$, then, it follows from I.20.22 that $T_{U,\boldsymbol{\xi}} \cong T_{\tilde{U},\boldsymbol{\xi}}$ and so the definition of $T_{V,\boldsymbol{\xi}}$ makes sense.

Definition 15.15 $\boldsymbol{\xi} \in V$ is a *simple point of* V if $\dim_k T_{V,\boldsymbol{\xi}} = \dim V$ (or, equivalently, if $\mathfrak{o}_{\boldsymbol{\xi},V}$ is a regular local ring). The *singular locus of* V, $\mathrm{Sing}(V) = \{\boldsymbol{\xi} \in V\colon \boldsymbol{\xi}$ is singular i.e., not a simple point of $V\}$.

We have (as in Part I):

Proposition 15.16 Sing(V) *is closed.*

Proof. $\mathrm{Sing}(V) = \bigcup_{i=0}^{N} \mathrm{Sing}(V \cap U_i)$, $U_i = \mathbb{P}^N - V(Y_i)$, and $\mathrm{Sing}(V \cap U_i)$ are closed by I.20.33.

Proposition 15.17 $\mathrm{Sing}(V) < V$ *and hence,* $\dim \mathrm{Sing}(V) < \dim V$.

Proof. Let U be a non-empty open affine in V. Then there is a non-empty open $U_1 \subset U$ with $\dim_k T_{V,\boldsymbol{\xi}} = \dim V$ for $\boldsymbol{\xi} \in U_1$. Thus, $\mathrm{Sing}(V) \subset V - U \cap U_1$ (which is closed).

The local approach also enables us to define the *differential,* $d\psi$, *of a morphism* $\psi\colon V \to W$. If $\psi\colon V \to W$ is a morphism (of quasi-projective varieties) with $\boldsymbol{\xi} \in V$, $\psi(\boldsymbol{\xi}) \in W$, then there are open affine neighborhoods $U_{\boldsymbol{\xi}}$ of $\boldsymbol{\xi}$ and $U_{\psi(\boldsymbol{\xi})}$ of $\psi(\boldsymbol{\xi})$ such that $\psi(U_{\boldsymbol{\xi}}) \subset U_{\psi(\boldsymbol{\xi})}$ and the restriction $\tilde{\psi}$ of ψ to $U_{\boldsymbol{\xi}}$ is an (affine) regular map. Let $\tilde{\psi}^*\colon k[U_{\psi(\boldsymbol{\xi})}] \to k[U_{\boldsymbol{\xi}}]$ be the comorphism and define

$$(d\psi)_{\boldsymbol{\xi}} = (d\tilde{\psi})_{\boldsymbol{\xi}}. \tag{15.18}$$

Then $(d\psi)_{\boldsymbol{\xi}}\colon T_{V,\boldsymbol{\xi}} = T_{U_{\boldsymbol{\xi}},\boldsymbol{\xi}} \to T_{U_{\psi(\boldsymbol{\xi})},\boldsymbol{\xi}} = T_{W,\psi(\boldsymbol{\xi})}$ is given by

$$(d\psi)_{\boldsymbol{\xi}}(D) = (d\tilde{\psi})_{\boldsymbol{\xi}}(D) = D \circ \tilde{\psi}^* \tag{15.19}$$

for $D \in \mathrm{Der}_k(k[U_{\boldsymbol{\xi}}], k_{\boldsymbol{\xi}})$. (This is independent of the choice of $U_{\boldsymbol{\xi}}$, $U_{\psi(\boldsymbol{\xi})}$ and so is well-defined). We can describe $d\psi$ in "coordinates". Let $\psi_a\colon V_a \to W_a$ be a regular map of (quasi) affine varieties with $V_a \subset \mathbb{A}^n$, $W_a \subset \mathbb{A}^m$, $\dim V_a = r$

and $\dim W_a = s$. Then $\psi_a(\boldsymbol{\xi}) = \psi_a(\xi_1, \ldots, \xi_n) = \boldsymbol{\eta} = (\eta_1, \ldots, \eta_m)$ with $\eta_i = \psi_i(\xi_1, \ldots, \xi_n)$ and $\psi_i \in k[X_1, \ldots, X_n]$. If $\boldsymbol{\xi} \in V_a$ and $\boldsymbol{\eta} \in W_a$, then

$$T_{V_a,\boldsymbol{\xi}} = V\left(\sum_{i=1}^{n} \frac{\partial f}{\partial X_i}(\boldsymbol{\xi})(X_i - \xi_i)\right)$$

where f runs through $I(V_a)$ and

$$T_{W_a,\boldsymbol{\eta}} = V\left(\sum_{j=1}^{m} \frac{\partial g}{\partial Y_j}(\boldsymbol{\eta})(Y_j - \eta_j)\right)$$

where g runs through $I(W_a)$. Since $\psi_a(V_a) \subset W_a$, $(g \circ \psi_a)(\boldsymbol{\xi}) = g(\psi_a(\boldsymbol{\xi})) = g(\boldsymbol{\eta}) = 0$ for all $g \in I(W_a)$. Let $\boldsymbol{\lambda} = (\lambda_1, \ldots, \lambda_n) \in T_{V_a,\boldsymbol{\xi}}$. Then

$$(d\psi_a)_{\boldsymbol{\xi}}(\boldsymbol{\lambda}) = \left(\sum_{i=1}^{n} \frac{\partial \psi_1}{\partial X_i}(\boldsymbol{\xi})\lambda_i, \ldots, \sum_{i=1}^{n} \frac{\partial \psi_m}{\partial X_i}(\boldsymbol{\xi})\lambda_i\right). \tag{15.20}$$

Note that $(d\psi_a)_{\boldsymbol{\xi}}(\boldsymbol{\lambda}) \in T_{W_a,\psi_a(\boldsymbol{\xi})}$ for if $g \in I(W_a)$ then

$$\sum_{j=1}^{m} \frac{\partial g}{\partial Y_j}(\psi_a(\boldsymbol{\xi}))\left(\sum_{i=1}^{n} \frac{\partial \psi_j}{\partial X_i}(\boldsymbol{\xi})\lambda_i\right) = \sum_{i=1}^{n}\left(\frac{\partial g \circ \psi_a}{\partial X_i}\right)(\boldsymbol{\xi})\lambda_i = 0.$$

Moreover, in view of (14.19), $(d\psi_a)_{\boldsymbol{\xi}}$ may be viewed as the $m \times n$ matrix

$$J(\psi_1, \ldots, \psi_m; X_1, \ldots, X_n)(\boldsymbol{\xi}) = \begin{pmatrix} \frac{\partial \psi_1}{\partial X_1}(\boldsymbol{\xi}) & \cdots & \frac{\partial \psi_1}{\partial X_n}(\boldsymbol{\xi}) \\ \vdots & & \vdots \\ \frac{\partial \psi_m}{\partial X_1}(\boldsymbol{\xi}) & \cdots & \frac{\partial \psi_m}{\partial X_n}(\boldsymbol{\xi}) \end{pmatrix}. \tag{15.21}$$

Then the rank of $(d\psi_a)_{\boldsymbol{\xi}}$, $\rho((d\psi_a)_{\boldsymbol{\xi}})$, is simply $\operatorname{rank}[J(\psi_1, \ldots, \psi_m;\ X_1, \ldots, X_n)(\boldsymbol{\xi})]$. It follows (linear algebra) that $\dim_k T_{V_a,\boldsymbol{\xi}} = \rho((d\psi_a)_{\boldsymbol{\xi}}) + \dim[\operatorname{Ker}(d\psi_a)_{\boldsymbol{\xi}}]$ and hence that

$$\begin{aligned} n + \dim \operatorname{Ker}(d\psi_a)_{\boldsymbol{\xi}} &= \operatorname{rank}[J(f_1, \ldots, f_\alpha; X_1, \ldots, X_n)(\boldsymbol{\xi})] \\ &\quad + \operatorname{rank}[J(\psi_1, \ldots, \psi_m; X_1, \ldots, X_n)(\boldsymbol{\xi})] \end{aligned} \tag{15.22}$$

where $I(V_a) = (f_1, \ldots, f_\alpha)$. If $\psi_a(\boldsymbol{\xi}) = \boldsymbol{\eta}$ is simple on W_a, then

$$\operatorname{rank}[J(\psi_1, \ldots, \psi_m; X_1, \ldots, X_n)(\boldsymbol{\xi})] \leq s.$$

Since $n - r \geq \operatorname{rank}[J(f_1, \ldots, f_\alpha;\ X_1, \ldots, X_n)(\boldsymbol{\xi})]$, it follows that

$$\operatorname{rank}[J(f_1, \ldots, f_\alpha, \psi_1, \ldots, \psi_m; X_1, \ldots, X_n)(\boldsymbol{\xi})] \leq n - r + s$$

whenever $\boldsymbol{\xi} \in V_a - \psi_a^{-1}(\mathrm{Sing}(W_a))$. If, for $\boldsymbol{\xi} \in V_a - \psi_a^{-1}(\mathrm{Sing}(W_a))$, we actually have

$$\mathrm{rank}[J(f_1, \ldots, f_\alpha, \psi_1, \ldots, \psi_m; X_1, \ldots, X_n)(\boldsymbol{\xi})] = n - r + s$$

then

$$\mathrm{rank}[J(f_1, \ldots, f_\alpha; X_1, \ldots, X_n)(\boldsymbol{\xi})] \\ + \mathrm{rank}[J(\psi_1, \ldots, \psi_m; X_1, \ldots, X_n)(\boldsymbol{\xi})] = n - r + s$$

and $\dim \mathrm{Ker}(d\psi_a)_{\boldsymbol{\xi}} = r - s$. Then $\mathrm{rank}[J(f_1, \ldots, f_\alpha;\ X_1, \ldots, X_n)(\boldsymbol{\xi})] = n - r$ and $\mathrm{rank}[J(\psi_1, \ldots, \psi_m;\ X_1, \ldots, X_n)(\boldsymbol{\xi})] = s$. In other words, $\boldsymbol{\xi}$ is simple on V_a and $(d\psi_a)_{\boldsymbol{\xi}}$ is surjective. We then say that ψ_a is a *smooth morphism at* $\boldsymbol{\xi}$ (or is *smooth at* $\boldsymbol{\xi}$). More precisely, we have the following:

Definition 15.23 Let $\psi\colon V \to W$ be a morphism of (quasi-) projective varieties. Let $\boldsymbol{\xi} \in V$. Then ψ is *smooth at* $\boldsymbol{\xi}$ if (i) $\boldsymbol{\xi}$ is simple on V and $\psi(\boldsymbol{\xi})$ is simple on W; and (ii) $(d\psi)_{\boldsymbol{\xi}}\colon T_{V,\boldsymbol{\xi}} \to T_{W,\psi(\boldsymbol{\xi})}$ is surjective.

Clearly smoothness is a local property. Thus, in view of Theorem 15.9, if ψ is smooth at $\boldsymbol{\xi}$, we may suppose that $V_a = V(f_1, \ldots, f_{n-r})$ where $d_{\boldsymbol{\xi}} f_1, \ldots, d_{\boldsymbol{\xi}} f_{n-r}$ are independent and that there are $g_1, \ldots, g_s$ with $g_j(\psi(\boldsymbol{\xi})) = 0$ and $d_{\psi(\boldsymbol{\xi})} g_1, \ldots,$ $d_{\psi(\boldsymbol{\xi})} g_s$ independent. Then $\psi^{-1}(\psi(\boldsymbol{\xi}))$ is defined "near" $\boldsymbol{\xi}$ by the equations $f_1 = \cdots = f_{n-r} = g_1 \circ \psi = \cdots = g_s \circ \psi = 0$ with independent differentials. It follows that

$$\mathrm{rank}[J(f_1, \ldots, f_{n-r}, \psi_1, \ldots, \psi_m; X_1, \ldots, X_n)(\boldsymbol{\xi})] = n - r + s$$

(and a fortiori, $\mathrm{rank}[J(f_1, \ldots, f_\alpha, \psi_1, \ldots, \psi_m;\ X_1, \ldots, X_n)(\boldsymbol{\xi})] = n - r + s$ for any basis $f_1, \ldots, f_\alpha$ of $I(V_a)$). In other words, if $\psi(\boldsymbol{\xi})$ is simple on W, then ψ is smooth at $\boldsymbol{\xi}$ if and only if

$$\mathrm{rank}[J(f_1, \ldots, f_\alpha, \psi_1, \ldots, \psi_m; X_1, \ldots, X_n)(\boldsymbol{\xi})] = n - r + s.$$

We shall soon show that smoothness has some interesting consequences. But first we note that the basic results of Part I Chapter 20 continue to hold. For example, we have:

Proposition 15.24 (cf. I.20.25) *Let V, W be varieties and let $(\boldsymbol{\xi}, \boldsymbol{\eta})$ be an element of $V \times W$. Then*

$$T_{V,\boldsymbol{\xi}} \times T_{W,\boldsymbol{\eta}} \simeq T_{V \times W,(\boldsymbol{\xi},\boldsymbol{\eta})} \tag{15.25}$$

and hence, if $\boldsymbol{\xi}$ is simple on V and $\boldsymbol{\eta}$ is simple on W, then $(\boldsymbol{\xi}, \boldsymbol{\eta})$ is simple on $V \times W$.

Proof. Let U_V, U_W be open affines in V, W with $\xi \in U_V$, $\eta \in U_W$ and $U_V \times U_W$ affine in $V \times W$. Then apply I.20.25.

Now let V_a be an affine variety and let $\xi \in V_a$. Let $R = k[V_a]$ and let $\mathfrak{m}_\xi = \{f \in R\colon f(\xi) = 0\}$. If M is an R-module, then $M_\xi = M/\mathfrak{m}_\xi M$ is a $k = k_\xi$ vector space. We write $\Omega[V_a]$ for the module $\Omega_{k[V_a]/k}$ of differentials of $k[V_a]$ over k (Appendix D). Then $\mathrm{Der}_k(k[V_a], k_\xi) \simeq \mathrm{Hom}_{k[V_a]}(\Omega[V_a], k_\xi) \simeq \mathrm{Hom}_k(\Omega[V_a]/\mathfrak{m}_\xi\Omega[V_a], k)$ so that

$$T_{V_a,\xi} = \mathrm{Hom}_k(\Omega[V_a]/\mathfrak{m}_\xi\Omega[V_a], k) \tag{15.26}$$

and

$$(T_{V_a,\xi})^* = \Omega[V_a]/\mathfrak{m}_\xi\Omega[V_a] \tag{15.27}$$

is the *cotangent space to V_a at ξ*. Again, if V is *any* variety and $\xi \in V$, then there is an open affine $U \subset V$ with $\xi \in U$ and we make the following:

Definition 15.28 The cotangent space to V at ξ, $(T_{V,\xi})^*$ is given by $(T_{U,\xi})^*$.

Proposition 15.29 *If ξ is a simple point of the r-dimensional variety V, then there is an open affine U with $\xi \in U$ such that $\Omega[U]$ is a free $k[U]$-module with basis of $df_1, \dots, df_r$, $f_1, \dots, f_r \in k[U]$.*

Proof. We may assume that V is affine and let $I(V) = (f_1, \dots, f_\alpha)$. Since ξ is simple, $\mathrm{rank}[J(f_1, \dots, f_\alpha; X_1, \dots, X_n)(\xi)] = N - r$. Let $x_1, \dots, x_r$ be local parameters at ξ (Appendix A). Since

$$\sum_{j=1}^{N} \frac{\partial f_i}{\partial X_j}\, dx_j = 0$$

it follows that all the dx_j can be expressed in terms of $dx_1, \dots, dx_r$ with coefficients in $\mathfrak{o}_{\xi,V}$. Thus, there is a neighborhood U_ξ of ξ on which all these coefficients are regular i.e., in $\mathfrak{o}_V(U_\xi)$. If $\eta \in U_\xi$, then $(dx_1)_\eta, \dots, dx_r)_\eta$ are a basis of $(T_{V,\eta})^*$. If $\omega \in \Omega[U_\xi]$, then $\omega = \sum_{i=1}^{r} h_i dx_i$, $h_i \in \mathfrak{o}_V(U_\xi)$ (since at any $\eta \in U_\xi$, $\omega = \sum g_i(dx_i)_\xi$ with $g_i \in \mathfrak{o}_{\eta,U_\xi} = \mathfrak{o}_{\eta,V}$). $\Omega[U]$ for a suitable U is free since $\sum g_i dx_i = 0$, $g_1 \neq 0$ (say) would imply that $\sum_{i=1}^{r} g_i(\eta)(dx_i)_\eta = 0$ on U_{g_1} (a contradiction).

Corollary 15.30 *If $u_1, \dots, u_r$ are a regular system of parameters at ξ, then $du_1, \dots, du_r$ generate $\Omega[U]$ on some neighborhood U of ξ.*

We are now prepared to prove the following theorem which includes Proposition I.23.52.

Theorem 15.31 *Let $\psi\colon V \to W$ be a morphism. If ψ is dominant and separable, then there is a non-empty open set $U \subset V$ such that ψ is smooth on U. Conversely, if ψ is smooth at $\boldsymbol{\xi} \in V$, then ψ is dominant and separable.*

Proof. We may suppose that V is affine with $I(V) = (f_1, \dots, f_\alpha)$ (so $V = V(f_1, \dots, f_\alpha)$) and that W is also affine with $W \subset \mathbb{A}_k^m$. Then $\psi = (\psi_1, \dots, \psi_m)$ with $\psi_i \in k[X_1, \dots, X_n]$. Consider $J(\cdot) = J(f_1, \dots, f_\alpha, \psi_1, \dots, \psi_m; X_1, \dots, X_n)(\cdot)$. If $\boldsymbol{\xi} \in V - \psi^{-1}(\text{Sing } W)$ (which is open in V), then $\rho(J(\boldsymbol{\xi})) \leq n - r + s$ ($r = \dim V$, $s = \dim W$) and ψ is smooth at $\boldsymbol{\xi}$ if and only if $\rho(J(\boldsymbol{\xi})) = n - r + s$. Now $J(\cdot)$ can be viewed as an $m + \alpha \times n$ matrix with entries in $k(V)$ i.e., as an element of $M(m + \alpha, n; k(V))$. The null space of $J(\cdot)$ can be identified with $\text{Der}_{\psi^*(k(W))}(k(V), k(V))$ via $D = 0$ on $\psi^*(k(W))$ if and only if

$$\sum_1^n \frac{\partial f_\alpha}{\partial X_j} D(X_j) = 0, \quad \sum_{j=1}^n \frac{\partial \psi_i}{\partial X_j} D(X_j) = 0.$$

Since ψ is separable, $\dim \text{Der}_{k(W)}(k(V), k(V)) = r - s = \text{tr. deg}\, k(V)/k(W)$ (Theorem 14.27) and $\rho(J(\cdot)) = n - r + s$. Consequently, $\rho(J(\boldsymbol{\xi})) = n - r + s$ for $\boldsymbol{\xi} \in$ some open U in $V - \psi^{-1}(\text{Sing } W)$ and ψ is smooth on U. As for the second assertion of the Theorem, if U_V, U_W are open affines in V, W respectively with $\psi\colon U_V \to U_W$, then $\overline{\psi(U_V)} = U_W$ would imply $\overline{\psi(V)} = W$. Thus we may assume that V, W are affine and, in view of Proposition 15.16, that V, W are nonsingular. Then $\Omega[V]$ is a free $k[V]$-module of rank r ($= \dim V$) and $\Omega[W]$ is a free $k[W]$-module of rank s ($= \dim W$). Let $\psi^*\colon k[W] \to k[V]$ be the comorphism. Then (cf. Appendix C with $k[W] = R$, $k[V] = S$) ψ^* induces a natural homomorphism $\tilde{\psi}^*\colon \Omega[W] \to \Omega[V]$ which, in turn, induces a homomorphism $\tilde{\Psi}^*\colon \Omega[W] \otimes_{k[W]} \Omega[V] \to \Omega[V]$ of free $k[V]$-modules. $\tilde{\Psi}^*$ may be described by an $r \times s$ matrix with entries in $k[V]$. If $(d\psi)_{\boldsymbol{\xi}}\colon \text{Hom}_{k[V]}(\Omega[V], k_\xi) \to \text{Hom}_{k[W]}(\Omega[W], k_{\psi(\boldsymbol{\xi})})$ is surjective, then the map $h \to h \circ \tilde{\psi}^*$ of $\text{Hom}_{k[V]}(\Omega[V], k) \to \text{Hom}_{k[W]}(\Omega[W], k)$ is surjective and the k-rank of $\tilde{\Psi}^*$ is s. But this implies $\rho(\tilde{\Psi}^*) = s$ and hence (Corollary C.14) that ψ^* is injective. In other words, ψ is dominant (I.7.18). Consequently, the map $\Omega_{k(W)/k} \otimes_{k(W)} k(V) \to \Omega_{k(V)/k}$ is also injective and its "co-map" $\text{Der}_k(k(V), k(V)) \to \text{Der}_k(k(W), k(V))$ is surjective. Since the kernel of the "co-map" is $\text{Der}_{k(W)}(k(V), k(V))$, $\dim \text{Der}_{k(W)}(k(V), k(V)) = r - s = \text{tr. deg}\, k(V)/k(W)$ and $k(V)$ is separable over $k(W)$ by Theorem 14.30.

Corollary 15.32 ([M-5]) *If $\psi\colon V \to W$ is dominant and k has characteristic 0, then there is an open $U_W \subset W$ such that ψ is smooth on $\psi^{-1}(U_W) - \text{Sing } V$.*

Proof. Let $V_s = \{\boldsymbol{\xi} \in V\colon \psi \text{ is smooth at } \boldsymbol{\xi}\}$. By the Theorem, V_s is non-empty and $V_1 = V - V_s - \psi^{-1}(\text{Sing } W) - \text{Sing } V$ is locally closed. We claim that $\overline{\psi(V_1)} < W$ (clearly this implies the corollary with $U_W = W - \overline{\psi(V_1)}$. If $\overline{\psi(V_1)} = W$, then $\psi_1 \colon V_1 \to W$ (ψ_1 the restriction of ψ) is dominant and $\overline{V}_1$ has a component X_1 with $\psi_1 \colon X_1 \to W$ dominant and separable (since k has characteristic 0). But then for some point $\boldsymbol{\xi} \in X_1$, $(d\psi_1)_{\boldsymbol{\xi}}$ is surjective. A fortiori $(d\psi)_{\boldsymbol{\xi}}$ is surjective and $\boldsymbol{\xi} \notin V_1$ (a contradiction).

We now wish to treat subvarieties. Let $W \subset V \subset \mathbb{A}_k^N$ be (irreducible) affine varieties with $s = \dim W$, $r = \dim V$. We recall (Chapter 7) that $\mathfrak{o}_{W,V} = k[V]_{\mathfrak{p}_W}$ and that $\mathfrak{m}_{W,V} = \mathfrak{p}_W k[V]_{\mathfrak{p}_W}$. Also, $\mathfrak{o}_{W,V}/\mathfrak{m}_{W,V} = k(W)$ and $K \dim \mathfrak{o}_{W,V} = \operatorname{codim}_V W = \dim V - \dim W = r - s$ so that $\mathfrak{o}_{W,V}$ is a regular local ring if and only if $\dim_{k(W)} \mathfrak{m}_{W,V}/\mathfrak{m}_{W,V}^2 = r - s$ i.e., if and only if $\mathfrak{m}_{W,V} = (f_1, \ldots, f_{r-s})$ with $f_1, \ldots, f_{r-s} \in \mathfrak{p}_W$ $(= \mathfrak{m}_{W,V} \cap k[V])$. Let $x = (x_1, \ldots, x_s, v_1, \ldots, v_{N-s})$ be a "generic point" of V and let $x' = (x_1, \ldots, x_s, w_1, \ldots, w_{N-s})$ be a "generic point" of W (see Appendix B) so that

$$\begin{aligned}\mathfrak{o}_{W,V} = \{ & f^*(v_1, \ldots, v_{N-s})/g^*(v_1, \ldots, v_{N-s}) : \\ & f^*, g^* \in K^*[v_1, \ldots, v_{N-s}], g^*(w_1, \ldots, w_{N-s}) \neq 0\}\end{aligned}$$

and $\mathfrak{m}_{W,V} = \{h^* \in \mathfrak{o}_{W,V} \colon h^*(w_1, \ldots, w_{N-s}) = 0\}$ (where $K^* = k(x_1, \ldots, x_s)$). The *tangent space to* V *at* W, $T_{V,W}$, is given by $(\mathfrak{m}_{W/V}/\mathfrak{m}_{W,V})^*$ so that W is *regular on* V if and only if $\mathfrak{o}_{W,V}$ is a regular local ring i.e., if and only if $\dim_{k(W)} T_{V,W} = \operatorname{codim}_V W = r - s$. We now make the following:

Definition 15.33 W is a *simple subvariety of* V (or is *nonsingular on* V) if

$$\operatorname{rank}[J(f_1, \ldots, f_\alpha; X_1, \ldots, X_N)(x')] = N - r \tag{15.34}$$

where $\mathfrak{p}_V = I(V) = (f_1, \ldots, f_\alpha)$ and $x' = (x_1, \ldots, x_s, w_1, \ldots, w_{N-s})$ is "the generic point of W".

Proposition 15.35 *W is a simple subvariety of V if and only if there is an open $U \subset W$ with $U \subset V - \operatorname{Sing} V$ (i.e., almost all points of W are simple points of V).*

Proof. If $\operatorname{rank}[J(f_1, \ldots, f_\alpha; X_1, \ldots, X_n)(\boldsymbol{\xi})] = N - r$ for some $\boldsymbol{\xi} \in W$, then $\boldsymbol{\xi}$ is a simple point of V and there is an $N - r \times N - r$ minor, h, of $J(f_1, \ldots, f_\alpha; X_1, \ldots, X_N)$ such that $h(\boldsymbol{\xi}) \neq 0$. But then $h \not\equiv 0$ on W and hence, $h(x') = h(x_1, \ldots, x_s, w_1, \ldots, w_{N-s}) \neq 0$. In other words, $\operatorname{rank}[J(f_1, \ldots, f_\alpha; X_1, \ldots, X_N)(x')] = N - r$. Conversely, if $\operatorname{rank}[J(f_1, \ldots, f_\alpha; X_1, \ldots, X_N)(x')] = N - r$, then there is an $N - r \times N - r$ minor h with $h(x') \neq 0$. If $\boldsymbol{\xi} \in W_h$, then $h(\boldsymbol{\xi}) \neq 0$ and $\operatorname{rank}[J(f_1, \ldots, f_\alpha; X_1, \ldots, X_N)(\boldsymbol{\xi})] = N - r$ so that $W_h \subset V - \operatorname{Sing} V$.

Proposition 15.36 *Let $V = \mathbb{A}_k^N$. Then W is simple on $\mathbb{A}^N$ if and only if $\mathfrak{o}_{W,\mathbb{A}^N}$ is regular.*

Proof. Since $I(\mathbb{A}^N) = (0)$, any W is simple on $\mathbb{A}^N$. So, we need only show that if W is simple on $\mathbb{A}^N$, then $\mathfrak{o}_{W,\mathbb{A}^N}$ is regular i.e., that $\mathfrak{m}_{W,\mathbb{A}^N} = (f_1, \ldots, f_{N-s})$. We claim that there are such $f_1, \ldots, f_{N-s} \in \mathfrak{p}_W = \mathfrak{m}_{W,\mathbb{A}^N} \cap K^*[v_1, \ldots, v_{N-s}]$. Now, w_{N-s} is algebraic over $K^*[w_1, \ldots, w_{N-s-1}]$ $(= K^*(w_1, \ldots, w_{N-s-1}))$ since tr. deg $k(W)/k = s$ and the w_i are algebraic over K^*. Let $f_{N-s}(x_1, \ldots, x_s, v_1, \ldots, v_{N-s-1}, Z)$ be a monic polynomial in Z such that $f_{N-s}(x_1, \ldots, x_s, w_1, \ldots, w_{N-s-1}, Z)$ is the irreducible minimal polynomial of w_{N-s} over $K^*[w_1, \ldots, w_{N-s-1}]$. Then $f_{N-s}(x') = 0$ and $f_{N-s} \in \mathfrak{p}_W = \mathfrak{m}_{W,\mathbb{A}^N} \cap K^*[v_1, \ldots, v_{N-s}]$. If h is any element of $\mathfrak{m}_{W,\mathbb{A}^N} \cap K^*[v_1, \ldots, v_{N-s}]$, then, by the Euclidean algorithm,

$$h(v_1, \ldots, v_{N-s}) = q(v_1, \ldots, v_{N-s}) f_{N-s}(v_1, \ldots, v_{N-s}) + r(v_1, \ldots, v_{N-s})$$

(with $q, r \in K^*(v_1, \ldots, v_{N-s-1})[v_{N-s}]$) where $\deg r < \deg f_{N-s}$ or $r = 0$ (as polynomials in v_{N-s}). If $\deg r < \deg f_{N-s}$, then $h(w_1, \ldots, w_{N-s}) = 0$ and $f_{N-s}(w_1, \ldots, w_{N-s}) = 0$ together would imply $r(w_1, \ldots, w_{N-s}) = 0$ and $f_{N-s}(w_1, \ldots, w_{N-s-1}, Z)$ would *not* be the minimal polynomial of w_{N-s} over $K^*[w_1, \ldots, w_{N-s-1}]$. Thus $r(w_1, \ldots, w_{N-s-1}, Z)$ is the zero polynomial and all the coefficients $r_j(w_1, \ldots, w_{N-s-1})$ of r are in $\mathfrak{p}_W$. Thus it would be enough to show that there are $f_1, \ldots, f_{N-s-1}$ which generate $\mathfrak{m}_{W,\mathbb{A}^N} \cap K^*[v_1, \ldots, v_{N-s-1}]$ over $K^*[v_1, \ldots, v_{N-s-1}]$. The result follows by induction.

Now let $\mathfrak{o}_{W,\mathbb{A}^N} = R^* = k[X_1, \ldots, X_N]_{I(W)}$, $\mathfrak{p} = I(V)$, and $\mathfrak{p}^* = R^*_{\mathfrak{p}}$. Then $R = \mathfrak{o}_{W,V} \simeq R^*/\mathfrak{p}^*$ and, setting $\mathfrak{m} = \mathfrak{m}_{W,V}$, we have $\mathfrak{m} \simeq \mathfrak{m}^*/\mathfrak{p}^*$ where $\mathfrak{m}^* = I(W)R^*$ is the maximal ideal of the local ring R^*. If $m_1, m_2 \in \mathfrak{m}$, then $m_1 = m_1^* + \mathfrak{p}^*$, $m_2 = m_2^* + \mathfrak{p}^*$ and $m_1 m_2 = m_1^* m_2^* + \mathfrak{p}^*$. In other words, $\mathfrak{m}^2 = (\mathfrak{p}^* + \mathfrak{m}^{*2})/\mathfrak{p}^*$. Moreover, $k(W) = R^*/\mathfrak{m}^* \simeq (R^*/\mathfrak{p}^*)/(\mathfrak{m}^*/\mathfrak{p}^*) = R/\mathfrak{m}$. We let $K = k(W)$. Then

$$\mathfrak{m}/\mathfrak{m}^2 \simeq \frac{(\mathfrak{m}^*/\mathfrak{p}^*)}{(\mathfrak{p}^* + \mathfrak{m}^{*2})/\mathfrak{p}^*} \simeq \frac{(\mathfrak{m}^*/\mathfrak{m}^{*2})}{(\mathfrak{p}^* + \mathfrak{m}^{*2}/\mathfrak{m}^{*2})} \tag{15.37}$$

and all are K-vector spaces. Then

$$\begin{aligned} \dim_K \mathfrak{m}/\mathfrak{m}^2 &= \dim_K \mathfrak{m}^*/\mathfrak{m}^{*2} - \dim_K (\mathfrak{p}^* + \mathfrak{m}^{*2})/\mathfrak{m}^{*2} \\ &= N - s - \dim_K (\mathfrak{p}^* + \mathfrak{m}^{*2})/\mathfrak{m}^{*2} \end{aligned} \tag{15.38}$$

(in view of Proposition 15.36 and (15.34)).

Corollary 15.39 $\mathfrak{o}_{W,V}$ *is regular if and only if* $\dim_k (\mathfrak{p}^* + \mathfrak{m}^{*2})/\mathfrak{m}^{*2} = N - r$.

Proposition 15.40 *If W is a simple subvariety of V, then $\mathfrak{o}_{W,V}$ is regular.*

Proof. If W is simple, then there are $f_1, \ldots, f_{N-r}$ (say) $\in \mathfrak{p}$ with

$$\det \left(\frac{\partial f_i}{\partial X_j} \right)(x') \neq 0, \quad j = 1, \ldots, N - r. \tag{15.41}$$

The f_i are a fortiori in $\mathfrak{p}^*$. Consider the elements $f_i + \mathfrak{m}^{*2}$ of $\mathfrak{p}^* + \mathfrak{m}^{*2}$. Since $\mathfrak{p}^* \subset \mathfrak{m}^*$, the f_i have no constant terms and we can write

$$f_i + \mathfrak{m}^{*2} = \sum_{j=1}^{N} \alpha_{ij} X_j + \mathfrak{m}^{*2} \tag{15.42}$$

with $\alpha_{ij} \in k$, $i = 1, \ldots, N - r$. Then $f_1 + \mathfrak{m}^{*2}, \ldots, f_{N-r} + \mathfrak{m}^{*2}$ will be linearly independent over K if and only if the vectors $\boldsymbol{\alpha}_i = (\alpha_{i1}, \ldots, \alpha_{iN})$ are independent over K i.e., if and only if $\operatorname{rank}(\boldsymbol{\alpha}_i) = N - r$. Since $\frac{\partial f_i}{\partial x_j}(x') = \alpha_{ij}$, the result follows from 15.41.

Proposition 15.43 *If $\mathfrak{o}_{W,V}$ is regular, then W is simple on V.*

Proof. By virtue of Corollary 15.39, $\dim_K(\mathfrak{p}^* + \mathfrak{m}^{*2})/\mathfrak{m}^{*2} = N - r$ and there are $g_1, \ldots, g_{N-r} \in \mathfrak{p}^*$ which give a basis $g_1 + \mathfrak{m}^{*2}, \ldots, g_{N-r} + \mathfrak{m}^{*2}$. Since $g_j = \sum r_j h_j / h$ with $h_j \in \mathfrak{p}$ and $h \notin I(W)$, there are $f_1, \ldots, f_{N-r} \in \mathfrak{p}$ such that the $f_i + \mathfrak{m}^{*2}$ give a basis. We can (as in the previous proposition) write $f_i + \mathfrak{m}^{*2} = \sum_{j=1}^{N} \alpha_{ij} X_j + \mathfrak{m}^{*2}$ with the vectors $\boldsymbol{\alpha}_i = (\alpha_{i1}, \ldots, \alpha_{iN})$ independent over K. Since $J(f_1, \ldots, f_{N-r};\, X_1, \ldots, X_N)(x') = (\boldsymbol{\alpha}_i)$, we have $\operatorname{rank}[J(f_1, \ldots, f_{N-r};\, X_1, \ldots, X_N)(x')] = N - r$ and W is simple on V. (These arguments are originally due to Zariski [Z-2].)

Since the notion of a simple subvariety is local, if $W \subset V \subset \mathbb{P}_k^N$ are projective varieties, then W is *regular* (*simple*) on V if $\mathfrak{o}_{W,V}$ is a regular local ring (or, equivalently, if an appropriate Jacobian has the right rank).

Example 15.44 (cf. Example 7.52) View ψ_F as a map of $\mathbb{A}^1 - (0)$ into $\mathbb{A}^4$ given by $\psi_F(z) = (-(z+1)/z,\ (z+1)(z-1)/z,\ -(z+1),\ (z+1)^2(z-1)/z)$ and use X_1, X_2, X_3, X_4 as coordinates on $\mathbb{A}^4$. Let C_F denote the curve $\psi_F(z)$ so that C_F is determined by the equations

$$\begin{aligned} f_1(X_1, X_2, X_3, X_4) &= (X_1 + 1)X_3 + X_3 = 0 \\ f_2(X_1, X_2, X_3, X_4) &= (X_2 + 1)X_3 + X_3(X_3 + 2) = 0 \\ f_3(X_1, X_2, X_3, X_4) &= (X_4 + 1)X_3 - X_3^2(X_3 - 2) = 0. \end{aligned} \tag{15.45}$$

Then $C_F \cap V(X_4) = \{P_1, P_2\}$ where $P_1 = (0,0,0,0)$ and $P_2 = (-2,0,-2,0)$. $C_F \cap V(X_4)$ is determined by the equations (15.45) and the equation $f_4(X_1, X_2, X_3, X_4) = X_4 = 0$. We observe that

$$J(f_1, f_2, f_3; X_1, X_2, X_3, X_4) = \begin{bmatrix} X_3 + 1 & 0 & X_1 + 1 & 0 \\ 0 & X_3 + 1 & 2X_3 + 2 + X_2 & 0 \\ 0 & 0 & X_4 - 3X_3^2 - 4X_3 & X_3 + 1 \end{bmatrix}$$

and that

$$J(f_1, f_2, f_3, f_4; X_1, X_2, X_3, X_4) = \begin{bmatrix} X_3+1 & 0 & X_1+1 & 0 \\ 0 & X_3+1 & 2X_3+2+X_2 & 0 \\ 0 & 0 & X_4-3X_3^2-4X_3 & X_3+1 \\ 0 & 0 & 0 & 1 \end{bmatrix}$$

It follows that $\mathcal{C}_F$ is nonsingular on $\mathbb{A}^4$, that P_1, P_2 are simple points of $\mathcal{C}_F$ (and also of $V(X_4)$), and that $\mathfrak{m}_1 = \mathfrak{m}_{P_1,\mathcal{C}_F} = (\overline{x}_1, \overline{x}_2, \overline{x}_3)$, $\mathfrak{m}_2 = \mathfrak{m}_{P_2,\mathcal{C}_F} = (\overline{x}_1+2, \overline{x}_2, \overline{x}_3+2)$. However, rank$[J(f_1, f_2, f_3, f_4; X_1, X_2, X_3, X_4)(P_1) = 3$ so that P_1 is a singular point of $\mathcal{C}_F \cap V(X_4)$. Since rank$[J(f_1, f_2, f_3, f_4; X_1, X_2, X_3, X_4)(P_2)] = 4$, P_2 is a simple point of $\mathcal{C}_F \cap V(X_4)$. This is relevant to the notion of simple and multiple poles for linear systems.

Exercises

(1) Show in detail that the differential of a morphism is well-defined.

(2) Show that the cotangent space is well-defined and give as many interpretations of this space as you can.

16
The Geometric Quotient Theorem

We recall from I.14 the following:

Definition 16.1 (See I.14.19) Let E be an equivalence on the affine variety V. A *geometric quotient of* V *modulo* E is a pair (W, ψ) consisting of an algebraic set W and a morphism ψ such that:

(i) for each $w \in W$, $\psi^{-1}(w)$ is a closed orbit;

(ii) for each invariant open set $U \subset V$, there is an open set $U_0 \subset W$ such that $\psi^{-1}(U_0) = U$;

(iii) the comorphism $\psi^*: k[W] \to k[V]$ is a surjective k-isomorphism between $k[W]$ and the ring of invariants $k[V]^E$.

If $E = E_G$ where G acts on V, then we speak of a *geometric quotient of* V *modulo* G.

We note that in this definition V and W are affine. We now recall some ideas from Chapter 13. Let $C_m^n = \{y = (A, B)\colon y \text{ is controllable}\}$, $\mathcal{O}_p^n = \{z = (A, C)\colon z \text{ is observable}\}$, $\mathbf{Y}\colon C_m^n \to M_*(n, (n+1)m)$, and $\mathbf{Z}\colon \mathcal{O}_p^n \to M_*((n+1)p, n)$ where

$$\mathbf{Y}(A, B) = [B\ AB \cdots A^n B]$$
$$\mathbf{Z}(A, C) = \begin{bmatrix} C \\ CA \\ \vdots \\ CA^n \end{bmatrix}. \tag{16.2}$$

P. Falb, *Methods of Algebraic Geometry in Control Theory: Part II*,
Modern Birkhäuser Classics, https://doi.org/10.1007/978-3-319-96574-1_16

$G = \mathrm{GL}(n,k)$ acts on $\mathcal{C}_m^n$, $\mathcal{O}_p^n$ via

$$g \cdot (A,B) = (gAg^{-1}, gB), \quad g \cdot (A,C) = (gAg^{-1}, Cg^{-1}) \tag{16.3}$$

and both are G-invariant. Consider $\mathcal{O}_p^n \times \mathcal{C}_m^n$ and let Δ_A be the A-diagonal i.e., $\Delta_A = \{(A_1,C) \times (A_2,B)\colon A_1 = A_2\}$. Clearly $S_{m,p}^n$ is G-isomorphic to Δ_A and so Δ_A is an irreducible affine variety of dimension $n^2 + n(m+p)$. The map $\psi\colon \Delta_A \to M_*((n+1)p,n) \times M_*(n,(n+1)m) \times \mathbb{A}^{n^2}$ given by

$$\psi((A,C) \times (A,B)) = (\mathbf{Z}(A,C), \mathbf{Y}(A,B), A)$$

is an injective G-morphism. Let $\mathbf{Z}^1,\ldots,\mathbf{Z}^{n+1}(p\times n)$, $\mathbf{Y}_1,\ldots,\mathbf{Y}_{n+1}(n\times m)$ and $\mathbf{X}(n\times n)$ be block coordinates on $M_*((n+1)p,n) \times M_*(n,(n+1)m) \times \mathbb{A}^{n^2}$. The equations

$$\mathbf{Z}^i = \mathbf{Z}^1\mathbf{X}^{i-1}, \quad \mathbf{Y}_j = \mathbf{X}^{j-1}\mathbf{Y}_1 \tag{16.4}$$

$i,j = 1,\ldots,n+1$ define the affine variety $\mathbf{V}_A(n,m,p)$ in $M_*((n+1)p,n) \times M_*(n,(n+1)m) \times \mathbb{A}^{n^2}$. ψ is a G-isomorphism between Δ_A and $\mathbf{V}_A(n,m,p)$ and a fortiori $\mathbf{V}_A(n,m,p)$ and $S_{m,p}^n$ are G-isomorphic (Proposition 13.46). Let $\psi_A\colon \mathbf{V}_A(n,m,p) \to \mathrm{Hank}(n,m,p)$ be given by

$$\begin{aligned}\psi_A(\mathbf{Z},\mathbf{Y},\mathbf{X}) &= (\mathbf{Z}^1\mathbf{Y}_1,\ldots,\mathbf{Z}^n\mathbf{Y}_1, -\chi_i(\mathbf{X}))\\ &= (\mathbf{Z}^1\mathbf{Y}_1, \mathbf{Z}^1\mathbf{X}\mathbf{Y}_1,\ldots,\mathbf{Z}^1\mathbf{X}^{n-1}\mathbf{Y}_1, -\chi_i(\mathbf{X})).\end{aligned} \tag{16.5}$$

Then ψ_A is a surjective G-morphism which, by the State Space Isomorphism Theorem, is injective on G-orbits. If we view G as acting trivially on $\mathrm{Hank}(n,m,p)$, $\mathcal{H}(n,m,p)$, and $\mathrm{Hankel}(n,m,p)$, then these are all G-isomorphic (Theorem 11.29 and Proposition 13.32). We further observe that in $X = \mathbb{A}^{\tau(n,m,p)} \times (\mathbb{P}^n)_{Y_{n+1}} \times M((n+1)p,(n+1)m) = \mathbb{A}^{\tau(n,m,p)} \times \mathbb{A}^n \times M((n+1)p,(n+1)m)$ (i.e., $\xi_{n+1} = 1$), the variety defined by the Block Symmetry, Hankel Structure, Dependence, and Grassmann equations (11.5, 11.7, 11.9 and 11.19 for instance) is an affine variety and the open subset where $\rho(\mathbf{M}) = n$ is a quasi-affine variety isomorphic to $\mathrm{Hank}(n,m,p)$ (Chapter 11). Let $\psi_h\colon \mathrm{Hank}(n,m,p) \to \mathrm{Hankel}(n,m,p)$ be given by

$$\psi_h\colon (\mathbf{H}_1,\ldots,\mathbf{H}_n,\xi_0,\ldots,\xi_{n-1}) = (\mathbf{H}_1,\ldots,\mathbf{H}_{2n}) \tag{16.6}$$

where

$$\mathbf{H}_{n+j} = -\xi_0\mathbf{H}_j - \cdots - \xi_{n-1}\mathbf{H}_{n+j-1} \tag{16.7}$$

for $j = 1,\ldots,n$. ψ_h is an isomorphism between $\mathrm{Hank}(n,m,p)$ and $\mathrm{Hankel}(n,m,p)$. The variety $\mathrm{Hankel}(n,m,p)$ is a normal, quasi-affine variety of dimension $n(m+p)$. We consider the map $\psi\colon V_A(n,m,p) \to \mathrm{Hankel}(n,m,p)$ given by $\psi = \psi_h \circ \psi_A$ or, equivalently, by

$$\psi(\mathbf{Z},\mathbf{Y},\mathbf{X}) = (\mathbf{Z}^1\mathbf{X}^{j-1}\mathbf{Y}_1)_{j=1,\ldots,2n}. \tag{16.8}$$

We shall show that $(\text{Hankel}(n, m, p), \psi)$ is a geometric quotient of $\mathbf{V}_A(n, m, p)$ modulo G.

Proposition 16.9 *If* $(\mathbf{H}_1, \dots, \mathbf{H}_{2n})$ *is an element of* $\text{Hankel}(n, m, p)$*, then* $\psi^{-1}((\mathbf{H}_1, \dots, \mathbf{H}_{2n}))$ *is a closed* G*-orbit.*

Proof. If $\mathbf{v} \in \mathbf{V}_A(n, m, p)$, then $\dim O_G(\mathbf{v}) = n^2$ (by the State Space Isomorphism Theorem), where $O_G(\mathbf{v})$ is the G-orbit of $\mathbf{v}$. It follows that $O_G(\mathbf{v})$ is closed (Lemma I.15.9). If $\mathbf{v} \in \psi^{-1}((\mathbf{H}_1, \dots, \mathbf{H}_{2n}))$, then $O_G(\mathbf{v}) \subset \psi^{-1}((\mathbf{H}_1, \dots, \mathbf{H}_{2n}))$ as $\mathbf{Z}^1\mathbf{X}^{j-1}\mathbf{Y}_1$ is G-invariant. On the other hand, if $\mathbf{v}_1 \in \psi^{-1}((\mathbf{H}_1, \dots, \mathbf{H}_{2n}))$, then $\psi(\mathbf{v}_1) = (\mathbf{H}_1, \dots, \mathbf{H}_{2n}) = \psi(\mathbf{v})$, and, by the State Space Isomorphism Theorem, $\mathbf{v}_1 \in O_G(\mathbf{v})$.

Thus, if $(\mathbf{H}_1, \dots, \mathbf{H}_{2n}) \in \text{Hankel}(n, m, p)$, then $\psi^{-1}((\mathbf{H}_1, \dots, \mathbf{H}_{2n}))$ is a closed G-orbit which is irreducible and has dimension n^2.

Proposition 16.10 *If* U *is an open* G*-invariant set in* $\mathbf{V}_A(n, m, p)$*, then there is an open set* U_0 *in* $\text{Hankel}(n, m, p)$ *with* $\psi^{-1}(U_0) = U$.

Proof. Since $\dim \mathbf{V}_A(n, m, p) = n^2 + n(m + p)$, $\text{Hankel}(n, m, p)$ is normal, and $\dim \psi^{-1}((\mathbf{H}_1, \dots, \mathbf{H}_{2n})) = n^2$ for all $(\mathbf{H}_1, \dots, \mathbf{H}_{2n}) \in \text{Hankel}(n, m, p)$, ψ is an open map by Corollary 14.14. Let $U_0 = \psi(U)$. Then U_0 is open and $U \subset \psi^{-1}(U_0)$. If $\mathbf{v} \in \psi^{-1}(U_0)$, then $\psi(\mathbf{v}) \in \psi(U)$ and there is a $\mathbf{v}_1$ in U with $\psi(\mathbf{v}) = \psi(\mathbf{v}_1)$. Since ψ is injective on G-orbits and U is G-invariant, $O_G(\mathbf{v}) = O_G(\mathbf{v}_1)$ and $\mathbf{v} \in U$.

Thus we have established properties (i) and (ii) of a geometric quotient. We now turn our attention to the ring of invariants property (iii).

Proposition 16.11 *Let* Γ *act on an affine variety* V*. If* $f \in k[V]^\Gamma$ *and* U *is an open,* Γ*-invariant subset of* V*, then* $f_U = f \mid U$ *is an element of* $k[U]^\Gamma$ *where* $f \mid U$ *is the restriction of* f *to* U*.*

Proof. Obvious.

Proposition 16.12 *Let* Γ *act on an affine variety* V*. If* $V = \bigcup_{i=1}^{r} U_i$ *with the* U_i *open and* Γ*-invariant, if* $f_i \in k[U_i]^\Gamma$*,* $i = 1, \dots, r$*, and if* $f_i = f_j$ *on* $U_i \cap U_j$*, then* f *given by* $f(v) = f_i(v)$ *for* $v \in U_i$ *is a well-defined regular function on* V *with* $f \in k[V]^\Gamma$*.*

Proof. If $v \in U_i \cap U_j$, then $f(v) = f_i(v) = f_j(v)$ and f is well-defined. If $v \in V$, then $v \in$ some U_i and $f(v) = f_i(v)$. Since U_i is invariant, $g \cdot v \in U_i$ for all $g \in \Gamma$ and $f(g \cdot v) = f_i(g \cdot v) = f_i(v)$ as $f_i \in k[U_i]^\Gamma$.

Proposition 16.13 *Let Γ act on the affine varieties V and W and $\psi\colon V \to W$ be a Γ-morphism. Suppose that $V = \bigcup_{i=1}^{r} U_i$, $W = \bigcup_{i=1}^{r} X_i$ where U_i, X_i are Γ-invariant open sets and that $\psi(U_i) = X_i$ (so that $\psi^*(k[X_i]) \subset k[U_i]$). If ψ is injective on Γ-orbits and if $\psi^*(k[X_i]^\Gamma) = k[U_i]^\Gamma$ for $i = 1, \ldots, r$, then $\psi^*(k[W]^\Gamma) = k[V]^\Gamma$.*

Proof. If $h \in k[W]^\Gamma$, then $h_i = h \mid X_i \in k[X_i]^\Gamma$ and $\psi^*(h_i) = f_i \in k[U_i]^\Gamma$. Is $f_i = f_j$ on $U_i \cap U_j$? If $v \in U_i \cap U_j$, then $\psi(v) \in \psi(U_i) \cap \psi(U_j) = X_i \cap X_j$ and $h_i(\psi(v)) = h_j(\psi(v))$ $(= h(\psi(v)))$. In other words, $f_i = \psi^*(h_i) = h_i \circ \psi = h_j \circ \psi = \psi^*(h_j) = f_j$ on $U_i \cap U_j$. By Proposition 16.12, there is an $f \in k[V]^\Gamma$ with $f = f_i$ on U_i and clearly, $f = \psi^*(h)$. In other words, $\psi^*(k[W]^\Gamma) \subset k[V]^\Gamma$. If $f \in k[V]^\Gamma$, then $f_i = f \mid U_i \in k[U_i]^\Gamma$ and there is an $h_i \in k[X_i]^\Gamma$ with $f_i = h_i \circ \psi = \psi^*(h_i)$. Is $h_i = h_j$ on $X_i \cap X_j$? Now $X_i \cap X_j = \psi(U_i) \cap \psi(U_j)$. Say $x = \psi(v_i) = \psi(v_j)$, then $\psi(O_\Gamma(v_i)) = \psi(O_\Gamma(v_j))$. Since ψ is injective on orbits, $O_\Gamma(v_i) = O_\Gamma(v_j)$ and $v_j = g \cdot v_i$. But then $v_j \in U_i$ (as U_i is Γ-invariant) and similarly, $v_i \in U_j$. Therefore, $X_i \cap X_j = \psi(U_i \cap U_j)$. But $f_i = f_j$ on $U_i \cap U_j$ so that $h_i = h_j$ on $\psi(U_i \cap U_j) = X_i \cap X_j$. Then, by Proposition 16.12, there is an $h \in k[W]^\Gamma$ with $\psi^*(h) = f$ and so, $k[V]^\Gamma \subset \psi^*(k[W]^\Gamma)$.

We shall use Proposition 16.13 to prove property (iii) by applying it to $V = \mathbf{V}_A(n, m, p)$, $W = \text{Hankel}(n, m, p)$ and the morphism ψ of (16.8) (with G acting trivially on Hankel(n, m, p)). Note that, by the State Space Isomorphism Theorem, ψ is injective on orbits.

Consider first $\mathbf{V}_A(n, m, p)$ and let $\boldsymbol{\alpha} = (e^1, \ldots, e^n)$, $1 \le e^1 < \cdots < e^n \le (n+1)p$ be a row index and $\boldsymbol{\beta} = (f_1, \ldots, f_n)$, $1 \le f_1 < \cdots < f_n \le (n+1)m$ be a column index (where the e^i do not include $(n+1)$, $(n+1) \cdot 2, \ldots, (n+1)p$ and the f_j do not include $(n+1)$, $(n+1)2, \cdots, (n+1)m$). Set

$$U_{\boldsymbol{\beta}}^{\boldsymbol{\alpha}} = \{\mathbf{v} = (\mathbf{Z}, \mathbf{Y}, \mathbf{X}) \in \mathbf{V}_A(n, m, p)\colon \det \mathbf{Z}^{\boldsymbol{\alpha}} \neq 0, \det \mathbf{Y}_{\boldsymbol{\beta}} \neq 0\} \tag{16.14}$$

and note that the $U_{\boldsymbol{\beta}}^{\boldsymbol{\alpha}}$ form an open, affine covering of $\mathbf{V}_A(n, m, p)$ by G-invariant sets. If $(\mathbf{H}_1, \ldots, \mathbf{H}_{2n})$ is an element of Hankel(n, m, p) then we let $\mathcal{H}(\mathbf{H}_1, \ldots, \mathbf{H}_{2n})$ be the corresponding Hankel matrix. Set

$$X_{\boldsymbol{\beta}}^{\boldsymbol{\alpha}} = \{\mathbf{H} = (\mathbf{H}_1, \ldots, \mathbf{H}_{2n}) \in \text{Hankel}(n, m, p)\colon \det \mathcal{H}_{\boldsymbol{\beta}}^{\boldsymbol{\alpha}}(\mathbf{H}) \neq 0\} \tag{16.15}$$

and note that the $X_{\boldsymbol{\beta}}^{\boldsymbol{\alpha}}$ form an open, affine covering of Hankel(n, m, p) by G-invariant sets. Clearly $\psi(U_{\boldsymbol{\beta}}^{\boldsymbol{\alpha}}) \subset X_{\boldsymbol{\beta}}^{\boldsymbol{\alpha}}$.

Proposition 16.16 $\psi(U_{\boldsymbol{\beta}}^{\boldsymbol{\alpha}}) = X_{\boldsymbol{\beta}}^{\boldsymbol{\alpha}}$ *and* $\psi^*(k[X_{\boldsymbol{\beta}}^{\boldsymbol{\alpha}}]) \subset k[U_{\boldsymbol{\beta}}^{\boldsymbol{\alpha}}]^G$.

Proof. If $\mathbf{H} \in X_{\boldsymbol{\beta}}^{\boldsymbol{\alpha}}$, then, in view of (say) the Propositions 13.38 and 13.46, there is a $\mathbf{v} = (\mathbf{Z}, \mathbf{Y}, \mathbf{X})$ with $\psi(\mathbf{v}) = \mathbf{H}$. Thus, $\mathbf{H}_j = \mathbf{Z}^1 \mathbf{X}^{j-1} \mathbf{Y}_1$ and $\mathcal{H}(\mathbf{H}) = \mathbf{ZY}$.

It follows that $\mathcal{H}^\alpha_\beta(\mathbf{H}) = \mathbf{Z}^\alpha\mathbf{Y}_\beta$ and hence that $\det\mathcal{H}^\alpha_\beta(\mathbf{H}) \neq 0$ if and only if $\det\mathbf{Z}^\alpha \neq 0$ and $\det\mathbf{Y}_\beta \neq 0$. Hence, $\mathbf{v} \in U^\alpha_\beta$ and $\psi(U^\alpha_\beta) = X^\alpha_\beta$. If $h \in k[X^\alpha_\beta]$, then $(h\circ\psi)(g\cdot v) = h(\psi(g\cdot v)) = h(\mathbf{Z}g^{-1}g\mathbf{Y}) = h(\mathbf{ZY}) = h(\psi(\mathbf{v})) = (h\circ\psi)(\mathbf{v})$ (for $\mathbf{v} \in U^\alpha_\beta$).

So, by virtue of Proposition 16.13, we need only prove that $\psi^*(k[X^\alpha_\beta]) = k[U^\alpha_\beta]^G$. We shall do this using the classical ideas of the third proof given in I.Chapter 19. Let

$$V^\alpha_\beta = \{\mathbf{v} = (\mathbf{Z},\mathbf{Y},\mathbf{X}) \in \mathbf{V}_A(n,m,p)\colon \mathbf{Z}^\alpha = I, \det\mathbf{Y}_\beta \neq 0\} \tag{16.17}$$

so that $V^\alpha_\beta \subset U^\alpha_\beta$. Let G act on $G\times V^\alpha_\beta$ via

$$g(g_1,\mathbf{v}) = (gg_1,\mathbf{v}). \tag{16.18}$$

Then

Proposition 16.19 $k[G\times V^\alpha_\beta]^G = k[V^\alpha_\beta]$.

Proof. (cf. I.19.50) Now $k[G\times V^\alpha_\beta] = k[G]\otimes_k k[V^\alpha_\beta]$ and $k[G]^G = k$. If $f \in k[G\times V^\alpha_\beta]$, then $f = \sum r_i\otimes s_i$ with $r_i \subset k[G]$, $s_i \in k[V^\alpha_\beta]$ and $f((g,\mathbf{v})) = \sum r_i(g)s_i(\mathbf{v})$. If f is G-invariant, then $f((g,\mathbf{v})) = f((g\cdot I,\mathbf{v})) = f((I,\mathbf{v})) = \sum r_i(I)s_i(\mathbf{v})$ and so $f = \sum r_i(I)s_i \in k[V^\alpha_\beta]$.

Next, define a morphism $\gamma\colon G\times V^\alpha_\beta \to U^\alpha_\beta$ via

$$\gamma((g,\mathbf{v})) = g\cdot\mathbf{v} \tag{16.20}$$

or, writing $\mathbf{v} \in V^\alpha_\beta$ in the form

$$\mathbf{v} = \left(\begin{bmatrix}\mathbf{Z}-\mathbf{Z}^\alpha\\ I^\alpha\end{bmatrix},\mathbf{Y},\mathbf{X}\right) \tag{16.21}$$

(where $\mathbf{Z}-\mathbf{Z}^\alpha$ represents the rows of $\mathbf{Z}$ excluding the α rows and I^α is in the α rows),

$$\gamma((g,\mathbf{v})) = \left(\begin{bmatrix}(\mathbf{Z}-\mathbf{Z}^\alpha)g^{-1}\\ I^\alpha g^{-1}\end{bmatrix},g\mathbf{Y},g\mathbf{X}g^{-1}\right). \tag{16.22}$$

We then have:

Lemma 16.23 *γ is a G-isomorphism between $G\times V^\alpha_\beta$ and U^α_β.*

Proof. (cf. I.19.52) First note that $\gamma(g_1\cdot(g,v)) = \gamma((g_1g,v)) = (g_1g)\cdot v = g_1(g\cdot v) = g_1\gamma((g,v))$ so that γ is a G-morphism. Define a morphism $\gamma^{-1}\colon U^\alpha_\beta \to G\times V^\alpha_\beta$ as follows:

$$\gamma^{-1}((\mathbf{Z},\mathbf{Y},\mathbf{X})) = ((\mathbf{Z}^\alpha)^{-1},\begin{bmatrix}(\mathbf{Z}-\mathbf{Z}^\alpha)(\mathbf{Z}^\alpha)^{-1}\\ I^\alpha\end{bmatrix},\mathbf{Z}^\alpha\mathbf{Y},\mathbf{Z}^\alpha\mathbf{X}(\mathbf{Z}^\alpha)^{-1}) \tag{16.24}$$

(note that $\mathbf{Z}(\mathbf{Z}^\alpha)^{-1} = \begin{bmatrix} (\mathbf{Z}-\mathbf{Z}^\alpha)(\mathbf{Z}^\alpha)^{-1} \\ I^\alpha \end{bmatrix}$). Then we *claim* that $\gamma^{-1}\circ\gamma = i$. If $\mathbf{v} \in V_\beta^\alpha$ is given by (16.21) and $g \in G$, then $\gamma((g,\mathbf{v})) = g\cdot\mathbf{v}$ is given by (16.22) and $\mathbf{Z}^\alpha$ for $g\cdot\mathbf{v}$ is precisely g^{-1} (so that $(\mathbf{Z}^\alpha)^{-1} = g$). It follows that $\gamma^{-1}(\gamma(g,\mathbf{v})) = \gamma^{-1}(g\cdot\mathbf{v})$ is given by

$$\begin{aligned}\gamma^{-1}(g\cdot\mathbf{v}) &= \left((g^{-1})^{-1}, \begin{bmatrix} (\mathbf{Z}-\mathbf{Z}^\alpha)g^{-1}g \\ I^\alpha \end{bmatrix}, g^{-1}g\mathbf{Y}, g^{-1}g\mathbf{X}gg^{-1}\right) \\ &= \left(g, \begin{bmatrix} \mathbf{Z}-\mathbf{Z}^\alpha \\ I^\alpha \end{bmatrix}, \mathbf{Y}, \mathbf{X}\right) = (g,\mathbf{v}). \qquad (16.25)\end{aligned}$$

Similarly, $\gamma\circ\gamma^{-1} = i$ and the lemma is established.

Corollary 16.26 $k[U_\beta^\alpha]^G = k[G\times V_\beta^\alpha]^G = k[V_\beta^\alpha]$.

Since we have shown in Proposition 16.16 that $\psi^*(k[X_\beta^\alpha]) \subset k[U_\beta^\alpha]^G$, we need only show that $k[U_\beta^\alpha]^G \subset \psi^*(k[X_\beta^\alpha])$. We have:

Proposition 16.27 $k[U_\beta^\alpha]^G \subset \psi^*(k[X_\beta^\alpha])$.

Proof. If $f \in k[U_\beta^\alpha]^G$, then f is a function of $(\mathbf{Z}-\mathbf{Z}^\alpha)(\mathbf{Z}^\alpha)^{-1}$, $\mathbf{Z}^\alpha\mathbf{Y}$, $\mathbf{Z}^\alpha\mathbf{X}(\mathbf{Z}^\alpha)^{-1}$. If $h \in k[X_\beta^\alpha]$, then $h\circ\psi = \psi^*(h)$ is a function of the (entries in) $\mathbf{ZY}$. Thus, $\mathbf{Z}^\alpha\mathbf{Y} \in \psi^*(k[X_\beta^\alpha])$. Since

$$(\mathbf{Z}^\alpha)^{-1} = \mathbf{Y}_\beta[\mathbf{Z}^\alpha\mathbf{Y}_\beta]^{-1} = \mathbf{Y}_\beta[\text{Adj } \mathbf{Z}^\alpha\mathbf{Y}_\beta]/\det[\mathbf{Z}^\alpha\mathbf{Y}_\beta]$$

(as $\det \mathbf{Z}^\alpha\mathbf{Y}_\beta \neq 0$ on X_β^α),

$$(\mathbf{Z}-\mathbf{Z}^\alpha)(\mathbf{Z}^\alpha)^{-1} = (\mathbf{Z}-\mathbf{Z}^\alpha)\mathbf{Y}_\beta[\text{Adj } \mathbf{Z}^\alpha\mathbf{Y}_\beta]/\det[\mathbf{Z}^\alpha\mathbf{Y}_\beta] \in \psi^*(k[X_\beta^\alpha])$$

(note $\det[\mathbf{Z}^\alpha\mathbf{Y}_\beta] = \det \mathcal{H}_\beta^\alpha$) and also,

$$\begin{aligned}\mathbf{Z}^\alpha\mathbf{X}(\mathbf{Z}^\alpha)^{-1} &= \mathbf{Z}^\alpha\mathbf{X}\mathbf{Y}_\beta[\text{Adj } \mathbf{Z}^\alpha\mathbf{Y}_\beta]/\det(\mathbf{Z}^\alpha\mathbf{Y}_\beta) \\ &= (\mathbf{Z}^\alpha)^{-1}\mathbf{Y}_\beta[\text{Adj } \mathbf{Z}^\alpha\mathbf{Y}_\beta]/\det(\mathbf{Z}^\alpha\mathbf{Y}_\beta) \in \psi^*(k[X_\beta^\alpha])\end{aligned}$$

and we are done.

Theorem 16.28 (Hankel(n,m,p), ψ) *(or* (Hank(n,m,p), ψ_A), *etc.) is a geometric quotient of* $S_{m,p}^n$ *modulo* G*. The geometric quotient is a non-singular variety of dimension* $n(m+p)$*.*

Now let us recall that in either the single input ($m = 1$) or single output ($p = 1$) case, there was a global canonical form modulo G. For example, if $m = 1$ and $x = (A,b,C)$, then $\psi_c(x) = x_c = (A^c, \epsilon^1, [CA^{j-1}b]_{j=1}^n)$ (I.19.21) gives this canonical form. Since $\psi_c(x) = g \cdot x$ where $g = \mathbf{Y}(x)^{-1}$, we see that (i) $\psi_c\colon S^n_{p,1} \to S^n_{p,1}$ is a morphism; (ii) given $x \in S^n_{p,1}$, there is a g $(= \mathbf{Y}(x)^{-1})$ with $g \cdot x = \psi_c(x)$; and, (iii) $\psi_c(x) = \psi_c(x_1)$ if and only if x and x_1 are equivalent modulo G. Such a morphism ψ_c is called a *continuous canonical form*. If $\psi_c(S^n_{p,1}) = V_c$, then $S^n_{p,1}$ is isomorphic to $G \times V_c$. In fact, if G acts on $G \times V_c$ via $g \cdot (g_1, v) = (gg_1, v)$, then $S^n_{p,1}$ is actually G-isomorphic to $G \times V_c$. This isomorphism is global in the sense that it applies on *all* of $S^n_{p,1}$. Now we note that $\mathbf{V}_A(n,m,p) = \cup U^{\alpha}_{\beta}$ and that U^{α}_{β} is G-isomorphic to $G \times V^{\alpha}_{\beta}$. Consider the map $\psi^{\alpha}_{c,\beta}\colon U^{\alpha}_{\beta} \to V^{\alpha}_{\beta}$ given by

$$\psi^{\alpha}_{c,\beta} = \pi_2 \circ \gamma^{-1} \tag{16.29}$$

with γ^{-1} given by (16.24) and $\pi_2\colon G \times V^{\alpha}_{\beta} \to V^{\alpha}_{\beta}$ the projection.

Proposition 16.30 *$\psi^{\alpha}_{c,\beta}$ is a morphism such that (a) given $\mathbf{v} \in U^{\alpha}_{\beta}$, there is a g with $g \cdot \mathbf{v} = \psi^{\alpha}_{c,\beta}(\mathbf{v})$; and, (b) $\psi^{\alpha}_{c,\beta}(\mathbf{v}) = \psi^{\alpha}_{c,\beta}(\mathbf{v}_1)$ if and only if $\mathbf{v}$ and $\mathbf{v}_1$ are equivalent modulo G. In other words, $\psi^{\alpha}_{c,\beta}$ is a continuous canonical form on U^{α}_{β}.*

Proof. If $\mathbf{v} = (\mathbf{Z}, \mathbf{Y}, \mathbf{X}) \in U^{\alpha}_{\beta}$, then $(\pi_2 \circ \gamma^{-1})(\mathbf{v}) = ((\mathbf{Z} - \mathbf{Z}^{\alpha})(\mathbf{Z}^{\alpha})^{-1}, \mathbf{Z}^{\alpha}\mathbf{Y}, \mathbf{Z}^{\alpha}\mathbf{X}(\mathbf{Z}^{\alpha})^{-1})$ (by abuse of notation in omitting I^{α}). But clearly $(\pi_2 \circ \gamma^{-1})(\mathbf{v}) = g \cdot \mathbf{v}$ where $g = \mathbf{Z}^{\alpha}$ and so (a) is established. If $(\pi_2 \circ \gamma^{-1})(\mathbf{v}) = (\pi_2 \circ \gamma^{-1})(\mathbf{v}_1)$, then $(\mathbf{Z} - \mathbf{Z}^{\alpha})(\mathbf{Z}^{\alpha})^{-1} = (\mathbf{Z}_1 - \mathbf{Z}^{\alpha}_1)(\mathbf{Z}^{\alpha}_1)^{-1}$, $\mathbf{Z}^{\alpha}\mathbf{Y} = \mathbf{Z}^{\alpha}_1\mathbf{Y}_1$, $\mathbf{Z}^{\alpha}\mathbf{X}(\mathbf{Z}^{\alpha})^{-1} = \mathbf{Z}^{\alpha}_1\mathbf{X}(\mathbf{Z}^{\alpha}_1)^{-1}$. Let $g^{-1} = (\mathbf{Z}^{\alpha}_1)^{-1}\mathbf{Z}^{\alpha}$ $(= \mathbf{Y}_{1,\beta}\mathbf{Y}^{-1}_{\beta})$. Then $g \cdot \mathbf{v} = \mathbf{v}_1$ and $\mathbf{v}$, $\mathbf{v}_1$ are equivalent modulo G. Conversely, if $\mathbf{v} = g_1 \cdot \mathbf{v}_1$ for some g_1, then $\mathbf{Z} = \mathbf{Z}_1 g_1^{-1}$, $\mathbf{Z}^{\alpha} = \mathbf{Z}^{\alpha}_1 g_1^{-1}$, $\mathbf{Y} = g_1\mathbf{Y}_1$, $\mathbf{X} = g_1\mathbf{X}_1 g_1^{-1}$ and it follows that $(\pi_2 \circ \gamma^{-1})(\mathbf{v}) = (\pi_2 \circ \gamma^{-1})(g_1 \cdot \mathbf{v}_1) = (\pi_2 \circ \gamma^{-1})(\mathbf{v}_1)$.

The issue, of course, in the multivariable case is whether or not the local continuous canonical forms can be pieced together. We shall see that this is *not* the case in general.

Proposition 16.31 *The open sets V^{α}_{β} and X^{α}_{β} are isomorphic and hence, $U^{\alpha}_{\beta} \simeq G \times X^{\alpha}_{\beta}$ and $\mathbf{V}_A(n,m,p) \simeq \cup(G \times X^{\alpha}_{\beta})$.*

Proof. Let us consider the map $(\psi \circ \gamma^{-1})\colon G \times V^{\alpha}_{\beta} \to X^{\alpha}_{\beta}$ and let $\hat{\psi}$ be its restriction to V^{α}_{β}. Since ψ is surjective and γ is a G-isomorphism, $(\psi \circ \gamma^{-1})$ is also surjective and a G-morphism. Let $\mathbf{H} \in X^{\alpha}_{\beta}$. Then there is a $\mathbf{v} \in U^{\alpha}_{\beta}$ with $\psi(\mathbf{v}) = \mathbf{H}$ and a g with $g \cdot \mathbf{v} \in V^{\alpha}_{\beta}$. But then $\hat{\psi}(g \cdot \mathbf{v}) = \mathbf{H}$ and $\hat{\psi}$ is surjective. If $\mathbf{v}, \mathbf{v}_1 \in V^{\alpha}_{\beta}$ and $\hat{\psi}(\mathbf{v}) = \hat{\psi}(\mathbf{v}_1) = \mathbf{H}$, then $\mathbf{v} = g \cdot \mathbf{v}_1$ and hence $\mathbf{Z}^{\alpha} = \mathbf{Z}^{\alpha}_1 g^{-1}$. But $\mathbf{v}, \mathbf{v}_1 \in V^{\alpha}_{\beta}$ so that $I^{\alpha} = I^{\alpha} g^{-1}$ and $g = I^{\alpha}$. Thus $\hat{\psi}$ is injective. As $\hat{\psi}^{-1} = \pi_2 \circ \gamma \circ \psi^{-1}$ is also a morphism the result follows.

So our situation is this: we have a variety V $(= \mathbf{V}_A(n,m,p))$ on which G acts, a variety B $(= \text{Hankel}(n,m,p))$ on which G acts trivially with $B \simeq V/G$, a G-morphism $\psi\colon V \to B$ with $\psi^{-1}(b) \simeq G$, and a covering U_i $(= X^{\alpha}_{\beta})$ of B by open sets such that $\psi^{-1}(U_i) \simeq G \times U_i$. The triple (V, ψ, B) is called a *principal G-bundle*. We shall now proceed to make this precise and relate it to continuous canonical forms.

Definition 16.32 A triple (V, ψ, B) consisting of varieties V and B and a morphism $\psi\colon V \to B$ is called a *bundle* (or *family*). B is the *base*; ψ is the *projection*, and $\psi^{-1}(b)$ is the *fiber over b*. If $V = X \times B$ and $\psi = \pi_B$, then V is the *product bundle* (over B) *with fiber X*. A morphism $s\colon B \to V$ is a *cross section* of (V, ψ, B) if $\psi \circ s = i$ i.e., if $s(b) \in \psi^{-1}(b)$ for all $b \in B$. If the fibers $\psi^{-1}(b)$ are all vector spaces, then (V, ψ, B) is a *family of vector spaces* (or *bundle of vector spaces*). If (V, ψ, B) and (V_1, ψ_1, B) are bundles over B, then a map $\phi\colon V \to V_1$ is a B-morphism if $\psi_1 \circ \phi = \psi$ (or, equivalently, if $\phi(\psi^{-1}(b)) \subset \psi_1^{-1}(b)$ for each $b \in B$). If (V, ψ, B) and (V_1, ψ_1, B_1) are bundles of vector spaces, then ϕ becomes a linear map of the fibers (over $k(b)$). A bundle (V, ψ, B) is *trivial* if it is (isomorphic to) a product bundle. If (V, ψ, B) is a bundle and U is open in B, then $(\psi^{-1}(U), \psi, U)$ (often written $(V \mid U, \psi, U)$) is a bundle called the *restriction to U*. A bundle (V, ψ, B) is *locally trivial* if there is a covering U_i of B such that $(\psi^{-1}(U_i), \psi, U_i)$ is trivial for all i. A *vector bundle* is a locally trivial family of vector spaces.

Suppose that (V, ψ, B) is a vector bundle with fibers $\psi^{-1}(b) = \mathbb{A}^n$ (i.e., $\dim \psi^{-1}(b) = n$ for all $b \in B$). Then (V, ψ, B) has *rank n* and we write $\rho(V) = n$. Let U_i be an open covering of B and let $\varphi_i\colon \psi^{-1}(U_i) \to \mathbb{A}^n \times U_i$ be the isomorphisms. Then $\varphi_{ij} = \varphi_i\big|_{\psi^{-1}(U_i \cap U_j)}$ and $\varphi_{ji} = \varphi_j\big|_{\psi^{-1}(U_i \cap U_j)}$ map $\psi^{-1}(U_i \cap U_j) \to \mathbb{A}^n \times (U_i \cap U_j)$. The map $\varphi_{ji}\varphi_{ij}^{-1}$ generates an automorphism of the bundle $(\mathbb{A}^n \times (U_i \cap U_j), \pi_2, U_i \cap U_j)$.

Example 16.33 Let $\mathbb{A}^m \times B$ and $\mathbb{A}^n \times B$ be trivial vector bundles and let $\varphi\colon \mathbb{A}^m \times B \to \mathbb{A}^n \times B$ be a (bundle) morphism. If $(x_1, \ldots, x_m)$, $(y_1, \ldots, y_n)$ are coordinates on $\mathbb{A}^m$, $\mathbb{A}^n$ respectively, then it is easy to see that $\varphi^*(\pi_1^* y_j)$, $j = 1, \ldots, n$ will determine φ. If $\epsilon_1, \ldots, \epsilon_m$ are a basis of $\mathbb{A}^m$, then the morphism $\varphi_i\colon B \to \mathbb{A}^m \times B$ given by composing $B \to \epsilon_i \times B$ and $\epsilon_i \times B \to \mathbb{A}^m \times B$ has the property that $\alpha_{ij} = \varphi_i^* \varphi^* y_j$ is regular. Moreover, $\varphi^* y_j = \sum a_{ij} x_i$. In other words, the morphism φ is determined by a linear map with a matrix of regular functions. Here φ is an isomorphism if and only if $m = n$ and $\det(a_{ij})$ is invertible.

From Example 16.33, we see that if (V, ψ, B) is a rank n vector bundle, then the automorphisms $\varphi_{ji}\varphi_{ij}^{-1}$ are represented by matrices $\mathbf{g}_{i,j}$ in $\text{GL}(n; k)$ which satisfy

$$\mathbf{g}_{i,i} = I, \quad \mathbf{g}_{i,\ell} = \mathbf{g}_{i,j}\mathbf{g}_{j,\ell}. \tag{16.34}$$

The $\mathbf{g}_{i,j}$ are the *transition matrices* of the bundle. On the other hand, given matrices satisfying (16.34), we can by appropriate "pasting" construct a rank n vector bundle.

Definition 16.35 A vector bundle of rank 1 is called a *line bundle.*

Line bundles are of considerable importance.

Example 16.36 Consider $\mathbb{A}^{n+1}$ and $\mathbb{P}^n = \text{Gr}(1, \mathbb{A}^{n+1})$. Let $V \subset \mathbb{A}^{n+1} \times \text{Gr}(1, \mathbb{A}^{n+1})$ be the incidence variety i.e., $V = \{(\boldsymbol{\xi}, L) \colon \boldsymbol{\xi} \in L\}$ and let π_2 be the projection on $\text{Gr}(1, \mathbb{A}^{n+1}) = \mathbb{P}^n$. Then $(V, \pi_2, \mathbb{P}^n)$ is a vector bundle. Consider the open set $U_0 = \mathbb{P}^n_{Y_0}$. Then $\pi_2^{-1}(U_0) = \pi_2^{-1}(\mathbb{P}^n_{Y_0}) \simeq k \times \mathbb{A}^n \simeq k \times U_0$ and similarly, $\pi_2^{-1}(U_i) \simeq k \times U_i$ for $U_i = \mathbb{P}^n_{Y_i}$. In other words, $(V, \pi_2, \mathbb{P}^n)$ is a line bundle which we write $(E_L, \pi, \mathbb{P}^n)$.

Suppose that (V, ψ, B) is a bundle and that $f \colon B_1 \to B$ is a morphism. Then f induces a bundle $f^*(V) = (V_1, \psi_1, B_1)$ over B_1, with $V_1 = \{(v, b_1) \in V \times B_1 \colon \psi(v) = f(b_1)\}$ and $\psi_1 \colon V_1 \to B_1$ the projection (i.e., $\psi_1 = \pi_2(V_1)$. We call $f^*(V)$ the *bundle induced by* f.

Example 16.37 Let $f \colon W \to \mathbb{P}^n$ be a closed imbedding. Then $f^*(E_L)$ is a line bundle over W.

Now let G be an algebraic group and V be a variety. Recall that a morphism $G \times V \to V$ is an *action* if $e \cdot \mathbf{v} = \mathbf{v}$ and $g \cdot (g_1 \cdot \mathbf{v}) = (gg_1) \cdot \mathbf{v}$ and that V/G is the set of equivalence classes modulo the action of G. If $\pi \colon V \to V/G$ is the natural projection and if V/G is given the *quotient topology,* then $(V, \pi, V/G)$ is a bundle.

Definition 16.38 Let G act on V and suppose that V/G is a variety and $\pi \colon V \to V/G$ is a morphism. Then $(V, \pi, V/G)$ is a *G-bundle.*

Definition 16.39 Let G act on V and suppose that (i) $g \cdot \mathbf{v} = \mathbf{v}$ if and only if $g = I$ (or equivalently, $g \cdot \mathbf{v} = g_1 \cdot \mathbf{v}$ if and only if $g = g_1$) and (ii) there is a *translation function* $\tau \colon V^* \to G$ where $V^* = \{(\mathbf{v}, g \cdot \mathbf{v}) \in V \times V \colon \mathbf{v} \in V, g \in G\}$ such that $\tau(\mathbf{v}, \mathbf{v}')\mathbf{v}' = \mathbf{v}$ for all $\mathbf{v}, \mathbf{v}' \in V^*$. V is *principal* if τ is a morphism. (V, ψ, B) is a *principal G-bundle* if V is principal and (V, ψ, B) is a G-bundle. (V, ψ, B) is a *trivial principal G-bundle* if it is a principal G-bundle and $V \simeq_G G \times X$ where G acts trivially on X. (The properties apply, of course, to any G-isomorphic bundles.)

Definition 16.40 Let G act on V and let $\psi_c \colon V \to V$ be a morphism. Then ψ_c is a *continuous canonical form* (for the action of G) if (i) given $\mathbf{v} \in V$, there is a $g \in G$ with $g \cdot \mathbf{v} = \psi_c(\mathbf{v})$; (ii) $\psi_c(\mathbf{v}) = \psi_c(\mathbf{v}_1)$ if and only if $\mathbf{v}$ is equivalent to $\mathbf{v}_1$ modulo G; and, (iii) $g \cdot \mathbf{v} = \mathbf{v}$ if and only if $g = I$.

Theorem 16.41 *Let G act on V. There is a continuous canonical form ψ_c on V if and only if V is a trivial principal G-bundle i.e., $V \simeq_G (G \times H)$ with G acting trivially on H.*

Proof. Suppose that $V \simeq_G (G \times H)$ is a trivial principal G-bundle with $\lambda(v) = (g_v, h_v)$ the G-isomorphism. If $\psi_c(v) = \pi_2(g_v, h_v) = h_v$, then $\psi_c = \pi_2 \circ \lambda$ is a morphism. Now $v \equiv_G h_v$ since $\lambda(g_v^{-1} \cdot v) = (I, h_v)$ implies $g_v^{-1} \cdot v = \lambda^{-1}(I, h_v) = h_v$. If $\psi_c(v) = \psi_c(v_1)$, then $h_v = h_{v_1}$ and $g_{v_1} \cdot g_v^{-1} \cdot v = g_{v_1} \cdot h_v = g_{v_1} \cdot h_{v_1} = g_{v_1} \cdot g_{v_1}^{-1} \cdot v_1 = v_1$ so that $v \equiv_G v_1$. If $v \equiv_G v_1$, then $g \cdot v = v_1$ implies $g \cdot g_v \cdot h_v = g_{v_1} \cdot h_{v_1}$ so that $(g_{v_1}^{-1} g g_v) \cdot h_v = h_{v_1}$. But G acts trivially on H so that $h_v = h_{v_1}$. Thus $\psi_c = \pi_2 \circ \lambda$ is a continuous canonical form. Conversely, if ψ_c is a continuous canonical form, we let $H = \psi_c(V)$ with G acting trivially on H. We claim that $V \simeq_G (G \times H)$. Let $\lambda(v) = (g_v, \psi_c(v))$ where $v = g_v \cdot \psi_c(v)$. Then $g \cdot v = g \cdot g_v \psi_c(v)$ and $\psi_c(v) = \psi_c(v_1)$ if $v \equiv_G v_1$ together give $\lambda(g \cdot v) = (g g_v, \psi_c(v)) = g \cdot \lambda(v)$ so that λ is a G-morphism. Given $(g, \psi_c(v))$ in $G \times H$, let $v_1 = g \cdot g_v^{-1} v$ so that $\lambda(v_1) = (g_{v_1}, \psi_c(v_1))$ where $g_{v_1} \cdot v_1 = \psi_c(v_1)$. But $\psi_c(v) = \psi_c(v_1)$ (as $v \equiv_G v_1$) so that $g_{v_1} \cdot \psi_c(v) = g_{v_1} \psi_c(v_1) = v_1 = g \cdot g_v^{-1} \cdot v = g \cdot \psi_c(v_1)$ and $g_{v_1} = g$ as G-acts trivially on $H = \psi_c(v)$. Thus $\lambda(v_1) = (g, \psi_c(v))$ and λ is surjective. If $\lambda(v) = \lambda(v_1)$, then $(g_v, \psi_c(v)) = (g_{v_1}, \psi_c(v_1))$ and hence, $g_v = g_{v_1}$, $\psi_c(v) = \psi_c(v_1)$ and $v = g_v \cdot \psi_c(v) = g_{v_1} \cdot \psi_c(v_1) = v_1$. In other words, λ is injective. Letting $\lambda^{-1}(g, \psi_c(v)) = g \cdot \psi_c(v)$, we see that the previous arguments give $\lambda \circ \lambda^{-1} = i$, $\lambda^{-1} \circ \lambda = i$. Clearly λ^{-1} is a morphism. Thus λ is a G-isomorphism and, since H is bijective with V/G, V is a trivial principal G-bundle.

We now note that $\mathbf{V}_A(n, m, p) = \cup U_\beta^\alpha$ where U_β^α is G-isomorphic with $G \times V_\beta^\alpha$ or $G \times X_\beta^\alpha$. In other words, $\mathbf{V}_A(n, m, p)$ is a principal G-bundle.

Corollary 16.42 *There is a continuous canonical form for equivalence of linear systems if and only if the principal G-bundle $\mathbf{V}_A(n, m, p)$ (or $S_{m,p}^n$) is trivial.*

Example 16.43 Let (V, ψ, B) be a principal G-bundle with $V = \cup(G \times U_i)$, $B = \cup U_i$, and $\psi^{-1}(U_i) = G \times U_i$. Let $k^* = \{a \in k\colon a \neq 0\}$ and let G act on k^* via $g \cdot a = (\det g)a$. Consider (V_d, ψ_d, B) where $V_d = \cup(k^* \times U_i)$, $\psi_d = \pi_2$. Then (V_d, ψ_d, B) is also a principal G-bundle and has rank 1 (i.e., is a line bundle). The map $\varphi\colon V \to V_d$ given by $\varphi(g, v) = (\det g, v)$ is clearly a G-morphism of bundles. If (V, ψ, B) is trivial, then so is (V_d, ψ_d, B).

Corollary 16.44 *If there is a continuous canonical form for equivalence of linear systems, then the principal line G-bundle* $\mathbf{V}_A(n,m,p)_{\det}$ $(= \mathbf{V}_A(n,m,p)_d)$ *is trivial.*

In effect, Corollary 16.44 can be viewed as the assertion that if there is a continuous canonical form, then the map $\det(\mathbf{Z}^\alpha \mathbf{Y}_\beta)^{-1}$ is defined *globally*. We shall show that this is not the case for $S^2_{2,2}$ using an example of Hazewinkel ([H-4]).

Consider $\mathbb{P}^1$ with homogeneous coordinates (ξ,η) and let $U_1 = \mathbb{P}^1_\eta \times (\mathbb{A}^2 - \{(0,0)\})$ with coordinates $(\xi,1,y_0,y_1)$ and $U_2 = \mathbb{P}^1_\xi \times (\mathbb{A}^2 - \{(0,0)\})$ with coordinates $(1,\eta,x_0,x_1)$. Set $U_{12} = U_{1\xi}$ and $U_{21} = U_{2\eta}$ i.e., $U_{12} = \{(\xi,1,y_0,y_1)\colon \xi \neq 0\}$ and $U_{21} = \{(1,\eta,x_0,x_1)\colon \eta \neq 0\}$. Let $\varphi\colon U_{12} \to U_{21}$ be the morphism given by

$$\varphi(\xi,1,y_0,y_1) = (1,\xi^{-1},y_0\xi,y_1\xi) \tag{16.45}$$

and let $\varphi^{-1}\colon U_{21} \to U_{12}$ be the morphism given by

$$\varphi^{-1}(1,\eta,x_0,x_1) = (\eta^{-1},1,x_0\eta,x_1\eta). \tag{16.46}$$

Then

$$\begin{aligned} (\varphi^{-1}\circ\varphi)(\xi,1,y_0,y_1) &= \varphi^{-1}(1,\xi^{-1},y_0\xi,y_1\xi) = (\xi,1,y_0,y_1) \\ (\varphi\circ\varphi^{-1})(1,\eta,x_0,x_1) &= \varphi(\eta^{-1},1,x_0\eta,x_1\eta) = (1,\eta,x_0,x_1) \end{aligned} \tag{16.47}$$

so that φ and φ^{-1} are isomorphisms. Let $W = U_1 \cup U_2$ "patched" by φ. Then W is a "variety". (This approach is discussed in Chapter 6.) Define maps $\psi_1\colon U_1 \to V$ and $\psi_2\colon U_2 \to V$ where $V = \mathbf{V}_A(2,2,2)$ by setting

$$\begin{aligned} \psi_1(\xi,1,y_0,y_1) &= \left(\begin{bmatrix} y_0 & 1 \\ \lambda_1 y_0 & \lambda_2 \\ y_1 & 1 \\ \lambda_1 y_1 & \lambda_2 \end{bmatrix}, \begin{bmatrix} \xi & \lambda_1\xi & 1 & \lambda_1 \\ 1 & \lambda_2 & 1 & \lambda_2 \end{bmatrix}, \begin{bmatrix} \lambda_1 & 0 \\ 0 & \lambda_2 \end{bmatrix} \right) \\ \psi_2(1,\eta,x_0,x_1) &= \left(\begin{bmatrix} x_0 & 1 \\ \lambda_1 x_0 & \lambda_2 \\ x_1 & 1 \\ \lambda_1 x_1 & \lambda_2 \end{bmatrix}, \begin{bmatrix} 1 & \lambda_1 & \eta & \lambda_1 \\ 1 & \lambda_2 & 1 & \lambda_2 \end{bmatrix}, \begin{bmatrix} \lambda_1 & 0 \\ 0 & \lambda_2 \end{bmatrix} \right) \end{aligned} \tag{16.48}$$

where $\lambda_1 \neq 0$, $\lambda_2 \neq 0$, and $\lambda_1 \neq \lambda_2$. These define a morphism of W into V. For if $\xi\eta = 1$, $x_0 = y_0\xi$, $x_1 = y_1\xi$ and

$$g_\xi = \begin{bmatrix} \xi^{-1} & 0 \\ 0 & 1 \end{bmatrix}.$$

Then $(g_\xi \cdot \psi_1)(\xi, 1, y_0, y_1) = \psi_2(1, \xi^{-1}, y_0\xi, y_1\xi) = (\psi_2 \circ \varphi)(\xi, 1, y_0, y_1)$. We now observe that

$$Z^{1,2}(\psi_1(\xi,1,y_0,y_1)) = \begin{bmatrix} y_0 & 1 \\ \lambda_1 y_0 & \lambda_2 \end{bmatrix}, \quad Z^{1,2}(\psi_2(1,\eta,x_0,x_1)) = \begin{bmatrix} x_0 & 1 \\ \lambda_1 x_0 & \lambda_2 \end{bmatrix}$$

$$Y_{1,2}(\psi_1(\xi,1,y_0,y_1)) = \begin{bmatrix} \xi & \lambda,\xi \\ 1 & \lambda_2 \end{bmatrix}, \quad Y_{1,2}(\psi_2(1,\eta,x_0,x_1)) = \begin{bmatrix} 1 & \lambda_1 \\ 1 & \lambda_2 \end{bmatrix}$$

$$Y_{2,4}(\psi_1(\xi,1,y_0,y_1)) = \begin{bmatrix} 1 & \lambda_1 \\ 1 & \lambda_2 \end{bmatrix}, \quad Y_{3,4}(\psi_2(1,\eta,x_0,x_1)) = \begin{bmatrix} \eta & \lambda_1\eta \\ 1 & \lambda_2 \end{bmatrix}$$

and hence that

$$\begin{aligned} \det(Z^{1,2}Y_{3,4})(\psi_1) &= (\lambda_2 - \lambda_1)^2 y_0 \\ \det(Z^{1,2}Y_{3,4})(\psi_2) &= (\lambda_2 - \lambda_1)^2 x_0\eta \\ \det(Z^{1,2}Y_{1,2})(\psi_1) &= (\lambda_2 - \lambda_1)^2 y_0\xi \\ \det(Z^{1,2}Y_{1,2})(\psi_2) &= (\lambda_2 - \lambda_1)^2 x_0 \end{aligned} \tag{16.49}$$

Everything is compatible with the "patching" by φ. *But* $\det(Z^{1,2}Y_{1,2})(\psi_1(0,1,y_0, y_1)) = 0$ and $\det(Z^{1,2}Y_{3,4})(\psi_2(1,0,x_0,x_1)) = 0$. Thus the associated line bundle $V_{\det}$ is not trivial and there is no continuous canonical form by Corollary 16.44. This example is universal in that it implies there is no continuous canonical form on $\mathbf{V}_A(n,m,p)$ for $n \geq 2$, $m \geq 2$, $p \geq 2$. We can see this by using the notion of a direct sum (see Chapter 4). Let $x_2 = (A_2, B_2, C_2)$ be an element of $S^2_{2,2}$ and let $(A_{n-2,n-2}, B_{n-2,m-2}, C_{p-2,n-2}) = y_{-2}$ be an element of $S^{n-2}_{m-2,p-2}$. Then $x = x_2 \oplus y_{-2} = (A,B,C)$ is an element of $S^n_{m,p}$ where

$$A = \begin{bmatrix} A_2 & O_{2,n-2} \\ O_{n-2,2} & A_{n-2,n-2} \end{bmatrix}, \quad B = \begin{bmatrix} B_2 & O_{2,m-2} \\ O_{n-2,2} & B_{n-2,m-2} \end{bmatrix},$$

$$C = \begin{bmatrix} C_2 & O_{2,n-2} \\ O_{p-2,2} & C_{p-2,n-2} \end{bmatrix}.$$

In other words (with the obvious notation), there is a natural embedding $S^2_{2,2} \oplus S^{n-2}_{m-2,p-2} \to S^n_{m,p}$ which is compatible with the similar imbedding $\mathrm{GL}(2;k) \oplus \mathrm{GL}(n-2,k) \to \mathrm{GL}(n;k)$. Moreover, there is also (clearly) the natural corresponding imbedding of $\mathrm{Hank}(2,2,2) \oplus \mathrm{Hank}(n-2, m-2, p-2)$ into $\mathrm{Hank}(n,m,p)$. Calling this embedding restricted to $\mathrm{Hank}(2,2,2)$, f_2, we get an induced bundle $f_2^*(\mathbf{V}_A(n,m,p)) = \mathbf{V}_A(2,2,2)$ with base $\mathrm{Hank}(2,2,2)$. But if $\mathbf{V}_A(n,m,p)$ were trivial (i.e., if there were a continuous canonical form on $\mathbf{V}_A(n,m,p)$) then $f_2^*(\mathbf{V}_A(n,m,p)) = \mathbf{V}_A(2,2,2)$ would be trivial as well. This would contradict the example of Hazewinkel.

Example 16.50 Consider $S^1_{m,p}$ with $m > 1$ and $p > 1$. Let $x = (a, \mathbf{b}, \mathbf{c}) \in S^1_{m,p}$ so that $\mathbf{b} \neq 0$, $\mathbf{c} \neq 0$ and write $\mathbf{b} = (b_1, \ldots, b_m)$, $\mathbf{c} = \begin{pmatrix} c^1 \\ \vdots \\ c^p \end{pmatrix}$. Let $\lambda = \lambda(\mathbf{b}) = \min\{i \colon b_i \neq 0\}$. Then set

$$\begin{aligned} \psi_c^\lambda(x) &= (a, b_\lambda^{-1}\mathbf{b}, \mathbf{c}b_\lambda) \\ &= (a, (0, \ldots, 0, 1, b_\lambda^{-1}b_{\lambda+1}, \ldots, b_\lambda^{-1}b_m), (c^i b_\lambda)). \end{aligned} \tag{16.51}$$

We *claim* that the map ψ_c^λ satisfies the condition (i), (ii), (iii) of Definition 16.40 for the action of $\mathrm{GL}(1, k) = k^*$ on $S^1_{m,p}$. Clearly, $t \cdot x = x$ if and only if $t = 1$. Moreover, if $x = (a, \mathbf{b}, \mathbf{c}) \in S^1_{m,p}$, then, for $t = b_\lambda^{-1}$, $t \cdot x = \psi_c^\lambda(x)$. Thus, (i) and (iii) of Definition 16.40 are satisfied. Suppose that $t \cdot x = x' = (a', \mathbf{b}', \mathbf{c}')$. Then $a' = tat^{-1} = a$. Since $tb_i = b_i'$, $t \neq 0$, $\lambda(\mathbf{b}) = \lambda(\mathbf{b}') = \lambda$. It follows from $tb_\lambda = b_\lambda'$ that $t = b_\lambda' b_\lambda^{-1}$ and hence that $b_\lambda'^{-1} b_{\lambda+j}' = b_\lambda'^{-1} t b_{\lambda+j} = b_\lambda'^{-1} b_\lambda' b_\lambda^{-1} b_{\lambda+j} = b_\lambda^{-1} b_{\lambda+j}$ and $c'^i b_\lambda' = c^i t^{-1} b_\lambda' = c^i b_\lambda b_\lambda'^{-1} b_\lambda' = c^i b_\lambda$. In other words, $\psi_c^\lambda(x) = \psi_c^\lambda(x')$. Conversely, if $\psi_c^\lambda(x) = \psi_c^\lambda(x')$, then $\lambda = \lambda(\mathbf{b}) = \lambda(\mathbf{b}')$ and, taking $t = b_\lambda' b_\lambda^{-1}$, we have $tb_{\lambda+j} = b_\lambda' b_\lambda^{-1} b_{\lambda+j} = b_\lambda'(b_\lambda'^{-1} b_{\lambda+j}')$ (since $\psi_c^\lambda(x) = \psi_c^\lambda(x')$) $= b_{\lambda+j}'$ and $c^i t^{-1} = c^i b_\lambda b_\lambda'^{-1} = c'^i b_\lambda' b_\lambda'^{-1} = c'^i$. Thus our claim is valid. However, if $U^\lambda = \{(a, \mathbf{b}, \mathbf{c}) \colon \lambda = \lambda(b)\}$, then $S^1_{p,m} = \bigcup_{\lambda=1}^{m} U^\lambda$ and $U^\lambda \cap U^\mu = \emptyset$ for $\lambda \neq \mu$. Thus, the ψ_c^λ can *not* be "patched" to give a morphism.

We have, in effect, established the following:

Theorem 16.52 *There is a continuous canonical form for equivalence on $S^n_{m,p}$ if and only if either $m = 1$ or $p = 1$.*

Note that G acts on $\mathbb{A}^{n^2+n(m+p)}$ and that $S^n_{m,p}$ is an open invariant subset. Let $W^n_{m,p} = \mathbb{A}^{n^2+n(m+p)} - S^n_{m,p}$. Then $W^n_{m,p}$ is a closed non-empty invariant subset of $S^n_{m,p}$ and $\dim W^n_{m,p} < n^2 + n(m+p)$. The effect of the action of G on $W^n_{m,p}$ can be quite complicated.

Example 16.53 Suppose that $k = \mathbb{C}$ and let $n = 2$, $p = 1$ and $m = 1$. Consider $x = (A, b, c)$ where

$$A = \begin{bmatrix} 1 & 1 \\ 0 & 1 \end{bmatrix}, \quad b = \begin{bmatrix} 1 \\ 0 \end{bmatrix}, \quad c = \begin{bmatrix} 0 & 1 \end{bmatrix}.$$

Then x is not controllable or observable. Let

$$g_\alpha = \begin{bmatrix} 1 & 0 \\ 0 & \alpha^{-1} \end{bmatrix}$$

for $\alpha \in \mathbb{C}^*$. Then $g_\alpha b = b = b_\alpha$, $cg_\alpha^{-1} = [0\ \alpha] = c_\alpha$ and

$$g_\alpha A g_\alpha^{-1} = \begin{bmatrix} 1 & \alpha \\ 0 & 1 \end{bmatrix} = A_\alpha.$$

Hence $(A_\alpha, b_\alpha, c_\alpha) \in O_G(x)$ for all $\alpha \in \mathbb{C}^*$. But $\lim_{\alpha \to 0} (A_\alpha, b_\alpha, c_\alpha) = (I_2, b, 0) \notin O_G(x)$ as $g \cdot I_2 g^{-1} = I_2$ for all $g \in G$ and also, $O \cdot g^{-1} = 0$ for all g in G. Consider next $x_0 = (A, b, c_0)$ where $c_0 = [1\ 1]$. Then x_0 is not controllable but is observable. Since $c_0 g_\alpha^{-1} = [1\ \alpha] = c_{0\alpha}$, $(A_\alpha, b_\alpha, c_{0\alpha}) \in O_G(x_0)$. But $\lim_{\alpha \to 0} (A_\alpha, b_\alpha, c_{0\alpha}) = (I_2, b, [1\ 0]) \notin O_G(x_0)$. Note that $(I_2, b, [1, 0])$ is not observable. Thus the limit of observable systems need not be observable.

Example 16.54 Suppose again that $k = \mathbb{C}$ and that $x = (A, b, c) \in \mathbb{A}_{\mathbb{C}}^{4+2(1+1)}$ where

$$A = \begin{bmatrix} 1 & 1 \\ 0 & 1 \end{bmatrix}, \quad b = \begin{bmatrix} 1 \\ 0 \end{bmatrix}, \quad c = [0\ 1].$$

Let $x_t = (A, b_t, c_t)$, $t \in \mathbb{C}^*$ where

$$b_t = \begin{bmatrix} 1 \\ t \end{bmatrix}, \quad c_t = [t\ 1].$$

Then $x_t \in S_{1,1}^2$ for all t in $\mathbb{C}^*$ and $\lim_{t \to 0} x_t = x$, which is consistent with the density of $S_{1,1}^2$ in $\mathbb{A}_{\mathbb{C}}^8$. In other words, the limit of a family of minimal systems need not be either controllable or observable. Suppose that $g \cdot x_t = x_s$. Then $gA = Ag$ so that g is of the form

$$g = \begin{bmatrix} \lambda & \alpha \\ 0 & \lambda \end{bmatrix}, \quad \lambda \in \mathbb{C}^*.$$

If $gb_t = b_s$, then $\lambda + \alpha t = 1$ and $\lambda t = s$. If $c_t g^{-1} = c_s$, then $t = \lambda s$ and $-\alpha t + \lambda = \lambda^2$. It follows that $g = I$ and $x_t = x_s$ and $s = t$. Thus x_t is not equivalent to x_s if $t \neq s$.

Let us now briefly examine the projective situation. Let E be an equivalence on a projective variety V. Then:

Definition 16.55 A *geometric quotient of* V *modulo* E is a pair (W, ψ) consisting of a projective variety W and a morphism $\psi: V \to W$ such that (i) $\psi^{-1}(w)$ is a closed orbit for each $w \in W$; (ii) for each invariant open set U in V, there is an open $U_0 \subset W$ such that $\psi^{-1}(U_0) = U$; and, (iii) if U_0 is open in W, then $f \in \mathfrak{o}_W(U_0)$ if and only if $\psi^*(f) \in \mathfrak{o}_V(\psi^{-1}(U_0))^E$ i.e., $(f \circ \psi)(v_1) = (f \circ \psi)(v)$ for all v in $\psi^{-1}(U_0)$ and $v_1 E v$.

Proposition 16.56 *Let (W, ψ) be a geometric quotient of V modulo E.* (a) *If U_0 is open in W, then $\psi^{-1}(U_0)$ is an open invariant subset of V;* (b) *If $U_1 \subset W$ with $\psi^{-1}(U_1)$ open invariant in V, then U_1 is open in W; and,* (c) (ii) *is equivalent to* (ii)$'$: *U_0 is open in W if and only if $\psi^{-1}(U_0)$ is an open invariant subset of V.*

Proof. (a) $\psi^{-1}(U_0)$ is open by continuity. If vEv_1, then $\psi(v) = w$, $\psi(v_1) = w_1$ but $\psi^{-1}(w) = O_E(v) = O_E(v_1) = \psi^{-1}(w_1)$ so that $w = w_1$. Thus $v \in \psi^{-1}(U_0)$ implies $\psi(v) = w \in U_0$ which in turn implies $O_E(v) = \psi^{-1}(w) \subset \psi^{-1}(U_0)$. (b) By (ii), there is a U_0 open in W with $\psi^{-1}(U_0) = \psi^{-1}(U_1)$ and, by (a), $\psi^{-1}(U_0)$ is open and invariant. We *claim* that $U_0 = U_1$. If $u_1 \in U_1$, then there is a $v \in \psi^{-1}(U_0)$ with $\psi(v) = u_1$. But $v \in \psi^{-1}(U_0) = \psi^{-1}(U_1)$ so that $\psi(v) = u_1 \in U_0$. Similarly, $U_0 \subset U_1$. (c) If (ii) holds, then (a), (b) give (ii)$'$. On the other hand, assume (ii)$'$ and let U be open, invariant in V and let $U_0 = \psi(U)$. Then $\psi^{-1}(U_0) = U$ would imply that U_0 is open in W by (ii)$'$. If $v \in U$, then $\psi(v) \in \psi(U) = U_0$ so that $U \subset \psi^{-1}(U_0)$. If $v \in \psi^{-1}(U_0)$, then $\psi(v) \in \psi(U)$ so that $\psi(v) = \psi(v_1) = w$ for some $v_1 \in U$. It follows that $v, v_1 \in \psi^{-1}(w) = O_E(v_1) = O_E(v)$. Since U is invariant, $O_E(v_1) \subset U$ and $v \in U$ i.e., $\psi^{-1}(U_0) \subset U$.

We note that (Hankel(n, m, p), ψ) is also a projective geometric quotient for linear systems under equivalence modulo the action of GL(n, k).

Exercises

(1) Interpret Propositions 16.12 and 16.13 in terms of sheaves.

(2) Show that the maps $\varphi_{ji}\varphi_{ij}^{-1}$ are automorphisms of the bundle $(\mathbb{A}^n \times (U_i \cap U_j), \pi_2, U_i \cap U_j)$ (p. 250).

(3) Show, in detail, how the transition matrices (16.34) can be used to construct a rank n vector bundle. Describe the situation for a line bundle (i.e., $n = 1$).

(4) Let $n = n_1 + n_2$, $m = m_1 + m_2$, $p = p_1 + p_2$. Show that there is a natural (injective) imbedding of $S^{n_1}_{m_1,p_1} \oplus S^{n_2}_{m_2,p_2} \to S^n_{m,p}$ which is compatible with the (similar) imbedding GL$(n_1, k) \oplus$GL$(n_2, k) \to$ GL(n, k). Show how this generates a map of Hank$(n_1, m_1, p_1) \oplus$ Hank$(n_2, m_2, p_2) \to$ Hank(n, m, p) which is consistent with equivalence. Analyze issues of surjectivity. For example, attempt to characterize the images of the various imbeddings.

(5) Consider $W_{1,1}^n$. Show that $\dim W_{1,1}^n = n^2 + 2n - 1$ by noting that $W_{1,1}^n = V_0 \cup V_c$ where $V_0 = V(\det \mathbf{Z}(A,c))$, $V_c = V(\det \mathbf{Y}(A,b))$. Is $W_{1,1}^n$ irreducible? Let $U_{1,1}^{n-1}$ be the open invariant subset where $\rho(\mathbf{Z}(A,c)) = n-1$ and $\rho(\mathbf{Y}(A,b)) = n-1$. Can you construct a geometric quotient for $U_{1,1}^{n-1}/G$? What about $W_{2,1}^n$? [Attempting this exercise will provide insight into the issues involved in quotients.] Do "nice" indices help?

17

Projective Algebraic Geometry VII: Divisors

A point ξ in a variety V may be viewed as the smallest proper subvariety of V while an irreducible subvariety W of codimension 1 in V may be viewed as the largest (type of) proper subvariety. We shall call such a W a *prime divisor on* V if V is normal (i.e., $\mathfrak{o}_{\xi,V}$ is integrally closed for all $\xi \in V$). For example, if $F \in k[Y_0, \ldots, Y_N]$ is an irreducible form, then $W = V(F)$ is a prime divisor on $\mathbb{P}^N$ and every prime divisor on $\mathbb{P}^N$ is of this form.

Example 17.1 Let $\mathbf{F}(z)$ be an element of $\text{Rat}(2,2,2)$ and let $\psi_{\mathbf{F}}\colon \mathbb{P}^1 \to \mathbb{P}^5$ be the associated regular map. Then $H_5 = V(Y_5)$ is a prime divisor and $\psi_{\mathbf{F}}(\mathbb{P}^1) \cap H_5$ consists of the poles of $\mathbf{F}$. (We shall study the structure of this intersection in the sequel.)

Now let $W \subset V$ be an irreducible subvariety of codim 1. Then $K \operatorname{Dim} \mathfrak{o}_{W,V} = 1$ and $\mathfrak{o}_{W,V}$ is an integral domain with $\mathfrak{o}_{W,V} \subset k(V)$, the function field of V. If V is affine, then

$$k[V] \subset k[V]_{\mathfrak{p}_W} = \mathfrak{o}_{W,V} \subset k(V) \tag{17.2}$$

(also for an affine representative in the projective case) and so, the quotient field of $\mathfrak{o}_{W,V}$ is $k(V)$. We observe that if V is *normal*, then $\mathfrak{o}_{W,V}$ is also integrally closed (Proposition I.16.31). So, in general, $(\mathfrak{o}_{W,V}, \mathfrak{m}_{W,V})$ is a Noetherian local domain with Krull dimension 1. When is it regular?

Definition 17.3 A subring R of a field K is a *valuation ring of* K if $x \in K$, $x \neq 0$, implies either x or x^{-1} (or both) is in R. A map $v\colon K^* \to \mathbb{Z}$ is a *discrete valuation of* K if $v(xy) = v(x) + v(y)$ and $v(x+y) \geq \min\{v(x), v(y)\}$. The set $K_v = \{x\colon v(x) \geq 0\} \cup \{0\}$ is a valuation ring of K and is called a *discrete valuation ring.*

P. Falb, *Methods of Algebraic Geometry in Control Theory: Part II*,
Modern Birkhäuser Classics, https://doi.org/10.1007/978-3-319-96574-1_17

We wish to prove the following theorem (the method used is that of [A-2]):

Theorem 17.4 *Let $(\mathfrak{o}, \mathfrak{m})$ be a Noetherian local domain with $K \dim \mathfrak{o} = 1$. Let $k = \mathfrak{o}/\mathfrak{m}$. Then the following are equivalent:*

(1) *$\mathfrak{o}$ is a discrete valuation ring;*

(2) *$\mathfrak{o}$ is integrally closed;*

(3) *$\mathfrak{m} = (m)$ is a principal ideal;*

(4) *$\dim_k(\mathfrak{m}/\mathfrak{m}^2) = 1$ i.e., $\mathfrak{o}$ is regular;*

(5) *every non-zero ideal $\mathfrak{a}$ in $\mathfrak{o}$ is a power of $\mathfrak{m}$; and,*

(6) *there is an m such that every non-zero ideal $\mathfrak{a}$ of $\mathfrak{o}$ is of the form $\mathfrak{a} = (m^\nu)$, some $\nu \geq 0$, and so, $\mathfrak{o}$ is a principal ideal domain.*

Corollary 17.5 *Any such $\mathfrak{o}$ is a unique factorization domain.*

Corollary 17.6 *If V is normal, then $\mathfrak{o}_{W,V}$ is regular for all irreducible $W < V$ with $\operatorname{codim}_V W = 1$. In such a case, V is called nonsingular in codimension 1.*

Before proving the theorem, we need to develop some preliminary results.

Definition 17.7 An ideal $\mathfrak{q}$ in a ring R is *primary if $xy \in \mathfrak{q}$* implies either $x \in \mathfrak{q}$ or $y \in \sqrt{\mathfrak{q}}$ (i.e., every zero divisor in $R/\mathfrak{q}$ is nilpotent).

Proposition 17.8 *If $\mathfrak{a}$ is an ideal in a Noetherian ring R, then $\mathfrak{a} \supset (\sqrt{\mathfrak{a}})^\nu$ for some $\nu > 0$.*

Proof. Let $\sqrt{\mathfrak{a}} = (f_1, \ldots, f_t)$. Then $f_i^{\nu_i} \in \mathfrak{a}$ and for large enough ν, all monomials of degree ν in the f_i are in $\mathfrak{a}$.

Proposition 17.9 *If $\sqrt{\mathfrak{a}} = \mathfrak{m}$ is a maximal ideal, then $\mathfrak{a}$ is primary.*

Proof. If $\sqrt{\mathfrak{a}} = \mathfrak{m}$, then $\overline{\mathfrak{m}}$ in $R/\mathfrak{a}$ is the set of all nilpotents in $R/\mathfrak{a}$ and so every zero-divisor in $R/\mathfrak{a}$ is nilpotent.

Proposition 17.10 *If R is Noetherian and $\mathfrak{m}$ is a maximal ideal, then $\sqrt{\mathfrak{a}} = \mathfrak{m}$ if and only if $\mathfrak{m}^\nu \subset \mathfrak{a} \subset \mathfrak{m}$ for some $\nu > 0$.*

Proof. If $\sqrt{\mathfrak{a}} = \mathfrak{m}$, then $\mathfrak{m}^\nu \subset \mathfrak{a} \subset \mathfrak{m}$ by Proposition 17.8. If $\mathfrak{m}^\nu \subset \mathfrak{a} \subset \mathfrak{m}$, then $\mathfrak{m} = \sqrt{\mathfrak{m}^\nu} \subset \sqrt{\mathfrak{a}} \subset \sqrt{\mathfrak{m}} = \mathfrak{m}$.

We can now prove Theorem 17.4.

Proof of Theorem 17.4 ([A-2]): (1) $\Rightarrow$ (2) Since $\mathfrak{o}$ is a discrete valuation ring, if $h \in K(\mathfrak{o})$, then either h or h^{-1} is in $\mathfrak{o}$. If h is integral over $\mathfrak{o}$ with $h^n + \alpha_1 h^{n-1} + \cdots + \alpha_n = 0$ and h were not in $\mathfrak{o}$, then $h^{-(n-1)} \in \mathfrak{o}$ and $h^{-(n-1)}(h^n + \cdots + \alpha_n) = 0$ or $h = -(\alpha_1 + \cdots + \alpha_n h^{-(n-1)})$ is in $\mathfrak{o}$ (a contradiction).

(2) $\Rightarrow$ (3) Let $x \neq 0$, $x \in \mathfrak{m}$. Then $\mathfrak{m}^\nu \subset (x)$ and $\mathfrak{m}^{\nu-1} \not\subset (x)$ for some ν. Let $y \in \mathfrak{m}^{\nu-1}$, $y \notin (x)$ and set $m = x/y$. Then $m^{-1} \notin \mathfrak{o}$ (otherwise $y \in (x)$) and $m^{-1}\mathfrak{m} \not\subset \mathfrak{m}$ (otherwise m^{-1} integral over $\mathfrak{o}$). Therefore, $m^{-1}\mathfrak{m} = \mathfrak{o}$ (since $\mathfrak{m}^\nu \subset (x)$) and $\mathfrak{m} = \mathfrak{o}m = (m)$.

(3) $\Rightarrow$ (4) By Corollary 7.46 to Nakayama's Lemma.

(4) $\Rightarrow$ (5) Let $\mathfrak{a} \neq (0)$ and $\mathfrak{m} = (m)$. Then there is a ν with $\mathfrak{a} \subset \mathfrak{m}^\nu$, $\mathfrak{a} \not\subset \mathfrak{m}^{\nu+1}$ (by Krull's Intersection Theorem, Corollary A.3). Let $y \in \mathfrak{a}$, $y = am^\nu$, $y \notin \mathfrak{m}^{\nu+1}$. Then $a \notin \mathfrak{m}$ and so is a unit in $\mathfrak{o}$. Thus $\mathfrak{m}^\nu = (m^\nu) \subset \mathfrak{a} \subset \mathfrak{m}^\nu = (m^\nu)$.

(5) $\Rightarrow$ (6) Since $\mathfrak{m} \neq \mathfrak{m}^2$, there is an $m \in \mathfrak{m}$, $m \notin \mathfrak{m}^2$. By hypothesis, $(m) = \mathfrak{m}^\nu$ for some ν and so $\nu = 1$ and $(m^\nu) = \mathfrak{m}^\nu$.

(6) $\Rightarrow$ (1) Since $(m) = \mathfrak{m}$ and $\mathfrak{m}^\nu = (m^\nu) \neq (m^{\nu+1}) = \mathfrak{m}^{\nu+1}$, if $h \in \mathfrak{o}$, then $(h) = (m^\nu)$ for a unique ν. Set $v_\mathfrak{o}(h) = \nu$ and extend $v_\mathfrak{o}$ to $k(\mathfrak{o})^*$ by setting $v_\mathfrak{o}(hg^{-1}) = v_\mathfrak{o}(h) - v_\mathfrak{o}(g)$. Clearly $v_\mathfrak{o}$ is a discrete valuation with $\mathfrak{o}$ as valuation ring.

Corollary 17.11 *If $h \neq 0$ is an element of $k(V)$ with V nonsingular in codimension* 1, *then there are only a finite number of (irreducible) subvarieties $W_1, \ldots, W_r$ with* codim 1 *such that $v_W(h) = 0$ if $W \neq W_i$, $i = 1, \ldots, r$.*

Proof. h is regular on an affine $U \subset V$ i.e., $h \in \mathfrak{o}_V(U)$ and U_h is open in V. If $W \cap U_h \neq \emptyset$, then $v_W(h) = 0$ (since $h, h^{-1} \notin \mathfrak{p}_W$). If $W \cap U_h = \emptyset$, then $W \subset V - U_h$ and so W would be an irreducible component of $V - U_h$. Let $W_1, \ldots, W_r$ be the components of $V - U_h$.

Corollary 17.12 *A non-constant rational function h on a normal variety V has only a finite number of poles and zeros where W (with* $\mathrm{codim}_V W = 1$*) is a pole (zero) of h if $v_W(h) < 0$ ($v_W(h) > 0$).*

We make use of this observation in studying divisors and intersections.

Definition 17.13 Let V be a normal variety. A closed, irreducible subvariety Γ of V with $\mathrm{codim}_V \Gamma = 1$ is called a *prime divisor of* V. The group $\{\sum n_i \Gamma_i\colon n_i \in \mathbb{Z},\ \Gamma_i$ a prime divisor on V, $n_i \neq 0$ for a finite number of $\Gamma_i\}$ (i.e., the free Abelian group generated by the prime divisors on V) is called the *divisor group*, Div(V), *on* V. If $D = \sum n_i \Gamma_i$ and $n_i \geq 0$ for all i, then D is *effective.* If $D = \sum n_i \Gamma_i$, then $\bigcup_{n_i \neq 0} \Gamma_i$ is called the *support*, Supp D, *of* D.

If $D_1, D_2 \in \mathrm{Div}(V)$, then $D_1 \geq D_2$ if $D_1 - D_2$ is effective. By Corollary 17.12, if $h \neq 0$ is an element of $k(V)$, then $\sum v_{\Gamma_i}(h)\Gamma_i$ is a divisor on V, is denoted by (h) and is called the *divisor of* h. If $D = \sum n_i \Gamma_i$ is any divisor, then $D = D_+ - D_-$ where $D_+ = \sum_{n_j > 0} n_j \Gamma_j$ and $D_- = -\sum_{n_j < 0} n_j \Gamma_j$ so that any divisor is the difference of effective divisors. A divisor (h) is called a *principal divisor*. If $(h)_0 = (h)_+$ and $(h)_\infty = (h)_-$, then $(h) = (h)_0 - (h)_\infty$. $(h)_0$ is the *divisor of zeros of* h and $(h)_\infty$ is the *divisor of poles of* h.

Definition 17.14 Divisors D_1 and D_2 are *linearly equivalent*, written $D_1 \sim D_2$, if $D_1 - D_2 = (h)$ for some $h \in k(V)^*$.

Proposition 17.15 (a) *The principal divisors form a group;* (b) $\sim$ *is an equivalence relation; and,* (c) $\mathrm{Pic}(V) = \mathrm{Div}(V)/\sim$ *is a group called the Picard (or Divisor Class) Group of* V.

Proof. (a) Since $v_\Gamma(h_1 h_2) = v_\Gamma(h_1) + v_\Gamma(h_2)$, we have $(h_1) + (h_2) = (h_1 h_2)$. Since $v_\Gamma(h_1 h_1^{-1}) = v_\Gamma(h_1) - v_\Gamma(h_1) = 0 = v_\Gamma(1)$, the principal divisors form a subgroup of $\mathrm{Div}(V)$. (b) $D \sim D$ as $D - D = 0 = (1)$. $D_1 \sim D_2$ means $D_1 - D_2 = (h)$ and so, $D_2 - D_1 = -(h) = (h^{-1})$. Finally, if $D_1 \sim D_2$, $D_2 \sim D_3$, then $D_1 - D_2 = (h)$, $D_2 - D_3 = (f)$ and $D_1 - D_3 = D_1 - D_2 + D_2 - D_3 = (h) + (f) = (hf)$. (c) follows immediately.

We now state a result whose proof we only sketch.

Proposition 17.16 *If* R *is an integrally closed Noetherian domain, then*

$$R = \cap R_{\mathfrak{p}} \tag{17.17}$$

where $\mathfrak{p}$ *runs through the prime ideals of* R *of height* 1 *i.e., the minimal prime ideals of* R ([Z-3]).

"Sketch of Proof". Let $K = K(R)$ be the quotient field of R. Clearly $R \subset \cap R_{\mathfrak{p}}$. Let $v_{\mathfrak{p}}$ be the valuation of K with valuation ring $R_{\mathfrak{p}}$ (Theorem 17.4). If $f \in R$, then $Rf = \mathfrak{p}_1^{(n_1)} \cap \cdots \cap \mathfrak{p}_r^{(n_r)}$ where the $\mathfrak{p}_i$ are minimal prime ideals in R and $\mathfrak{p}^{(n)} = (\mathfrak{p}^n R_{\mathfrak{p}}) \cap R$ (the so-called nth symbolic power of $\mathfrak{p}$, [Z-3]). Then $v_{\mathfrak{p}_i}(f) = n_i$ and $v_{\mathfrak{p}}(f) = 0$ if $\mathfrak{p} \neq$ any $\mathfrak{p}_i$, $i = 1, \ldots, r$. Let $h = f/g$ be an element of K^* and suppose $h \in \cap R_{\mathfrak{p}}$. Then $v_{\mathfrak{p}}(h) \geq 0$ for all $\mathfrak{p}$. But $Rf = \cap \mathfrak{p}^{(v_{\mathfrak{p}}(f))}$ and $Rg = \cap \mathfrak{p}^{(v_{\mathfrak{p}}(g))}$ and $0 \leq v_{\mathfrak{p}}(h) = v_{\mathfrak{p}}(f) - v_{\mathfrak{p}}(g)$. It follows that $Rf \subset Rg$ and $h \in R$.

Corollary 17.18 *If* V_a *is a normal affine variety, then* $(f) \geq 0$ *if and only if* $f \in k[V_a]$. *Hence, if* V *is a normal projective variety, then* $(g) \geq 0$ *if and only if* $g \in k$.

Proof. $(f) \geq 0$ if and only if $v_\Gamma(f) \geq 0$ for all prime divisors Γ on V_a if and only if $f \in \mathfrak{o}_{\Gamma,V_a}$ for all Γ. Thus, $(f) \geq 0$ if and only if

$$f \in \cap \mathfrak{o}_{\Gamma,V_a} = \bigcap_{\mathfrak{p}_\Gamma} k[V_a]_{\mathfrak{p}_\Gamma} = k[V_a].$$

Proposition 17.19 *Let V_a be a normal affine variety and let $\Gamma_1, \ldots, \Gamma_t$ be prime divisors and $\nu_1, \ldots, \nu_t$ be non-negative integers. Then there is an f in $k[V_a]$ with $v_{\Gamma_i}(f) = \nu_i$.*

Proof. Let $\mathfrak{p}_i = I(\Gamma_i)$. Since $\mathfrak{p}_i \not\subset \mathfrak{p}_j$ for $i \neq j$, there are $u_1, \ldots, u_t$ in $k[V_a]$ such that u_i is a unit in $\mathfrak{o}_{\Gamma_i,V_a}$ and $u_i \in \mathfrak{m}_{\Gamma_j,V_a}^\rho$ for $j \neq i$ and ρ large. Let $\mathfrak{m}_{\Gamma_i,V_a} = (m_i)$, $i = 1, \ldots, t$ and let $\omega_i = m_i^{\nu_i}$, $i = 1, \ldots, t$ with $m_i \in k[V_a]$. Then $v_{\Gamma_i}(\omega_i) = \nu_i$ and if $f = \sum_{i=1}^{t} u_i\omega_i$, then $v_{\Gamma_i}(f) = \nu_i$ (by the properties of valuations).*

Example 17.20 Let $V = \mathbb{P}^N$. The prime divisors on $\mathbb{P}^N$ are of the form $\Gamma = V(F)$ where $F \in k[Y_0, \ldots, Y_N]$ is an irreducible homogeneous polynomial. Hence, any divisor D on $\mathbb{P}^N$ is of the form $D = \sum n_i V(F_i)$. If $h \in k(\mathbb{P}^N)^*$, then $h = h_1/h_2$ where h_1, h_2 are relatively prime forms of the same degree q (say). Thus $h_1 = F_1^{n_1} \cdots F_r^{n_r}$, $h_2 = F_1^{m_1} \cdots F_r^{m_r}$ where the F_i are irreducible forms of degree q_i and $q = \sum n_i q_i$, $q = \sum m_j q_j$. In other words, $h = \prod_{i=1}^{r} F_i^{n_i - m_i}$ with

$$0 = \deg h = \sum (n_i - m_i) q_i = \sum r_i \deg F_i$$

and $(h) = \sum r_i \deg F_i V(F_i)$. But, if $D = \sum r_i \Gamma_i$ and $\sum r_i \deg F_i = 0$, then $D = (F_1^{r_1} \cdots F_t^{r_t})$. Let us set $\deg D = \sum n_i \deg F_i$ when $D = \sum n_i V(F_i)$. Then the sequence

$$0 \longrightarrow k(\mathbb{P}^N)^*/k^* \longrightarrow \mathrm{Div}(\mathbb{P}^N) \stackrel{\text{degree}}{\longrightarrow} \mathbb{Z} \longrightarrow 0 \tag{17.21}$$

is exact and so $\mathrm{Pic}(\mathbb{P}^n) \simeq \mathbb{Z}$. This can be interpreted as saying that $D \sim rH$ where H is a hyperplane in $\mathbb{P}^N$ and $r \in \mathbb{Z}$. Note that if $H_1 = V(\ell_1)$, $H_2 = V(\ell_2)$ where ℓ_1, ℓ_2 are linear forms then $H_1 - H_2 = (\ell_1/\ell_2)$ so that any two hyperplanes are linearly equivalent. Let $D = \sum n_i V(F_i)$ be a divisor and let $H_0 = V(Y_0)$ be the hyperplane $Y_0 = 0$. If we show that $V(F_i) \sim r_i H_0$, then $D \sim (\sum n_i r_i) H_0$ and the interpretation is clear. Let $U_0 = \mathbb{P}^N - H_0 \simeq \mathbb{A}^N$. If $\Gamma = V(F)$ and $\Gamma \cap U_0 = \emptyset$,

---

*In particular, $v(x+y) = v(x)$ if $v(x) < v(y)$ and $v(\sum x_i) = v(x_1)$ if $v(x_1) < v(x_i)$, $i = 2, \ldots$.

then $\Gamma \subset H_0$ and $\Gamma = H_0$ (both irreducible of dimension $N-1$). If $\Gamma = V(F)$ and $\Gamma \cap U_0 \neq \emptyset$, then $\Gamma \cap U_0 = V(f)$ where f is an irreducible polynomial (i.e., $\Gamma \cap U_0$ is a prime divisor on $\mathbb{A}^N$), $f(X_1, \ldots, X_N) = F(1, X_1, \ldots, X_N)$. Then $f(Y_1/Y_0, \ldots, Y_N/Y_0)Y_0^r = F(Y_0, \ldots, Y_N)$ and $f(Y_1/Y_0, \ldots, Y_N/Y_0) = F(Y_0, \ldots, Y_N)/Y_0^r$. Viewing f as a rational function on $\mathbb{P}^N$, we have $(f) = V(F) - rH_0 = \Gamma - rH_0$ and $\Gamma \sim rH_0$.

Definition 17.22 Let $D \in \mathrm{Div}(V)$ and let $\mathcal{L}_V(D) = \{f \in k(V)\colon f = 0 \text{ or } f \neq 0 \text{ and } (f) + D \geq 0\}$. If $\mathcal{L}_V(D) \neq 0$, then $|D| = \{(f) + D\colon f \in \mathcal{L}_V(D)\}$ is called the *complete linear system of* D.

If $(f) + D \geq 0$ and $D = \sum_{i=1}^{r} n_i\Gamma_i$, then $v_{\Gamma_i}(f) \geq -n_i$ for $i = 1, \ldots, r$ and $v_\Gamma(f) \geq 0$ for all other prime divisors Γ. Conversely, if $v_{\Gamma_i}(f) \geq -n_i, i = 1, \ldots, r$ and $v_\Gamma(f) \geq 0$ all other prime divisors Γ, then $(f) + D \geq 0$. In other words,

$$\mathcal{L}_V(D) = \{f\colon v_{\Gamma_i}(f) \geq -n_i, v_\Gamma(f) \geq 0, \Gamma \neq \Gamma_i, D = \sum n_i\Gamma_i\}. \qquad (17.23)$$

Thus, if $f_1, f_2 \in \mathcal{L}_V(D)$, then $v_{\Gamma_i}(f_1 + f_2) \geq \min\{v_{\Gamma_i}(f_1), v_{\Gamma_i}(f_2)\}$ so that $f_1 + f_2 \in \mathcal{L}_V(D)$. Since $v_{\Gamma_i}(\alpha f) = v_{\Gamma_i}(f)$ for $\alpha \in k$, $\alpha f \in \mathcal{L}_V(D)$ if $f \in \mathcal{L}_V(D)$. In other words, $\mathcal{L}_V(D)$ is a k-vector space. Since $(f_1) + D = (f_2) + D$ if and only if $f_1 = \lambda f_2$ with $\lambda \in k^*$, $|D| \simeq \mathbb{P}(\mathcal{L}_V(D))$ and $\dim |D| = \dim \mathcal{L}_V(D) - 1$. If $D \sim E$, then $\mathcal{L}_V(D) \simeq \mathcal{L}_V(E)$ for if $D - E = (\varphi)$, $\varphi \in k(V)$, then $(h\varphi) + E = (h) + D$ and so, $h \in \mathcal{L}_V(D)$ if and only if $h\varphi \in \mathcal{L}_V(E)$.

Definition 17.24 A *linear system of divisors*[†] is any (projective) linear subspace of a complete linear system $|D|$.

Let $h \in k[Y_0, \ldots, Y_N]/I_h(V) = S(V)$ and suppose that $h \in S(V)_q$ i.e., h is a form of degree q. Then $V(h) \cap V = \Gamma_1 \cup \cdots \cup \Gamma_r$ where the Γ_i are prime divisors on V. For each i, there is a $\lambda(i)$ such that $Y_{\lambda(i)} \not\equiv 0$ on Γ_i (i.e., $\Gamma_i \cap U_{\lambda(i)} \neq \emptyset$) and so $v_{\Gamma_i}(h/Y_{\lambda(i)}^q) = n_i$ is defined. Thus, $(h) = V \cdot V(h) = \sum n_i\Gamma_i \in \mathrm{Div}(V)$. This divisor is called the *hypersurface section of* V *by* h. If $h_1, h_2 \in S(V)_q$, then $h_1/h_2 \in k(V)$ and $(h_1) = (h_2) + (h_1/h_2)$. We let

$$\mathfrak{o}_V(q) = \{D \in \mathrm{Div}(V)\colon D \text{ is a hypersurface section of } V \text{ by } h \in S(V)_q\}. \qquad (17.25)$$

We have:

Proposition 17.26 *$\mathfrak{o}_{\mathbb{P}^N}(q)$ is a complete linear system of dimension $\binom{N+q}{q} - 1$ for all $q \geq 1$.*

[†]The standard terminology is *linear system* but we make the distinction to avoid confusion.

Proof. Let $D_0 = qH_0$ and consider $\mathcal{L}_{\mathbb{P}^N}(D_0)$. Then $f \in \mathcal{L}_{\mathbb{P}^N}(D_0)$ if and only if $(f)+qH_0 \geq 0$ i.e., if and only if $Y_0^q f$ is a polynomial and, hence, if and only if f is a homogeneous polynomial of degree q. In other words, $\mathcal{L}_{\mathbb{P}^N}(D_0)$ is spanned by the monomials $Y_0^{r_0} \cdots Y_N^{r_N}$, $r_i \geq 0$, $\sum r_i = q$. Hence $\dim |qH_0| = \binom{N+q}{q} - 1$. But $\mathfrak{o}_{\mathbb{P}^N}(q) = \{D \in \mathrm{Div}(\mathbb{P}^N)\colon D = V(h)$, degree $h = q$, $h \in S_q$ and $D \geq 0\}$. Since any such $D \sim qH_0$ and $qH_0 \in \mathfrak{o}_{\mathbb{P}^N}(q)$, we have the result.

We can view $\mathfrak{o}_V(q)$ as the set $\{D \cdot V\colon D \in \mathfrak{o}_{\mathbb{P}^N}$ and $V \not\subset \mathrm{Supp}(D)\}$ and can call $\mathfrak{o}_V(q)$ the "*trace of* $\mathfrak{o}_{\mathbb{P}^N}(q)$ *on* V". $\mathfrak{o}_V(q)$ is thus a linear system of divisors on V. It need not be complete. However, we do have:

Theorem 17.27 *If V is a normal projective variety, then $\mathfrak{o}_V(q)$ is complete if q is large.*

Proof ([M-5]). We may assume that $V \not\subset$ any hyperplane (replacing $\mathbb{P}^N$ by a smaller linear space). Let $\Gamma_i = (Y_i)$, $i = 0, \ldots, N$ be elements of $\mathfrak{o}_V(1)$. Consider $\mathfrak{o}_V(q)$ and let D be an element of the complete linear system (i.e., the projective space, perhaps of infinite dimension) containing $\mathfrak{o}_V(q)$. Then $D \sim q\Gamma_i$ for all i and there are $f_0, \ldots, f_N$ in $k(V)$ with $(f_i) = D - q\Gamma_i$. Then $(f_i/f_j) = q(\Gamma_j - \Gamma_i) = (Y_j^q/Y_i^q)$. Thus, we may assume that

$$Y_i^q f_i = Y_j^q f_j \tag{17.28}$$

for all i, j (modulo changing the f_i's by constants). In other words, $h = Y_i^q f_i \in k(V)$. But f_i has no poles in $V \cap U_i$ ($U_i = \mathbb{P}^N_{Y_i}$) and so $f_i \in k[V \cap U_i]$. Then $f_i = F_i/Y_i^n$ with $F_i \in S(V)_n$ and

$$Y_i^n h = Y_i^q F_i \in S(V)_{n+q} \tag{17.29}$$

for all i. So we have the following situation: $S(V) = k[Y_0, \ldots, Y_N]/\mathfrak{p}_h$, $\mathfrak{p}_h$ a homogeneous prime ideal, $h \in$ quotient field of $S(V)$ and $Y_0^n h, \ldots, Y_n^n h \in S(V)_{n+q}$. We *claim* there is a q_0 such that if $q \geq q_0$, then $h \in S(V)_q$. Assuming the claim, then $D = (h) \in \mathfrak{o}_V(q)$.

Claim *Let $S = k[Y_0, \ldots, Y_N]/\mathfrak{p}_h$. Then there is a q_0 such that if $q \geq q_0$, then $h \in K(S)$ and $Y_0^n h, \ldots, Y_N^n h \in S_{n+q}$ together imply $h \in S_q$.*

Proof of Claim. Let $r = \dim V$. Then S is a finitely generated torsion-free graded $k[Z_0, \ldots, Z_r]$ module where $\{Z_i\}$ is a transcendence basis of S, $Z_i = \sum_{j=0}^{N} \alpha_{ij} Y_j$ (by the Noether Normalization Lemma). Then S is a graded submodule of a free module $S' = \oplus \sum_{j=1}^{t} k[Z_0, \ldots, Z_r]\omega_j$, ω_j homogeneous of degree d_j. Let $\tilde{S} =$

$\{f \in S' \colon \text{there is a } d \text{ with } Z_i^d f \in S \text{ all } i\}$. $\tilde{S} \subset S'$ and so $\tilde{S}$ is finitely generated. Let $\tilde{\omega}_s$ be generators with degrees m_s. Then $Z_i^d \tilde{\omega}_s \in S$ for all i, s and some d. Hence, $(Z_0, \ldots, Z_r)^{d(r+1)}\tilde{\omega}_s \subset S$ for all s. If $q_0 = \max(m_s)+d(r+1)$ and $q \geq q_0$, then $\tilde{S}_q = S_q$. For, clearly $S_q \subset \tilde{S}_q$ for all q, and if $g \in \tilde{S}_q$, then $g = \sum g_s \tilde{\omega}_s$, $g_s \in (Z_0, \ldots, Z_r)^{q-m_s}$ so that $q \geq q_0$ implies $g_s \in (Z_0, \ldots, Z_r)^{d(r+1)}$ and $\sum g_s \tilde{\omega}_s \in S_q$.

Corollary 17.30 $\dim_k |D|$ *is finite for any* D.

Proof. Let D be any divisor on V and let $\varphi \neq 0$ be an element of $|D|$. Then $E = (\varphi) + D \geq 0$ and $\mathcal{L}_V(D) \simeq \mathcal{L}_V(E)$. In other words, we may suppose that $D = \sum_{i=1}^{t} n_i \Gamma_i$ with $n_i > 0$ (if $D = 0$, then $|D| = \mathbb{P}(k)$). Let $f_i \in I(\Gamma_i)$, $f_i \notin I(V)$, f_i a form of degree q_i. Consider $g = f_1^{n_1} \cdots f_t^{n_t}$ and let $\tilde{q} = \sum n_i q_i$. Then $(g) = D + E_1$ with $E_1 \geq 0$ (since $v_{\Gamma_i}(g) \geq n_i$ and $v_\Gamma(g) \geq 0$, $\Gamma \neq \Gamma_i$). Let $h = g^m$ with $g = m\tilde{q}$ such that $\mathfrak{o}_V(q)$ is complete. Then $\mathcal{L}_V(D) \subseteq \mathcal{L}_V((g))$ and $|D| + E_1 \subseteq |(g)| = \mathfrak{o}_V(q)$ so that $\dim_k |D| \leq \dim_k \mathfrak{o}_V(q)$ is finite.

We can also prove the corollary directly by induction. We begin with:

Lemma 17.31 *Let* $D = D_+ - D_-$ *with* $D_+ \geq 0$, $D_- \geq 0$ *and suppose that* Γ *does not occur in* D_-. *Then there is a divisor* D^* *on* Γ *such that if* $f \in \mathcal{L}_V(D)$, *then* f *induces a function* f' *on* Γ *and* $f' \in \mathcal{L}_\Gamma(D^*)$.

Proof. Since Γ is not in D_-, $v_\Gamma(f) \geq 0$ and $f \in \mathfrak{o}_{\Gamma,V}$. Let $f' = \operatorname{Tr}_\Gamma f = \overline{f}$ be the $\mathfrak{m}_{\Gamma,V}$ residue of f. f' is the *trace of* f *on* Γ and is *induced by* f *on* Γ. Note that $\mathfrak{o}_{\Gamma,V}/\mathfrak{m}_{\Gamma,V} = k(\Gamma)$, the function field of Γ. Let $V_i = V - H_i \cap V = V \cap U_i$, $H_i = V(Y_i)$, for $i = 0, 1, \ldots, N$. Then there are f_i (say $Y_i^{d_i}$) such that f_i is a unit in $\mathfrak{o}_{\Gamma,V}$ and $f_i f$ has no poles on V_i. Since f_i, f, $f_i f$ are defined at Γ, so are the traces f_i', f', $f_i' f'$. Since $f_i f$ is defined at all points of V_i, $f_i' f'$ is defined at all points of $\Gamma \cap V_i$. Since the poles of f' are bounded by the zeros of f_i', we can take $D^* = (f_i')_0$ (note on $V_i \cap V_j$, $f_i' f' = f_j' f'$).

Theorem 17.32 $\mathcal{L}_V(D)$ *is finite dimensional.*

Proof. We use induction on $\dim V$. If $D = 0$, then $\mathcal{L}_V(D) = k$. So, for $r = 1$, we must show that if $\mathcal{L}_V(D)$ is finite dimensional, then so is $\mathcal{L}_V(D + P)$ where P is a point. Suppose $f, g \in \mathcal{L}_V(D + P)$ which are independent modulo $\mathcal{L}_V(D)$. Let $n = v_P(D)$ so that $v_P(f) = v_P(g) = n + 1$. Then $v_P(f/g) = 0$ and f/g is a unit α at P. It follows that $g(f/g - \alpha) = f - \alpha g$ has a pole of order no greater than n at P. In other words, $f - \alpha g \in \mathcal{L}_V(D)$, a contradiction. Thus, $\dim \mathcal{L}_V(D + P) \leq \dim \mathcal{L}_V(D) + 1$. We now assume the theorem for varieties of dimension $< r$. Clearly, it suffices to show that if $D \geq 0$ and $\mathcal{L}_V(D)$ is finite dimensional, then so is $\mathcal{L}_V(D+\Gamma)$. Suppose $D = n\Gamma + \cdots$. If there is no $f \in \mathcal{L}_V(D+\Gamma)$ with $v_P(f) = -(n+1)$, then $\mathcal{L}_V(D+\Gamma) = \mathcal{L}_V(D)$. So we can assume that there is an $f \in \mathcal{L}_V(D+\Gamma)$ with $v_\Gamma(f) = -(n+1)$. Consider $\mathcal{L}_V(D+\Gamma+(f)) = f^{-1}\mathcal{L}_V(D+\Gamma)$ and note that if $f^{-1}g$ has a zero at Γ, then $v_\Gamma(g) \geq -n$ and $g \in \mathcal{L}_V(D)$. If $h \in \mathcal{L}_V(D+\Gamma+(f))$, then h has no pole at Γ and the map $h \to h' = \mathrm{Tr}_\Gamma h$ is a homomorphism. By the lemma, the image is finite dimensional. Since the kernel is $\{f^{-1}g\colon f^{-1}g \text{ has a zero at } \Gamma\} \simeq \mathcal{L}_V(D)$, the result follows.

If $\mathbf{L}$ is a linear subspace of $\mathcal{L}_V(D)$, then $\Lambda(\mathbf{L}, D) = \{(f)+D\colon f \in \mathbf{L}\}$ is called the *linear system of divisors defined by* $\mathbf{L}$‡ and is clearly a linear subspace of $|D|$. If $D \sim D_1$ and $\mathbf{L} = \mathbf{L}_1 f$ where $D_1 = (f) + D$, then $\Lambda(\mathbf{L}, D) = \Lambda(\mathbf{L}_1, D_1)$ and conversely. If $f_0, \ldots, f_n$ are a basis of the space $\mathbf{L}$ and we let $D_i = (f_i) + D$, then $\Lambda(\mathbf{L}, D) = \Lambda(\mathbf{L}_i, D_i)$ where $\mathbf{L}_i = \mathbf{L}f_i^{-1}$. Let $\Lambda = \Lambda(\mathbf{L}, D) = \Lambda(\mathbf{L}_i, D_i)$, $i = 0, \ldots, n$. Since the f_i are in $k(V)$, we can define maps $\psi_i\colon V - D_i \to \mathbb{P}^n$ by setting

$$\psi_i(\xi) = (f_0(\xi)/f_i(\xi), \ldots, f_n(\xi)/f_i(\xi)). \tag{17.33}$$

Clearly, $\psi_i = \psi_j$ on $V - (D_i \cup D_j)$ and so the map $\psi = (\psi_0, \ldots, \psi_n)$ of V into $\mathbb{P}^n$ is a rational map.

Example 17.34 Let $V = \mathbb{P}^N$ and $H_i = V(Y_i)$. Then

$$|H_0| = \left\{ \left(\sum_{j=0}^{N} \alpha_j Y_j / Y_0 \right) + H_0\colon (\alpha_0, \ldots, \alpha_N) \in \mathbb{P}^N \right\}.$$

If F is a form of degree q, then $(F) = (F/Y_0^q) + qH_0$. If

$$\mathbf{L} = \left\{ \left(\sum_{j=0}^{M} \alpha_j Y_j \right) : (\alpha_0, \ldots, \alpha_M) \in \mathbb{P}^M \subset \mathbb{P}^N \right\},$$

then $\psi = (Z_0, \ldots, Z_M, 0, \ldots, 0)$ is the projection of $\mathbb{P}^N$ onto $\mathbb{P}^M$ with center $V(Y_0) \cap \cdots \cap V(Y_M) = H_0 \cap \cdots \cap H_M$ (an L_{N-M}).

Example 17.35 Let $\psi\colon \mathbb{P}^1 \to \mathbb{P}^N$ be a regular map and let $V = \psi(\mathbb{P}^1)$. Let $\Gamma = V(Y_N)$ as a prime divisor on $\mathbb{P}^N$. Suppose that $V \not\subset \Gamma$. Then $V \cdot \Gamma \in \mathfrak{o}_V(1)$ and has a finite set of points as support. These points are the elements of $V \cap \Gamma$. Let $V \cap \Gamma = \{P_1, \ldots, P_t\}$. Then $V \cdot \Gamma = \sum_{j=1}^{t} n_j P_j$ is a divisor on V. If $\mathbf{F}(z)$ is an

‡Again the standard mathematical terminology is *linear system*.

element of $\text{Rat}(n, 2, 2)$ and $\psi_{\mathbf{F}}\colon \mathbb{P}^1 \to \mathbb{P}^5$ is the associated regular map, then, letting $\Gamma = H_5 = V(Y_5)$, we have

$$\psi_{\mathbf{F}}(\mathbb{P}^1) \cdot \Gamma = \sum_{i=1}^{t} n_i P_i \tag{17.36}$$

where $\psi_{\mathbf{F}}(\mathbb{P}^1) \cap \Gamma = \{P_1, \ldots, P_t\}$. The P_i are the poles of $\mathbf{F}$ and the n_i are their "multiplicities". The "degree" of the divisor $\psi_{\mathbf{F}}(\mathbb{P}^1) \cdot \Gamma$ on $\psi_{\mathbf{F}}(\mathbb{P}^1)$ is $n = \sum_{i=1}^{t} n_i$. [We shall prove this in general in Chapter 18.] Let us reexamine Example 2.11 in this light. If $\mathbb{P}^1 = \{(s,t)\colon \text{either } s \neq 0 \text{ or } t \neq 0\}$, then $\mathbf{F}(z) \in \text{Rat}(3, 2, 2)$ with $\psi_{\mathbf{F}}(s,t) = (\psi_0, \ldots, \psi_5)$ where $\psi_0(s,t) = s^2 t$, $\psi_1(s,t) = -s^2(t+s)$, $\psi_2(s,t) = s(t+s)(t-s)$, $\psi_3(s,t) = -st(t+s)$, $\psi_4(s,t) = 0$, $\psi_5(s,t) = (t+s)^2(t-s)$. We observe that $\psi_{\mathbf{F}}(0,1) = (0, \ldots, 0, 1)$ and that $\psi_{\mathbf{F}}(1,0) = (0, -1, -1, 0, 0, -1)$. Thus the points where $\psi_{\mathbf{F}}(s,t)$ intersects H_5 all lie in the affine space $\mathbb{A}^5 = \mathbb{P}^5_{Y_0}$ with affine coordinates $(1, X_1, X_2, X_3, X_4, X_5)$ and we can consider the (rational) map $\psi_{\mathbf{F}}\colon \mathbb{A}^1 - (0) \to \mathbb{A}^5$ given by $\psi_{\mathbf{F}}(z) = (-(z+1)/z, (z+1)(z-1)/z, -(z+1), 0, (z+1)^2(z-1)/z)$. Since $\psi_{\mathbf{F}}(z) \subset V(X_4) \simeq \mathbb{A}^4$, we shall view $\psi_{\mathbf{F}}$ as the map of $\mathbb{A}^1 - (0)$ into $\mathbb{A}^4$ given by

$$\psi_{\mathbf{F}}(z) = (-(z+1)/z, (z+1)(z-1)/z, -(z+1), (z+1)^2(z-1)/z) \tag{17.37}$$

and we let X_1, X_2, X_3, X_5 be the coordinates on $\mathbb{A}^4$. Let $C_{\mathbf{F}}$ be the curve $\psi_{\mathbf{F}}(z)$ and let $\mathfrak{a}_{\mathbf{F}} = I(C_{\mathbf{F}})$ so that $k[C_{\mathbf{F}}] = k[X_1, X_2, X_3, X_5]/\mathfrak{a}_{\mathbf{F}}$. $C_{\mathbf{F}}$ is determined by the equations

$$\begin{aligned} f_1(X_1, X_2, X_3, X_5) &= (X_3+1)X_1 + X_3 = 0 \\ f_2(X_1, X_2, X_3, X_5) &= (X_3+1)X_2 + X_3(X_3+2) = 0 \\ f_3(X_1, X_2, X_3, X_5) &= (X_3+1)X_5 - X_3^2(X_3+2) = 0 \end{aligned} \tag{17.38}$$

(and the f_i are in $\mathfrak{a}_{\mathbf{F}}$) and $C_{\mathbf{F}} \cap V(X_5) = \{P_1, P_2\}$ where $P_1 = (0,0,0,0)$ and $P_2 = (-2, 0, -2, 0)$. If $\Gamma = V(X_5)$, then the divisor $C_{\mathbf{F}} \cdot \Gamma = 2P_1 + P_2$ which has degree 3 (= degree $\mathbf{F}(z)$). Let $\mathfrak{m}_1 = \mathfrak{m}_{P_1, C_{\mathbf{F}}}$, $\mathfrak{m}_2 = \mathfrak{m}_{P_2, C_{\mathbf{F}}}$ and let $\bar{\ }$ denote $\mathfrak{a}_{\mathbf{F}}$-residue. Then, at P_1, $\overline{x}_3 \in \mathfrak{m}_1$ and $\overline{x}_3 + 1 \notin \mathfrak{m}_1$ and $\overline{x}_3 + 2 \notin \mathfrak{m}_1$ so that $\overline{x}_5 \in \mathfrak{m}_1^2$. In other words, $v_{P_1}(\overline{x}_5) = 2$. At P_2, $\overline{x}_3 + 2 \in \mathfrak{m}_2$ but $\overline{x}_3$ and $\overline{x}_3 + 1$ are not in $\mathfrak{m}_2$ (otherwise $1 \in \mathfrak{m}_2$). It follows that $\overline{x}_5 = u(\overline{x}_3 + 2)$ with u a unit in $\mathfrak{o}_{P_2, C_{\mathbf{F}}}$ and that $v_{P_2}(\overline{x}_5) = 1$. Thus, $C_{\mathbf{F}} \cdot \Gamma = 2P_1 + P_2 = v_{P_1}(\overline{x}_5)P_1 + v_{P_2}(\overline{x}_5)P_2$. Note also that $\overline{x}_5$ is a local equation of $H_5 = V(X_5)$ at P_1 and P_2.

Example 17.39 Let $\mathbf{L}$ be a linear system of divisors with $\dim \mathbf{L} = M$. Then $\mathbf{L}$ is an L_M i.e., a linear subspace of a projective space. So L_M is isomorphic to $(\mathbb{P}^M)^*$, the space of hyperplanes, $\sum_{i=0}^{M} a_i Z_i = 0$, on $\mathbb{P}^M$. Let λ be such an isomorphism and let $D_0 \in \mathbf{L}$ with $\lambda(D_0) = (1, 0, \ldots, 0)$ i.e., $\lambda(D_0)$ is the hyperplane

$Z_0 = 0$. Set $V_{\mathbf{L}} = \{f\colon (f) + D_0 \geq 0,\ (f) + D_0 \in \mathbf{L}, \text{ or } f = 0\}$. Then $V_{\mathbf{L}}$ is a vector space of dimension $M+1$ and $\mathbb{P}(V_{\mathbf{L}}) \simeq (\mathbb{P}^M)^*$ via an isomorphism $\boldsymbol{\lambda}$ of $V_{\mathbf{L}}$ and $\mathbb{A}^{M+1}$ induced by λ. Let $f_i = \boldsymbol{\lambda}^{-1}(Z_i)$, $i = 1, \ldots, M$, so that $f_i \in k(V)$. Let $\psi\colon V - \operatorname{supp} D_0 \to \mathbb{P}^M$ be given by

$$\psi(\xi) = (1, f_1(\xi), \ldots, f_M(\xi)). \tag{17.40}$$

Then ψ is clearly a rational map.

Example 17.41 Let $f(z) = p(z)/q(z)$ be a transfer function of degree n. Then $f(x_0, x_1) = a_0 x_0^n + \cdots + a_{n-1} x_0^{n-1} x_1 / b_0 x_0^n + \cdots + b_{n-1} x_0^{n-1} x_1^{n-1} + x_1^n \in k(\mathbb{P}^1)$ and $(f) \in \operatorname{Div}(\mathbb{P}^1)$. Moreover, $(f) = (f)_0 - (f)_\infty$.

Now let V be a nonsingular variety. If Γ is a prime divisor on V and $\xi \in V$, then there is an open (affine) neighborhood U_ξ of ξ such that $\Gamma \cap U_\xi = V(f_\Gamma)$ i.e., f_Γ is a local equation of Γ at ξ. If $D = \sum n_i \Gamma_i$ is a divisor, then $D \cap U_\xi = (\Pi f_{\Gamma_i}^{n_i})$. In other words, near ξ, D is a principal divisor. It follows that there is a finite open covering $U_1, \ldots, U_t$ of V such that $D \cap U_i = (h_i)$ is principal i.e., D is "locally principal". We also observe that $h_1, \ldots, h_t$ satisfy the following:

(a) the h_i are not zero; and,

(b) $(h_i) \cap U_j = (h_j) \cap U_i = D \cap (U_i \cap U_j)$.

It follows from (b) that $(h_i h_j^{-1}) = 0$ on $U_i \cap U_j$ and that $h_i h_j^{-1}$, $h_j h_i^{-1} \in \mathfrak{o}_V(U_i \cap U_j)$ (i.e., are regular on $U_i \cap U_j$).

Definition 17.42 *A locally principal* (*or Cartier*) *divisor on a variety* V (not necessarily nonsingular) is a set $\{(h_i, U_i)\colon h_i \in \mathfrak{o}_V(U_i)\}$ of rational functions such that $V = \bigcup_{i=1}^{t} U_i$ and

(a) the h_i are not zero (i.e., do not vanish everywhere); and,

(b) $h_i h_j^{-1}$, $h_j h_i^{-1} \in \mathfrak{o}_V(U_i \cap U_j)$ for all i, j.

Two sets $\{(h_i, U_i)\}$, $\{(h_i', U_i')\}$ define the same Cartier divisor if $h_i h_j'^{-1}$, $h_i^{-1} h_j'$ are in $\mathfrak{o}_V(U_i \cap U_j')$ for all i, j.

If V is a nonsingular variety and $\{(h_i, U_i)\}$ is a Cartier divisor on V, then $\{(h_i, U_i)\}$ determines a divisor on V. For, let Γ be a prime divisor on V. If $\Gamma \cap U_i \neq \emptyset$, set $n_\Gamma = v_\Gamma(h_i)$ (since $(h_i) \cap U_j = (h_j) \cap U_i$, n_Γ is well-defined). Then $D = \sum n_\Gamma \Gamma$ is a divisor on V.

If $g \in k(V)$, then g determines a Cartier divisor by (say) taking $\{(g, U_i)\colon U_i = (\mathbb{P}^N_{Y_i}) \cap V\}$.

Theorem 17.43 *Let $\psi\colon V \to W$ be a dominant morphism of nonsingular varieties. Then ψ defines a homomorphism $\psi_D^*\colon \mathrm{Div}(W) \to \mathrm{Div}(V)$ such that $\psi_D^*((g)) = (\psi^*(g)) = ((g \circ \psi))$ for $g \in k(W)$, and, hence, defines a homorphism of divisors (abuse of notation) $\psi_D^*\colon \mathrm{Pic}(W) \to \mathrm{Pic}(V)$.*

Proof. Let $D \in \mathrm{Div}(W)$. Then $D = \{(g_i, U_i)\}$ with $g_i \in o_W(U_i)$. Since ψ is a morphism, $V_i = \psi^{-1}(U_i)$ is open in V and $h_i = \psi^*(g_i) \in o_V(V_i)$. [Note that $U_i \cap \psi(V) \neq \emptyset$ for all i since ψ is dominant and W is irreducible.] Since $\psi^{-1}(W) = V$, the V_i cover V. Since ψ is dominant, ψ^* is injective and so the $h_i \not\equiv 0$. But $h_i h_j^{-1} = \psi^*(g_i)\psi^*(g_j^{-1}) = \psi^*(g_i g_j^{-1})$ and $V_i \cap V_j = \psi^{-1}(U_i \cap U_j)$ together imply that $\{(h_i, V_i)\}$ is a Cartier divisor on V which determines a divisor $\psi_D^*(D)$ in $\mathrm{Div}(V)$. If $D = \{(g_i, U_i)\} = \{(g_i', U_i')\}$, then $\psi^{-1}(U_i \cap U_j') = \psi^{-1}(U_i) \cap \psi^{-1}(U_j') = V_i \cap V_j'$ and $\psi^*(g_i g_j'^{-1}) = \psi^*(g_i)\psi^*(g_j'^{-1}) = h_i h_j'^{-1}$ on $V_i \cap V_j'$ so that $\psi_D^*(D)$ is well-defined. Suppose that $D^1 = \{(g_i^1, U_i^1)\}$ and $D^2 = \{(g_j^2, U_j^2)\}$ are elements of $\mathrm{Div}(W)$, then $D^1 + D^2 = \{(g_i^1 g_j^2, U_i^1 \cap U_j^2)\}$ so that $\psi_D^*(D^1 + D^2) = \{(h_i^1 h_j^2, V_i^1 \cap V_j^2)\} = \psi_D^*(D^1) + \psi_D^*(D^2)$. In other words, ψ_D^* is a homomorphism. The remaining assertions are clear.

Exercises

(1) Determine the primary ideals in $\mathbb{Z}$.

(2) Show that $\mathrm{Div}(V)$ is a group.

(3) Consider the rational function $h = x_0/x_0 + x_1$ on $\mathbb{P}^1$. Find $(h)_0$, $(h)_\infty$. Let $p(z) = b_0 + \cdots + b_{n-1}z^{n-1}$ and $q(z) = a_0 + \cdots + a_{n-1}z^{n-1} + z^n$ be relatively prime polynomials and let $h(x_0, x_1) = b_0 x_0^n + \cdots + b_{n-1}x_0 x_1^{n-1}/a_0 x_0^n + \cdots + a_{n-1}x_0 x_1^{n-1} + x_1^n$ be a rational function on $\mathbb{P}^1$. Find $(h)_0$, $(h)_\infty$ and (h).

(4) Let V be a nonsingular variety and let $\{(h_i, U_i)\}$, $\{(h_j', U_j')\}$ define the same Cartier divisor on V. Show that they determine the same divisor on V.

18

Projective Algebraic Geometry VIII: Intersections

We shall examine in a brief elementary way the notion of intersection of varieties ([F-5], [H-3]). We shall eventually prove Bezout's Theorem which plays a role in pole placement.

Example 18.1 Consider the curve $\psi_F(x_0, x_1) = (x_1 x_0^2, (x_1 + x_0)x_0^2, x_0(x_1^2 - x_0^2), -x_1 x_0(x_1 + x_0), 0, (x_1 + x_0)(x_1^2 - x_0^2))$ which is a rational cubic C_F lying in the hyperplanes $Y_4 = 0$, $Y_1 + Y_2 + Y_3 = 0$. In addition, the equations

$$\begin{aligned} F_1(Y_0, \ldots, Y_5) &= Y_0 Y_5 - Y_1 Y_4 + Y_2 Y_3 = 0 \\ F_2(Y_0, \ldots, Y_5) &= Y_1 Y_2 + Y_1 Y_4 - Y_2 Y_3 - Y_1 Y_5 = 0 \\ F_3(Y_0, \ldots, Y_5) &= Y_0 Y_1 - Y_0 Y_3 + Y_1 Y_3 = 0 \end{aligned} \tag{18.2}$$

define C_F in $\mathbb{P}^5$. We have

$$\begin{aligned} &\psi_F(0,1) = (0, \ldots, 0, 1) \in C_F \\ &\psi_F(1,z) = (z, z+1, (z-1)(z+1), -z(z+1), 0, (z+1)(z^2-1)) \in C_F \end{aligned}$$

and we note that

$$\begin{aligned} \psi_F(1,-1) &= (-1, 0, \ldots, 0) \\ \psi_F(1,1) &= (1, 2, 0, -2, 0, 0). \end{aligned}$$

What is the set $V(Y_5) \cap C_F$? If $\boldsymbol{\xi} \in V(Y_5) \cap C_F$, then $\xi_5 = 0$ and $\xi_2 \xi_3 = 0$ (as $\xi_4 = 0$) and $\xi_1 \xi_2 = 0$. If $\xi_2 \neq 0$, then $\xi_1 = 0$, $\xi_3 = 0$ and $\xi_1 + \xi_2 + \xi_2 = 0$ would give the contradiction $\xi_2 = 0$. Hence we have $\xi_2 = 0$. If $\xi_1 = 0$, then $\xi_3 = 0$ and $\boldsymbol{\xi} = (\xi_0, 0, \ldots, 0)$ with $\xi_0 \neq 0$ i.e., $\boldsymbol{\xi} = \psi_F(1, -1)$. If $\xi_1 \neq 0$, then $\xi_1 = -\xi_3 \neq 0$

P. Falb, *Methods of Algebraic Geometry in Control Theory: Part II*,
Modern Birkhäuser Classics, https://doi.org/10.1007/978-3-319-96574-1_18

and $2\xi_0\xi_1 - \xi_1^2 = 0$ so that $2\xi_0 = \xi_1 = -\xi_3$ and $\boldsymbol{\xi} = (\xi_0, 2\xi_0, 0, -2\xi_0, 0, 0)$ with $\xi_0 \neq 0$ i.e., $\boldsymbol{\xi} = \psi_F(1,1)$. Thus, as a *set*,

$$V(Y_5) \cap C_F = \{\psi_F(1,-1), \psi_F(1,1)\} = \{(-1,0,\dots,0), (1,2,0,-2,0,0)\}.$$

Let $P_1 = (-1,0,\dots,0)$, $P_2 = (1,2,0,-2,0,0)$. We want to look at the structure of the intersection of $H_5 = V(Y_5)$ and C_F at P_1 and P_2. Consider T_{C_F,P_1}. The equations

$$\begin{aligned} d_{P_1}(Y_4 = 0) &= z_4 \\ d_{P_1}(Y_1 + Y_2 + Y_3 = 0) &= z_1 + z_2 + z_3 \\ d_{P_1}(F_1 = 0) &= -z_5 \\ d_{P_1}(F_2 = 0) &= 0 \\ d_{P_1}(F_3 = 0) &= -z_1 + z_3 \end{aligned}$$

define T_{C_F,P_1}. Writing these in matrix form, we have

$$\begin{bmatrix} 0 & 0 & 0 & 0 & 1 & 0 \\ 0 & 1 & 1 & 1 & 0 & 0 \\ 0 & 0 & 0 & 0 & 0 & -1 \\ 0 & 0 & 0 & 0 & 0 & 0 \\ 0 & -1 & 0 & 1 & 0 & 0 \end{bmatrix} \begin{bmatrix} z_0 \\ z_1 \\ \vdots \\ z_5 \end{bmatrix} = 0 = A_{P_1}\mathbf{z}$$

so that $\dim T_{C_F,P_1} = \dim \operatorname{Ker} A_{P_1} = 2 > 1$ and P_1 is a singular point of C_F. Consider next T_{C_F,P_2}. The equations

$$\begin{aligned} d_{P_2}(Y_4 = 0) &= z_4 \\ d_{P_2}(Y_1 + Y_2 + Y_3 = 0) &= z_1 + z_2 + z_3 \\ d_{P_2}(F_1 = 0) &= -2z_2 - 2z_4 + z_5 \\ d_{P_2}(F_2 = 0) &= 4z_2 + 2z_4 - z_5 \\ d_{P_2}(F_3 = 0) &= 4z_0 + z_3 \end{aligned}$$

define T_{C_F,P_2}. Writing these in matrix form, we have

$$\begin{bmatrix} 0 & 0 & 0 & 0 & 1 & 0 \\ 0 & 1 & 1 & 1 & 0 & 0 \\ 0 & 0 & -2 & 0 & -2 & 1 \\ 0 & 0 & 4 & 0 & 2 & -1 \\ 4 & 0 & 0 & 1 & 0 & 0 \end{bmatrix} \begin{bmatrix} z_0 \\ z_1 \\ \vdots \\ z_5 \end{bmatrix} = 0 = A_{P_2}\mathbf{z}$$

so that $\dim T_{C_F,P_2} = \dim \operatorname{Ker} A_{P_2} = 1$ and P_2 is a simple point of C_F. Now $\dim H_5 = 4$ and $\dim C_F = 1$. Every irreducible component of $H_5 \cap C_F$ (i.e., the points P_1, P_2) has dim 0 so that $\dim H_5 + \dim C_F - 5 = 0 = \dim P_i$. We say that H_5 and C_F *intersect properly*. At P_1, $\dim T_{C_F,P_1} + \dim T_{H_5,P_1} = 6 > 5$ so that the vector spaces T_{C_F,P_1} and T_{H_5,P_1} have a non-trivial intersection. On the other hand, at P_2, $\dim T_{C_F,P_2} + \dim T_{H_5,P_2} = 5$ and $T_{C_F,P_2} \cap T_{H_5,P_2} = (0)$. In this case, we say that H_5 and C_F *intersect transversally* (or are *transversal*) *at* P_2. At P_1, C_F and H_5 have a common tangent and so there is a "multiple" intersection. (Cf. Example 17.35.)

We want to generalize this example to the case of an element $\psi_{\mathbf{F}}(x_0, x_1)$ in $\mathrm{Rat}(n, m, p)$ and the hyperplane $H_N = V(Y_N)$. In fact, we will show that $\deg(\psi_{\mathbf{F}} \cdot H_N) = n$ viewing $\psi_{\mathbf{F}} \cap H_N$ as the "cycle" $\psi_{\mathbf{F}} \cdot H_N$. Let $\gamma_{\mathbf{F}} = \psi_{\mathbf{F}}(\mathbb{P}^1)$ be the "curve" corresponding to $\mathbf{F} \in \mathrm{Rat}(n, m, p)$. Since $\mathbb{P}^1$ is irreducible and $\psi_{\mathbf{F}}$ is a morphism, $\gamma_{\mathbf{F}}$ is an irreducible curve. Since $\psi_{\mathbf{F}}(0,1) = (0, \dots, 0, 1)$, $\gamma_{\mathbf{F}} \not\subset H_N$ and so $\dim(\gamma_{\mathbf{F}} \cap H_N) = 1 + N - 1 - N = 0$. In other words, $\gamma_{\mathbf{F}}$ and H_N intersect properly. Thus, $\gamma_{\mathbf{F}} \cap H_N = \{P_1, \dots, P_t\}$ where the P_i are points. We have $\gamma_{\mathbf{F}} \cdot H_N = \sum_{i=1}^{t} n_i P_i$ as a divisor on $\gamma_{\mathbf{F}}$ and degree $(\gamma_{\mathbf{F}} \cdot H_N) = \sum_{i=1}^{t} n_i$. We wish to characterize n_i as an "intersection multiplicity", and so show that $\sum_{i=1}^{t} n_i = n = \deg \gamma_{\mathbf{F}}$. Since $\gamma_{\mathbf{F}}$ and H_N intersect properly, we have, by Bezout's Theorem (18.17), which shall be proved shortly, that

$$\deg(\gamma_{\mathbf{F}} \cdot H_N) = \deg \gamma_{\mathbf{F}} \deg H_N. \tag{18.3}$$

But $\deg H_N = 1$ and so $\deg(\gamma_{\mathbf{F}} \cdot H_N) = \deg \gamma_{\mathbf{F}}$. We claim that $\deg \gamma_{\mathbf{F}} = n$. Since $\gamma_F = \psi_{\mathbf{F}}(\mathbb{P}^1)$, $\boldsymbol{\xi} = (\xi_0, \dots, \xi_N) \in \gamma_{\mathbf{F}}$ if and only if $\xi_i = \psi_{\mathbf{F},i}(\eta_0, \eta_1)$ for some $(\eta_0, \eta_1) \in \mathbb{P}^1$. Now, $\deg \gamma_{\mathbf{F}}$ is the number of points of intersection of a general L_{N-1} with $\gamma_{\mathbf{F}}$. Let $L_{N-1} = V(\sum a_j Y_j)$ be a general L_{N-1} with $a_N \neq 0$ so that $\boldsymbol{\xi} \in \gamma_{\mathbf{F}} \cap L_{N-1}$ if and only if $\sum a_j \psi_{\mathbf{F},j}(\eta_0, \eta_1) = 0$. If $\boldsymbol{\eta} = (0, 1)$, then $\sum a_j \psi_{\mathbf{F},j}(0, 1) = a_N \neq 0$ so that no $\boldsymbol{\xi}$ in $\gamma_{\mathbf{F}} \cap L_{N-1}$ has $\xi_i = \psi_{\mathbf{F},i}(0, 1)$. In other words, we may suppose that $\xi_i = \psi_{\mathbf{F},i}(1, t)$ and hence that t is a root of the polynomial $\sum a_j \psi_{\mathbf{F},j}(1, T) = G(\mathbf{a}, T)$. Since $G(\mathbf{a}, T)$ has degree n, the polynomial has, for general $\mathbf{a}$, n distinct roots (as the $\psi_{\mathbf{F},i}(1, z)$ are relatively prime). Hence, $\gamma_{\mathbf{F}} \cap L_{N-1}$ will contain n points for a general L_{N-1} and $\deg \gamma_{\mathbf{F}} = n$.

Example 18.4 Let $P \in \gamma_{\mathbf{F}} \cap H_N$ and let n_P be the "multiplicity" of P in $\gamma_{\mathbf{F}} \cdot H_N$. Since P is a simple point of H_N, $\dim T_{H_N,P} = N - 1$. Then $\gamma_{\mathbf{F}}$ and H_N will intersect transversally at P if and only if $\dim T_{\gamma_{\mathbf{F}},P} = 1$ i.e., if and only if P is a simple point of $\gamma_{\mathbf{F}}$ i.e., if and only if $\mathfrak{o}_{P,\gamma_{\mathbf{F}}}$ is a regular local ring. Since $P \subset \gamma_{\mathbf{F}}$ is an irreducible subvariety of codimension 1, $K\,\mathrm{Dim}\, \mathfrak{o}_{P,\gamma_{\mathbf{F}}} = 1$ and $\mathfrak{o} = \mathfrak{o}_{P,\gamma_{\mathbf{F}}}$ will be regular if and only if $\mathfrak{m} = \mathfrak{m}_{P,\gamma_{\mathbf{F}}} = (m)$ is a principal ideal (Theorem 17.4). If $\mathfrak{o}' = \mathfrak{o}_{P,H_N}$, then $\mathfrak{o}'$ is regular and $\mathfrak{m}' = \mathfrak{m}_{P,H_N} = (\ell_1, \dots, \ell_{N-1})$ (the ℓ_i

are "linear"). Since the sum $T_{\gamma_{\mathbf{F}},P} + T_{H_N,P}$ is direct, $dm, d\ell_1, \ldots, d\ell_{N-1}$ may be asumed independent. It follows that $\mathfrak{m}_{P,\mathbf{P}^N} = (m, \ell_1, \ldots, \ell_{N-1}) = (I(\gamma_{\mathbf{F}}), I(H_n))\mathfrak{o}_{P,\mathbf{P}^N}$ and that

$$\dim_k \mathfrak{o}_{P,\mathbf{P}^N}/(I(\gamma_{\mathbf{F}}), I(H_n))\mathfrak{o}_{P,\mathbf{P}^N} = \ell(\mathfrak{o}_{P,\mathbf{P}^N}/(I(\gamma_{\mathbf{F}}), I(H_N))\mathfrak{o}_{P,\mathbf{P}^N}) = 1$$

where $\ell(\cdot)$ is length). In this case $n_P = 1$ is the "intersection multiplicity" of $\gamma_{\mathbf{F}}$ and H_N at P.

Let $V \subset \mathbb{A}^N$ be a variety with affine coordinate ring $k[V]$ and let W be a closed subset of V. The irreducible components Z_i of W correspond to the minimal prime ideals $\mathfrak{p}_i = I(Z_i)$ in $k[V]$ containing $I(W)$.

Definition 18.5 The (*geometric*) *multiplicity* of Z_i in W on V, $m(Z_i; W, V)$, is given by

$$m(Z_i; W, V) = \ell(\mathfrak{o}_{Z_i,V}/I(W)\mathfrak{o}_{Z_i,V}) \tag{18.6}$$

(which is finite since the ring is Artinian [A-2]).

In the case of a projective variety V and a closed $W \subset V$, we can, as usual, use an affine open $U \subset V$ with $U \cap W \neq \emptyset$ to define $m(Z_i; W, V)$. If $Z_1, \ldots, Z_t$ are the irreducible components of W, then the *cycle*, $[W]$, is given by

$$[W] = \sum_{i=1}^{t} m(Z_i; W, V) Z_i \tag{18.7}$$

as a formal sum. If W is pure s-dimensional (i.e., every component Z_i is s-dimensional), then the *degree of* $[W]$, $\deg[W]$, is given by

$$\deg[W] = \sum_{i=1}^{t} m(Z_i; W, V) \deg Z_i. \tag{18.8}$$

If we let $\mathfrak{p}_i = I_V(Z_i)$, then $\mathfrak{p}_1, \ldots, \mathfrak{p}_t$ are the minimal prime ideals of $k[V]$ containing $I_V(W)$ and are the minimal primes of the annihilator, $\mathrm{Ann}(k[V]/I_V(W))$, of the $k[V]$-module, $k[V]/I_V(W)$. We observe that

$$\begin{aligned} m(Z_i; W, V) &= \ell(k[V]_{\mathfrak{p}_i}/I_V(W)k[V]_{\mathfrak{p}_i}) \\ &= \ell(R_i/I_V(W)R_i) \end{aligned} \tag{18.9}$$

where $R_i = k[V]_{\mathfrak{p}_i}$. $M_i = R_i/I_V(W)R_i$ is an R_i-module and $\mathfrak{m}_i = \mathfrak{p}_i R_i$ is the minimal prime ideal of Ann M_i. But M_i being a module of finite length has a decomposition $(0) = N_0 < N_1 < \cdots < N_{\ell_i} = M_i$ with $N_j/N_{j-1} \simeq R_i/\mathfrak{m}_i$ and $\ell_i =$ length of M_i. In general ([A-2], [M-1]) if M is a module of finite type over a Noetherian ring, then M has a decomposition (not unique) $(0) = M_0 < \cdots < M'_\lambda = M$ with $M'_j/M'_{j-1} \simeq R/\mathfrak{q}_j$ where $\mathfrak{q}_j$ is a prime ideal which contains

Ann M. If $h_M(t)$ is the Hilbert polynomial of M, then $h_M(t) = \sum_{j=1}^{\lambda} h_{M'_j/M'_{j-1}}(t)$. If $V(\mathfrak{q}_j) = V_j$ is a variety of dimension s_j and degree d_j, then

$$h_{M'_j/M'_{j-1}}(t) = \frac{d_j}{s_j!} t^{s_j} + \cdots . \tag{18.10}$$

So, if $M = k[V]/I_V(W)$, then

$$h_M(t) = \frac{\deg[W]}{s!} t^s + \cdots = \sum_{j=1}^{\lambda} \frac{d_j}{s_j!} t^{s_j} + \cdots . \tag{18.11}$$

Since $W = Z_1 \cup \cdots \cup Z_\sigma$ and $\mathfrak{p}_1, \dots, \mathfrak{p}_\sigma$ are the *minimal* prime ideals of Ann M, we need only consider the $h_{M'_j/M'_{j-1}}(t)$ for the $\mathfrak{p}_i$. Since the number of times $\mathfrak{p}_1$ (say) occurs for some M'_j/M'_{j-1} is by localization $\ell(R_1/I_V(W)R_1)$ and similarly for the other $\mathfrak{p}_i$, and since

$$h_{\mathfrak{p}_i}(t) = \frac{(\deg Z_i)}{s!} t^s + \cdots \tag{18.12}$$

we have

$$\deg[W] = \sum_{i=1}^{t} m(Z_i; W, V) \deg Z_i \tag{18.13}$$

where

$$m(Z_i; W, V) = \ell(R_i/I_V(W)R_i). \tag{18.14}$$

We shall, from now on, deal with the situation where $V = \mathbb{P}^N$ (or $\mathbb{A}^N$) so that $\mathfrak{o}_{Z_i,V} = \mathfrak{o}_{Z_i,\mathbb{P}^N}$ is a regular local ring.

If X is an r-dimensional subvariety of $\mathbb{P}^N$ and $H = V(F)$ is a hypersurface not containing X, then

$$\deg(X \cap H) = (\deg X)(\deg H) = m \deg X \tag{18.15}$$

where $m = \deg F$. If $W = X \cap H$ with components $Z_1, \dots, Z_t$, then $[W] = [X \cdot H] = \sum_{j=1}^{t} m(Z_j; W, \mathbb{P}^N) Z_j$ and

$$\deg[W] = \sum_{j=1}^{t} m(Z_j; W, \mathbb{P}^N) \deg Z_j = (\deg X)(\deg H).$$

In particular, if $H = V(L)$ is a hyperplane, then $\deg(X \cap H) = \deg X$.

Definition 18.16 Let V_1, V_2 be subvarieties of $\mathbb{P}^N$ with $\dim V_1 = r$, $\dim V_2 = s$. Then V_1 and V_2 *intersect properly* if $\dim(V_1 \cap V_2) = r + s - N$. If Z is an irreducible component of $V_1 \cap V_2$, then V_1 and V_2 *intersect transversally at* Z if $m(Z; V_1 \cap V_2, \mathbb{P}^N) = 1$.

If V_1 and V_2 intersect properly, then every irreducible component of $V_1 \cap V_2$ has dimension $r + s - N$ so that $V_1 \cap V_2$ is pure $r + s - N$ dimensional.

We now have:

Theorem 18.17 (Bezout's Theorem) *Let V_1 and V_2 be subvarieties of $\mathbb{P}^N$ with $\dim V_1 = r$, $\dim V_2 = s$. Suppose that V_1 and V_2 intersect properly and that $Z_1, \ldots, Z_t$ are the irreducible components of $V_1 \cap V_2$. Then*

$$\sum_{j=1}^{t} m(Z_j; V_1 \cap V_2, \mathbb{P}^N) \deg Z_j = (\deg V_1)(\deg V_2). \tag{18.18}$$

Proof. Let $(X_0, \ldots, X_N, Y_0, \ldots, Y_N)$ be homogeneous coordinates on $\mathbb{P}^{2N+1}$. Let $\mathbb{P}_1^N = L_{1,N} = V(Y_0, \ldots, Y_N)$ and $\mathbb{P}_2^N = L_{2,N} = V(X_0, \ldots, X_N)$. Then $V_1 \subset \mathbb{P}_1^N$ and $V_2 \subset \mathbb{P}_2^N$ (or better isomorphic images). Consider the join, $J(V_1, V_2)$, of V_1 and V_2 in $\mathbb{P}^{2N+1}$. Then (Example 12.26) $\deg J(V_1, V_2) = (\deg V_1)(\deg V_2)$. Let Δ be the linear subspace (the "diagonal") defined by $X_i - Y_i = 0$, $i = 0, 1, \ldots, N$ i.e., $\Delta = V(X_0 - Y_0, \ldots, X_N - Y_N)$. Then $\Delta \simeq \mathbb{P}^N$ and

$$V_1 \cap V_2 = \Delta \cap J(V_1, V_2). \tag{18.19}$$

In other words, $\deg(V_1 \cap V_2) = \deg(\Delta \cap J(V_1, V_2))$. Since Δ is an intersection of hyperplanes, it will be enough to show that if L is a hyperplane and V is a variety, then

$$\deg(L \cap V) = (\deg L)(\deg V) = \deg V = \sum_{j=1}^{t} m(Z_j; L \cap V, \mathbb{P}^N) \deg Z_j.$$

Now, either $V \subset L$ and $L \cap V = V = Z_1$ and $\deg V = \deg Z_1$ or, $V \not\subset L$ and the result follows from the hypersurface case. Thus,

$$\begin{aligned} \deg(V_1 \cap V_2) &= \sum_{j=1}^{t} m(Z_j; V_1 \cap V_2, \mathbb{P}^N) \deg Z_j \\ &= \deg(L \cap J(V_1, V_2)) = (\deg V_1)(\deg V_2). \end{aligned}$$

Corollary 18.20 (Classical Bezout's Theorem) *Let C_1, C_2 be distinct curves in $\mathbb{P}^2$ of degrees d_1, d_2 respectively. Let $C_1 \cap C_2 = \{P_1, \ldots, P_t\}$. Then*

$$\deg(C_1 \cap C_2) = \sum_{j=1}^{t} m(P_i; C_1 \cap C_2, \mathbb{P}^2) = d_1 d_2. \tag{18.21}$$

Corollary 18.22 *Let $\psi_{\mathbf{F}}(x_0, x_1) \in \mathrm{Rat}(n, m, p)$ and let $\gamma_{\mathbf{F}} = \psi_{\mathbf{F}}(\mathbb{P}^1)$. Let $N = \binom{m+p}{m} - 1$ and $H_N = V(Y_N)$ in $\mathbb{P}^N$. Then*

$$\deg(\gamma_{\mathbf{F}} \cap H_N) = \sum_{j=1}^{t} m(P_j; \gamma_{\mathbf{F}} \cap H_N, \mathbb{P}^N) = n \tag{18.23}$$

and

$$[\gamma_{\mathbf{F}} \cdot H_N] = \sum_{j=1}^{t} m(P_j; \gamma_{\mathbf{F}} \cap H_N, \mathbb{P}^N) P_j \tag{18.24}$$

where $\gamma_{\mathbf{F}} \cap H_N = \{P_1, \dots, P_t\}$ and

$$m(P_j; \gamma_{\mathbf{F}} \cap H_N, \mathbb{P}^N) = \ell(\mathfrak{o}_{P,\mathbb{P}^N} / I(\gamma_{\mathbf{F}} \cap H_N)\mathfrak{o}_{P_j,\mathbb{P}^N}).$$

While this is satisfactory for our current purposes, some additional examples seem worthwhile.

Example 18.25 Let $C = V(f)$ be an irreducible curve in $\mathbb{P}^2$ with degree $f = d$. Let $P = \xi = (\xi_0, \xi_1, \xi_3)$ be a point of C. Assume that $\xi_0 \neq 0$ so that $P = (1, s, t) \in \mathbb{A}^2$ $(= \mathbb{P}^2 - V(Y_0))$. We can change coordinates so that $P = (0,0)$ in $\mathbb{A}^2$ and $f = f(x,y) = f_0(x,y) + f_1(x,y) + \cdots + f_d(x,y)$ with f_i homogeneous of degree i. Since $(0,0) \in C$, $f_0(x,y) \equiv 0$. Let us call $m(P;C) = \min\{i \colon f_i \not\equiv 0\}$, the *multiplicity of P on C*. Let $f_1(x,y) = a_1 x + b_1 y$. Then $(\partial f_1/\partial x)(0,0) = a_1$ and $(\partial f_1/\partial y)(0,0) = b_1$. If $f_1 \not\equiv 0$, then P is a simple point of C and conversely. Thus, P is a simple point of C if and only if $m(P;C) = 1$. Let $L = V(ax + by)$ be a line through P. Then $m(P;L) = 1$. Suppose that $m(P;C) = \mu \geq 1$ and let $L_1, \dots, L_\mu$ be the linear factors of $f_\mu(x,y)$. If L is distinct from the L_i, then $V(L,f) = \{P\}$ and, by the Nullstellensatz, $(x,y)^t \subset (L,f) \subset (x,y)$ for some t (in fact for $t \geq \mu$). Let $\mathfrak{o} = \mathfrak{o}_{P,\mathbb{A}^2} = k[X,Y]_M$, $M = (X,Y)$ and let $\mathfrak{m} = \mathfrak{m}_{P,\mathbb{A}^2}$ so that $(\mathfrak{o}, \mathfrak{m})$ is a regular local ring with $\mathfrak{m} = (x,y)$. Since $\mathfrak{m}^\mu \subset (L,f)$, the natural map $\pi \colon \mathfrak{o}/(L,f) \to \mathfrak{o}/(\mathfrak{m}^{\mu+1}, L, f)$ is an isomorphism and so, we have $\dim_k \mathfrak{o}/(L,f) = \dim_k \mathfrak{o}/(\mathfrak{m}^{\mu+1}, L, f)$. But $V(\mathfrak{m}^{\mu+1}, L, f) = \{P\}$ implies $\mathfrak{o}/(\mathfrak{m}^{\mu+1}, L, f)$ is isomorphic to $k[X,Y]/(M^{\mu+1}, L, F)$ so that $\dim_k \mathfrak{o}/(L,f) = \dim_k k[X,Y]/(M^{\mu+1}, L, F)$ where $F = F(X,Y) = f_\mu(X,Y) + \cdots + f_d(X,Y)$. Consider the exact sequence

$$\frac{k[X,Y]}{M} \times \frac{k[X,Y]}{M^\mu} \xrightarrow{\varphi_1} \frac{k[X,Y]}{M^{\mu+1}} \xrightarrow{\varphi_2} \frac{k[X,Y]}{(M^{\mu+1}, L, F)} \longrightarrow 0$$

where φ_2 is the natural map and φ_1 is given by

$$\varphi_1(\overline{G}, \overline{H}) = \overline{HL + GF}.$$

If $\varphi_1(\overline{G},\overline{H}) = 0$, then $\deg(HL+GF) \geq \mu+1$. If $G = G_r + \cdots$, $H = H_s + \cdots$, then $HL+GF = H_sL+G_rF_\mu + \cdots$ so that $s+1 = r+\mu$ and $H_sL+G_rF_\mu = 0$. Since L, F_μ have no common factors, $s \geq \mu$ and $r \geq 1$ so that $\overline{G} = 0$, $\overline{H} = 0$. In other words, φ_1 is injective. It follows that $\dim_k k[X,Y]/(M^{\mu+1}, L, F) = \dim_k k[X,Y]/M^{\mu+1} - \dim_k k[X,Y]/M - \dim_k k[X,Y]/M^\mu = \frac{(\mu+1)(\mu+2)}{2} - 1 - \frac{\mu(\mu+1)}{2} = \mu$. So $m(P;C) = \dim_k \mathfrak{o}/(L,f)$ for almost all L.

Example 18.26 ([F-4]) Let $C_1 = V(F_1)$, $C_2 = V(F_2)$ be distinct irreducible curves in $\mathbb{P}^2$ of degrees d_1, d_2 respectively. C_1 and C_2 intersect properly with $C_1 \cap C_2 = \{P_1, \ldots, P_t\}$. Let $P = P_1$ (say) and consider $m(P; C_1 \cap C_2, \mathbb{P}^2) = \ell(\mathfrak{o}/(f_1,f_2))$ where $\mathfrak{o} = \mathfrak{o}_{P,\mathbb{P}^2}$ and the f_i are local equations of the C_i at P. We can, by a change of coordinates assume that $P = (0,0)$. Let $m_i = m(P;C_i)$. We claim that $m(P;C_1 \cap C_2, \mathbb{P}^2) \geq m_1m_2$ and that equality holds if and only if C_1, C_2 have distinct tangent directions at P i.e., if and only if f_{1m_1}, f_{2m_2} have no common linear factors. Let $M = (X,Y)$ and $\mathfrak{m} = \mathfrak{m}_{P,\mathbb{P}^2} = (x,y)$ so that $(\mathfrak{o},\mathfrak{m})$ is a regular local ring. Since P is an isolated point of $C_1 \cap C_2$ and f_1, f_2 are local equations of C_1, C_2 at P, we have $\mathfrak{m}^t \subset (f_1,f_2)$ for some t. If C_1, C_2 have distinct tangent directions at P, then $\mathfrak{m}^t \subset (f_1,f_2)$ for $t \geq m_1+m_2-1$ and the natural map $\pi\colon \mathfrak{o}/(f_1,f_2) \to \mathfrak{o}/(\mathfrak{m}^{m_1+m_2}, f_1, f_2)$ is an isomorphism. Again consider the exact sequence

$$\frac{k[X,Y]}{M^{m_1}} \times \frac{k[X,Y]}{M^{m_2}} \xrightarrow{\varphi_1} \frac{k[X,Y]}{M^{m_1+m_2}} \xrightarrow{\varphi_2} \frac{k[X,Y]}{(M^{m_1+m_2}, F_1, F_2)} \longrightarrow 0$$

where φ_2 is the natural map and φ_1 is given by

$$\varphi_1(\overline{G},\overline{H}) = \overline{HF_1 + GF_2}.$$

Since $V(M^{m_1+m_2}, F_1, F_2) = \{P\}$, φ_2 is surjective and $k[X,Y]/(M^{m_1+m_2}, F_1, F_2)$ is isomorphic to $\mathfrak{o}/(\mathfrak{m}^{m_1+m_2}, f_1, f_2)$. It follows that

$$\begin{aligned} m(P;C_1 \cap C_2, \mathbb{P}^2) &= \dim_k \mathfrak{o}/(f_1,f_2) \\ &\geq \dim_k \mathfrak{o}/(\mathfrak{m}^{m_1+m_2}, f_1, f_2) = \dim_k \frac{k[X,Y]}{(M^{m_1+m_2}, F_1, F_2)} \end{aligned}$$

and that

$$\begin{aligned} \dim_k \frac{k[X,Y]}{(M^{m_1+m_2}, F_1, F_2)} &= \dim_k \frac{k[X,Y]}{M^{m_1+m_2}} - \dim_k \mathrm{Ker}\, \varphi_1 \\ \dim_k \frac{k[X,Y]}{M^{m_1}} + \dim_k \frac{k[X,Y]}{M^{m_2}} &\geq \dim_k \mathrm{Ker}\, \varphi_1. \end{aligned}$$

Thus, we have

$$\begin{aligned} m(P; C_1 \cap C_2, \mathbb{P}^2) &\geq \dim_k \frac{k[X,Y]}{M^{m_1+m_2}} - \dim_k \frac{k[X,Y]}{M^{m_1}} - \dim_k \frac{k[X,Y]}{M^{m_2}} \\ &\geq \frac{(m_1+m_2)(m_1+m_2+1)}{2} - \frac{m_1(m_1+1)}{2} - \frac{m_2(m_2+1)}{2} \\ &\geq m_1 m_2. \end{aligned}$$

If f_1, f_2 have distinct tangents, then φ_1 is injective (arguing as in the previous example) and $\mathfrak{o}/(f_1, f_2)$ is isomorphic to $\mathfrak{o}/(\mathfrak{m}^{m_1+m_2}, f_1, f_2)$ so that

$$\begin{aligned} \dim_k \mathfrak{o}/(f_1, f_2) &= \dim_k \mathfrak{o}/(\mathfrak{m}^{m_1+m_2}, f_1, f_2) \\ &= \dim_k k[X,Y]/(M^{m_1+m_2}, F_1, F_2) \\ &= \dim_k k[X,Y]/M^{m_1+m_2} - \dim_k k[X,Y]/M^{m_1} - \dim_k k[X,Y]/M^{m_2} \\ &= m_1 m_2. \end{aligned}$$

On the other hand, if $m(P; C_1 \cap C_2, \mathbb{P}^2) = m_1 m_2$, then φ_1 is injective and f_1, f_2 have no common tangents at P. For if L were such a tangent, then $F_{1m_1} = LF_1'$, $F_{2m_2} = LF_2'$ with $\deg F_i' = m_i - 1$ and $\varphi_1(\overline{F}_1' - \overline{F}_2') = 0$.

Example 18.27 Let $V_1 = V(X_1, X_2)$, $V_2 = V(X_3, X_4)$ in $\mathbb{A}^4$. Then V_1, V_2 are the irreducible components of $V = V_1 \cup V_2$ and $I(V) = I(V_1) \cap I(V_2) = (X_1, X_2) \cap (X_3, X_4) = (X_1X_3, X_1X_4, X_2X_3, X_2X_4)$. Let $Z = V(X_1 - X_3, X_2 - X_4)$. Then $Z \cap V = (Z \cap V_1) \cup (Z \cap V_2) = \{(0,0,0,0)\}$. Since the notion of intersection multiplicity should be additive, we should have $m(P; Z \cap V, \mathbb{A}^4) = m(P; Z \cap V_1, \mathbb{A}^4) + m(P; Z \cap V_2, \mathbb{A}^4) = 2$ (for $P = (0,0,0,0)$). But $m(P; Z \cap V, \mathbb{A}^4) = \dim_k \mathfrak{o}_{P,\mathbb{A}^4}/I(Z \cap V)\mathfrak{o}_{P,\mathbb{A}^4} = \dim_k k[X_1, X_2, X_3, X_4]/(X_1X_3, X_1X_4, X_2X_3, X_2X_4, X_1 - X_3, X_2 - X_4) = \dim_k k[u,v]/(u^2, uv, v^2) = 3$. [Thus, for a truly general theory Definition 18.5 is not adequate.]

Example 18.28 Let V be the surface in $\mathbb{A}^4$ given parametrically by $\gamma(s,t) = (s^4, s^3t, st^3, t^4)$ so that $I(V) = (x_1x_4 - x_2x_3, x_1^2x_3 - x_2^3, x_2x_4^2 - x_3^3, x_2^2x_4 - x_3^2x_1)$. Let $P = (0,0,0,0)$ and let $V_1 = V(X_1)$, $V_2 = V(X_4)$. Then $W = V_1 \cap V_2 = \{P\}$ in V and $m(P; W, V) = \dim_k \mathfrak{o}_{P,V}/I(V)\mathfrak{o}_{P,V} = \dim_k k[x_1, x_2, x_3, x_4]/(x_1x_4 - x_2x_3, x_1^2x_3 - x_2^3, x_2x_4^2 - x_3^3, x_2^2x_4 - x_3^2x_1, x_1, x_4) = \dim_k k[u,v]/(u^3, uv, u^2v, uv^2, v^3) = 5$. However, if $x_1 = s^4$ and $x_4 = t^4$, then the multiplicity should be 4.

If V is a nonsingular variety of dimension r, if $H_1, \ldots, H_r$ are hypersurfaces in V which intersect properly (so $\dim H_1 \cap \ldots \cap H_r = 0$), if P is a point of $H_1 \cap \cdots \cap H_r$, and if $\mathfrak{a}_{H_1 \cap \cdots \cap H_r} = (h_1, \ldots, h_r)$ in $\mathfrak{o}_{P,V}$ with h_i a local equation of H_i in $\mathfrak{o}_{P,V}$, then

$$\begin{aligned} i(P; H_1, \ldots, H_r, V) &= \dim_k \mathfrak{o}_{P,V}/(f_1, \ldots, f_r) \\ &= \dim_k \mathfrak{o}_{P,V}/\mathfrak{a}_{H_1 \cap \cdots \cap H_r} \mathfrak{o}_{P,V} \end{aligned} \tag{18.29}$$

is a reasonable definition of the intersection multiplicity (V is then a local complete intersection at P) and agrees with the geometric multiplicity.

Let V be a nonsingular variety of dimension r and let $D_1, \ldots, D_r$ be divisors on V. Suppose that

$$D_i = \sum n_{ij} \Gamma_j \tag{18.30}$$

and let $X = \bigcap_{i=1}^{r} \operatorname{Supp} D_i$. If $\xi \in X$, then we let $\dim_\xi X = \max\{\dim Z_\ell \colon \xi \in Z_\ell$ and Z_ℓ a component of $X\}$.

Definition 18.31 $D_1, \ldots, D_r$ *intersect properly at* ξ if $\dim_\xi X = 0$ i.e., ξ is an isolated point of X. $D_1, \ldots, D_r$ *intersect properly* if $\dim X = 0$ (i.e., the D_i intersect properly at every $\xi \in X$).

If $D_1, \ldots, D_r$ are effective and intersect properly at ξ, then there are local equations f_i of the D_i at ξ for $i = 1, \ldots, r$ and $(f_1, \ldots, f_r) \supset \mathfrak{m}_\xi^t$ for some t (by the Nullstellensatz) where $(\mathfrak{o}_\xi, \mathfrak{m}_\xi)$ is the local ring of ξ on V. It follows that $\dim_k \mathfrak{o}_\xi/(f_1, \ldots, f_r)\mathfrak{o}_\xi$ is finite and does not depend on the choice of local equations.

Definition 18.32 The *intersection multiplicity of divisors* $D_1, \ldots, D_r$ *at* ξ, is given by

$$i(\xi\,; D_1, \ldots, D_r) = \dim_k(\mathfrak{o}_\xi/(f_1, \ldots, f_r)\mathfrak{o}_\xi). \tag{18.33}$$

Suppose that (say) $D_1 = D_{1+} - D_{1-}$ is not necessarily effective, then we set

$$i(\xi; D_1, \ldots, D_r) = i(\xi; D_{1+}, D_2, \ldots, D_r) - i(\xi; D_{1-}, D_2, \ldots, D_r) \tag{18.34}$$

and so can define the intersection multiplicity for *any* divisors $D_1, \ldots, D_r$ which intersect properly at ξ. If $\Gamma_1, \ldots, \Gamma_r$ are prime divisors, then $i(\xi; \Gamma_1, \ldots, \Gamma_r) = 1$ if and only if $(f_1, \ldots, f_r) = \mathfrak{m}_\xi$ ($\Gamma_i = V(f_i)$ locally) i.e., if and only if $\Gamma_1, \ldots, \Gamma_r$ intersect transversally at ξ. It can also be shown ([S-2]) that if $D_1, D_2, \ldots, D_r$ and $D_1', D_2, \ldots, D_r$ intersect properly and $D_1 = D_1' + (f)$ (i.e., $D_1 \sim D_1'$), then

$$\sum_{\xi \in \cap \operatorname{Supp} D_i} i(\xi; D_1, \ldots, D_r) = \sum_{\xi \in \operatorname{Supp} D_1' \cap \bigcap_{i=1}^{r} \operatorname{Supp} D_i} i(\xi; D_1', D_2, \ldots, D_r) \tag{18.35}$$

or, equivalently,

$$\sum i(\xi; (f), D_2, \dots, D_r) = 0 \tag{18.36}$$

when $(f), D_2, \dots, D_r$ intersect properly. This allows us to define an *intersection index* $(D_1 \cdot D_2 \cdot \dots \cdot D_r)$ of elements of $\operatorname{Pic}(V)$ by setting

$$(D_1 \cdot D_2 \cdot \dots \cdot D_r) = \sum_{\xi \in \cap \operatorname{Supp} D_i'} i(\xi; D_1', D_2', \dots, D_r') \tag{18.37}$$

where $D_i' \sim D_i$ and the D_i' intersect properly. The existence of such D_i' can be shown by induction using the fact that given a divisor D, then $D \sim \lambda H$ where H is a hyperplane so that if $\xi \in \operatorname{Supp} D$, there is a $D' \sim D$ with $\xi \notin \operatorname{Supp} D'$. For instance, if $V = \mathbb{P}^2$ and L is a line, then $(L \cdot L) = 1$ since there are distinct lines L_1, L_2 with $L_1 \sim L$, $L_2 \sim L$ and $(L_1 \cdot L_2) = 1$ (as $L_1 \cap L_2 = \xi$ a unique point). Similarly, if $V = V(F(Y_0, Y_1, Y_2, Y_3))$ is a smooth hypersurface in $\mathbb{P}^3$ of degree m, then $(\Theta \cdot \Theta) = m$ where Θ is a hyperplane section of V.

For much more on intersections, we refer to [F-5].

19

State Feedback

We now turn our attention to the study of state feedback. Let $G_n = \mathrm{GL}(n,k)$, $G_m = \mathrm{GL}(m,k)$ and $H_{mn} = \mathrm{Hom}_k(\mathbb{A}^n, \mathbb{A}^m) \simeq \mathbb{A}^{nm} \simeq M(m,n,k)$. Consider the set $\Gamma_f(n,m) = G_n \times H_{mn} \times G_m$ and *define* a multiplication in $\Gamma_f(n,m)$ as follows:

$$[g, K, \alpha][g_1, K_1, \alpha_1] = [gg_1, Kg_1 + \alpha K_1, \alpha\alpha_1] \tag{19.1}$$

where $g, g_1 \in G_n$, $K, K_1 \in H_{mn}$, and $\alpha, \alpha_1 \in G_m$. Then

Theorem 19.2 (i) *$\Gamma_f(n,m)$ is a group and a quasi-affine variety;* (ii) *the mappings $\mu\colon \Gamma_f(n,m) \times \Gamma_f(n,m) \to \Gamma_f(n,m)$ and $i\colon \Gamma_f(n,m) \to \Gamma_f(n,m)$ given by $\mu(\gamma,\gamma_1) = \gamma\gamma_1$ and $i(\gamma) = \gamma^{-1}$ are morphisms;* (iii) *the mapping $\tau\colon \Gamma_f(n,m) \to \mathrm{GL}(n+m,k)$ given by*

$$\tau([g,K,\alpha]) = \begin{bmatrix} g & O_{n,m} \\ K & \alpha \end{bmatrix} \tag{19.3}$$

is an isomorphism onto a closed subgroup $\tilde{\Gamma}_f(n,m)$; (iv) *$\tilde{\Gamma}_f(n,m)$ is a linear subvariety (hence nonsingular) of dimension $n^2 + nm + m^2$; and* (v) *$\Gamma_f(n,m)$ acts on $\mathbb{A}^{n^2+n(m+p)}$ via the morphism φ given by*

$$\varphi([g,K,\alpha],(A,B,C)) = (g(A - B\alpha^{-1}K)g^{-1}, gB\alpha^{-1}, Cg^{-1}) \tag{19.4}$$

or, equivalently, $\tilde{\Gamma}_f(n,m)$ acts via the morphism $\tilde{\varphi}$ given by

$$\tilde{\varphi}\left(\begin{bmatrix} g & 0 \\ K & \alpha \end{bmatrix}, (A,B,C)\right) = \begin{bmatrix} g & O_{n,p} \\ O_{p,n} & I_p \end{bmatrix} \begin{bmatrix} A & B \\ C & O_{p,m} \end{bmatrix} \begin{bmatrix} g^{-1} & O \\ -\alpha^{-1}Kg^{-1} & \alpha^{-1} \end{bmatrix}. \tag{19.5}$$

P. Falb, *Methods of Algebraic Geometry in Control Theory: Part II*,
Modern Birkhäuser Classics, https://doi.org/10.1007/978-3-319-96574-1_19

Proof. Entirely analogous to the proof of Theorem I.21.2.

Definition 19.6 $\Gamma_f(n,m)$ is the *state feedback group* (of type (n,m)) and the action of $\Gamma_f(n,m)$ on $\mathbb{A}^{n^2+n(m+p)}$ is called *state feedback.*

Proposition 19.7 *If* $\mathbf{Y}(A,B,C) = [B\ AB \cdots A^{n-1}B]$, *then the rank of* $\mathbf{Y}$, $\rho(\mathbf{Y})$, *is invariant under state feedback and so controllability is invariant under state feedback.*

Proof. $\mathbf{Y}(\gamma \cdot x) = g[B\ (A - B\alpha^{-1}KB)B \cdots]\alpha^{-1}$ so that range $\mathbf{Y}(\gamma \cdot x) = g[\text{range}\,\mathbf{Y}(x)]\alpha^{-1}$.

Observability and minimality are, of course, not invariant.

Proposition 19.8 *Let* $N_f = \{[I,K,I] \in \Gamma_f(n,m)\}$ *and let* $G_f = \{[g,0,\alpha] \in \Gamma_f(n,m)\}$. *Then* (i) N_f *is a normal subgroup;* (ii) G_f *is a subgroup which acts on* N_f *via inner automorphisms; and* (iii) $\Gamma_f = N_f G_f = G_f N_f$.

Proof. Simply note that $\gamma = [g,K,\alpha] = [I,Kg^{-1},I] \cdot [g,0,\alpha] = [g,0,\alpha] \cdot [I,\alpha^{-1}K,I]$ and that $[g,K,\alpha] \cdot [I,K,I][g^{-1},-\alpha^{-1}Kg^{-1},\alpha^{-1}] = [I,\alpha^{-1}Kg^{-1},I]$.

Now let $x = (A,B,C)$ and consider the map $\Delta(\gamma,x)\colon \Gamma_f(n,m) \times \mathbb{A}^{n^2+n(m+p)} \to$ polynomials of degree n given by

$$\Delta(\gamma,x) = \det(zI - g(A - B\alpha^{-1}K)g^{-1}). \tag{19.9}$$

Then $\Delta(\gamma_e,x) = \det(zI - A)$, $\Delta(\gamma_e, \gamma \cdot x) = \Delta(\gamma,x)$, and

$$\Delta(\gamma_1\gamma_2, x) = \Delta(\gamma_1, \gamma_2 \cdot x) \tag{19.10}$$

for all $\gamma_1, \gamma_2 \in \Gamma_f(n,m)$. If $\gamma_g \in G_f$, then $\Delta(\gamma_g,x) = \Delta(\gamma_e,\gamma_g \cdot x) = \det(zI - A) = \Delta(\gamma_e,x)$ so that Δ is a G_f-invariant. Thus, the range $\Delta(\Gamma_f(n,m),x)$ is the range of $\Delta(N_f,x)$ which is the range of $\Delta(\gamma_K,x)$ where $\gamma_K = [I,K,I]$ runs through N_f. If we view the (monic) polynomials of degree n as the vector space of their coefficients, then we have the map $\Delta_x\colon H_{mn} \to \mathbb{A}_k^n$ given by

$$\Delta_x(K) = \Delta(K,x) = (\chi_i(A - BK)) \tag{19.11}$$

and we want to know when it is surjective. In fact, we shall prove the following:

Theorem 19.12 (Coefficient Assignment Theorem) $\Delta_x(\cdot) = \Delta(\cdot,x)$ *is surjective if and only if* x *is controllable).*

Before proving the theorem, we note that we cannot directly use the method of Part I, Chapter 21. We start with Heymann's Lemma.

Lemma 19.13 (Heymann's Lemma, Theorem I.23.38) *The pair (A, B) is controllable if and only if there is $\gamma_{K_0} = [I, K_0, I] \in N_f$ and $w \in \text{Hom}_k(\mathbb{A}^m, k) = M(m, 1, k)$ such that $(A + BK_0, Bw)$ is controllable (or, equivalently, there is $\gamma_{K_0} = [I, K_0, I] \in N_f$ and $b \in \text{Col}(B)$ such that $(A + BK_0, b)$ is controllable).*

Let us, for the moment, assume the lemma. Then, we have:

Proof (of Theorem 19.12). If x is controllable, then $(A - BK_0, Bw) = x_1$ is a controllable scalar system. Then $\Delta(K_1, x_1) = (\chi_i(A - BK_0 - BwK_1))$ with $K_1 \in M(1, n, k)$ is surjective. But $\chi_i(A - BK_0 - BwK_1) = \chi_i(A - B(K_0 + wK_1))$ and $wK_1 \in M(m, n, k)$ so that $\Delta(\cdot, x)$ is surjective. If, on the other hand, $\Delta(\cdot, x)$ is surjective, then there is a K_0 with $A + BK_0$ having distinct eigenvalues and $\Delta(\cdot, \gamma_{K_0} \cdot x)$ surjective. So we may suppose A is diagonal with distinct eigenvalues. Then (A, B) is controllable if and only if there is a $w \in \mathbb{A}^m$ with $w \notin \bigcup_{i=1}^{n} \text{Ker } B^i$. If $V_j = \text{Ker } B^j$ and if *all* $B^j \neq 0$, then done. If, say, $B^1 = 0$, then

$$A = \begin{bmatrix} \lambda_1 & & & 0 \\ & \lambda_2 & & \\ & & \ddots & \\ 0 & & & \lambda_n \end{bmatrix}, \quad A + BK = \begin{bmatrix} A^1 + B^1 K \\ \cdots\cdots \\ \tilde{A} + \tilde{B}K \end{bmatrix} = \begin{bmatrix} \lambda_1 \; 0 \cdots 0 \\ \tilde{A} + \tilde{B}K \end{bmatrix} \tag{19.14}$$

and there is a fixed pole at λ_1 so $\Delta(\cdot, x)$ is not surjective. Alternatively, if $\sigma(A + BK_0) = \{\lambda_1, \ldots, \lambda_n\}$, λ_i distinct, then there are eigenvectors $v_1, \ldots, v_n$ with $(A + BK_0)v_i = \lambda_i v_i$ and

$$v_i = (\lambda_i I - A)^{-1} BK_0 v_i. \tag{19.15}$$

Since

$$(\lambda I - A)^{-1} = \sum \phi_j(\lambda) A^{j-1} / \det[\lambda I - A],$$

we get

$$\det[\lambda I - A] v_i = \sum_{j=1}^{n} \phi_j(\lambda) A^{j-1} BK_0 v_i$$

and (for $\lambda \neq \lambda_i$) $v_1, \ldots, v_n \in \text{range } \mathbf{Y}(A, B, C)$.

Now, we have established in Part I (I.23.39 and I.23.40) that if $(A + BK_0, Bw)$ is controllable, then so is (A, B). Consider now the map $\phi_x \colon \Gamma_f(n, m) \to M(n, n)$ given by

$$\phi_x([g, K, \alpha]) = g(A + B\alpha^{-1}K)g^{-1}. \tag{19.16}$$

Then ϕ_x is a morphism and $\Gamma_f(n, m)$ and $M(n, n)$ are non-singular varieties. If ϕ_x is dominant, then $\phi_x(\Gamma_f(n, m)) \supset U$ open in $M(n, n)$ and, since $M(n, n)_\Delta$

(Δ = discriminant) is open in $M(n,n)$, there is an element $A+BK_0$ with distinct eigenvalues.

Proposition 19.17 *ϕ_x is dominant if and only if x is controllable.*

Proof. Since ϕ_x may be viewed as a morphism of projective varieties, ϕ_x will be dominant if and only if ϕ_x is surjective. If ϕ_x is surjective, then $\Delta(\cdot, x)$ is surjective and by Theorem 19.12, x is controllable. If we show that ϕ_x is dominant if x is controllable, then ϕ_x will also be surjective. Now, in view of Theorem 15.31, we need only show that ϕ_x is smooth at $\gamma_e = [I, 0, I]$. Let $\gamma = [g, K, \alpha]$. Then

$$\begin{aligned}\phi_x(\gamma_e + t\gamma) &= (I+tg)(A+B(I+t\alpha^{-1})tK)(I+tg)^{-1} \\ &= [A+B(I+t\alpha^{-1})tK + tg(A+B(I+t\alpha^{-1})tK](I+tg)^{-1}. \quad (19.18)\end{aligned}$$

Now $(I+tg)^{-1} = I - tg + t^2g^2 - \cdots$ (viewed as an element of the completion $\mathfrak{o}^*_{\gamma_e}$ of $\mathfrak{o}_{\gamma_e}$) and so

$$\phi_x(\gamma_e + t\gamma) = A + tBK - tAg + tgA + \text{higher order.}$$

It follows that

$$(d\phi_x)_{\gamma_e}(\gamma) = gA - Ag + BK \quad (19.19)$$

and hence, that

$$(d\phi_x)^*_{\gamma_e}(M) = \operatorname{tr}((gA - Ag + BK)M). \quad (19.20)$$

Since $(d\phi_x)_{\gamma_e}$ will be surjective if and only if $(d\phi_x)^*_{\gamma_e}$ is injective, we see that ϕ_x is smooth at γ_e if and only if

$$\operatorname{tr}((AM - MA)g) = 0, \quad \operatorname{tr}(MBK) = 0 \quad (19.21)$$

for all g, K i.e., if and only if $AM - MA = 0$, $MB = 0$ i.e., if and only if $M\mathbf{Y}(A,B,C) = 0$ implies $M = 0$ i.e., if and only if $x = (A,B,C)$ is controllable.

Let $\mathcal{C}_*(n,m,p) = \{x = (A,B,C)$: x is controllable and C has full rank$\}$. $\mathcal{C}_*(n,m,p)$ is open in $\mathbb{A}^{n^2+n(m+p)}$ and is invariant under state feedback. Let $x = (A,B,C)$ be an element of $\mathcal{C}_*(n,m,p)$ and let $\mathbf{f}(x) = (f_1(x), \ldots, f_m(x))$ be the Hermite indices of x [i.e., $[b_1 \cdots A^{f_1-1}b_1 \cdots A^{f_n-1}b_n]$ span $\mathbf{Y}(A,B,C)$]. If $\gamma_g = [g, 0, \alpha] \in G_f$, then $x_1 = \gamma_g \cdot x = (gAg^{-1}, gB\alpha^{-1}, Cg^{-1})$ and, clearly, $\mathbf{Y}(x_1) = g\mathbf{Y}(x)\alpha^{-1}$.

Proposition 19.22 $\mathbf{f}(x) = \mathbf{f}(x_1)$ *so that the Hermite indices are invariant under G_f.*

Proof. Let $gB\alpha^{-1} = [\beta_1, \dots, \beta_m]$, $gAg^{-1} = A_1$. Then $[\beta_1 \cdots A_1^{f_1-1}\beta_1] = g[b_1 \cdots A^{f_1-1}b_1]\alpha^{-1}$ so that $f_1(x_1) \geq f_1(x)$. But $A^{f_1}b_1 = \sum \alpha_{j1}A^{f_1-j_1}b_1$ so that $gA^{f_1}b_1\alpha^{-1} = \sum \alpha_{j1}gA^{f_1-j_1}b_1\alpha^{-1}$. Hence

$$A_1^{f_1}\beta_1 = \sum \alpha_{j1}A_1^{f_1-j_1}\beta_1 \tag{19.23}$$

and $f_1(x_1) = f_1(x)$. Assume that $f_j(x) = f_j(x_1)$ for $j = 1, \dots, t-1$, $t = 2, \dots$. Then $g[b_1 \cdots A^{f_j-1}b_j]\alpha^{-1} = [\beta_1 \cdots A_1^{f_j-1}\beta_j]$ gives independence and an argument similar to that used to derive (19.23) gives dependence.

Example 19.24 Let

$$A = \begin{bmatrix} 0 & 0 & 0 \\ 0 & 0 & 1 \\ \alpha & \beta & \gamma \end{bmatrix}, \quad b_1 = \begin{bmatrix} 0 \\ 0 \\ 1 \end{bmatrix}, \quad b_2 = \begin{bmatrix} 1 \\ 0 \\ 0 \end{bmatrix}, \ C \text{ any.}$$

Then

$$\mathbf{Y}(A, B, C) = \begin{bmatrix} 0 & 0 & 0 & 1 & \\ 0 & 1 & \gamma & 0 & \cdots \\ 1 & \gamma & \beta + \gamma^2 & 0 & \end{bmatrix}$$

so $A^2b_1 = \beta b_1 + \gamma Ab_1$ and $\mathbf{f}(x) = (2, 1)$. Let

$$K = \begin{bmatrix} 0 & 0 & 0 \\ 0 & -1 & 0 \end{bmatrix}.$$

Then

$$A - BK = \begin{bmatrix} 0 & 1 & 0 \\ 0 & 0 & 1 \\ \alpha & \beta & \gamma \end{bmatrix} = A_1$$

and

$$\tilde{\mathbf{Y}}(A_1, B, C) = \begin{bmatrix} 0 & 0 & 1 & 1 & \\ 0 & 1 & \gamma & 0 & \cdots \\ 1 & \gamma & \alpha + \beta + \gamma^2 & 0 & \end{bmatrix}$$

so that $\mathbf{f}(x_1) = (3, 0) \neq \mathbf{f}(x)$. Thus, the Hermite indices are *not* invariant under N_f (a fortiori under $\Gamma_f(n, m)$).

So we must use the Kronecker indices of $\mathbf{Y}(x)$. In view of the analysis in Chapter 4, the Kronecker indices $\boldsymbol{\kappa}(x) = (\kappa_1, \dots, \kappa_m)$ correspond to determining the first n independent columns of $[B\ AB \cdots]$ and putting them in the form $[b_1 \cdots A^{\kappa_1-1}b_1\ b_2 \cdots A^{\kappa_m-1}b_m]$. We may assume $\rho(B) = r$ and that $b_1 \cdots b_r$ are independent. Then $\kappa_{r+1} = \cdots = \kappa_m = 0$ and $\sum \kappa_j = \sum_{j=1}^{r} \kappa_j = n$. The

Kronecker indices are again invariant under G_f (by a similar argument to Proposition 19.22).

Example 19.25 Let

$$A = \begin{bmatrix} 0 & 0 & 0 \\ 0 & 0 & 1 \\ \alpha & \beta & \gamma \end{bmatrix}, \quad b_1 = \begin{bmatrix} 0 \\ 0 \\ 1 \end{bmatrix}, \quad b_2 = \begin{bmatrix} 1 \\ 0 \\ 0 \end{bmatrix}, \ C \text{ any}$$

then

$$\mathbf{Y}(A,B,C) = \begin{bmatrix} b_1 \ b_2 \ Ab_1 \ Ab_2 \cdots \end{bmatrix} = \begin{bmatrix} 0 & 1 & 0 & \\ 0 & 0 & 1 & \cdots \\ 1 & 0 & \gamma & \end{bmatrix}$$

so that $\kappa_1 = 2$, $\kappa_2 = 1$ are the Kronecker indices. If K is given by

$$K = \begin{bmatrix} -k_1 & -k_2 & -k_3 \\ -k_4 & -k_5 & -k_6 \end{bmatrix}$$

then

$$A_1 = A - BK = \begin{bmatrix} k_1 & k_5 & k_6 \\ 0 & 0 & 1 \\ \alpha + k_1 & \beta + k_2 & \gamma + k_3 \end{bmatrix}$$

and

$$\mathbf{Y}(A_1,B,C) = \begin{bmatrix} b_1 \ b_2 \ A_1b_1 \ A_1b_2 \cdots \end{bmatrix} = \begin{bmatrix} 0 & 1 & k_6 & \\ 0 & 0 & 1 & \cdots \\ 1 & 0 & \gamma + k_3 & \end{bmatrix}$$

so that again the Kronecker indices are $\kappa_1 = 2$, $\kappa_2 = 1$. Note also that if $k_1 = -\alpha + \chi_1$, $k_2 = -\beta + \chi_2$, $k_3 = -\gamma + \chi_3$, then $\det(zI - (A - BK)) = z^3 - \chi_3 z^2 - \chi_2 z - \chi_1$ and *any* characteristic polynomial can be obtained by state feedback.

Proposition 19.26 *Let $B \in M(n,m)$ and let $L \in M(m,t)$. Then* $\mathbf{Col}(BL) \subset \mathbf{Col}(B)$.

Proof. $(BL)_j = B \cdot L_j$ and $L_j = (\lambda_j^i)$, $i = 1, \ldots, m$ so that

$$B \cdot L_j = \sum_{i=1}^{m} \lambda_j^i B_i \in \mathbf{Col}(B).$$

Corollary 19.27 *The Kronecker indices are invariant under state feedback.*

Proof. $(A - BK)^j B = A^j B + B \cdot L$ for some $L \in M(m, m)$.

Let $x = (A, B, C) \in \mathcal{C}_*(n, m, p)$ and let $\boldsymbol{\kappa}(x) = (\kappa_1, \dots, \kappa_t, 0, \dots, 0)$ be the Kronecker indices of x. Then $[b_1 \cdots A^{\kappa_1 - 1} b_1 \; b_2 \cdots A^{\kappa_t - 1} b_t]$ is nonsingular and (Chapter 4), there is a g_x in $\mathrm{GL}(n, k)$ such that $g_x \cdot x$ is in the controllable companion form with indices $\boldsymbol{\kappa}(x)$. Let

$$d_0 = 0, \quad d_j = \sum_{i=1}^{j} \kappa_j, \quad j = 1, \dots, t \tag{19.28}$$

so that

$$\begin{aligned} d_{j+1} - d_j &= \kappa_j, \quad j = 0, \dots, t-1 \\ d_t &= \sum_{j=1}^{t} \kappa_j = n. \end{aligned} \tag{19.29}$$

Then $g_x A g_x^{-1} = A_x$, $g_x B = B_x$, $C g_x^{-1} = C_x$ where $A_x = [(A_x)_j^i]$, $i, j = 1, \dots, t$ is in block companion form with $(A_x)_j^i$ a $\kappa_i \times \kappa_j$ matrix

$$(A_x)_i^i = \begin{bmatrix} 0 & 1 & & \\ 0 & 0 & 1 & \\ \vdots & \vdots & \ddots & \\ 0 & 0 & & 1 \\ a_{d_{i-1}+1}^{d_i} & \vdots & & a_{d_i}^{d_i} \end{bmatrix}, \quad (A_x)_j^i = \begin{bmatrix} O_{\kappa_i - 1, \kappa_j} \\ a_{d_{j-1}+1}^{d_i} \cdots a_{d_j}^{d_i} \end{bmatrix} \tag{19.30}$$

and

$$B_x = \begin{bmatrix} O_{\kappa_1 - 1, m} \\ B^1 \\ \vdots \\ O_{\kappa_t - 1, m} \\ B^t \end{bmatrix} = \begin{bmatrix} & & O_{\kappa_1 - 1, m} & & & \\ 1 & * & \cdots & * & & \\ 0 & 0 & \cdots & & & \\ \vdots & & & & & \\ 0 & \cdots & 1 & * & \cdots & * \end{bmatrix} \tag{19.31}$$

$$\begin{matrix} \uparrow \\ t \end{matrix}$$

(so that $\rho(B_x) = \rho(B) = t$). Note that B^i is the d_ith row of $g_x B$. Now, in general, we can find an $\alpha_x^{-1} \in \mathrm{GL}(m, k)$ such that

$$B_x \alpha_x^{-1} = \left[E_{\boldsymbol{\kappa}, t} \; *_{n, m-t} \right] \tag{19.32}$$

where

$$E_{\kappa,t} = \begin{bmatrix} O_{\kappa_1-1,t} \\ \epsilon^1 \\ O_{\kappa_2-1,t} \\ \epsilon^2 \\ \vdots \\ \epsilon^t \end{bmatrix} \tag{19.33}$$

is an $n \times t$-matrix. Let K be an element of $\mathrm{Hom}_k(\mathbb{A}^n, \mathbb{A}^m) = M(m,n)$ and let $K_n^t = [K^j]_{j=1,\dots,t}$ and $K_n^{m-t} = [K^j]_{j=t+1,\dots,m}$. Then

$$B_x\alpha_x^{-1}K = E_{\kappa,t}K_n^t + *_{n,m-t}K_n^{m-t}. \tag{19.34}$$

Setting $K_n^{m-t} = O_n^{m-t}$, we have

$$B_x\alpha_x^{-1}K = E_{\kappa,t}K_n^t = \begin{bmatrix} O_{\kappa_1-1,n} \\ K^1 \\ O_{\kappa_2-1,n} \\ \vdots \\ K^t \end{bmatrix}. \tag{19.35}$$

Then $A_x - B_x\alpha_x^{-1}K$ is a block matrix with d_ith row $A_x^{d_i} - K^i$. If we choose $K^i = A_x^{d_i}$, then $A_x - B_xK = [L_j^i]$ where L_j^i is $\kappa_i \times \kappa_j$ and

$$L_i^i = \begin{bmatrix} 0 & 1 & & \\ & & \ddots & \\ & & & 1 \\ 0 & & \cdots & 0 \end{bmatrix}, \quad L_j^i = O_{\kappa_j}^{\kappa_i} \quad (i \neq j). \tag{19.36}$$

Moreover, since the $n \times m - t$ matrix $*_{n,m-t}$ in (19.32) is of the form

$$*_{n,m-t} = \begin{bmatrix} O_{\kappa_1-1,m-t} \\ *^{d_1} \\ O_{\kappa_2-1,m-t} \\ \vdots \\ *^{d_t} \end{bmatrix} \tag{19.37}$$

we can take $\alpha_1^{-1} \in \mathrm{GL}(m,k)$ such that $B_x\alpha_x^{-1}\alpha_1^{-1} = [E_{\kappa,t}\ O_{n,m-t}]$ (i.e., we let

$$\alpha_1^{-1} = \begin{bmatrix} I_t & X_{t,m-t} \\ O_{m-t,t} & I_{m-t} \end{bmatrix}$$

where $X^i_{t,m-t} = -*^{d_i}$). Thus, if (A,B) is controllable, then the Kronecker indices are a complete invariant for the action of state feedback on C^n_m and if $A_{x,c} = [L^i_j]$ and $B_{x,c} = [E_{\kappa,t}\ O_{n,m-t}]$, then $(A_{x,c}, B_{x,c})$ is a canonical form for the action of state feedback on the space C^n_m of controllable systems. Since $B_{x,c} = [E_{\kappa,t}\ O_{n,m-t}]$, we have

$$C(zI - A_{x,c})^{-1}B_{x,c} = \left[C(zI - A_{x,c})E_{\kappa,t}\ \ O_{p,m-t}\right];$$

and, since $\rho(B) = t$ is invariant under state feedback, we may as well suppose that B is of full rank $m \leq n$.

We now want to examine the orbits $O_{\Gamma_f}(x) = O_{\Gamma_f}((A,B))$ for the action of $\Gamma_f(n,m)$ on C^n_m with $\rho(B) = m \leq n$.

Proposition 19.38 *If $B_{x,c}\alpha^{-1} = B_{x,c}$, then $\alpha = I_m$. If $gB_{x,c} = B_{x,c}$, then $g\epsilon_{d_i} = \epsilon_{d_i}$. If $\gamma_K = [I, K, I]$ and $\gamma_K \cdot (A_{x,c}, B_{x,c}) = (A_{x,c}, B_{x,c})$, then $K = 0$.*

Proof. Let $B^m_{x,c} = [B^{d_i}_{x,c}]^m_{i=1} = I_m$. Then $B_{x,c}\alpha^{-1} = B_{x,c}$ implies $B^m_{x,c}\alpha^{-1} = B^m_{x,c}$ and $\alpha = I_m$. The second assertion is obvious. Since $A_{x,c} - B_{x,c}K = A_{x,c}$, we have $B^m_{x,c}K = 0$ and $K = 0$.

Now $A_{x,c}$ = block-diagonal $[L^i_i]$ where L^i_i is a $\kappa_i \times \kappa_i$ matrix of the form (19.36). Let $g = [\mathbf{g}^i_j] \in \mathrm{GL}(n,k)$ where $\mathbf{g}^i_j$ is a $\kappa_i \times \kappa_j$ block. Then

$$gA_{x,c} = \begin{bmatrix} \mathbf{g}^1_1 L_{11} & \mathbf{g}^1_2 L_{22} & \cdots & \mathbf{g}^1_m L_{mm} \\ \vdots & \vdots & & \\ \mathbf{g}^m_1 L_{11} & & \cdots & \mathbf{g}^m_m L_{mm} \end{bmatrix}$$

$$A_{x,c}g = \begin{bmatrix} L_{11}\mathbf{g}^1_1 & L_{11}\mathbf{g}^1_2 & \cdots & L_{11}\mathbf{g}^1_m \\ \vdots & & & \\ L_{mm}\mathbf{g}^m_1 & \cdots & & L_{mm}\mathbf{g}^m_m \end{bmatrix}. \tag{19.39}$$

If $gB_{x,c} = B_{x,c}$, then the d_jth column of g is ϵ_j.

Lemma 19.40 (Lemma 1) *Let*

$$L = \begin{bmatrix} 0 & 1 & \cdots & 0 \\ 0 & \cdots & & 1 \\ 0 & \cdots & & 0 \end{bmatrix} \tag{19.41}$$

be $t \times t$ and let $h \in \mathrm{GL}(t,k)$ be such that $hL = Lh$ and with tth column $h_t = \epsilon_t$. Then $h = I_t$.

Proof. Note that

$$h = \begin{bmatrix} h_1^1 & \cdots & h_{t-1}^1 & 0 \\ \vdots & & & \vdots \\ h_1^t & \cdots & h_{t-1}^t & 1 \end{bmatrix}$$

$$hL = \begin{bmatrix} 0 & h_1^1 & \cdots & h_{t-1}^1 \\ 0 & h_1^2 & \cdots & h_{t-1}^2 \\ \vdots & \vdots & & \\ 0 & h_1^t & \cdots & h_{t-1}^t \end{bmatrix}, \quad Lh = \begin{bmatrix} h_1^2 & \cdots & h_{t-1}^2 & 0 \\ \vdots & & & \\ h_1^t & \cdots & h_{t-1}^t & 1 \\ 0 & & 0 & 0 \end{bmatrix}.$$

It follows that $h_1^t = \cdots = h_{t-1}^t = 0$, $h_1^2 = \cdots = h_1^t = 0$, and $h_{t-1}^1 = \cdots = h_{t-1}^{t-2} = 0$, $h_{t-1}^{t-1} = 1$ i.e.,

$$h = \begin{bmatrix} h_1^1 & h_2^1 & \cdots & 0 & 0 \\ 0 & \vdots & & 0 & 0 \\ \vdots & \vdots & & \vdots & \vdots \\ & h_2^{t-1} & & 1 & 0 \\ 0 & 0 & & 0 & 1 \end{bmatrix}$$

and the result holds by induction.

Lemma 19.42 (Lemma 2) *Let* $\mathbf{h} = (h_j^i)$ *be an* $r \times s$ *matrix with* s*th column* $\mathbf{h}_s = 0$ *and let* L_r, L_s *be* $r \times r$, $s \times s$ *matrices of the form (19.41). If* $\mathbf{h}L_s = L_r\mathbf{h}$, *then* $\mathbf{h} = 0$.

Proof. (Exercise).

Corollary 19.43 *If* $gA_{x,c}g^{-1} = A_{x,c}$, *then* $g = I$ *and*

$$S_{\mathrm{GL}(n,k)\times\mathrm{GL}(m,k)}(x_c) = \{(I, 0, I)\}.$$

Example 19.44 Let $m = 1$, $n = 3$. Then K is 1×3, B is 3×1 and $\mathrm{GL}(1,k) = k^*$. Let

$$A_0 = \begin{bmatrix} 0 & 1 & 0 \\ 0 & 0 & 1 \\ 0 & 0 & 0 \end{bmatrix}, \quad b_0 = \epsilon_3 = \begin{bmatrix} 0 \\ 0 \\ 1 \end{bmatrix}.$$

If $\gamma = [g, K, \alpha] \in \Gamma_f(3,1)$, what conditions insure that $\gamma \cdot (A_0, b_0) = (A_0, b_0)$? We have $gb_0\alpha^{-1} = b_0$ and $g(A_0 - b_0\alpha^{-1}K)g^{-1} = A_0$. Let $g = (g_j^i)$. Then $gb_0\alpha^{-1} = b_0 = \epsilon_3$ implies $g_3^1 = 0$, $g_3^2 = 0$, $g_3^3 = \alpha$. Then

$$gA_0 = \begin{bmatrix} 0 & g_1^1 & g_2^1 \\ 0 & g_1^2 & g_2^2 \\ 0 & g_1^3 & g_2^3 \end{bmatrix}, \quad A_0 g = \begin{bmatrix} g_1^2 & g_2^2 & 0 \\ g_1^3 & g_2^3 & \alpha \\ 0 & 0 & 0 \end{bmatrix},$$

$$gb_0\alpha^{-1}K = \begin{bmatrix} 0 & 0 & 0 \\ 0 & 0 & 0 \\ \alpha k_1 & \alpha k_2 & \alpha k_3 \end{bmatrix}$$

and $gA_0 - gb_0\alpha^{-1}K = A_0 g$ together imply $g_1^2 = 0$, $g_2^1 = 0$, $g_1^3 = 0$, $g_2^3 = 0$, $K_1 = 0$, $K_2 = 0$, $K_3 = 0$ and so $g = \alpha I$. Thus, $S_{\Gamma_f}((A_0, b_0)) = \{[\alpha I, 0, \alpha]\}$. (In fact, this is true for any n.)

Example 19.45 Let $m = 1$, $p \leq n$ and $x = (A, b, C) \in \mathcal{C}_*(n, 1, p)$. Let $b_0 = \epsilon_n$ and

$$A_0 = \begin{bmatrix} 0 & 1 & \cdots & 0 \\ \vdots & & & \\ & & \cdots & 1 \\ 0 & & & 0 \end{bmatrix}.$$

Then there is a $\gamma_x = [g_x, K_x, \alpha_x]$ with $\gamma_x \cdot x = (A_0, \epsilon_n, Cg_x^{-1})$. Let $Cg_x^{-1} = C_{\gamma_x}$. If $\gamma_x' \in \Gamma_f(n,1)$ with $\gamma_x' \cdot x = (A_0, \epsilon_n, C_{\gamma_x'})$, then $\gamma_x'\gamma_x^{-1}(A_0, \epsilon_n, C_{\gamma_x}) = (A_0, \epsilon_n, C_{\gamma_x'})$ and $\gamma_x'\gamma_x^{-1} = \{[\alpha I, 0, \alpha]\} = S_{\Gamma_f}((A_0, \epsilon_n))$. Hence $C_{\gamma_x} = C_{\gamma_x'} \cdot \alpha$. Moreover, $[\alpha I, 0, \alpha]\gamma_x = [\alpha g_x, \alpha K_x, \alpha\alpha_x] = \gamma_x \cdot [\alpha I, 0, \alpha]$ and $[\alpha I, 0, \alpha]\gamma_x \cdot x = (A_0, \epsilon_n, C_{\gamma_x}\alpha^{-1})$. We observe also that there is a γ_x whose entries are regular functions of x. Let $\psi \colon \mathcal{C}_*(n, 1, p) \to \mathbb{P}^{np-1}$ be given by

$$\psi(x) = [C_{\gamma_x}].$$

ψ is a morphism into $\cup(\mathbb{P}^{np-1})_{\delta^i}$ (δ^i the $p \times p$ minors) which is open in $\mathbb{P}^{np-1}$. If $xE_{\Gamma_f}y$, then $\psi(x) = \psi(y)$ (by the argument given earlier) and ψ is invariant. If $U = \cup(\mathbb{P}^{np-1})_{\delta^i}$ and $[C] \in U$, then $(A_0, \epsilon_n, C) \in \mathcal{C}_*(n, 1, p)$ and hence ψ is surjective (onto U). If $\gamma \cdot x = x$, then $\gamma\gamma_x^{-1}(A_0, \epsilon_n, C_{\gamma_x}) = \gamma_x^{-1}(A_0, \epsilon_n, C_{\gamma_x})$ and hence, $\gamma_x\gamma\gamma_x^{-1}(A_0, \epsilon_n, C_{\gamma_x}) = (A_0, \epsilon_n, C_{\gamma_x})$. Then $\gamma_x\gamma\gamma_x^{-1} = [\alpha I, 0, \alpha]$ and $\alpha C_{\gamma_x} = C_{\gamma_x}$. Since $C_{\gamma_x} \neq 0$, $\alpha = 1$ and $\gamma =$ identity. In other words, $S_{\Gamma_f}(x) = \{\gamma_e\}$. It follows that $\dim O_{\Gamma_f}(x) = \dim \Gamma_f(n,1) = n^2 + n + 1$ for *all* x and hence that the orbits are closed. If $\mathbf{u} \in U$, then $\psi^{-1}(\mathbf{u})$ is a closed orbit which is irreducible of dimension $n^2 + n + 1 = n^2 + n + np - (np - 1)$. Since U is normal (open in $\mathbb{P}^{np-1}$), ψ is an open map. We *claim* that (U, ψ) is in fact a geometric quotient. Consider $\psi^*\colon k[U_{\delta^1}] \to k[\mathcal{C}_*(n,1,p)]$. If $f \in k[\mathcal{C}_*(n,1,p)]^{\Gamma_f}$, then $f(x) = f(\gamma_x \cdot x) = f_0(A_0, \epsilon_n, C_{\gamma_x}) = \tilde{f}(C_{\gamma_x})$ (and $\tilde{f}(C_{\gamma_x}\alpha^{-1}) = \tilde{f}(C_{\gamma_x})$) i.e., $\tilde{f} \in k[U_{\delta^1}]$ and $\psi^*(\tilde{f}) = \tilde{f} \circ \psi = f$. By the results in Chapter 16, (U, ψ) is a geometric quotient. This is special to the case $m = 1$ as we shall see in the sequel.

Now let $\boldsymbol{\partial} = (\partial_1, \ldots, \partial_m)$, $\partial_1 \geq \cdots \geq \partial_m \geq 1$, $\sum_{j=1}^{m} \partial_j = n$ be an m-partition of n. Let $x = (A, B, C) \in \mathcal{C}_m^n \times M_*(p,n) = \mathcal{C}_*(n,m,p)$ with Kronecker indices $\boldsymbol{\partial}(x) = \boldsymbol{\partial}$. Then there is a $\gamma \in \Gamma_f(n,m)$ with $\gamma \cdot x = (A_{\boldsymbol{\partial}}, B_{\boldsymbol{\partial}}, \tilde{C})$ where $A_{\boldsymbol{\partial}} = A_{x,c}$, $B_{\boldsymbol{\partial}} = B_{x,c}$. Let $d_0 = 0$, $d_j = \sum_1^j \partial_i$ so that

$$B_{\boldsymbol{\partial}} = \left[\epsilon_{d_1} \cdots \epsilon_{d_m}\right] \tag{19.46}$$

$$A_{\boldsymbol{\partial}} = [L_j^i], \quad L_i^i = \begin{bmatrix} 0 & 1 & \cdots & \\ & & & 1 \\ & 0 & \cdots & 0 \end{bmatrix}, \quad L_j^i = O_{\partial_i,\partial_j}, \quad i \neq j. \tag{19.47}$$

We first wish to determine $S_{\Gamma_f}(A_{\boldsymbol{\partial}}, B_{\boldsymbol{\partial}}) = \Gamma_f(\boldsymbol{\partial})$ (the stabilizer of $(A_{\boldsymbol{\partial}}, B_{\boldsymbol{\partial}})$). Then $\gamma = [g, K, \alpha] \in \Gamma_f(\boldsymbol{\partial})$ if and only if

$$gB_{\boldsymbol{\partial}} = B_{\boldsymbol{\partial}}\alpha, \quad gA_{\boldsymbol{\partial}} - B_{\boldsymbol{\partial}}K = A_{\boldsymbol{\partial}}g. \tag{19.48}$$

Now

$$B_{\boldsymbol{\partial}}\alpha = \begin{bmatrix} O_{\partial_1-1,m} \\ \epsilon^1 \\ O_{\partial_2-1,m} \\ \epsilon^2 \\ \vdots \\ \epsilon^m \end{bmatrix} \quad \alpha = \begin{bmatrix} O_{\partial_1-1,m} \\ \alpha^1 \\ O_{\partial_2-1,m} \\ \alpha^2 \\ \vdots \\ \alpha^m \end{bmatrix} \tag{19.49}$$

so that $gB_{\boldsymbol{\partial}} = B_{\boldsymbol{\partial}}\alpha$ if and only if

$$\begin{aligned} g_{d_j}^i &= 0, \qquad i \neq d_1, \ldots, d_m \\ g_{d_j}^{d_i} &= \alpha_j^i \end{aligned} \tag{19.50}$$

for $i, j = 1, \dots, m$. We also have

$$B_{\partial}K = \begin{bmatrix} O_{\partial_1-1,n} \\ K^1 \\ O_{\partial_2-1,n} \\ K^2 \\ \vdots \\ K^m \end{bmatrix}. \tag{19.51}$$

In view of (19.39), we shall prove a series of lemmas.

Lemma 19.52 *Let L_t be a $t \times t$ matrix of the form*

$$L_t = \begin{bmatrix} 0 & 1 & \cdots & 0 \\ \vdots & & & \vdots \\ 0 & & \cdots & 1 \\ 0 & & \cdots & 0 \end{bmatrix} \tag{19.53}$$

and let K_t^1 be $1 \times t$ and let θ_t^t be a $t \times t$ matrix

$$\theta_t^t = \begin{bmatrix} \theta_{11} & \cdots & \theta_{1t-1} & 0 \\ \vdots & & & \vdots \\ \theta_{t1} & \cdots & \theta_{tt-1} & \alpha \end{bmatrix}. \tag{19.54}$$

Then

$$\theta_t^t L_t - \begin{bmatrix} O_{t-1,t} \\ K_t^1 \end{bmatrix} = L_t \theta_t^t \tag{19.55}$$

if and only if $\theta_t^t = \alpha I_t$, $K_t^1 = 0$.

Proof. Clear if $\theta_t^t = \alpha I_t$, $K_t^1 = 0$. Let $t = 2$. Then

$$\theta_t^t L_2 - \begin{bmatrix} 0 & 0 \\ K^1 & K^2 \end{bmatrix} = \begin{bmatrix} 0 & \theta_{11} \\ -K^1 & \theta_{21} - K^2 \end{bmatrix}$$

$$L_2 \theta_t^t = \begin{bmatrix} \theta_{21} & \alpha \\ 0 & 0 \end{bmatrix}$$

so that $\theta_{21} = 0$, $K^1 = 0$, $K^2 = 0$ and $\theta_2^2 = \alpha I_2$. Now we use induction. Let

$$\theta_{t+1}^{t+1} = \begin{bmatrix} \theta_{t+1}^1 & \\ \theta_1^t & \theta_t^t \end{bmatrix}, \qquad L_{t+1} = \begin{bmatrix} O_{t+1,1} & \begin{matrix} \epsilon_t^1 \\ L_t \end{matrix} \end{bmatrix}$$

$$\theta_{t+1}^1 = (\theta_{11} \ \theta_t^1), \qquad K_{t+1}^1 = (K^1 \ K_t^1).$$

Then

$$\theta_{t+1}^{t+1} L_{t+1} - \begin{bmatrix} O_{t,t+1} \\ K_{t+1}^1 \end{bmatrix} = \begin{bmatrix} 0 & \theta_{11}\epsilon_t^1 + \theta_t^1 L_t \\ \vdots & \\ -K^1 & \theta_1^t \epsilon_t^1 + \theta_t^t L_t - K_t^1 \end{bmatrix}$$

$$L_{t+1}\theta_{t+1}^{t+1} = \begin{bmatrix} \epsilon_t^1 \theta_1^t & \epsilon_t^1 \theta_t^t \\ L_t \theta_1^t & L_t \theta_t^t \end{bmatrix}$$

so that $\epsilon_t^1 \theta_1^t = \theta_{21} = 0$, $\theta_1^t \epsilon_t^1 = (\theta_1^t \ O_{t-1}^t)$,

$$L_t \theta_1^t = \begin{bmatrix} \theta_{3\,1} \\ \vdots \\ \theta_{t+1\,1} \\ 0 \end{bmatrix}.$$

These imply $\theta_1^t \epsilon_t^1 = O_{t,t}$, $L_t \theta_1^t = 0$ and $K^1 = 0$. By the induction hypothesis, $\theta_t^t = \alpha I_t$, $K_t^1 = 0$. But $\epsilon_t^1 \theta_t^t = [\theta_{22}\ \theta_{23} \cdots \theta_{2t+1}] = [\alpha\ 0 \cdots 0]$ and $\theta_{11}\epsilon_t^1 + \theta_t^1 L_t = [\theta_{11}\ \theta_{12} \cdots \theta_{1t}]$ and so, $\theta_{11} = \alpha$ and $\theta_t^1 = O_t^1$. In other words, $\theta_{t+1}^{t+1} = \alpha I_{t+1}$.

Corollary 19.56 *Let θ_t^t be of the form*

$$\theta_t^t = \begin{bmatrix} \theta_{11} & \cdots & \theta_{1t-1} & 0 \\ \vdots & & \vdots & \vdots \\ & & & \alpha_1 \\ \theta_{t1} & & \theta_{tt-1} & \alpha_2 \end{bmatrix}. \tag{19.57}$$

Then

$$\theta_t^t L_t - \begin{bmatrix} O_{t-1,t} \\ K_t^1 \end{bmatrix} = L_t \theta_t^t$$

if and only if $\theta_t^t = \alpha_2 I_t + \alpha_1 L_t$, $K_t^1 = 0$.

Proof. Let $t = 2$. Then

$$\theta_2^2 L_2 - \begin{bmatrix} 0 & 0 \\ K_1^1 & K_1^2 \end{bmatrix} = \begin{bmatrix} 0 & \theta_{11} \\ -K_1^1 & \theta_{21} - K_1^2 \end{bmatrix}, \quad L_2\theta_2^2 = \begin{bmatrix} \theta_{21} & \alpha_2 \\ 0 & 0 \end{bmatrix}$$

so that $\theta_{21} = 0$, $\theta_{11} = \alpha_2$ and

$$\theta_2^2 = \begin{bmatrix} \alpha_2 & \alpha_1 \\ 0 & \alpha_2 \end{bmatrix} = \alpha_2 I_2 + \alpha_1 L_2.$$

Now use induction on t. If

$$\theta_{t+1}^{t+1} = \begin{bmatrix} \theta_{11} & \theta_t^1 \\ \theta_1^t & \theta_t^t \end{bmatrix}$$

with θ_t^t of the form (19.57), then

$$\theta_{t+1}^{t+1} L_{t+1} - \begin{bmatrix} O_{t,t+1} \\ K_{t+1}^1 \end{bmatrix} = \begin{bmatrix} 0 & \theta_{11}\epsilon_t^1 + \theta_t^1 L_t \\ -K^1 & \theta_1^t \epsilon_t^1 + \theta_t^t L_t - K_t^1 \end{bmatrix}$$

and

$$L_{t+1}\theta_{t+1}^{t+1} = \begin{bmatrix} \epsilon_t^1 \theta_1^t & \epsilon_t^1 \theta_t^t \\ L_t \theta_1^t & L_t \theta_t^t \end{bmatrix}$$

imply as in the lemma, that $K^1 = 0$, $\theta_1^t = 0$. By the induction hypothesis, $\theta_t^t = \alpha_2 I_t + \alpha_1 L_t$ and $K_t^1 = 0$. Then $\epsilon_t^1 \theta_t^t = \alpha_2 \epsilon_t^1 + \alpha_1 \epsilon_t^1 L_t = (\alpha_2, \alpha_1, 0, \ldots, 0)$ and $\theta_{11}\epsilon_t^1 + \theta_t^1 L_t = (\theta_{11}, \theta_{12}, \ldots, \theta_{1t})$. It follows that $\theta_{11} = \alpha_2$, $\theta_{12} = \alpha_1$ and $\theta_{13} = \cdots = \theta_{1t} = \theta_{1t+1} = 0$. Thus $\theta_{t+1}^{t+1} = \alpha_2 I_{t+1} + \alpha_1 L_{t+1}$.

Corollary 19.58 *Let θ_{s+1}^s be an $s \times s+1$ matrix of the form $\theta_{s+1}^s = [X_s^s \ \alpha\epsilon_s]$ with $X_s^s = (x_j^i)$. Set $\alpha_1 = x_s^{s-1}$, $\alpha_2 = x_s^s$. Then*

$$\theta_{s+1}^s L_{s+1} - \begin{bmatrix} 0 \\ \mathbf{K}_{s+1}^1 \end{bmatrix} = L_s \theta_{s+1}^s$$

where $\mathbf{K}_{s+1}^1 = (\mathbf{K}_s^1, K_{s+1}^1)$ if and only if $X_s^s = \alpha_2 I_s + \alpha_1 L_s$, $\mathbf{K}_s^1 = \mathbf{0}$, $K_{s+1}^1 = \alpha_2$.

Proof. Simply note that

$$\theta_{s+1}^s L_{s+1} - \begin{bmatrix} 0 \\ \mathbf{K}_{s+1}^2 \end{bmatrix} = \left(X_s^s L_s - \begin{bmatrix} 0 \\ \mathbf{K}_s^1 \end{bmatrix}, X_s^s \epsilon_s - \begin{bmatrix} 0 \\ K_{s+1}^1 \end{bmatrix} \right)$$

and that

$$L_s \theta_{s+1}^s = (L_s X_s^s \ \ \alpha L_s \epsilon_s).$$

Since $X_s^s \epsilon_s = (x_s^i)$, we can apply the previous corollary.

Lemma 19.59 *Let $r = s + t$ and K_t^1 be a $1 \times t$ matrix. Let θ_t^r be an $r \times t$ matrix of the form*

$$\theta_t^{s+t} = \begin{bmatrix} \theta_{t-1}^s & O_1^s \\ & \theta_t^t \end{bmatrix} \tag{19.60}$$

where θ_t^t is a $t \times t$ matrix of the form (19.54). Then

$$\theta_t^{s+t} L_t - \begin{bmatrix} O_t^{s+t-1} \\ K_t^1 \end{bmatrix} = L_{s+t} \theta_t^{s+t} \tag{19.61}$$

if and only if $\theta_t^{s+t} = 0$ and $K_t^1 = 0$.

Proof. Clear if $\theta_t^{s+t} = 0$ and $K_t^1 = 0$. Let $\theta_t^s = [\theta_{t-1}^s \ O_1^s]$. Then

$$\theta_t^{s+t} L_t - \begin{bmatrix} 0 \\ K_t^1 \end{bmatrix} = \begin{bmatrix} \theta_t^s L_t \\ \theta_t^t L_t - \begin{bmatrix} O_t^{t-1} \\ K_t^1 \end{bmatrix} \end{bmatrix}$$

and

$$L_{s+t}\theta_t^{s+t} = \begin{bmatrix} L_s\theta_t^s - E_t^{1,s}\theta_t^t \\ L_t\theta_t^t \end{bmatrix}$$

where

$$E_t^{1,s} = \begin{bmatrix} O_t^{s-1} \\ \epsilon_t^1 \end{bmatrix}$$

and noting that

$$L_{s+t} = \begin{bmatrix} L_s & E_t^{1,s} \\ O_s^t & L_t \end{bmatrix}.$$

By Lemma 19.52, $K_t^1 = 0$, $\theta_t^t = \alpha I_t$ and so $\theta_t^s L_t = L_s\theta_t^s - \alpha E_t^{1,s}$ or

$$\begin{bmatrix} \theta_{11} & \cdots & \theta_{1t-1} & 0 \\ \vdots & & & \vdots \\ \theta_{s1} & \cdots & \theta_{st-1} & \theta_{st} \end{bmatrix} L_t = L_s \begin{bmatrix} \theta_{11} & \cdots & \theta_{1t-1} & 0 \\ \vdots & & & \vdots \\ \theta_{s1} & \cdots & \theta_{s+1} & \theta_{st} \end{bmatrix}$$

$$- \begin{bmatrix} 0 & 0 & \cdots & 0 \\ \vdots & \vdots & & \\ \alpha & 0 & \cdots & 0 \end{bmatrix}.$$

In view of the form of L_t, L_s, $\alpha = 0$ and

$$\theta_t^s = \begin{bmatrix} \theta_{t-1}^1 & 0 \\ \vdots & \vdots \\ \theta_{t-1}^s & 0 \end{bmatrix}$$

which implies

$$\theta_t^s L_t = \begin{bmatrix} 0 & \theta_{t-1}^1 \\ \vdots & \vdots \\ 0 & \theta_{t-1}^s \end{bmatrix}, \quad L_s\theta_t^s = \begin{bmatrix} \theta_{t-1}^2 & 0 \\ \vdots & \vdots \\ \theta_{t-1}^s & \\ 0 & 0 \end{bmatrix}.$$

It follows that $\theta_t^s = 0$.

Lemma 19.62 *Let $r = s + t$ with $t < s$. Let K_r^1 be a $1 \times r$ matrix with $K_r^1 = (K_t^1\ K_s^1)$ where K_t^1, K_s^1 are $1 \times t$ and $1 \times s$ matrices. Let θ_r^s be an $s \times r$ matrix of the form*

$$\theta_r^s = [\theta_t^s\ \ \theta_s^s]$$

where θ_t^s is of the form (19.60) and θ_s^s is of the form (19.54). Then

$$\theta_r^s L_r - \begin{bmatrix} O_r^{s-1} \\ K_r^1 \end{bmatrix} = L_s \theta_r^s \tag{19.63}$$

if and only if $\theta_t^s = 0$, $K_t^1 = 0$, $\theta_s^s = \alpha I_s$, $K_s^1 = 0$.

Proof. Note that

$$L_r = L_{t+s} = \begin{bmatrix} L_t & E_s^{1,t} \\ O_t^s & L_s \end{bmatrix}$$

where $E_s^{1,t} = [e_t\ \ O_{s-1}^t]$. Then

$$\begin{aligned} \theta_r^s L_r &= [\theta_t^s L_t\ \ \theta_t^s E_s^{1,t} + \theta_s^s L_s] \\ \theta_r^s L_r - \begin{bmatrix} O_r^{s-1} \\ K_r^1 \end{bmatrix} &= \left(\theta_t^s L_t - \begin{bmatrix} O_t^{s-1} \\ K_t^1 \end{bmatrix}, \theta_t^s E_s^{1,t} + \theta_s^s L_s - \begin{bmatrix} O_s^{s-1} \\ K_s^1 \end{bmatrix} \right) \\ L_s \theta_r^s &= \left[L_s \theta_t^s, L_s \theta_s^s \right]. \end{aligned}$$

Thus, it is clear that if $\theta_t^s = 0$, $K_t^1 = 0$, $\theta_s^s = \alpha I_s$, $K_s^1 = 0$, the relation (19.63) holds. On the other hand, if (19.63) holds, then Lemma (19.59) gives $\theta_t^s = 0$, $K_t^1 = 0$ $(s > t)$ and Lemma (19.52) gives $\theta_s^s = \alpha I_s$.

Corollary 19.64 (of Proof) *If $r = 2s$, then*

$$\theta_{2s}^s L_{2s} - \begin{bmatrix} O_{2s}^{s-1} \\ K_{2s}^1 \end{bmatrix} = L_s \theta_{2s}^s$$

where

$$\theta_{2s}^s = \begin{bmatrix} & O_1^{s-1} & & O_1^{s-1} \\ \theta_{s-1}^{1,s} & & \theta_{s-1}^{2,s} & \\ & \beta & & \alpha \end{bmatrix} \qquad K_{2s}^1 = (K_s^1, K_s^2)$$

if and only if $\theta_{s-1}^{1,s} = \beta I_s$, $K_s^1 = 0$, $K_s^2 = (\beta, O_{s-1}^1)$ and $\theta_s^{2,s} = \alpha I_s$.

Example 19.65 Let $s = 2$. Then

$$\theta_4^2 = \begin{bmatrix} \theta_{11} & 0 & \theta_{13} & 0 \\ \theta_{21} & \beta & \theta_{23} & \alpha \end{bmatrix}$$

$$\theta_4^2 L_4 - \begin{bmatrix} 0 \\ K_4^1 \end{bmatrix} = \begin{bmatrix} 0 & \theta_{11} & 0 & \theta_{13} \\ -K_1^1 & \theta_{21} - K_2^1 & \beta - K_1^2 & \theta_{23} - K_2^2 \end{bmatrix}$$

$$L_2\theta_4^2 = \begin{bmatrix} \theta_{21} & \beta & \theta_{23} & \alpha \\ 0 & 0 & 0 & 0 \end{bmatrix}$$

so that

$$\theta_4^2 = \begin{bmatrix} \beta & 0 & \alpha & 0 \\ 0 & \beta & 0 & \alpha \end{bmatrix}$$

$$K_4^1 = (0 \; 0 \; \beta \; 0).$$

Lemma 19.66 *Let $r = \lambda s + t$ with $s > t > 0$ and $\lambda > 1$. Let $K_r^1 = (K_t^1, K_{1,s}^1 \cdots K_{\lambda,s}^1)$ where K_t^1, $K_{j,s}^1$ are $1 \times t$ and $1 \times s$ matrices, respectively. Let θ_r^s be an $s \times r$ matrix of the form*

$$\theta_r^s = \left[\theta_t^s \; \theta_{1,s}^s \cdots \theta_{\lambda,s}^s\right]$$

where

$$\theta_t^s = \left[\theta_{t-1}^s \;\; O_1^s\right], \quad \theta_{j,s}^s = \left[\theta_{j,s-1}^s \;\; \alpha_j \epsilon_s\right]$$

for $j = 1, \ldots, \lambda$. Then

$$\theta_r^s L_r - \begin{bmatrix} O_r^{s-1} \\ K_r^1 \end{bmatrix} = L_s \theta_r^s \tag{19.67}$$

if and only if $\theta_t^s = 0$, $K_t^1 = 0$, $\theta_{j,s}^s = \alpha_j I_s$ and $K_{j,s}^1 = \alpha_{j-1}\epsilon^{1,s}$, $j = 2, \ldots, \lambda$.

Proof. Note that

$$L_r = \begin{bmatrix} L_t & E_s^{1,t} & O_s^t & \cdots & O_s^t \\ O_t^s & L_s & E_s^{1,s} & & O_s^s \\ \vdots & & L_s & & \vdots \\ & & & & E_s^{1,s} \\ O_t^s & O_s^s & O_s^s & \cdots & L_s \end{bmatrix}$$

where $E_s^{1,t} = \left[\epsilon_t \; O_{s-1}^t\right]$, $E_s^{1,s} = \left[\epsilon_s \; O_{s-1}^s\right]$. Then

$$\theta_r^s L_r = \left[\theta_t^s L_t, \theta_t^s E_s^{1,t} + \theta_{1,s}^s L_s, \cdots, \theta_{j-1,s}^s E_s^{1,s} + \theta_{j,s}^s L_s, \cdots\right]$$
$$L_s \theta_r^s = \left[L_s \theta_t^s, L_s \theta_{1,s}^s, \ldots, L_s \theta_{j,s}^s \ldots\right].$$

By repeated applications of prior lemmas, we can see that $\theta_t^s = 0$, $K_t^1 = 0$, $\theta_{j,s}^s = \alpha_j I_s$ and $K_{j,s}^1 = \alpha_{j-1}\epsilon^{1,s}$.

Corollary 19.68 (of Proof) *If $r = \lambda s$, $\lambda > 1$, then, for $\theta_{\lambda s}^s = [\theta_{1,s}^s \cdots \theta_{\lambda,s}^s]$, we have*

$$\theta_{\lambda s}^s L_{\lambda s} - \begin{bmatrix} O_{\lambda s}^{s-1} \\ K_{\lambda s}^1 \end{bmatrix} = L_s \theta_{\lambda s}^s$$

for $K_{\lambda s}^1 = (K_{1,s}^1 \cdots K_{\lambda,s}^1)$, if and only if $\theta_{j,s}^s = \alpha_j I_s$, $K_{1,s}^1 = 0$ and $K_{j,s}^1 = \alpha_{j-1}\epsilon^{1,s}$ for $j = 2, \ldots, \lambda$.

Now suppose that $\gamma = [\mathbf{g}, \mathbf{K}, \boldsymbol{\alpha}]$ is an element of $\Gamma_{f,\partial}$. Let

$$\mathbf{g} = \begin{bmatrix} \theta_1^1 & \cdots & \theta_m^1 \\ \theta_1^m & & \theta_m^m \end{bmatrix}$$

where θ_j^i is a $\partial_i \times \partial_j$-matrix with

$$\theta_j^i = [\theta_{\partial_j - 1}^{\partial_i} \;\; \alpha_j^i \epsilon_{\partial_j}].$$

Let

$$\mathbf{K} = \begin{bmatrix} \mathbf{K}_1^1 & \cdots & \mathbf{K}_m^1 \\ \mathbf{K}_1^m & \cdots & \mathbf{K}_m^m \end{bmatrix}$$

where $\mathbf{K}_j^i$ is a $1 \times \partial_j$ matrix. Let $L_{ii} = L_{O_i} - L_i$ so that

$$\mathbf{g}A_\partial = \begin{bmatrix} \theta_1^1 L_1 & \theta_2^1 L_2 & \cdots & \theta_m^1 L_m \\ \vdots & \vdots & & \vdots \\ \theta_1^m L_1 & \theta_2^m L_2 & \cdots & \theta_m^m L_m \end{bmatrix}$$

$$B_\partial \mathbf{K} = \begin{bmatrix} O_{\partial_1}^{\partial_1 - 1} & \cdots & O_{\partial_m}^{\partial_1 - 1} \\ K_1^1 & & K_m^1 \\ \vdots & & \\ K_1^m & \cdots & K_m^m \end{bmatrix}$$

$$A_\partial \mathbf{g} = \begin{bmatrix} L_1\theta_1^1 & L_1\theta_2^1 & \cdots & L_1\theta_m^1 \\ \vdots & \vdots & & \vdots \\ L_m\theta_1^m & L_m\theta_2^m & \cdots & L_m\theta_m^m \end{bmatrix}.$$

The equations become

$$\theta_j^i L_j - F_j^i = L_i \theta_j^i, \quad F_j^i = \begin{bmatrix} O_{\partial_j}^{\partial_i - 1} \\ \mathbf{K}_j^i \end{bmatrix}.$$

Repeated applications of the lemmas give the structure of $\mathbf{g}$ and a fortiori of $\Gamma_{f,\partial}$. The $\mathbf{g}$ are in "block stripe" form.

Example 19.69 Let $n = 8$, $m = 3$ so that $n = 2 \cdot 3 + 2 = qm + r$ with $q = 2$, $r = 2$. Let $\partial_1 = \partial_2 = 3$ and $\partial_3 = 2$. Since

$$\big[B\ AB \cdots\big] = \big[B_1\ B_2\ AB_1\ AB_2\ AB_3, A^2B_1, A^2B_2, \dots\big],$$

here $\det\big[\mathbf{Y}(A,B)_{1\cdots n}^{1\cdots n}\big] \neq 0$ so that in $\mathbb{A}^{n^2+nm} = \mathbb{A}^{64+24}$, the set of such (A,B) is open (hence "generic"). The equations for $\boldsymbol{\partial} = (3,3,2)$ are, for $i = 1$,

$$\theta_1^1 L_1 - \begin{bmatrix} O \\ \mathbf{K}_1^1 \end{bmatrix} = L_1 \theta_1^1$$

$$\theta_2^1 L_2 - \begin{bmatrix} O \\ \mathbf{K}_2^1 \end{bmatrix} = L_1 \theta_2^1$$

$$\theta_3^1 L_3 - \begin{bmatrix} O \\ \mathbf{K}_3^1 \end{bmatrix} = L_1 \theta_3^1.$$

Since $\partial_1 = \partial_2 > \partial_3$, we have

(1) $\theta_1^1 = \alpha_1^1 I_3$, $\mathbf{K}_1^1 = 0$ (Lemma 19.52)

(2) $\theta_2^1 = \alpha_2^1 I_3$, $\mathbf{K}_2^1 = 0$ (Lemma 19.52)

(3) $\theta_3^1 = 0$, $\mathbf{K}_3^1 = 0$ (Lemma 19.59)

Similarly, for $i = 2$, we have $\theta_1^2 = \alpha_1^2 I_3$, $\mathbf{K}_1^2 = 0$, $\theta_2^2 = \alpha_2^2 I_3$, $\mathbf{K}_2^2 = 0$, $\theta_3^2 = 0$, $\mathbf{K}_3^2 = 0$. For $i = 3$, the equations are

$$\theta_1^3 L_1 - \begin{bmatrix} O \\ \mathbf{K}_1^3 \end{bmatrix} = L_3 \theta_1^3$$

$$\theta_2^3 L_2 - \begin{bmatrix} O \\ \mathbf{K}_2^3 \end{bmatrix} = L_3 \theta_2^3$$

$$\theta_3^3 L_3 - \begin{bmatrix} O \\ \mathbf{K}_3^3 \end{bmatrix} = L_3 \theta_3^3$$

and we have, by Corollaries 19.64 and 19.68 and by Lemma 19.52,

(1) $\theta_1^3 = \big[(\mathbf{K}_1^3)_3 I_2 + \alpha_1^3 L_2\ \alpha_1^3 \epsilon_2\big]$, $\mathbf{K}_1^3 = (0\ 0\ (\mathbf{K}_1^3)_3)$

(2) $\theta_2^3 = \big[(\mathbf{K}_2^3)_3 I_2 + \alpha_2^3 L_2\ \alpha_2^3 \epsilon_2\big]$, $\mathbf{K}_2^3 = (0\ 0\ (\mathbf{K}_2^3)_3]$

(3) $\theta_3^3 = \alpha_3^3 I_2$, $\mathbf{K}_3^3 = 0$.

Hence, $\mathbf{g}$ is of the form

$$\mathbf{g} = \begin{bmatrix} \alpha_1^1 & 0 & 0 & \alpha_2^1 & 0 & 0 & 0 & 0 \\ 0 & \alpha_1^1 & 0 & 0 & \alpha_2^1 & 0 & 0 & 0 \\ 0 & 0 & \alpha_1^1 & 0 & 0 & \alpha_2^1 & 0 & 0 \\ \alpha_1^2 & 0 & 0 & \alpha_2^2 & 0 & 0 & 0 & 0 \\ 0 & \alpha_1^2 & 0 & 0 & \alpha_2^2 & 0 & 0 & 0 \\ 0 & 0 & \alpha_1^2 & 0 & 0 & \alpha_2^2 & 0 & 0 \\ (\mathbf{K}_1)_3^3 & \alpha_1^3 & 0 & (\mathbf{K}_2^3)_3 & \alpha_2^3 & 0 & \alpha_3^3 & 0 \\ 0 & (\mathbf{K}_1^3)_3 & \alpha_1^3 & 0 & (\mathbf{K}_2^3)_3 & \alpha_2^3 & 0 & \alpha_3^3 \end{bmatrix}$$

and $\boldsymbol{\alpha}$ is of the form

$$\boldsymbol{\alpha} = \begin{bmatrix} \alpha_1^1 & \alpha_2^1 & 0 \\ \alpha_1^2 & \alpha_2^2 & 0 \\ \alpha_1^3 & \alpha_2^3 & \alpha_3^3 \end{bmatrix}$$

(and $\mathbf{K} = 0$). If $\Delta = \alpha_1^1\alpha_2^2 - \alpha_2^1\alpha_1^2$, then $0 \neq \det \boldsymbol{\alpha} = \alpha_3^3 \cdot \Delta$ and $\det \mathbf{g} = (\alpha_3^3)^2 \cdot \Delta^3 \neq 0$. We observe that $\dim \Gamma_{f,\partial} = 9$ and that $\Gamma_{f,\partial}$ contains the subgroup $[k^* I_8, 0, k^* I_3]$ of diagonal elements. Let

$$\boldsymbol{\alpha}^1 = \begin{bmatrix} \alpha_1^1 & \alpha_2^1 \\ \alpha_1^2 & \alpha_2^2 \end{bmatrix}, \quad \boldsymbol{\alpha}^2 = \begin{bmatrix} \alpha_1^3 & \alpha_2^3 \end{bmatrix}, \quad \boldsymbol{\alpha}^3 = \alpha_3^3 \cdot 1$$

$$\mathbf{K} = \begin{bmatrix} (\mathbf{K}_1^3)_3 & (\mathbf{K}_2^3)_3 \end{bmatrix}$$

then

$$\mathbf{g} = \begin{bmatrix} I_3 \otimes \boldsymbol{\alpha}^1 & \mathbf{O}_2^6 \\ E_1 \otimes \mathbf{K} + E_2 \otimes \boldsymbol{\alpha}^2 & I_2 \otimes \boldsymbol{\alpha}^3 \end{bmatrix}$$

where $E_1 = \begin{bmatrix} I_2 & O_1^2 \end{bmatrix}$, $E_2 = \begin{bmatrix} O_1^2 & I_2 \end{bmatrix}$.

Example 19.70 (The "Generic" Case) Let $n = qm + r$ with $0 \leq r < m$. Let $\mathbf{Y}(A, B) = \begin{bmatrix} B & AB \cdots A^{n-1}B \end{bmatrix}$ and $y(A, B) = \det\left[\mathbf{Y}(A, B)\big|_{1\cdots n}^{1\cdots n}\right]$. Then $(\mathbb{A}^{n^2+nm})_y \times M_*(p, n)$ is open and if $x = (A, B, C) \in (\mathbb{A}^{n^2+nm})_y \times M_*(p, n)$, then $\boldsymbol{\partial}(x) = (\partial_1, \ldots, \partial_r, \partial_{r+1}, \ldots, \partial_m)$ with $\partial_1 = \cdots = \partial_r = q + 1$, $\partial_{r+1} =$

$\cdots = \partial_m = q$. Such an x is called "generic." What is $\Gamma_{f,\boldsymbol{\partial}}$? If we let

$$\mathbf{g} = \begin{bmatrix} \theta_1^1 & \cdots & \theta_r^1 & \theta_{r+1}^1 & \cdots & \theta_m^1 \\ \vdots & & \vdots & \vdots & & \vdots \\ \theta_1^r & & \theta_r^r & \theta_{r+1}^r & \cdots & \theta_m^r \\ \theta_1^{r+1} & & \theta_r^{r+1} & \theta_{r+1}^{r+1} & \cdots & \theta_m^{r+1} \\ \vdots & & \vdots & \vdots & & \vdots \\ \theta_1^m & \cdots & \theta_r^m & \theta_{r+1}^m & \cdots & \theta_m^m \end{bmatrix}, \quad F_j^i = \begin{bmatrix} O_{\partial_j}^{\partial_i - 1} \\ \\ \mathbf{K}_j^i \end{bmatrix}$$

with θ_j^i a $\partial_i \times \partial_j$-matrix, then repeated application of the lemmas and corollaries gives:

(a) $\theta_{j_1}^{i_1} = \alpha_{j_1}^{i_1} I_{q+1}$, $F_{j_1}^{i_1} = 0$ so $\mathbf{K}_{j_1}^{i_1} = 0$ for $i_1 = 1, \ldots, r$, $j_1 = 1, \ldots, r$;

(b) $\theta_{j_2}^{i_1} = 0$, $\mathbf{K}_{j_2}^{i_1} = 0$, $\alpha_{j_2}^{i_1} = 0$ for $i_1 = 1, \ldots, r$, $j_2 = r+1, \ldots, m$;

(c) $\theta_{j_2}^{i_2} = \alpha_{j_2}^{i_2} I_q$, $F_{j_2}^{i_2} = 0$ so $\mathbf{K}_{j_2}^{i_2} = 0$ for $i_2 = r+1, \ldots, m$, $j_2 = r+1, \ldots, m$; and

(d) $\theta_{j_1}^{i_2} = \left[(\mathbf{K}_{j_1}^{i_2})_{q+1} I_q + \alpha_{j_1}^{i_2} L_q \;\; \alpha_{j_1}^{i_2} \epsilon_q \right]$, $F_{j_1-1}^{i_2} = 0$ so $\mathbf{K}_{j_2}^{i_2} = (O_q^1 \; \mathbf{K}_j^{i_2})$ with $(\mathbf{K}_j^{i_2})_{q+1} = K_{j_1}^{i_2}$ for $i_2 = r+1, \ldots, m$, $j_1 = 1, \ldots, r$.

In other words, if $\gamma = [\mathbf{g}, \mathbf{K}, \boldsymbol{\alpha}] \in \Gamma_{f,\boldsymbol{\partial}}$, then $\mathbf{K} = (\mathbf{K}^i)$ with $\mathbf{K}^1 = 0, \ldots, \mathbf{K}^r = 0$, $\mathbf{K}^i = (K_j^i)$, $i = r+1, \ldots, m$ and $K_j^i = 0$, $j = r+1, \ldots, m$ i.e.,

$$\mathbf{K} = \begin{bmatrix} O_q^1 & K_1^{r+1} & O_q^1 & \cdots & K_r^{r+1} & O_{q(m-r)}^1 \\ \vdots & \vdots & \vdots & & \vdots & \\ O_q^1 & K_1^m & O_q^1 & & K_r^m & O_{q(m-r)}^1 \end{bmatrix}$$

$$\boldsymbol{\alpha} = \begin{bmatrix} \alpha_1^1 & \cdots & \alpha_r^1 & 0 & & 0 \\ \vdots & & \vdots & \vdots & & \vdots \\ \alpha_1^r & & \alpha_r^r & 0 & & 0 \\ \alpha_1^{r+1} & & \alpha_r^{r+1} & \alpha_{r+1}^{r+1} & \cdots & \alpha_m^{r+1} \\ \vdots & & \vdots & \vdots & & \\ \alpha_1^m & & \alpha_r^m & \alpha_{r+1}^m & \cdots & \alpha_m^m \end{bmatrix} = \begin{bmatrix} \boldsymbol{\alpha}^1 & O \\ \boldsymbol{\alpha}^2 & \boldsymbol{\alpha}^3 \end{bmatrix}$$

with $\boldsymbol{\alpha}^1$ $r \times r$, $\boldsymbol{\alpha}^2$ $m - r \times r$, $\boldsymbol{\alpha}^3$ $m - r \times m - r$ and

$$\mathbf{g} = \begin{bmatrix} I_{q+1} \otimes \boldsymbol{\alpha}^1 & \mathbf{O}_{q(m-r)}^{(q+1)r} \\ E_1^q \otimes \tilde{K} + E_2^q \otimes \boldsymbol{\alpha}^2 & I_q \otimes \boldsymbol{\alpha}^3 \end{bmatrix}$$

where $\tilde{K} = (K_{j_1}^{i_2})$, $i_2 = r+1,\ldots,m$, $j_1 = 1,\ldots,r$ and $E_1^q = [I_q\ O_1^q]$, $E_2^q = [O_1^q\ I_q]$. We observe that $\dim \Gamma_{f,\boldsymbol{\partial}} = r^2+(m-r)m+(m-r)r = r^2+m^2-r^2 = m^2$ and that $\Gamma_{f,\boldsymbol{\partial}}$ contains the subgroup $[k^* I_n, 0, k^* I_m]$ of diagonal elements.

Example 19.71 Let $n = 5$, $m = 2$, $\partial_1 = 3$, $\partial_2 = 2$ so that $q = 2$, $r = 1$. Let $C = [C_1\ C_2]$ with $C_1\ p \times 3$ and $C_2\ p \times 2$. Then $\gamma = [\mathbf{g}, \mathbf{K}, \boldsymbol{\alpha}]$ is an element of $\Gamma_{f,(3,2)}$ if

$$\boldsymbol{\alpha} = \begin{bmatrix} \alpha_1^1 & 0 \\ \alpha_1^2 & \alpha_2^2 \end{bmatrix}, \quad \mathbf{K} = \begin{bmatrix} 0 & 0 & 0 & 0 & 0 \\ 0 & 0 & K & 0 & 0 \end{bmatrix}$$

$$\mathbf{g} = \begin{bmatrix} \alpha_1^1 & 0 & 0 & 0 & 0 \\ 0 & \alpha_1^1 & 0 & 0 & 0 \\ 0 & 0 & \alpha_1^1 & 0 & 0 \\ K & \alpha_1^2 & 0 & \alpha_2^2 & 0 \\ 0 & K & \alpha_1^2 & 0 & \alpha_2^2 \end{bmatrix} = \begin{bmatrix} I_3 \otimes \alpha_1^1 & 0 \\ E_1^2 \otimes K + E_2^2 \otimes \alpha_1^2 & I_2 \otimes \alpha_2^2 \end{bmatrix}.$$

Call $C = [C_1\ C_2]$ *equivalent to* $D = [D_1\ D_2]$ if there is a $\mathbf{g}$ with $C\mathbf{g} = D$ ($\mathbf{g} \in \Gamma_{f,(3,2)}$). Let $C_1 = [C_{1,1}\ C_{1,2}\ C_{1,3}]$, $C_2 = [C_{2,1}\ C_{2,2}]$ with $C_{i,j}\ p \times 1$ and let

$$\tilde{C}_1 = \begin{bmatrix} C_{1,1} \\ C_{1,2} \\ C_{1,3} \end{bmatrix}, \quad \tilde{C}_2 = \begin{bmatrix} C_{2,1} \\ C_{2,2} \\ 0 \end{bmatrix}, \quad \tilde{C}_3 = \begin{bmatrix} 0 \\ C_{2,1} \\ C_{2,2} \end{bmatrix}$$

be elements of $k^{3 \cdot p} = k^{(q+1)p} = k^{\partial_1 p}$. Consider the "flag" $V_1 \subset V_2 \subset V_3 \subset k^{3 \cdot p}$ where

$$\begin{aligned} V_1 &= \operatorname{span} \tilde{C}_3 \\ V_2 &= \operatorname{span}[V_1, \tilde{C}_2] = \operatorname{span}[\tilde{C}_2, \tilde{C}_3] \\ V_3 &= \operatorname{span}[V_2, \tilde{C}_1] = \operatorname{span}[\tilde{C}_1, \tilde{C}_2, \tilde{C}_3]. \end{aligned}$$

Suppose that $C\mathbf{g} = D$ and let $W_1 \subset W_2 \subset W_3 \subset k^{3p}$ be the "flag" corresponding to D. We *claim* that $V_1 = W_1$, $V_2 = W_2$, $V_3 = W_3$. Since

$$C\mathbf{g} = [\alpha_1^1 C_1 + C_2(E_1^2 \otimes K + E_2^2 \otimes \alpha_1^2)\ \alpha_2^2 C_2]$$

clearly $V_1 = W_1$, $V_2 = W_2$ as $\alpha_2^2 \neq 0$. Writing

$$C_2 = \begin{bmatrix} C_{2,1}^1 & & C_{2,2}^1 \\ \vdots & & \vdots \\ C_{2,1}^p & \cdots & C_{2,2}^p \end{bmatrix}$$

we have

$$C_2(E_1^2 \otimes K + E_2^2 \otimes \alpha_1^2) = \begin{bmatrix} KC_{2,1}^1 & \alpha_1^2 C_{2,1}^1 + KC_{2,2}^1 & \alpha_1^2 C_{2,2}^1 \\ \vdots & \vdots & \vdots \\ KC_{2,1}^p & \alpha_1^2 C_{2,1}^p + KC_{2,2}^p & \alpha_1^2 C_{2,2}^p \end{bmatrix}$$

so that

$$\tilde{D}_1 = \alpha_1^1 \tilde{C}_1 + K\tilde{C}_2 + \alpha_1^2 \tilde{C}_3$$

is an element of V_3, and, since $\alpha_1^1 \neq 0$, $\tilde{C}_1$ is an element of W_3. Hence $V_3 = W_3$. Now suppose that D is given with "flag" W_1, W_2, W_3 and that $V_1 = W_1$, $V_2 = W_2$, $V_3 = W_3$. Then $\alpha\tilde{C}_3 = \tilde{D}_3$ for some $\alpha \neq 0$ and (consequently) $\alpha\tilde{C}_2 = \tilde{D}_2$. If $\tilde{D}_1 \in V_3$, then $\tilde{D}_1 = a\tilde{C}_1 + K\tilde{C}_2 + \lambda\tilde{C}_3$. If $a \neq 0$, then, for

$$\mathbf{g} = \begin{bmatrix} aI_3 & 0 \\ E_1^2 \otimes K + E_2^2 \otimes \lambda & \alpha I_2 \end{bmatrix}$$

we have $C\mathbf{g} = D$ and C, D are equivalent. If $a = 0$, then $\tilde{D}_1 \in V_2$ implies $W_3 = V_2$ and $\tilde{C}_1 \in V_2$ (as $V_3 = W_3$). In other words, $\tilde{C}_1 = a_1\tilde{C}_2 + b_1\tilde{C}_3$ and $\tilde{D}_1 = 1 \cdot \tilde{C}_1 + (K - a_1)\tilde{C}_2 + (\lambda - b_1)\tilde{C}_3$ so that, for

$$\mathbf{g} = \begin{bmatrix} I_3 & 0 \\ E_1^2 \otimes (K - a_1) + E_2^2 \otimes (\lambda - b_1) & \alpha I_2 \end{bmatrix}$$

$C\mathbf{g} = D$. Thus, the "flag" $[V_1, V_2, V_3]$ characterizes the equivalence class of C modulo the action of $\Gamma_{f,\boldsymbol{\partial}}$.

Example 19.72 Let $n = 10$, $m = 4$, $\partial_1 = 3$, $\partial_2 = 3$, $\partial_3 = 2$, $\partial_4 = 2$ so that $q = 2$, $r = 2$. Let $C = \begin{bmatrix} C_1 \; C_2 \; C_3 \; C_4 \end{bmatrix}$ with C_1, C_2 $p \times 3$ and C_3, C_4 $p \times 2$. Then $\gamma = [\mathbf{g}, \mathbf{K}, \boldsymbol{\alpha}] \in \Gamma_{f,(3,3,2,2)}$ if

$$\boldsymbol{\alpha} = \begin{bmatrix} \boldsymbol{\alpha}^1 & 0 \\ \boldsymbol{\alpha}^2 & \boldsymbol{\alpha}^3 \end{bmatrix}, \quad \mathbf{K} = \begin{bmatrix} & 0 & & 0 \\ O_4^4 & K_1 & O_4^4 & K_2 \\ & 0 & & 0 \\ & K_3 & & K_4 \end{bmatrix}$$

$$\mathbf{g} = \begin{bmatrix} I_3 \otimes \boldsymbol{\alpha}^1 & 0 \\ E_1^2 \otimes \tilde{K} + E_2^2 \otimes \boldsymbol{\alpha}^2 & I_2 \otimes \boldsymbol{\alpha}^3 \end{bmatrix}$$

where $\boldsymbol{\alpha}^i$ are 2×2, $\det \boldsymbol{\alpha}^1 \neq 0$, $\det \boldsymbol{\alpha}^3 \neq 0$ and $\tilde{K} = (\mathbf{K}^i)$ $i = 1, 2$, $\mathbf{K}^1 = (K_1 \; K_2)$, $\mathbf{K}^2 = (K_3 \; K_4)$. Call C *equivalent to* $D = [D_1 \; D_2 \; D_3 \; D_4]$ if there is a $\mathbf{g}$

with $C\mathbf{g} = D$. Let $C_1 = [C_{1,1}\ C_{1,2}\ C_{1,3}]$, $C_2 = [C_{2,1}\ C_{2,2}\ C_{2,3}]$, $C_3 = [C_{3,1}\ C_{3,2}]$, $C_4 = [C_{4,1}\ C_{4,2}]$ with $C_{i,j}$ $p \times 1$ and let

$$\tilde{C}_1 = \begin{bmatrix} C_{1,1} \\ C_{1,2} \\ C_{1,3} \end{bmatrix}, \quad \tilde{C}_2 = \begin{bmatrix} C_{2,1} \\ C_{2,2} \\ C_{2,3} \end{bmatrix}, \quad \tilde{C}_{3,1} = \begin{bmatrix} C_{3,1} \\ C_{3,2} \\ 0 \end{bmatrix},$$

$$\tilde{C}_{3,2} = \begin{bmatrix} 0 \\ C_{3,1} \\ C_{3,2} \end{bmatrix}, \quad \tilde{C}_{4,1} = \begin{bmatrix} C_{4,1} \\ C_{4,2} \\ 0 \end{bmatrix}, \quad \tilde{C}_{4,2} = \begin{bmatrix} 0 \\ C_{4,1} \\ C_{4,2} \end{bmatrix}$$

be elements of $k^{3p} = k^{(q+1)p} = k^{\partial_1 p}$. Consider the "flag" $\mathcal{F}(C) = [V_1, V_2, V_3]$ where $V_1 \subset V_2 \subset V_3 \subset k^{3p}$ and

$$\begin{aligned} V_1 &= \operatorname{span}\left[\tilde{C}_{3,2}\ \tilde{C}_{4,2}\right] \\ V_2 &= \operatorname{span}\left[\tilde{C}_{3,1}\ \tilde{C}_{4,}, V_1\right] \\ V_3 &= \operatorname{span}\left[\tilde{C}_1, \tilde{C}_2, V_2\right]. \end{aligned}$$

Suppose that $C\mathbf{g} = D$ and let $\mathcal{F}(D) = [W_1, W_2, W_3]$ be the "flag" corresponding to D. We *claim* that $\mathcal{F}(C) = \mathcal{F}(D)$ i.e., $V_1 = W_1$, $V_2 = W_2$, $V_3 = W_3$. Now

$$C\mathbf{g} = \left[\mathbf{C}_1(I_3 \otimes \boldsymbol{\alpha}^1) + \mathbf{C}_2(E_1^2 \otimes \mathbf{K} + E_2^2 \otimes \boldsymbol{\alpha}^2)\ \mathbf{C}_2(I_2 \otimes \boldsymbol{\alpha}^3)\right]$$

where $\mathbf{C}_1 = [C_1\ C_2]$, $\mathbf{C}_2 = [C_3\ C_4]$ so that

$$\begin{aligned} D_{3,1} &= (\boldsymbol{\alpha}^3)_1^1 C_{3,1} + (\boldsymbol{\alpha}^3)_1^2 C_{4,1} \\ D_{3,2} &= (\boldsymbol{\alpha}^3)_1^1 C_{3,2} + (\boldsymbol{\alpha}^3)_1^2 C_{4,2} \\ D_{4,1} &= (\boldsymbol{\alpha}^3)_2^1 C_{3,1} + (\boldsymbol{\alpha}^3)_2^2 C_{4,1} \\ D_{4,2} &= (\boldsymbol{\alpha}^3)_2^1 C_{3,2} + (\boldsymbol{\alpha}^3)_2^2 C_{4,2}. \end{aligned}$$

It follows that $V_1 = W_1$, $V_2 = W_2$ as $\det \boldsymbol{\alpha}^3 \neq 0$. Since $\mathbf{C}_2 = \left[C_{3,1}\ C_{3,2}\ C_{4,1}\ C_{4,2}\right]$ and (say)

$$\boldsymbol{\alpha}^2 = \begin{bmatrix} \alpha_1 & \alpha_2 \\ \alpha_3 & \alpha_4 \end{bmatrix}$$

we can readily see that $\mathbf{C}_2(E_1^2 \otimes \mathbf{K} + E_2^2 \otimes \boldsymbol{\alpha}^2) = [K_1 C_{3,1} + K_3 C_{4,1} \vdots \alpha_1 C_{3,1} + K_1 C_{3,2} + \alpha_3 C_{4,1} + K_3 C_{4,2} \vdots \alpha_1 C_{3,2} + \alpha_3 C_{4,2} \vdots K_2 C_{3,1} + K_4 C_{4,1} \vdots \alpha_2 C_{3,1} + K_2 C_{3,2} + \alpha_4 C_{4,1} + K_4 C_{4,2} \vdots \alpha_2 C_{3,2} + \alpha_4 C_{4,2}]$ and that, consequently,

$$\begin{aligned} \tilde{D}_1 &= (\boldsymbol{\alpha}^1)_1^1 \tilde{C}_1 + (\boldsymbol{\alpha}^1)_1^2 \tilde{C}_2 + K_1 \tilde{C}_{3,1} + K_3 \tilde{C}_{4,1} + \alpha_1 \tilde{C}_{3,2} + \alpha_3 \tilde{C}_{4,2} \\ \tilde{D}_2 &= (\boldsymbol{\alpha}^1)_2^1 \tilde{C}_1 + (\boldsymbol{\alpha}^1)_2^2 \tilde{C}_2 + K_2 \tilde{C}_{3,1} + K_4 \tilde{C}_{4,1} + \alpha_2 \tilde{C}_{3,2} + \alpha_4 \tilde{C}_{4,2} \end{aligned}$$

i.e., $W_3 \subset V_3$. Since $\det \boldsymbol{\alpha}^1 \neq 0$, $V_3 \subset W_3$ and $\mathcal{F}(C) = \mathcal{F}$ $(D = C\mathbf{g})$. Now, suppose that D is given with "flag" $\mathcal{F}(D) = [W_1, W_2, W_3]$ and that $\mathcal{F}(D) = \mathcal{F}(C)$ i.e., $W_1 = V_1$, $W_2 = V_2$, $W_3 = V_3$. Then $\mathcal{R}([\tilde{C}_{3,2}\ \tilde{C}_{4,2}]) = \mathcal{R}([\tilde{D}_{3,2}\ \tilde{D}_{4,2}])$ which implies that

$$\mathcal{R}\left(\begin{bmatrix} D_{3,1} & D_{4,1} \\ D_{3,2} & D_{4,2} \end{bmatrix}\right) = \mathcal{R}\left(\begin{bmatrix} C_{3,1} & C_{4,1} \\ C_{3,2} & C_{4,2} \end{bmatrix}\right)$$

where $\mathcal{R}(\ ) =$ "range of ". It follows that there is a non-singular 2×2 matrix $\boldsymbol{\alpha}^3$ such that

$$\begin{bmatrix} D_{3,1} & D_{4,1} \\ D_{3,2} & D_{4,2} \end{bmatrix} = \begin{bmatrix} C_{3,1} & C_{4,1} \\ C_{3,2} & C_{4,2} \end{bmatrix} \boldsymbol{\alpha}^3$$

and hence that $\mathbf{C}_2(I_2 \otimes \boldsymbol{\alpha}^3) = \mathbf{D}_2$. Since $W_3 = V_3$, there are K_1, K_2, K_3, K_4 and a 2×2 matrix $\boldsymbol{\alpha}^2$ such that

$$\begin{aligned} \tilde{D}_1 &= a\tilde{C}_1 + b\tilde{C}_2 + K_1\tilde{C}_{3,1} + K_3\tilde{C}_{4,1} + (\boldsymbol{\alpha}^2)_1^1\tilde{C}_{3,2} + (\boldsymbol{\alpha}^2)_1^2\tilde{C}_{4,2} \\ \tilde{D}_2 &= c\tilde{C}_1 + d\tilde{C}_2 + K_2\tilde{C}_{3,1} + K_4\tilde{C}_{4,1} + (\boldsymbol{\alpha}^2)_2^1\tilde{C}_{3,2} + (\boldsymbol{\alpha}^2)_2^2\tilde{C}_{4,2}. \end{aligned}$$

Setting $\mathbf{K} = (\mathbf{K}^i)$, $\mathbf{K}^1 = (K_1, K_2)$, $\mathbf{K}^2 = (K_3, K_4)$, we have $\mathbf{C}_2(E_1^2 \otimes \mathbf{K} + E_2^2 \otimes \boldsymbol{\alpha}^2) = [D_{1,1} - aC_{1,1} - bC_{2,1} \vdots D_{1,2} - aC_{1,2} - bC_{2,2} \vdots D_{1,3} - aC_{1,3} - bC_{2,3} \vdots D_{2,1} - cC_{1,1} - dC_{2,1} \vdots D_{2,2} - cC_{1,12} - dC_{2,2} \vdots D_{2,3} - cC_{1,3} - dC_{2,3}]$. Thus, $\bmod W_2$ $(= V_2)$ $\mathcal{R}([\tilde{D}_1\ \tilde{D}_2]) = \mathcal{R}([\tilde{C}_1\ \tilde{C}_2])$ and there is a 2×2 non-singular matrix $\boldsymbol{\alpha}^1$ with

$$\begin{bmatrix} D_{1,1} & D_{2,1} \\ D_{1,2} & D_{2,2} \\ D_{1,3} & D_{2,3} \end{bmatrix} = \begin{bmatrix} C_{1,1} & C_{2,1} \\ C_{1,2} & C_{2,2} \\ C_{1,3} & C_{2,3} \end{bmatrix} \boldsymbol{\alpha}^1$$

$(\bmod V_2)$. It follows that there is a $\mathbf{g} \in \Gamma_{f,(3,3,2,2)}$ with $C\mathbf{g} = D$. So again the "flag" $\mathcal{F}(C)$ characterizes the equivalence class of C.

Example 19.73 (The "Generic" Case) Let $n = qm + r$ with $0 \leq r < m$ and let $x = (A, B, C)$ be "generic" as in Example 19.70. Let $C = [C_1, \ldots, C_r, C_{r+1}, \ldots, C_m]$ with $C_1, \ldots, C_r$ $p \times q+1$ and $C_{r+1}, \ldots, C_m$ $p \times q$. Let $\mathbf{g}$ be an element of $\Gamma_{f,\partial}$ so that

$$\mathbf{g} = \begin{bmatrix} I_{q+1} \otimes \boldsymbol{\alpha}^1 & \mathbf{O}_{q(m-r)}^{(q+1)r} \\ E_1^q \otimes \tilde{K} + E_2^q \otimes \boldsymbol{\alpha}^2 & I_q \otimes \boldsymbol{\alpha}^3 \end{bmatrix}.$$

Then $D = C\mathbf{g} = \left[\mathbf{C}_1(I_{q+1} \otimes \alpha^1) + \mathbf{C}_2(E_1^q \otimes \tilde{K} + E_2^q \otimes \alpha^2)\ \mathbf{C}_2(I_q \otimes \alpha^3)\right]$. As in the prior examples, let

$$\tilde{C}_i = \begin{bmatrix} C_{i,1} \\ \vdots \\ C_{i,q+1} \end{bmatrix}, \quad \tilde{C}_{j,1} = \begin{bmatrix} C_{j,1} \\ \vdots \\ C_{j,q} \\ 0 \end{bmatrix}, \quad \tilde{C}_{j,2} = \begin{bmatrix} 0 \\ C_{j,1} \\ \vdots \\ C_{j,q} \end{bmatrix}$$

for $i = 1, \ldots, r$, $j = r+1, \ldots, m$ (and similarly for $\tilde{D}_\ell$). Let $V_1 = \text{span}[\tilde{C}_{j,2}\colon j = r+1, \ldots, m]$, $V_2 = \text{span}[\tilde{C}_{j,1}, V_1\colon j = r+1, \ldots, m]$ and $V_3 = \text{span}[\tilde{C}_1, \ldots, \tilde{C}_r, V_2]$. Let $W_1 = \text{span}[\tilde{D}_{j,2}\colon j = r+1, \ldots, m]$, $W_2 = \text{span}[\tilde{D}_{j,1}, W_1\colon j = r+1, \ldots, m]$, and $W_3 = \text{span}[\tilde{D}_1, \ldots, \tilde{D}_r, W_2]$. Then $V_1 \subset V_2 \subset V_3$, $W_1 \subset W_2 \subset W_3$ are the "flags" $\mathcal{F}(C)$, $\mathcal{F}(D)$ corresponding to C, D respectively. Since

$$D_{j,s} = \sum_{t=r+1}^{m} (\alpha^3)_{j-r}^{t-r} C_{t,s}$$

for $j = r+1, \ldots, m$, $s = 1, \ldots, q$, we have

$$\tilde{D}_{j,1} = \sum_{t=r+1}^{m} (\alpha^3)_{j-r}^{t-r} \tilde{C}_{t,1}, \quad \tilde{D}_{j,2} = \sum_{t=r+1}^{m} (\alpha^3)_{j-r}^{t-r} \tilde{C}_{t,2}$$

for $j = r+1, \ldots, m$. Therefore, $W_1 = V_1$, $W_2 = V_2$. It is also easy to se that

$$\tilde{D}_i = \sum_{\ell=1}^{r} (\alpha^1)_i^\ell \tilde{C}_\ell + \sum_{t=r+1}^{m} \{(\tilde{K})_i^{t-r} \tilde{C}_{t,1} + (\alpha^2)_i^{t-r} \tilde{C}_{t,2}\}$$

for $i = 1, \ldots, r$ and hence, that $V_3 = W_3$. In other words, $\mathcal{F}(C) = \mathcal{F}(D)$ when $D = C\mathbf{g}$. Now suppose that D is given with $\mathcal{F}(D) = \mathcal{F}(C)$. Then

$$\mathcal{R}([\tilde{C}_{j,2}]_{j=r+1,\ldots,m}) = \mathcal{R}([\tilde{D}_{j,2}]_{j=r+1,\ldots,m})$$

which implies

$$\mathcal{R}\left(\begin{bmatrix} D_{r+1,1} & \cdots & D_{m,1} \\ \vdots & & \vdots \\ D_{r+1,q} & \cdots & D_{m,q} \end{bmatrix}\right) = \mathcal{R}\left(\begin{bmatrix} C_{r+1,1} & \cdots & C_{m,1} \\ \vdots & & \vdots \\ C_{r+1,q} & \cdots & C_{m,q} \end{bmatrix}\right)$$

It follows that there is an $\alpha^3 \in \text{GL}(m-r, k)$ such that

$$\begin{bmatrix} D_{r+1,1} & \cdots & D_{m,1} \\ \vdots & & \vdots \\ D_{r+1,q} & \cdots & D_{m,q} \end{bmatrix} = \begin{bmatrix} C_{r+1,1} & \cdots & C_{m,1} \\ \vdots & & \vdots \\ C_{r+1,q} & \cdots & C_{m,q} \end{bmatrix} \alpha^3$$

and hence, that $\mathbf{C}_2(I_q \otimes \boldsymbol{\alpha}^3) = \mathbf{D}_2$. Since $V_3 = W_3$, we have

$$\tilde{D}_i = \sum_{\ell=1}^{r} (\mathbf{a})_i^{\ell} \tilde{C}_{\ell} + \sum_{t=r+1}^{m} \{(\tilde{K})_i^{t-r} \tilde{C}_{t,1} + (\boldsymbol{\alpha}^2)_i^{t-r} \tilde{C}_{t,2}\}$$

for some $\mathbf{a}$, $\tilde{K}$, $\boldsymbol{\alpha}^2$. Then $\operatorname{mod} W_2$ $(= V_2)$, $\mathcal{R}([\tilde{D}_1, \ldots, \tilde{D}_r]) = \mathcal{R}([\tilde{C}_1, \ldots, \tilde{C}_r])$ and there is an $\boldsymbol{\alpha}^1 \in \mathrm{GL}(v, k)$ with

$$\begin{bmatrix} D_{1,1} & \cdots & D_{r,1} \\ \vdots & & \vdots \\ D_{1,q+1} & \cdots & D_{r,q+1} \end{bmatrix} = \begin{bmatrix} C_{1,1} & \cdots & C_{r,1} \\ \vdots & & \vdots \\ C_{1,q+1} & \cdots & C_{r,q+1} \end{bmatrix} \boldsymbol{\alpha}^1$$

$(\operatorname{mod} V_2)$. It follows that there is a $\mathbf{g} \in \Gamma_{f,\partial}$ with $C\mathbf{g} = D$. Thus, the "flag" $\mathcal{F}(C)$ characterizes the class of $C \operatorname{mod} \Gamma_{f,\partial}$ in the "generic" case. (Analysis of the general situation is left to the reader ([D-1]).) Call C *full* if the $\dim V_i$ are maximal i.e., $\dim V_1 = m - r$, $\dim V_2 = m - r + m - r = 2m - 2r$, and $\dim V_3 = 2m - 2r + r = 2m - r$. Let $u_1, \ldots, u_{m-r}, v_1, \ldots, v_{2m-2r}, w_1, \ldots, w_{2m-2r}, x_1, \ldots, x_r$ be elements of $\mathbb{A}^{p(q+1)}$. Let $[\ ,\ldots,\ ]$ = span of. Let $X_1, \ldots, X_{p(q+1)}$ be the coordinate functions on $\mathbb{A}^{p(q+1)}$. Set $L^1 = V(X_1, \ldots, X_p)$ and $L^2 = V(X_{qp+1}, \ldots, X_{qp+p})$. In an appropriate (product of) Grassmann variety, let

$$\mathbf{X}_{\Delta} = \{[u_1, \ldots, u_{m-r}] \times [v_1, \ldots, v_{2m-2r}] \times [w_1, \ldots, w_{2m-2r}, x_1, \ldots, x_r]\}$$

such that

(a) $u_i \in L^1$, $i = 1, \ldots, m - r$

(b) $v_i - u_i = 0$, $i = 1, \ldots, m - r$

(c) $v_{m-r+j} \in L^2$, $j = 1, \ldots, m - r$

(d) $(u_i)^{p+t} - (v_{i+m-r})^t = 0$, $t = 1, \ldots, pq$, $i = 1, \ldots, m - r$

(e) $w_{\ell} - v_{\ell} = 0$, $\ell = 1, \ldots, 2m - 2r$.

Let $\Phi = \{C \colon C \text{ is full}\}$ and let $\psi \colon \Phi \to \mathbf{X}_{\Delta}$ be given by

$$\psi(C) = \{V_1(C) \times V_2(C) \times V_3(C)\}.$$

ψ is clearly injective on $\Gamma_{f,\partial}$-orbits. If

$$\begin{aligned} \gamma = [\gamma_1, \ldots, \gamma_{m-r}] &\times [\gamma_1, \ldots, \gamma_{m-r}, \gamma_{m-r+1}, \ldots, \gamma_{2m-2r}] \\ &\times [\gamma_1, \ldots, \gamma_{2m-2r}, \lambda_1, \ldots, \lambda_r] \end{aligned}$$

is an element of $\mathbf{X}_\Delta$ and we let

$$C_i = \left[\lambda_i^{1\cdots p}\lambda_i^{p+1\cdots 2p}\cdots\lambda_i^{qp+1\cdots qp+p}\right]$$

for $i = 1, \ldots, r$ and

$$C_j = \left[\gamma_{j-r}^{p+1\cdots 2p}\cdots\gamma_{j-r}^{qp+1\cdots qp+p}\right] = \left[\gamma_{m-r+j}^{1\cdots p}\cdots\gamma_{m-r+j}^{(q-1)p+1\cdots qp}\right]$$

for $j = r+1, \ldots, m$, then $C = \left[C_1 \cdots C_r \; C_{r+1} \cdots C_m\right]$ is full and $\psi(C) = \gamma$. In other words, ψ is surjective. Since ψ is a $\Gamma_{f,\partial}$-morphism, $(\mathbf{X}_\Delta, \psi)$ is a quotient for $\Phi \operatorname{mod} \Gamma_{f,\partial}$. It is *not*, in general, a geometric quotient.

20
Output Feedback

We now examine the notion of output feedback. Our approach (cf. [W-2], [W-3]) is quite naturally geometric. Let $G = GL(n, k)$ and let $H_{mp} = \mathrm{Hom}_k(\mathbb{A}^p, \mathbb{A}^m) \simeq M(m, p)$. Consider the set $\Gamma_0(n, m, p) = G \times H_{mp}$ and *define* a multiplication in $\Gamma_0(n, m, p)$ as follows:

$$[g, K][g_1, K_1] = [gg_1, K + K_1] \tag{20.1}$$

where $g, g_1 \in G$ and $K, K_1 \in H_{mp}$. Then

Theorem 20.2 (i) *$\Gamma_0(n, m, p)$ is a group and a quasi-affine variety;* (ii) *the mappings $\mu\colon \Gamma_0(n, m, p) \times \Gamma_0(n, m, p) \to \Gamma_0(n, m, p)$ and $i\colon \Gamma_0(n, m, p) \to \Gamma_0(n, m, p)$ given by $\mu(\gamma\gamma_1) = \gamma\gamma_1$ and $i(\gamma) = \gamma^{-1}$ are morphisms;* (iii) *$\Gamma_0(n, m, p)$ is non-singular of dimension $n^2 + mp$; and,* (iv) *$\Gamma_0(n, m, p)$ acts on $\mathbb{A}^{n^2+n(m+p)}$ via the morphism φ given by*

$$\varphi([g, K], (A, B, C)) = (g(A + BKC)g^{-1}, gB, Cg^{-1}). \tag{20.3}$$

Proof. (i), (ii) and (iii) are clear. As for (iv), note that $[g, K]\cdot[g_1, K_1]\cdot(A, B, C) = [g, K](g_1(A+BK_1C)g_1^{-1}, g_1B, Cg_1^{-1}) = (g[g_1(A+BK_1C)g_1^{-1} + g_1BKCg_1^{-1}]g^{-1}, gg_1B, Cg_1^{-1}g^{-1}) = [gg_1, K + K_1] \cdot (A, B, C)$.

Definition 20.4 $\Gamma_0(n, m, p)$ is the (*pure*) *output feedback group* of type (n, m, p) and the action of $\Gamma_0(n, m, p)$ on $\mathbb{A}^{n^2+n(m+p)}$ is called *output feedback*.

Proposition 20.5 *If $\mathbf{Y}(A, B, C)$ and $\mathbf{Z}(A, B, C)$ are the controllability and observability matrices, then $\rho(\mathbf{Y})$ and $\rho(\mathbf{Z})$ are invariant under output feedback.*

P. Falb, *Methods of Algebraic Geometry in Control Theory: Part II*,
Modern Birkhäuser Classics, https://doi.org/10.1007/978-3-319-96574-1_20

Thus, controllability, observability and minimality are invariant under output feedback.

Proof. $\mathbf{Y}(\gamma \cdot x) = g[B\ (A+BKC)B \cdots]$ so that $\mathcal{R}(\mathbf{Y}(\gamma \cdot x)) = g\mathcal{R}(\mathbf{Y}(x))$. Similarly,

$$\mathbf{Z}(\gamma \cdot x) = \begin{bmatrix} Cg^{-1} \\ C(A+BKC)g^{-1} \\ \vdots \end{bmatrix}$$

and $\mathcal{R}(\mathbf{Z}(\gamma \cdot x)) = \mathcal{R}(\mathbf{Z}(x))g^{-1}$.

Proposition 20.6 *Let $x = (A,B,C)$ and $F_x(z) = C(zI-A)^{-1}B$. Then $F_{\gamma \cdot x}(z) = C(zI-(A+BKC))^{-1}B = F_x(z)(I-KF_x(z))^{-1}$. If $F_x(z) = P(z)Q^{-1}(z)$ with P, Q coprime, then $F_{\gamma \cdot x}(z) = P(z)(Q(z)-KP(z))^{-1}$ and P, $Q-KP$ are coprime.*

Proof.

$$\begin{aligned} F_{\gamma \cdot x}(z) &= Cg^{-1}(zI - g(A+BKC)g^{-1})^{-1}gB \\ &= C(zI-(A+BKC))^{-1}B \\ &= C[(zI-A)(I-(zI-A)^{-1}BKC)]^{-1}B \\ &= C[I-(zI-A)^{-1}BKC]^{-1}(zI-A)^{-1}B \\ &= C(zI-A)^{-1}B[I+KF_x(z)+(KF_x(z))^2+\cdots] \\ &= F_x(z)\cdot[I-KF_x(z)]^{-1}. \end{aligned}$$

Since

$$\begin{aligned} F_{\gamma \cdot x}(z) &= F_x(z)\cdot[I-KF_x(z)]^{-1} = PQ^{-1}[I-KPQ^{-1}]^{-1} \\ &= PQ^{-1}[(Q-KP)Q^{-1}] = P[Q-KP]^{-1}. \end{aligned}$$

Finally, $XP+YQ=I$ implies $(X+YK)P+Y(Q-KP)=I$ and, conversely, $X_1P+Y_1(Q-KP)=I$ implies $(X_1-Y_1K)P+Y_1Q=I$.

Corollary 20.7 $\Delta_\gamma(z) = \Delta_K(z) = \det(zI-A+BKC) = \det(Q(z)+KP(z))$ *and so*

$$\Delta_K(z) = \det\left([K\ I_m]\begin{bmatrix} P(z) \\ Q(z) \end{bmatrix}\right). \tag{20.8}$$

We wish to study the question of "assigning" the roots of $\Delta_K(z)$. We observe that $[K\ I_m]$ is $m \times (p+m)$ and that $\begin{bmatrix} P \\ Q \end{bmatrix}$ is $p+m \times m$. Let

$$N = \binom{m+p}{m} - 1 = \binom{m+p}{p} - 1 \tag{20.9}$$

and let $[K \ I_m] = \mathbf{K}$ be viewed as an element of $\mathrm{Gr}(m, m+p)$. Then $\pi(\mathbf{K}) = (\pi_{\boldsymbol{\alpha}}(\mathbf{K}))$, $\boldsymbol{\alpha} = (\alpha_1, \ldots, \alpha_m)$, $1 \leq \alpha_1 < \cdots < \alpha_m \leq m+p$, with

$$\pi_{\boldsymbol{\alpha}}(K) = \det\big[\mathbf{K}\big|_{\alpha_1\cdots\alpha_m}^{1\cdots m}\big] \tag{20.10}$$

are the Plücker coordinates of $\mathbf{K}$ in $\mathbb{P}^N$. For each z, $\psi_F(z)$ may be viewed as an element of the "dual" Grassmannian $\mathrm{Gr}(n, m+p)^*$ with Plücker coordinates

$$\psi_{F,\boldsymbol{\alpha}}(z) = \det\big[\psi_F(z)\big|_{1\cdots m}^{\alpha_1\cdots\alpha_m}\big] \tag{20.11}$$

in $\mathbb{P}^N$. Then, in view of (20.8), we have

$$\Delta_K(z) = \sum_{\boldsymbol{\alpha}} \pi_{\boldsymbol{\alpha}}(\mathbf{K})\psi_{F,\boldsymbol{\alpha}}(z) \tag{20.12}$$

which is a polynomial of degree n. If we view polynomials of degree n as points in $\mathbb{P}^n$, then $\sum_{j=0}^{n} x_j z^j$ is a typical polynomial with $x_0, \ldots, x_n$ as coordinates in $\mathbb{A}^{n+1}$. Consider the map $\Lambda_F \colon \mathbb{P}^N \to \mathbb{A}^{n+1}$ given by

$$\Lambda_F(\xi = (\xi_{\boldsymbol{\alpha}})) = \sum_{\boldsymbol{\alpha}} \xi_{\boldsymbol{\alpha}}\psi_{F,\boldsymbol{\alpha}}(z). \tag{20.13}$$

If $\boldsymbol{\alpha} \neq \boldsymbol{\alpha}_* = (p+1, \ldots, p+m)$, then

$$\psi_{F,\boldsymbol{\alpha}}(z) = \sum_{j=0}^{n-1} a_{j,\boldsymbol{\alpha}} z^j = (a_{0,\boldsymbol{\alpha}}, \ldots, a_{n-1,\boldsymbol{\alpha}}, 0) \tag{20.14}$$

and

$$\psi_{F,\boldsymbol{\alpha}_*}(z) = \det Q(z) = \sum_{j=0}^{n-1} \chi_{j+1} z^j + z^n = (\chi_1, \ldots, \chi_n, 1). \tag{20.15}$$

It follows that

$$\begin{aligned} \Lambda_F(\boldsymbol{\xi}) &= \sum_{\boldsymbol{\alpha}} \left(\sum_{j=0}^{n-1} \xi_{\boldsymbol{\alpha}} a_{j,\boldsymbol{\alpha}} z^j \right) + \xi_{\boldsymbol{\alpha}_*} \left(\sum_{j=0}^{n-1} \chi_{j+1} z^j + z^n \right) \\ &= \left(\sum_{\boldsymbol{\alpha}} \xi_{\boldsymbol{\alpha}} a_{j,\boldsymbol{\alpha}} z^j + \xi_{\boldsymbol{\alpha}_*}\chi_{j+1}, \xi_{\boldsymbol{\alpha}_*} \right). \end{aligned} \tag{20.16}$$

Let $Y_0, \ldots, Y_N$ be coordinates on $\mathbb{P}^N$ and let

$$\begin{aligned} L_j(Y_0, \ldots, Y_N) &= \sum_{t=0}^{N-1} Y_t a_{j,t} + Y_N \chi_{j+1}, \quad j = 0, 1, \ldots, n-1 \\ L_n(Y_0, \ldots, Y_N) &= Y_N. \end{aligned} \tag{20.17}$$

Let $E_F = E = V(L_0, L_1, \dots, L_{n-1}, L_n)$ in $\mathbb{P}^N$. Then E is a linear subspace of $\mathbb{P}^N$ with $\dim E \geq N - (n+1)$. $\mathrm{Gr}(m, m+p)$ is a subvariety of $\mathbb{P}^N$ and Λ_F is a *projection* from $\mathbb{P}^N - E$ into $\mathbb{P}^n$.

Definition 20.18 $F(z)$ is *assignable* if the map $\Lambda_F\colon \mathrm{Gr}(m, m+p) - E_F \to \mathbb{P}^n$ is surjective.

If $\boldsymbol{\xi} = (\xi_{\boldsymbol{\alpha}}) \in \mathrm{Gr}(m, m+p) - E$, then $\xi_{\boldsymbol{\alpha}_*} \neq 0$ and if $\mathbf{Z} \in M(m, m+p)$ with $\pi_{\boldsymbol{\alpha}}(\mathbf{Z}) = \xi_{\boldsymbol{\alpha}}$, then $\det\left[\mathbf{Z}\big|_{p+1\cdots p+m}^{1\cdots m}\right] = \xi_{\boldsymbol{\alpha}_*} \neq 0$, $g = \mathbf{Z}_{p+1\cdots p+m}^{1\cdots m} \in \mathrm{GL}(m, k)$ and $g^{-1}\mathbf{Z} = [K_{\mathbf{Z}}\ I_m]$. In other words, $\Lambda_F(\mathbf{Z}) = \Delta_{F, K_{\mathbf{Z}}}(z)$ comes from an output feedback $K_{\mathbf{Z}}$.

Let $X = \mathrm{Gr}(m, m+p) - E$ so that $\dim X = mp = \dim \mathrm{Gr}(m, m+p)$ and let $\psi_F\colon X \to \mathbb{P}^n$ be the (restriction of the) projection Λ_F. Let $\psi_F(X) = Y$ so that the map $\psi_F\colon X \to Y$ is a dominant morphism and $\dim X \geq \dim Y$.

Proposition 20.19 *If $F(z)$ is assignable, then $mp \geq n$. In other words, $mp \geq n$ is a necessary condition for assignability.*

Proof. If $F(z)$ is assignable, then $Y = \psi_F(X) = \mathbb{P}^n$ and $\dim X = mp \geq \dim Y = n$.

Example 20.20 Let $n = 2$, $m = p = 1$ and let

$$A = \begin{bmatrix} 0 & 1 \\ 0 & 0 \end{bmatrix}, \quad b = \begin{bmatrix} 0 \\ 1 \end{bmatrix}, \quad c = \begin{bmatrix} 1 & 0 \end{bmatrix}$$

so that $x = (A, b, c) \in S_{1,1}^2$. Then

$$A + bKc = \begin{bmatrix} 0 & 1 \\ 0 & 0 \end{bmatrix} + \begin{bmatrix} 0 \\ 1 \end{bmatrix} \begin{bmatrix} K & 0 \end{bmatrix} = \begin{bmatrix} 0 & 1 \\ K & 0 \end{bmatrix}$$

and $\det(zI = (A + bKC)) = z^2 - K$. Thus, F_x is not assignable.

Now, let

$$\psi_{F,\boldsymbol{\alpha}}(z) = (a_{0,\boldsymbol{\alpha}}, \dots, a_{n-1,\boldsymbol{\alpha}}, a_{n,\boldsymbol{\alpha}}) = v_{\boldsymbol{\alpha}} \in \mathbb{A}^{n+1}$$

and let $V_F = \mathrm{span}[v_{\boldsymbol{\alpha}}]$. If $\dim V_F = n + 1$, then $\dim E_F = N - (n+1)$ (as technically $v_{\boldsymbol{\alpha}} \in (\mathbb{P}^N)^*$ and $E_F = \mathrm{Ker}\ V_F$). Let $\mathbf{L}_n$ be a linear subspace disjoint from E_F. Then $\mathbf{L}_n \simeq \mathbb{P}^n$ and if $\boldsymbol{\xi} = (\xi_{\boldsymbol{\alpha}}) \in \mathrm{Gr}(m, m+p) - E_F$, $L_{\boldsymbol{\xi}} = J(E_F, \boldsymbol{\xi})$ (join of), then $\dim L_{\boldsymbol{\xi}} = N - n$ and $L_{\boldsymbol{\xi}} \cap \mathbf{L}_n$ is a (unique) point $\Lambda_F(\boldsymbol{\xi})$. If $\mathbf{v} \in \mathbf{L}_n$ $(= \mathbb{P}^n)$, then $L_{\mathbf{v}} = J(E_F, \mathbf{v})$ has dimension $N - n$. It follows that if $mp \geq n$, then $\dim X + \dim L_{\mathbf{v}} \geq 0$, $L_{\mathbf{v}} \cap X \neq \emptyset$, and Λ_F is surjective. Thus, we have shown:

Proposition 20.21 *If $\dim V_F = n + 1$ and $mp \geq n$, then $F(z)$ is assignable.*

Proposition 20.22 *If $mp \geq n$, then* $\dim V_{F_x} = n+1$ *for almost all x and assignability is generic.*

Proof. Recall that if $F_x(z) = \sum_{t=0}^{n-1} \mathbf{B}_t z^t/z^n + \cdots$, then $F^i_{xj}(z) = \sum_{t=0}^{n-1} b^i_{tj} z^t/z^n + \cdots$ where $\mathbf{B}_t = (b^i_{tj})$ and

$$(-1)^{\sigma_{ij}} \psi_{F_x,\alpha_{ij}}(z) = \sum_{t=0}^{n-1} b^i_{tj} z^t$$

with $\alpha_{ij} = (i, p+1, \ldots, p+j-1, p+j+1, \ldots, p+m)$, $i = 1, \ldots, p$, $j = 1, \ldots, m$. In other words,

$$v_{\alpha_{ij}} = (b^i_{0j}, \ldots, b^i_{n-1,j}, 0)$$

and it will be sufficient to show that $\dim_k$ span $[v_{\alpha_{ij}}\colon i = 1, \ldots, p,\ j = 1, \ldots, m] = n$. If $x = (A, B, C)$, then

$$\mathbf{B}_t = CA^{n-t-1}B + \sum_{\ell=t+1}^{n-1} \chi_\ell(A) CA^{\ell-t-1}B, \quad t = 0, \ldots, n-2$$

$$\mathbf{B}_{n-1} = CB$$

and

$$b^i_{tj} = C^i(A^{n-t-1})B_j + \sum_{\ell=t+1}^{n-1} \chi_\ell(A) C^i(A^{\ell-t-1})B_j$$

$$b^i_{n-1_j} = C^i B_j.$$

Then $\dim_k \operatorname{span}[v_{\alpha_{ij}}] = \dim_k \operatorname{span}[w_{ij}]$ where $w_{ij} = (C^i A^{n-1} B_j, \ldots, C^i B_j)$ $(\in \mathbb{A}^n)$. But if $\#\{w_{ij}\} = mp \geq n$, then $\dim_k \operatorname{span}[w_{ij}] = n$ for almost all $x = (A, B, C)$ in $S^n_{m,p}$.

Since $\dim E_F = \dim \operatorname{Ker} V_F = \dim V_F^\perp$ (the annihilator of V_F), $\dim V_F + \dim E_F = N$. Since $\dim V_F \leq n+1$, $\dim E_F \geq N - (n+1)$. Also,

$$\begin{aligned} \dim E_F \cap \operatorname{Gr}(m, m+p) &\geq \dim E_F + \dim \operatorname{Gr}(m, m+p) - N \\ &\geq \dim E_F + mp - N \\ &\geq N - (n+1) + mp - N = mp - (n+1) \end{aligned}$$

(dimension of intersections in projective space).

Proposition 20.23 *If $mp \geq n$ and* $\dim E_F \cap \operatorname{Gr}(m, m+p) = mp - (n+1)$, *then Λ_F is surjective.*

Proof. If $\dim E_F \cap \mathrm{Gr}(m, m+p) = mp - (n+1)$, then $mp - (n+1) = \dim E_F + mp - N$ and $\dim E_F = N - (n+1)$. Since $\dim V_F + \dim E_F = N$, $\dim V_F = n+1$ and Λ_F is surjective by Proposition 20.21.

Example 20.24 Let $n = 2$, $m = 2$, $p = 2$ and $x = (A, B, C)$ where

$$A = \begin{bmatrix} 0 & 1 \\ 0 & 0 \end{bmatrix}, \quad B = \begin{bmatrix} 0 & 0 \\ 1 & 1 \end{bmatrix}, \quad C = \begin{bmatrix} 1 & 0 \\ 1 & 0 \end{bmatrix}.$$

Then $x \in S_{2,2}^2$ and $4 = mp > 2 = n$. We have

$$\begin{aligned} C(zI - A)^{-1}B &= \begin{bmatrix} 1 & 0 \\ 1 & 0 \end{bmatrix} \begin{pmatrix} z & -1 \\ 0 & z \end{pmatrix}^{-1} \begin{bmatrix} 0 & 0 \\ 1 & 1 \end{bmatrix} \\ &= \begin{bmatrix} 1 & 0 \\ 1 & 0 \end{bmatrix} \begin{bmatrix} 1/z & 1/z^2 \\ 0 & 1/z \end{bmatrix} \begin{bmatrix} 0 & 0 \\ 1 & 1 \end{bmatrix} \\ &= \begin{bmatrix} 1/z^2 & 1/z^2 \\ 1/z^2 & 1/z^2 \end{bmatrix}. \end{aligned}$$

If

$$\mathbf{K} = \begin{bmatrix} K_1 & K_2 \\ K_3 & K_4 \end{bmatrix}$$

and $K = \sum_{j=1}^{4} K_i$, then

$$zI - (A + BKC) = \begin{bmatrix} z & -1 \\ -K & z \end{bmatrix}$$

and $\det(zI - (A + BKC)) = z^2 - K = (z - \lambda)(z + \lambda)$, $\lambda = \sqrt{K}$. Thus F is *not* assignable. If

$$P = \begin{bmatrix} 1 & 0 \\ 1 & 0 \end{bmatrix}, \quad Q = \begin{bmatrix} z^2 & -1 \\ 0 & 1 \end{bmatrix}$$

then $PQ^{-1} = F$ and

$$\begin{bmatrix} P(z) \\ Q(z) \end{bmatrix} = \begin{bmatrix} 1 & 0 \\ 1 & 0 \\ z^2 & -1 \\ 0 & 1 \end{bmatrix}$$

so that P, Q are coprime. We have

$$\psi_F(z) = (0, -1, 1, -1, 1, z^2)$$

and, if $\boldsymbol{\xi} = (\xi_0, \xi_1, \xi_2, \xi_3, \xi_4, \xi_5) \in \mathrm{Gr}(2,4)$, then $\boldsymbol{\xi} \in E_F$ if and only if

$$-\xi_1 + \xi_2 - \xi_3 + \xi_4 + \xi_5 z^2 = 0.$$

The v_α are given by $v_{12} = (0,0,0)$, $v_{13} = (-1,0,0)$, $v_{14} = (1,0,0)$, $v_{23} = (-1,0,0)$, $v_{24} = (1,0,0)$, $v_{34} = (0,0,1)$ and $\dim V_F = 2 < 3 = n+1$. The equations which define E_F are

$$\begin{aligned} L_0(Y_0, \ldots, Y_5) &= -Y_1 + Y_2 - Y_3 + Y_4 = 0 \\ L_1(Y_0, \ldots, Y_5) &= 0 \\ L_2(Y_0, \ldots, Y_5) &= Y_5 = 0 \end{aligned}$$

so that $E_F = V(-Y_1 + Y_2 - Y_3 + Y_4, Y_5)$ and $\dim E_5 = 3$. Thus, $\dim E_F \cap \mathrm{Gr}(2,4) \geq \dim E_F + \dim \mathrm{Gr}(2,4) - 5 = 3+4-5 = 2 > 1 = 4-3 = mp-(n+1)$.

Proposition 20.25 *If $n = mp$ and $E_F \cap \mathrm{Gr}(m, m+p) = \emptyset$, then Λ_F is surjective.*

Proof. $\dim E_F \geq N-(n+1) = N-mp-1$ and $\dim E_F \cap \mathrm{Gr}(m, m+p) \geq -1$. Since $E_F \cap \mathrm{Gr}(m, m+p) = \emptyset$, $-1 \geq \dim E_F + n - N$ and $N-n-1 \geq \dim E_F \geq N-n-1$. Thus, $\dim E_F = N-(n+1)$ and $\dim V_F = n+1$. Λ_F is surjective by Proposition 20.21.

Definition 20.26 $F(z)$ is *non-degenerate* if $E_F \cap \mathrm{Gr}(m, m+p) = \emptyset$ and $F(z)$ is *degenerate* if $\dim E_F \cap \mathrm{Gr}(m, m+p) \geq 0$ (i.e., is non-empty).

Proposition 20.27 *If $mp > n$, then all $x \in S^n_{m,p}$ are degenerate.*

Proof. $\dim E_{F_x} \cap \mathrm{Gr}(m, m+p) \geq \dim E_{F_x} + mp - N \geq N-(n+1) + mp - N \geq mp-(n+1) \geq 0$.

Proposition 20.28 *If $mp = n$ and $F(z)$ is non-degenerate then Λ_F is surjective.*

If $mp = n$, then it is natural to ask if there are non-degenerate $F(z)$ and if the property is generic.

Example 20.29 Let $p = n$, $m = 1$ and

$$P(z) = \begin{bmatrix} 1 \\ z \\ \vdots \\ z^{n-1} \end{bmatrix}, \quad Q(z) = z^n, \quad F(z) = PQ^{-1} = \begin{bmatrix} 1/z^n \\ \vdots \\ 1/z \end{bmatrix}.$$

Then $\psi_F(z) = (1, z, \dots, z^{n-1}, z^n)$ (the so-called "rational curve") and, after homogenizing, $\psi_F(x_0, x_1) = (x_0^n, x_0^{n-1}x_1, \dots, x_1^n)$. Let $L(Y_0, \dots, Y_N) = \alpha_0 Y_0 + \dots + \alpha_N Y_N = 0$ be any hyperplane in $\mathbb{P}^N$ with $N = \binom{n+1}{n} - 1 = n$. Then $\psi_F(x_0, x_1) \not\subset L$ as $\sum_{j=0}^{n} \alpha_j x_0^{n-j} x_1^j \equiv 0$ implies all $\alpha_j = 0$.

Example 20.30 Let $p = 1$, $m = n$ and

$$P(z) = [1\ 0 \cdots 0], \quad Q(z) = \begin{bmatrix} z & -1 & 0 & \cdots & 0 \\ 0 & z & -1 & & \\ & & & & \vdots \\ 0 & \cdots & & 0 & z \end{bmatrix}$$

$F(z) = PQ^{-1} = [1/z, \dots, 1/z^n]$. Then $\psi_F(z) = ((-1)^{n-1}, z^{n-1}, -z^{n-2}, \dots, z^n)$ (also a "rational curve") and $\psi_F(x_0, x_1) = ((-1)^{n-1}x_0^n,\ x_0 x_1^{n-1}, \dots, x_1^n)$. If $V(L)$ is *any* hyperplane, then $\psi_F(x_0, x_1) \not\subset V(L)$.

Example 20.31 Let $m = 2$, $p = 2$, $n = 4$ and

$$P(z) = \begin{bmatrix} z & 1 \\ 1 & z \end{bmatrix}, \quad Q(z) = \begin{bmatrix} z^2 & -z \\ 0 & z^2 \end{bmatrix}, \quad F(z) = PQ^{-1} = \begin{bmatrix} 1/z & 2/z^2 \\ 1/z^2 & z^2 + 1/z^3 \end{bmatrix}.$$

Then $\psi_F(z) = (z^2 - 1, -2z^2, z^3, -z - z^3, z^2, z^4) \in \mathrm{Gr}(2,4)$ and $\psi_F(x_0, x_1) = (x_1^2 x_0^2 - x_0^4, -2x_1^2 x_0^2, x_1^3 x_0, -x_1 x_0^3 - x_1^3 x_0, x_1^2 x_0^2, x_1^4) \in \mathrm{Gr}(2,4) \subset \mathbb{P}^5$. Now let $\ell = \alpha_0 Y_0 + \dots + \alpha_5 Y_5$ be a linear form so that $V(\ell) \in (\mathbb{P}^5)^*$ is a hyperplane and $V(\ell) \in \mathrm{Gr}(2,4)^*$ i.e., $(\alpha_0, \dots, \alpha_5) \in \mathrm{Gr}(2,4)$ so that

$$\alpha_0 \alpha_5 - \alpha_1 \alpha_4 + \alpha_2 \alpha_3 = 0. \tag{20.32}$$

We *claim* that $\psi_F(x_0, x_1) \not\subset V(\ell)$ for any such ℓ. If $\psi_F(x_0, x_1) \subset V(\ell)$, then $-\alpha_0 x_0^4 - \alpha_3 x_1 x_0^3 + (\alpha_0 - 2\alpha_1 + \alpha_4) x_1^2 x_0^2 + (\alpha_2 - \alpha_3) x_1^3 x_0 + \alpha_5 x_1^4 \equiv 0$ so that $\alpha_0 = 0$, $\alpha_3 = 0$ (hence), $\alpha_2 = 0$, $\alpha_5 = 0$ and $\alpha_4 - 2\alpha_1 = 0$. But if $(\alpha_0, \dots, \alpha_5)$ satisfies (20.32) as well, then $\alpha_1 \alpha_4 = 0$ and both $\alpha_1 = 0$ and $\alpha_4 = 0$ (which contradicts $(\alpha_0, \dots, \alpha_5) \in \mathbb{P}^5$). In other words, $F(z)$ is non-degenerate. We observe that the points

$$\begin{aligned} \psi_F(1,0) &= (-1, 0, 0, 0, 0, 0) \\ \psi_F(1,1) &= (0, -2, 1, -2, 1, 1) \\ \psi_F(1,-1) &= (0, -2, -1, 2, 1, 1) \\ \psi_F(1,2) &= (3, -8, 8, -10, 4, 16) \\ \psi_F(0,1) &= (0, 0, 0, 0, 0, 1) \end{aligned}$$

are in general position and thus span an L_4 ($= L_n = L_{mp}$). We can also look at this example in a somewhat different way. Let $F(z)$ be *any* element of $\mathrm{Rat}(4,2,2)$ and let

$$\psi_{F,t}(z) = \sum_{j=0}^{3} a_{j,t} z^j, \quad t = 0,1,2,3,4$$

$$\psi_{F,5}(z) = \sum_{j=0}^{3} \chi_j z^j + z^4.$$

If $Y_0, \ldots, Y_5$ are coordinates on $\mathbb{P}^5$ and if

$$L_{F,j}(Y_0, \ldots, Y_5) = \sum_{t=0}^{4} a_{j,t} Y_t, \quad j = 0,1,2,3$$

$$L_{F,4}(Y_0, \ldots, Y_5) = Y_5$$

then $E_F = V(L_{F,0}, \ldots, L_{F,3}, L_{F,4})$ is a linear subspace of $\mathbb{P}^5$. $F(z)$ will be non-degenerate if and only if there is no $\boldsymbol{\xi} = (\xi_0, \ldots, \zeta_5) \in \mathrm{Gr}(2,4)$ such that $\boldsymbol{\xi} \in E_F$. In other words, if and only if the equations

$$\begin{aligned}
&\xi_5 = 0\\
&\xi_1\xi_4 - \xi_2\xi_3 = 0\\
&\xi_0 a_{0,0} + \xi_1 a_{0,1} + \xi_2 a_{0,2} + \xi_3 a_{0,3} + \xi_4 a_{0,4} = 0\\
&\xi_0 a_{1,0} + \xi_1 a_{1,1} + \xi_2 a_{1,2} + \xi_3 a_{1,3} + \xi_4 a_{1,4} = 0\\
&\xi_0 a_{2,0} + \xi_1 a_{2,1} + \xi_2 a_{2,2} + \xi_3 a_{2,3} + \xi_4 a_{2,4} = 0\\
&\xi_0 a_{3,0} + \xi_1 a_{3,1} + \xi_2 a_{3,2} + \xi_3 a_{3,3} + \xi_4 a_{3,4} = 0
\end{aligned}$$

have only the zero solution. If $\mathbf{A} = (\mathbf{a}_i)$ where $\mathbf{a}_i = (a_{i,0},\ a_{i,1},\ a_{i,2},\ a_{i,3},\ a_{i,4})$, then the system becomes (in $\mathbb{P}^4 \simeq V(Y_5) \subset \mathbb{P}^5$)

$$\mathbf{A}\boldsymbol{\xi} = \mathbf{0}, \quad \xi_1\xi_4 - \xi_2\xi_3 = 0$$

where here $\boldsymbol{\xi} = (\xi_0, \xi_1, \xi_2, \xi_3, \xi_4)$. For a general 4×5 $\mathbf{A}$ this system will have a unique solution in $\mathbb{A}^5$ and so, non-degeneracy is "generic".

Let $x \in S_{m,p}^n$ with transfer matrix F_x. Then $E_{F_x} = \{(\xi_{W,\alpha}) \in \mathrm{Gr}(m, m+p)^* \simeq \mathrm{Gr}(p, m+p)\colon \sum_{\alpha} \xi_{W,\alpha} \psi_{F_x,\alpha}(z) \equiv 0\}$ (viewing $\xi_{W,\alpha}$ as "dual" coordinates and W as the subspace in $\mathrm{Gr}(p, m+p)$). Let $Z \subset \mathbf{S}_{m,p}^n \times \mathrm{Gr}(p, m+p)$ be given by

$$Z = \{(x, W)\colon W \in E_{F_x}\} \tag{20.33}$$

where $\mathbf{S}_{m,p}^n = S_{m,p}^n / G$. Then Z is an algebraic set.

Theorem 20.34 *Z is irreducible and* $\dim Z = n(m+p) - n + mp - 1 = n(m+p) - 1$ *for $n = mp$.*

Proof. Let $V_\infty = \text{span}[\epsilon^{p+1}, \ldots, \epsilon^{m+p}] \in \text{Gr}(m, m+p)$. Clearly, $\psi_{F_x}(0,1) = \begin{bmatrix} 0 \\ I_m \end{bmatrix} = V_\infty$ for *all* $x \in \mathbf{S}^n_{m,p}$. If $\pi_1 \colon \mathbf{S}^n_{m,p} \times \mathbb{G}\text{r}(p, m+p) \to \mathbf{S}^n_{m,p}$, $\pi_2 \colon \mathbf{S}^n_{m,p} \times \mathbb{G}\text{r}(p, m+p) \to \text{Gr}(p, m+p)$ are the projections (restricted to Z), then $\pi_2^{-1}(W) = \{(x, W) \colon W \in E_{F_x}\} = \{(x, W) \colon \psi_{F_x}(x_0, x_1) \subset \text{Ker}\, W\}$. Then $\pi_2(Z) \subset \sigma_1(V_\infty) = \{W \in \mathbb{G}\text{r}(p, m+p) \colon \dim(V_\infty \cap W) \geq 1\}$ (see Example 12.17). But $\{W \in \text{Gr}(p, m+p) \colon \dim(W \cap V_\infty) \geq 2\} = \sigma_2(V_\infty) \subset \sigma_1(V_\infty)$. Since $\sigma_2(V_\infty)$ is closed (Example 12.17), $\sigma_1(V_\infty) - \sigma_2(V_\infty)$ is open (and non-empty) in $\sigma_1(V_\infty)$. It follows that $\sigma_1(V_\infty) - \sigma_2(V_\infty) \subset \pi_2(z) \subset \sigma_1(V_\infty)$ and

$$\overline{\pi_2(z)} = \sigma_1(V_\infty), \quad Z = \overline{\pi_2^{-1}(\sigma_1(V_\infty) - \sigma_2(V_\infty))}.$$

Since output feedback is transitive (as a group action) all the fibers $\pi_2^{-1}(W)$ are isomorphic for $W \in \sigma_1(V_\infty) - \sigma_2(V_\infty)$ and consequently, have the same dimension. Consider

$$W_{\mathbf{i},j} = \text{span}[\epsilon^{i_1}, \ldots, \epsilon^{i_{p-1}}, \epsilon^{p+j}]$$

where $\mathbf{i} = (i_1, \ldots, i_{p-1})$, $1 \leq i_1 < \cdots < i_{p-1} \leq p$, $j = 1, \ldots, m$. Then $W_{\mathbf{i},j} \in \sigma_1(V_\infty) - \sigma_2(V_\infty)$ and $(x, W_{\mathbf{i},j}) \in \pi_2^{-1}(W_{\mathbf{i},j})$ if and only if

$$F_{x,(\mathbf{p}-\mathbf{i},j)}(z) = 0 \tag{20.35}$$

where $\mathbf{p} = (1, \ldots, p)$. These equations give mp conditions and so $\dim \pi_2^{-1}(W) = n(m+p) - mp$. By ([H-2]), $\dim Z = \dim \pi_2^{-1}(W) + \dim \sigma_1(V_\infty) = n(m+p) - mp + mp - 1 = n(m+p) - 1 = n(m+p) - n + mp - 1$ for $n = mp$.

Corollary 20.36 *If $n = mp$, then non-degeneracy is generic.*

Proof. Since $\dim Z = n(m+p) - 1$, $\dim \pi_1(Z) \leq n(m+p) - 1 < n(m+p)$.

An alternative argument ([B-7]) can be developed along the following lines. Let $\psi_F(z_1), \ldots, \psi_F(z_n)$, $\psi_F(\xi_\infty)$ be $n+1$ points on the curve. If $F(z)$ is degenerate, then there is a W in $\bigcap_{i=1}^{n} \sigma_1(\psi_F(z_i)) \cap \sigma_1(V_\infty)$. Since the $\sigma_1(\psi_F(\cdot))$ are Schubert hypersurfaces (Example 12.17)

$$\dim \bigcap_{i=1}^{n} \sigma_1(\psi_F(z_i)) \cap \sigma_1(V_\infty) = mp - (n+1) \tag{20.37}$$

if the points $\psi_F(z_1), \ldots, \psi_F(z_n)$, $\psi_F(\xi_\infty) = V_\infty$ are in general position. Thus, if $mp = n$, then $F(z)$ will be degenerate if and only if any n (finite) points on the curve are dependent. This gives appropriate algebraic constraints on $F(z)$.

However, there is an even more elementary argument. Let $\boldsymbol{\xi} = (\xi_0, \dots, \xi_{N-1}, \xi_N)$ and let $(Y_0, \dots, Y_{N-1}, Y_N)$ be the coordinates on $\mathbb{P}^N$. If $x \in S^n_{m,p}$ with transfer function F_x, then $E_x = E_{F_x} = V(L^x_0, \dots, L^x_{n-1}, L^x_n)$ where

$$L^x_j(Y_0, \dots, Y_N) = \sum_{t=0}^{N-1} a^x_{j,t} Y_t, \quad j = 0, 1, \dots, n-1$$
$$L^x_n(Y_0, \dots, Y_N) = Y_N.$$

Thus, $\boldsymbol{\xi} \in \mathrm{Gr}(m, m+p) \cap E_x$ if and only of $\boldsymbol{\xi} \in \mathrm{Gr}(m, m+p) \cap V(L^x_0, \dots, L^x_{n-1}, L^x_n)$. In particular, we must have $\xi_N = 0$. So, we consider $\tilde{\boldsymbol{\xi}} = (\xi_0, \dots, \xi_{N-1})$ ("=" $(\xi_0, \dots, \xi_{N-1}, 0)$) $\in \mathbb{P}^{N-1}$ and $\mathbf{X} = \mathrm{Gr}(m, m+p) \cap H_N$ (where $H_N = V(Y_N)$) as a subvariety of $\mathbb{P}^{N-1}$. We observe that $\dim \mathbf{X} = mp - 1$. Let $\mathbf{E}_x = V(L^x_0, \dots, L^x_{n-1}) \subset \mathbb{P}^{N-1}$ so that $\boldsymbol{\xi} \in \mathrm{Gr}(m, m+p) \cap E_x$ if and only if $\xi_N = 0$ and $\tilde{\boldsymbol{\xi}} \in \mathbf{X} \cap \mathbf{E}_x$. Let $\mathbf{A}^x$ be the $n \times N$ matrix with entries $(a^x_{j,t})$, $j = 0, 1, \dots, n-1$, $t = 0, 1, \dots, N-1$. Then $\tilde{\boldsymbol{\xi}} \in \mathbf{X}$ will be an element of $\mathbf{E}_x$ if and only if

$$\mathbf{A}^x \tilde{\boldsymbol{\xi}} = 0 \tag{20.38}$$

i.e., if and only if $\boldsymbol{\xi}$ (viewed as a column) $\in \mathrm{Ker}\, \mathbf{A}^x$. But, for almost all x, Ker $\mathbf{A}^x$ is an $L_{N-(n+1)} = L_{N-1-n}$ and, almost all L_{N-1-n} come from a Ker $\mathbf{A}^x$. Let $r = \dim \mathbf{X} = mp - 1$ so that for, $r \leq (N-1) - 2 = N - 3$ almost all $L_{N-1-r-1} = L_{N-1-mp+1-1} = L_{N-(mp+1)}$ do not meet $\mathbf{X}$ (Example 12.9). Thus, for $mp \leq N - 2$* and $n = mp$, almost all E_x do not meet $\mathrm{Gr}(m, m+p)$ and so, non-degeneracy is generic.

Finally, if $X \subset \mathbb{P}^N - L$, $\dim X = r$, $\dim L = N - r - 1$ and $L \cap X = \emptyset$, then $\pi_L \colon X \to \mathbb{P}^r$ is a finite, surjective morphism (i.e., $\pi_L(x) = \mathbb{P}^r$). Since $k(X)$ is separable over $k(\mathbb{P}^r)$, π_L is a separable morphism and $\#\pi_L^{-1}(\eta) = \deg X$ for $\eta \in \mathbb{P}^r$. This implies the following:

Proposition 20.39 *If $F(z)$ is non-degenerate and $mp = n$, then*

$$\begin{aligned} \#\pi^{-1}_{E_F}(\eta) &= \deg\, \mathrm{Gr}(m, m+p) \\ &= \frac{1! \cdots (p-1)! 1! \cdots (m-1)! mp!}{m! \cdots (m+p-1)!} \end{aligned} \tag{20.40}$$

for $\eta \in \mathbb{P}^n$. (See [H-9].)

This is relevant to the question of what can be done with real feedback for real data.

---

*It is easy to see that for $m > 2$, $p > 2$, $mp + 2 \leq N = \binom{m+p}{m} - 1$ i.e., $mp + 3 \leq \binom{m+p}{m}$ (Exercise).

Example 20.41 Let $m = 2$, $p = 2$. Then (20.40) gives $d_{2,2} = \deg Gr(2,4) = 4!/2!3! = 2$. Let us "prove" this. Consider $\mathbb{P}^5$ with coordinates $Y_0, \ldots, Y_5$. Then $\mathrm{Gr}(2,4) = V(Y_0Y_5 - Y_1Y_4 + Y_2Y_3)$ has dimension 4. If $L_1 = V(H_1, H_2, H_3, H_4)$ where $H_i = \sum_{j=0}^{5} a_j^i Y_j$, $i = 1,2,3,4$, then, for general a_j^i $\dim L_1 = 1$ and $\dim \mathrm{Gr}(2,4) \cap L_1 \geq 0$. For almost all L_1, $\dim \mathrm{Gr}(2,4) \cap L_1 = 0$ and $\# \mathrm{Gr}(2,4) \cap L_1 = \text{degree } \mathrm{Gr}(2,4)$. We want to show that this number is 2. If we write

$$\mathbf{A} \begin{bmatrix} Y_0 \\ \vdots \\ Y_5 \end{bmatrix} = 0 \tag{20.42}$$

where $\mathbf{A} = (a_j^i)$, $i = 1, \ldots, 4$, $j = 0, 1, \ldots, 5$ is a 4×6 matrix, then we may assume that (say) $Y_0 \neq 0$ so that $Y_0 = 1$ (as $\dim \mathrm{Gr}(2,4) \cap V(Y_0) = 3$ and so almost all L_1 with $Y_0 = 1$ do not meet $\mathrm{Gr}(2,4) \cap V(Y_0)$) and we may assume that $\mathbf{A}$ is of full rank 4 and is in "echelon" form (this is "generic") i.e., for example,

$$\mathbf{A} = \begin{bmatrix} 1 & * & * & * & * & * \\ 0 & 1 & * & * & * & * \\ 0 & 0 & 1 & * & * & * \\ 0 & 0 & 0 & 1 & * & * \end{bmatrix} \quad \text{or } \mathbf{A} = \begin{bmatrix} 1 & * & * & * & * & * \\ 0 & 0 & 1 & * & * & * \\ 0 & 0 & 0 & 1 & * & * \\ 0 & 0 & 0 & 0 & 1 & * \end{bmatrix} \tag{20.43}$$

etc. If we examine (20.42) for (say) the first matrix $\mathbf{A}$ in (20.43), then, by elimination, it leads to an equation

$$\alpha Y_4 + \beta Y_5 - 1 = 0. \tag{20.44}$$

If we apply the elimination to the equation $Y_5 = Y_1Y_4 - Y_2Y_3$ of $\mathrm{Gr}(2,4)$ on $Y_0 \neq 0$, then we obtain an equation,

$$\gamma Y_5^2 + \delta Y_4 Y_5 + \varphi Y_4^2 - Y_5 = 0. \tag{20.45}$$

Thus, the points in $\mathrm{Gr}(2,4) \cap L_1$ correspond to the solutions (ξ_4, ξ_5) of (20.44) and (20.45) of which, in general, there are 2.

There is still another approach to output feedback. Let $\mathbf{y}(z) = \mathbf{F}(z)\mathbf{u}(z)$ and set $\mathbf{u}(z) = -\mathbf{K}\mathbf{y}(z)$. [This is the "traditional" frequency domain point of view.] The equations can be written

$$\begin{bmatrix} \mathbf{K} & I_m \\ -I_p & \mathbf{F}(z) \end{bmatrix} \begin{bmatrix} \mathbf{y}(z) \\ \mathbf{u}(z) \end{bmatrix} = 0. \tag{20.46}$$

Let $\psi_F(z) = \mathrm{Ker}[-I_p \; \mathbf{F}(z)]$ so that $\psi_F(z) \in \mathrm{Gr}(m, m+p)$.

Lemma 20.47 *Let* $\mathbf{x}(z)$ *be a* $(p+m)\times 1$ *vector with* $[x^i(z)]_1^p = \mathbf{y}(z)$ *and* $[x^j(z)]_{p+1}^{p+m} = \mathbf{u}(z)$. *Then* $\mathbf{x}(z) \in \operatorname{Ker}[-I_p \ \mathbf{F}(z)]$ *if and only if*

$$\mathbf{x}(z) \in \text{Image}\begin{bmatrix}\mathbf{F}(z)\\ I_m\end{bmatrix}. \tag{20.48}$$

Proof. Obvious.

In other words, if $\mathbf{F}(z) = \mathbf{P}(z)\mathbf{Q}^{-1}(z)$, then

$$\psi_{\mathbf{F}}(z) = \operatorname{Ker}[-I_p \ \mathbf{F}(z)] = \operatorname{Im}\begin{bmatrix}\mathbf{F}(z)\\ I_m\end{bmatrix} = \operatorname{Im}\begin{bmatrix}\mathbf{P}(z)\\ \mathbf{Q}(z)\end{bmatrix}$$

is an element of $\operatorname{Gr}(m, m+p)$. Consider now the group $\tilde{\Gamma}_0(m,p)$ of $GL(m+p,k)$ of matrices

$$\gamma_K = \begin{bmatrix}I_p & 0\\ K & I_m\end{bmatrix} \tag{20.49}$$

with $K \in M(m,p)$. Then $\tilde{\Gamma}_0(m,p)$ acts on the system matrix via

$$\gamma_K \cdot \begin{bmatrix}\mathbf{P}\\ \mathbf{Q}\end{bmatrix} = \begin{bmatrix}I_p & 0\\ K & I_m\end{bmatrix}\begin{bmatrix}\mathbf{P}\\ \mathbf{Q}\end{bmatrix} = \begin{bmatrix}\mathbf{P}\\ K\mathbf{P}+\mathbf{Q}\end{bmatrix}. \tag{20.50}$$

If $\mathbf{P}$, $\mathbf{Q}$ are co-prime, then $\mathbf{P}$, $K\mathbf{P}+\mathbf{Q}$ are also co-prime. The action of $\tilde{\Gamma}_0(m,p)$ is called *output feedback*. We observe that

$$\det[K\mathbf{P}+\mathbf{Q}] = \det\begin{bmatrix}I & \mathbf{P}\\ 0 & K\mathbf{P}+\mathbf{Q}\end{bmatrix} = \det\begin{bmatrix}I & 0\\ K & I\end{bmatrix}\begin{bmatrix}I & \mathbf{P}\\ -K & \mathbf{Q}\end{bmatrix}$$

and hence that

$$\det[K\mathbf{P}(z)+\mathbf{Q}(z)] = \det\begin{bmatrix}I & \mathbf{P}(z)\\ -K & \mathbf{Q}(z)\end{bmatrix}. \tag{20.51}$$

It follows that ξ is a zero of $\det[K\mathbf{P}(z)+\mathbf{Q}(z)]$ if and only if

$$\det\begin{bmatrix}I & \mathbf{P}(\xi)\\ -K & \mathbf{Q}(\xi)\end{bmatrix} = 0. \tag{20.52}$$

If

$$W_K = \operatorname{Im}\begin{bmatrix}I\\ -K\end{bmatrix}, \quad Z_\xi = \operatorname{Im}\begin{bmatrix}\mathbf{P}(\xi)\\ \mathbf{Q}(\xi)\end{bmatrix}$$

when $W_K \in \operatorname{Gr}(p, m+p)$, $Z_\xi \in \operatorname{Gr}(m, m+p)$ and ξ is a zero of $\det[K\mathbf{P}(z)+\mathbf{Q}(z)]$ if and only if $W_K \in \sigma_1(Z_\xi)$ $(= \{W \in \operatorname{Gr}(p, m+p)\colon \dim(W \cap Z_\xi) \geq 1\})$. So, the *output feedback assignment problem* becomes: given $\xi_1, \ldots, \xi_n$ and Schubert hypersurfaces $\sigma_1(Z_\xi), \ldots, \sigma_1(Z_{\xi_n})$, does there exist a W_K in $\bigcap_{i=1}^{n} \sigma_1(Z_{\xi_i})$? (See, [B-7].) Since $\sigma_1(Z_\xi)$ is the intersection of $\mathrm{Gr}(p, m+p)$ with a hyperplane, we have, in general, that

$$\dim \bigcap_{i=1}^{n} \sigma_1(Z_{\xi_i}) = mp - n. \tag{20.53}$$

It follows that if $mp < n$, then this intersection will be empty and generic pole placement will be impossible. In other words, $mp \geq n$ is a necessary condition. Let $\xi_\infty = (0, 1)$ so that $Z_{\xi_\infty} = \begin{bmatrix} 0 \\ I \end{bmatrix}$ and *no* $W_K \in \sigma_1(Z_{\xi_\infty})$. If $mp > n$, then

$$\dim \bigcap_{i=1}^{n} \sigma_1(Z_{\xi_i}) = mp - n \geq 1$$

and

$$\dim \bigcap_{i=1}^{n} \sigma_1(Z_{\xi_i}) \cap \sigma_1(Z_\infty) = mp - n - 1 \geq 0$$

so that

$$\bigcap_{i=1}^{n} \sigma_1(Z_{\xi_i}) - \left[\bigcap_{i=1}^{n} \sigma_1(Z_{\xi_i}) \cap \sigma_1(Z_\infty) \right]$$

is, in general, open and non-empty and so contains a W_K. Thus, if $mp > n$, assignability is generic. If $mp = n$, then a non-degeneracy condition such as

$$\left[\bigcap_{i=1}^{n} \sigma_1(Z_\xi) \right] \cap \sigma_1(Z_{\xi_\infty}) \neq \emptyset$$

is required. (See [B-7].)

Exercises

(1) Determine the conditions of (20.35) explicitly for $m = 2$, $p = 2$.

(2) Show that $mp + 3 \leq (m+p)!/m!p!$ for $p > 2$, $m > 2$ and that $mp + 3 \not\leq (m+p)!/m!p!$ for $p \leq 2$ or $m \leq 2$. [*Hint*: may assume $m = p + t$, $t \geq 0$. **case 1**: $t = 0$, do for $p = 3$, then use induction. **case 2**: use induction and a relation for $X(t) = (p + t + p)!/(p + t)!p!$]

(3) (Kimura) If $m + p - 1 \geq n$, then assignability by output feedback is generic. [Since $mp - m - p + 1 = (m-1)(p-1) \geq 0$, $m + p - 1 \geq n$ implies

$mp \geq n$.] Show the result directly without using Proposition 20.22. [*Hint*: note that $m+p-1 = \dim M_{1*}(m,p)$ (rank 1 $m \times p$ matrices). If $K = wv$, w $m \times 1$, v $1 \times p$, then consider the morphism $\psi\colon \mathbb{P}^{m-1} \times \mathbb{P}^{p-1} \times \mathbb{A}^{nmp} \to \mathbb{A}^n$ given by $\psi(w, v, \{H_1, \dots, H_n\}) = (\mathrm{Tr}(vH_jw))_{j=1}^n = (vH_jw)_{j=1}^n$ and the variety $X = \{(w, v, \{H_1, \dots, H_n\})\colon vH_jw = 0,\ j = 1, \dots, n\}$. Look at $\mathbb{P}^{m-1} \times \mathbb{P}^{p-1} \times \mathbb{A}^{nmp} - X = U$ and $\psi\colon U \to \mathbb{A}^n$.]

(4) Show that $\tilde{\Gamma}_0$ is an algebraic group, the action of $\tilde{\Gamma}_0(m,p)$ is by morphism, and that n is invariant under $\tilde{\Gamma}_0(m,p)$.

Appendix A

Formal Power Series, Completions, Regular Local Rings, and Hilbert Polynomials

We gather in this Appendix various results on commutative algebra ([A-2], [M-1], [Z-3]) which are used in Chapters 7, 10, 12, 14 and 17. We shall show that (geometric) regular local rings are unique factorization domains. The method of proof will involve demonstrating that the completion is a formal power series ring and that a formal power series ring is a UFD. We also introduce the Hilbert-Samuel function and examine its properties.

Definition A.1 Let R be a Noetherinan ring, $\mathfrak{a} \subset R$ an ideal, and M a finitely generated R-module. The *Rees ring*, $R_{\mathfrak{a}}(R)$, is the graded ring $\oplus \sum_{n=0}^{\infty} \mathfrak{a}^n$, and the *Rees module*, $R_{\mathfrak{a}}(M)$, is the $R_{\mathfrak{a}}(R)$-graded module $\oplus \sum_{n=0}^{\infty} \mathfrak{a}^n M$ ($\mathfrak{a}^0 = R$). The *associated graded ring* $G_{\mathfrak{a}}(R)$ is the graded ring $\oplus \sum_{n=0}^{\infty} \mathfrak{a}^n/\mathfrak{a}^{n+1}$ (where $\overline{r}_m \overline{r}_n = \overline{r}_{m+n}$) and the *associated graded module*, $G_{\mathfrak{a}}(M)$, is the $G_{\mathfrak{a}}(R)$ graded module $\oplus \sum_{n=0}^{\infty} \mathfrak{a}^n M/\mathfrak{a}^{n+1} M$.

We note that a graded ring G is Noetherian if and only if G_0 is Noetherian and G is a finitely generated G_0-algebra (i.e., $G = G_0[g_1, \dots, g_\nu]$ with $g_i \in G$). [See I.B.19.]

Theorem A.2 (Artin-Rees Theorem) *Let R be a Noetherian ring, $\mathfrak{a} \subset R$ an ideal, and M a finitely generated R-module. If $N \subset M$ is a submodule, then there is a t such that*

$$\mathfrak{a}^{n+t} M \cap N = \mathfrak{a}^n (\mathfrak{a}^t M \cap N) \tag{A.3}$$

for all n.

P. Falb, *Methods of Algebraic Geometry in Control Theory: Part II*, Modern Birkhäuser Classics, https://doi.org/10.1007/978-3-319-96574-1

Proof. If $\mathfrak{a} = (a_1, \ldots, a_s)$, then $R_{\mathfrak{a}}(R) = R[a_1, \ldots, a_s]$ (with the usual R-grading) is a finitely generated R-algebra and $R_{\mathfrak{a}}(M)$ is a finitely generated module. Let $N_n = \mathfrak{a}^n M \cap N$ so that $\tilde{N} = \oplus \sum N_n$ is a submodule of $R_{\mathfrak{a}}(M)$ (in fact, $\tilde{N} = R_{\mathfrak{a}}(M) \cap N$). Then $\tilde{N} = (\mathbf{n}_1, \ldots, \mathbf{n}_t)$ is finitely generated with $\mathbf{n}_i$ homogeneous of degree ν_i. If $\tau \geq \max \deg \mathbf{n}_i = \max \nu_i$, then $N_{n+\tau} = \mathfrak{a}^n N_\tau$ for all n. For, clearly, $\mathfrak{a}^n N_\tau \subset N_{n+\tau}$ and if $\alpha \in N_{n+\tau}$, then $\alpha = \sum_{i=1}^{t} r_i \mathbf{n}_i$, $r_i \in \mathcal{R}_{\mathfrak{a}}(R)$, $\deg r_i = n + \tau - \nu_i$, r_i homogeneous. Thus, $r_i \in \mathfrak{a}^{n+\tau-\nu_i}$ and $\alpha \in \mathfrak{a}^n N_\tau$.

Corollary A.4 (Krull's Intersection Theorem)

$$\mathfrak{a}^r \bigcap_{n=1}^{\infty} \mathfrak{a}^n M = \bigcap_{n=1}^{\infty} \mathfrak{a}^n M \tag{A.5}$$

for all r.

Proof. Let $N = \bigcap_{n=1}^{\infty} \mathfrak{a}^n M$. Then there is a t with $\mathfrak{a}^{n+t} M \cap N = \mathfrak{a}^n \; (\mathfrak{a}^t M \cap N)$ for all n. But $N \subseteq \mathfrak{a}^s M$ so that $\mathfrak{a}^s M \cap N = N$ for all s.

Corollary A.6 *If $(\mathfrak{o}, \mathfrak{m})$ is a Noetherian local ring and $\mathfrak{a} \subseteq \mathfrak{m}$ is an ideal in $\mathfrak{o}$, then*

$$\bigcap_{n=1}^{\infty} (\mathfrak{a} + \mathfrak{m}^n) = \mathfrak{a} \tag{A.7}$$

and, in particular,

$$\bigcap_{n=1}^{\infty} \mathfrak{m}^n = (0). \tag{A.8}$$

Proof. By passage to $\mathfrak{o}/\mathfrak{a}$, we may assume $\mathfrak{a} = (0)$. But then $\mathfrak{m} \cdot \cap \mathfrak{m}^n = \cap \mathfrak{m}^n$ and the Nakayama lemma applies.

Let R be a ring and let $S^* = R[[X_1, \ldots, X_N]]$ be the ring of formal power series over R and let $S = R[X_1, \ldots, X_N]$. We let ν be the order function on S^* (i.e., if $f = (f_0, \ldots, f_q, \ldots)$, then $\nu(f) = \min\{q\colon f_q \neq 0\}$).

Proposition A.9 *If R is an integral domain, then $R[[X_1, \ldots, X_N]] = S^*$ is an integral domain.*

Proof. If $f, g \neq 0$, then $f_{\nu(f)} g_{\nu(g)} \neq 0$ and $\nu(fg) = \nu(f) + \nu(g)$ so $fg \neq 0$.

Proposition A.10 *An element f of S^* is a unit if and only if f_0 is a unit in R.*

Proof. If $fg = (1, 0, \dots)$, then $f_0 g_0 = 1$ and f_0 is a unit. Conversely, if $f_0 g_0 = 1$ so that $g_0 = 1/f_0$ in R, then set $g_1 = -f_0^{-1}(g_0 f_1)$, $g_q = -f_0^{-1}(g_{q-1} f_1 + \cdots + g_0 f_q), \dots$. Then $fg = 1$.

Corollary A.11 *If* $(\mathfrak{o}, \mathfrak{m})$ *is a local ring, then* $\mathfrak{o}[[X_1, \dots, X_N]]$ *is a local ring with maximal ideal* $\mathfrak{M}$ *generated by* $(\mathfrak{m}, X_1, \dots, X_N)$ *and* $\mathfrak{o}[[X_1, \dots, X_n]]/\mathfrak{M} = \mathfrak{o}/\mathfrak{m}$.

Proof. By induction, we may take $N = 1$. But $(\mathfrak{m}, X_1)$ is the set of all non-units in $\mathfrak{o}[[X_1]]$ for $a_0 + a_1 X_1 + \cdots \notin (\mathfrak{m}, X_1)$ if and only if $a_0 \notin \mathfrak{m}$ (i.e., a_0 is a unit in $\mathfrak{o}$).

Consider the ring $S^* = R[[X_1, \dots, X_N]]$ and let $\mathfrak{m}^* = (X_1, \dots, X_N)$. Then $\cap \mathfrak{m}^{*t} = (0)$ and we can define a metric $d(f, g)$ on S^* by setting

$$d(f, g) = 2^{-\nu(f-g)} \tag{A.12}$$

where ν is the order on S^*.

Proposition A.13 (S^*, d) *is a complete metric space.*

Proof. We first show that $d(f, g)$ is a metric. Clearly $d(f, f) = 0$ (since $\nu(0) = +\infty$). If $d(f, g) = 0$ then $f - g \in \mathfrak{m}^{*t}$ for all t and so, $f = g$. Noting that $d(f, g) = 2^{-\nu}$ if and only if $f - g \in \mathfrak{m}^{*\nu}$ and $f - g \notin \mathfrak{m}^{*\nu+1}$, we have $d(f, g) = d(g, f)$. If f, g, h are distinct, then (say) $d(f, h) = 2^{-\nu}$, $d(g, h) = 2^{-\mu}$ with $\mu \geq \nu$ and so, $f - h \in \mathfrak{m}^{*\nu}$ and $h - g \in \mathfrak{m}^{*\mu} \subset \mathfrak{m}^{*\nu}$. It follows that $d(f - g) \leq 2^{-\nu} \leq 2^{-\nu} + 2^{-\mu}$. Thus, $d(f, g)$ is a metric. (Hence (S^*, d) is Hausdorff.) To show that S^* is complete we must show that every Cauchy sequence $\{f^i\}$ in S^* converges in S^*. The sequence $\{f^i\}$ is Cauchy if (and only if) $\epsilon > 0$ implies there is an $n(\epsilon)$ with $d(f^i, f^j) < \epsilon$ for $i, j \geq n(\epsilon)$. Taking $\epsilon = 2^{-\nu}$, we have $f^i - f^j \in \mathfrak{m}^{*\nu+1}$ for $i, j \geq n(2^{-\nu})$ and so $f^i_\nu = f^j_\nu$ for $i, j \geq n(2^{-\nu})$. Let $f_\nu = f_\nu^{n(2^{-\nu})}$ and let $f = (f_0, f_1, \dots)$. Then $d(f, f^i) < 2^{-\nu}$ if $i \geq \max\{n(1), n(1/2), \dots, n(1/2^\nu)\}$ and f^i converges to f.

S^* is a topological ring with the topology defined by the neighborhood base $\mathfrak{m}^{*t}$ (of 0). If $\{f^i\}$ is a sequence in S^* with $\nu(f^i) \to \infty$ as $i \to \infty$, then $\{f^i\}$ is Cauchy and $\lim\{f^i\} = 0$. The partial sums $\sum_{i=0}^{r} f^i = g^r$ are also Cauchy and we let $\sum_{i=0}^{\infty} f^i = \lim_{r \to \infty} g^r$. Allowing infinite sums, we can write, for any f in S^*,

$$f = \sum_{q=0}^{\infty} f_q \tag{A.14}$$

where $f_q \in S_q = R[X_1, \dots, X_N]_q$. In other words, S is dense in S^*.

Theorem A.15 *If R is Noetherian, then $S^* = R[[X_1, \ldots, X_N]]$ is also Noetherian.*

Proof. By induction, we may take $N = 1$ and consider $R[[X]]$. We use a proof similar to that of Theorem I.5.3. Let $\mathfrak{a}$ be an ideal in $R[[X]]$ and let $\ell_i(\mathfrak{a}) = \{c$: there is an $f \in \mathfrak{a}$ of order i with leading coefficient c or $c = 0\}$. Then $\ell_i(\mathfrak{a})$ is an ideal in R, $\ell_i(\mathfrak{a}) \subset \ell_{i+1}(\mathfrak{a})$, and $\ell = \cup \ell_i(\mathfrak{a})$ is an ideal in R. Let $\ell = (r_1, \ldots, r_n)$ and let $f_1(X), \ldots, f_n(X)$ be elements of $\mathfrak{a}$ with initial coefficient r_i. Let $d = \max\{\nu(f_i)\}$ and let $\ell_i(\mathfrak{a}) = (r_{i1}, \ldots, r_{i\mu(i)})$ for $i < d$ and let $f_{ij}(X)$ be corresponding elements of $\mathfrak{a}$. We *claim* that $\mathfrak{a} = (f_1, \ldots, f_n, f_{11}, \ldots, f_{d-1\mu(d-1)})$. If $f \in \mathfrak{a}$ with $\nu(f) < d$, then $f = cX^{\nu(f)} + \cdots$ and $c = \sum_1^{\mu(f)} c_j r_{\nu(f)j}$ so that $f - \sum c_j f_{\nu(f)j}$ has order $\geq \nu(f)+1$. Thus, it will be enough to show that, if $f \in \mathfrak{a}$ with $\nu(f) \geq d$, then $f \in (f_1, \ldots, f_n, f_{11}, \ldots, f_{d-1\mu(d-1)})$. If $f = cX^{\nu(f)} + \cdots$, then $c = \sum_{j=1}^{n} c_j r_j$ and $f(X) - \sum_{j=1}^{n} c_j X^{j-\nu(f_j)} f_j(X)$ has order $\geq \nu(f) + 1$. By continuing, we get sequences $\{c_j^m\}$ $j = 1, \ldots, n$, $m = \nu(f), \nu(f)+1, \ldots$ such that, for every ν,

$$f(X) - \sum_{j=1}^{n} \left(\sum_{m=\nu(f)}^{\nu(f)+\nu} c_j^m X^{m-\nu(f_j)} f_j(X) \right)$$

has order $> m$. Since $m - \nu(f_j) \to \infty$ as $m \to \infty$, the sum $\sum_{m=\nu(f)}^{\infty} c_j^m X^{m-\nu(f_j)}$ converges to an element $r_j(X)$ of $R[[X]]$ and $f(X) - \sum_{j=1}^{n} r_j(X) f_j(X)$ has order $> m$ for all m. In other words, $f - \sum r_j f_j \in \cap \mathfrak{m}^{*m} = (0)$.

Corollary A.16 *If $(\mathfrak{o}, \mathfrak{m})$ is a Noetherian local ring, then $\mathfrak{o}[[X_1, \ldots, X_N]]$ is a Noetherian local ring with maximal ideal $\mathfrak{M} = (\mathfrak{m}, X_1, \ldots, X_N)$ and $\mathfrak{o}/\mathfrak{m} = \mathfrak{o}[[X_1, \ldots, X_N]]/\mathfrak{M}$. In particular, if K is a field, then $K[[X_1, \ldots, X_N]]$ is a Noetherian local domain with maximal ideal $M = (X_1, \ldots, X_N)$.*

We observe that $(0) < (X_1) < \cdots < (X_1, \ldots, X_{N-1}) < M$ is a chain of prime ideals in $K[[X_1, \ldots, X_N]]$ so that the height of M, $h(M)$, is no less than N. In other words, the Krull dimension of the power series ring $\mathfrak{o} = K[[X_1, \ldots, X_N]]$, $K\,\mathrm{Dim}\,\mathfrak{o}$, $\geq N$. We have, using ideal theory:

Proposition A.17 $K\,\mathrm{Dim}\,\mathfrak{o} = N$ *where* $\mathfrak{o} = K[[X_1, \ldots, X_N]]$.

Proof. We need only show that $h(M) \leq N$. We use induction on N. First, suppose that $N = 1$ and hence $M = (X)$. Let $\mathfrak{p}$ be a prime ideal with $(0) \subset$

$\mathfrak{p} < M$ and let $\mathfrak{a}_n = \mathfrak{p}^n\mathfrak{o}_\mathfrak{p} \cap \mathfrak{o}$. Then $\mathfrak{a}_n + M \supset \mathfrak{a}_{n+1} + M \supset$ is a descending chain of ideals. Since $\mathfrak{o}/X\mathfrak{o} = \mathfrak{o}/\mathfrak{m} = K$ satisfies (trivially) the descending chain condition for ideals, there is an n_0 with $\mathfrak{a}_n + M = \mathfrak{a}_{n+1} + M = \cdots$ for $n \geq n_0$. If $x \in \mathfrak{a}_n$, then $x = y + Xz$ with $y \in \mathfrak{a}_{n+1}$, $z \in \mathfrak{o}$ and so $x - y = zX \in \mathfrak{a}_n$. But $\sqrt{\mathfrak{a}_n} = \mathfrak{p}$ and $X \notin \mathfrak{p}$ together give $z \in \mathfrak{a}_n$. Hence $\mathfrak{a}_n = \mathfrak{a}_{n+1} + \mathfrak{a}_n X$. By Nakayama's Lemma, $\mathfrak{a}_{n+1} = \mathfrak{a}_n$ for $n \geq n_0$. But $\cap\mathfrak{a}_n = (\cap\mathfrak{p}^n\mathfrak{o}_\mathfrak{p}) \cap \mathfrak{o} = (0)$ (by Krull's Intersection Theorem A.4) and so $\mathfrak{a}_n = \mathfrak{a}_{n+1} = \cdots = (0)$ for $n \geq n_0$. If $y \in \mathfrak{p} = \sqrt{\mathfrak{a}_n} = \sqrt{(0)}$, then $y^{n_0} = 0$ and $\mathfrak{p} = (0)$. In other words, there is no prime ideal between (0) and M. Thus, $h(M) = 1$. Assume the result for values $< N$. Let $(0) < \mathfrak{p} < M$ be a prime chain with no prime ideal between $\mathfrak{p}$ and M. We may suppose that X_N (say) is not in $\mathfrak{p}$. Then $\sqrt{\mathfrak{p} + X_N\mathfrak{o}} = M$ and $X_j^m = c_j + X_N d_j$ with $c_j \in \mathfrak{p}$, $j = 1, \ldots, N-1$. Thus $\mathfrak{p}$ is a minimal prime ideal of $\mathfrak{a} = X_1\mathfrak{o} + \cdots + X_{N-1}\mathfrak{o}$. For if $\mathfrak{a} \subset \mathfrak{p}' \subset \mathfrak{p}$, then, in $\mathfrak{o}/\mathfrak{a}$, we have $M/\mathfrak{a} = (\overline{X}_N)$ where $\overline{X}_N = X_N (\operatorname{mod} \mathfrak{a})$, and, by our earlier argument, $\mathfrak{p}' = \mathfrak{p}$. Since $h(\mathfrak{p}) \leq N - 1$ by the induction hypothesis, $h(M) \leq N$.

An alternative proof can be developed using a "normalization" approach ([Z-3]).

Now let us consider a ring R and an ideal $\mathfrak{a}$ in R with $\mathfrak{a} \neq (0), R$.

Proposition A.18 *The sets $x + \mathfrak{a}^n$, $n = 0, 1, \ldots$, form a neighborhood base for a topology on R which makes R a topological ring.*

Proof. Suppose $x + \mathfrak{a}^n \cap y + \mathfrak{a}^m \neq \emptyset$ and, say, $m \geq n$. Then there is a z with $z - x \in \mathfrak{a}^n$, $z - y \in \mathfrak{a}^m$ and so $x - y \in \mathfrak{a}^n$ (as $\mathfrak{a}^m \subset \mathfrak{a}^n$ for $m \geq n$). It follows that $x + \mathfrak{a}^n = y + \mathfrak{a}^n$ and $x + \mathfrak{a}^n \cap y + \mathfrak{a}^m = y + \mathfrak{a}^n \cap y + \mathfrak{a}^m = y + \mathfrak{a}^m$. Thus we have a topology. To show that addition is continuous, let $x - x_0 \in \mathfrak{a}^n$ and $y - y_0 \in \mathfrak{a}^n$. Then $x + y - (x_0 + y_0) \in \mathfrak{a}^n$. Similarly, if $x - x_0 \in \mathfrak{a}^n$ and $y - y_0 \in \mathfrak{a}^n$, then $xy - x_0y_0 = (x - x_0)y + x_0(y - y_0) \in \mathfrak{a}^n$.

Definition A.19 The topology of A.18 is called the $\mathfrak{a}$-*topology* (or the $\mathfrak{a}$-*adic topology*) on R.

Proposition A.20 (i) $\cap\mathfrak{a}^n = \overline{\{0\}}$ *(i.e., $\cap\mathfrak{a}^n$ is the $\mathfrak{a}$-adic closure of $\{0\}$);* (ii) *$R/\cap\mathfrak{a}^n$ is Hausdorff; and,* (iii) *R is Hausdorff if and only if $\cap\mathfrak{a}^n = (0)$.*

Proof. Clearly $r \in \cap\mathfrak{a}^n$ if and only if $0 \in r + \mathfrak{a}^n$ for all n i.e., if and only if $r \in \overline{\{0\}}$. As for (ii), $\overline{r}$ is closed for all r so that all points of $R/\cap\mathfrak{a}^n$ are closed sets. (iii) is an immediate consequence of (ii).

Corollary A.21 *If $(\mathfrak{o}, \mathfrak{m})$ is a Noetherian local ring, then the $\mathfrak{m}$-topology is Hausdorff.*

We can define the $\mathfrak{a}$-topology of an R-module M via the neighborhood base $m + \mathfrak{a}^n M$ and similar results apply.

Proposition A.22 *If $\mathfrak{b}$ is a subset of R, then the closure $\overline{\mathfrak{b}}$ of $\mathfrak{b}$ is $\bigcap_0^\infty(\mathfrak{b} + \mathfrak{a}^n)$. Hence, $\mathfrak{b}$ is closed if and only if $\mathfrak{b} = \cap(\mathfrak{b} + \mathfrak{a}^n)$.*

Proof. If $x \in \overline{\mathfrak{b}}$, then, for each n, there is a $b_n \in \mathfrak{b}$ with $b_n \in x + \mathfrak{a}^n$ i.e., $x \in \mathfrak{b} + \mathfrak{a}^n$ for all n. If, on the other hand, $x \in \cap(\mathfrak{b} + \mathfrak{a}^n)$, then there is a $b_n \in \mathfrak{b}$ with $b_n \in x + \mathfrak{a}^n$ i.e., $x \in \overline{\mathfrak{b}}$.

Corollary A.22 *If $N \subset M$, then the closure $\overline{N}$ of N is $\bigcap_0^\infty(N + \mathfrak{a}^n M)$. Hence, N is closed if and only if $N = \cap(N + \mathfrak{a}^n M)$.*

Suppose now that the $\mathfrak{a}$-adic topology on R (or M) is Hausdorff i.e., $\cap \mathfrak{a}^n = (0)$. Then, following the procedure used for formal power series, we have a metric $d(x, y)$ $(= d_\mathfrak{a}(x, y))$ defined on R by

$$d(x, y) = 2^{-\nu} \tag{A.23}$$

where $x - y \in \mathfrak{a}^\nu$, $x - y \notin \mathfrak{a}^{\nu+1}$. Clearly, (R, d) is a topological ring. A sequence $\{x_n\}$ is *Cauchy* if, for each n, there is a $t_0(n)$ such that $x_t - x_s \in \mathfrak{a}^n$ for all $s, t \geq t_0(n)$, or, equivalently, if, given $\epsilon > 0$, there is an $n_0(\epsilon)$ such that $d(x_t, x_s) < \epsilon$ for all $s, t \geq n_0(\epsilon)$. Two sequences $\{x_n\}$, $\{y_n\}$ are *equivalent*, $\{x_n\} \equiv \{y_n\}$, if, given n, there is a $t_0(n)$ such that $x_t - y_t \in \mathfrak{a}^n$ for $t \geq t_0(n)$, or, equivalently, if $d(x_t, y_t) \to 0$. Let R^* be the set of all equivalence classes $x^* = \{x_n\}$ of Cauchy sequences and set

$$d^*(x^*, y^*) = \lim_{n\to\infty} d(x_n, y_n) \tag{A.24}$$

$x^* = \{x_n\}$, $y^* = \{y_n\}$.

Proposition A.25 (1) *(R^*, d^*) is a metric space;* (2) *φ: $R \to R^*$ given by $\varphi(x) = x^* = \{x_n = x\}$ is an isometry so that R may be viewed as embedded in R^*;* (3) *R is dense in R^*; and,* (4) *R^* is complete (i.e., every Cauchy sequence in R^* has a limit in R^*).*

Proof. (1) d^* is well-defined for, if $x^* = \{x_n\} = \{x_n'\}$ and $y^* = \{y_n\} = \{y_n'\}$, then $d(x_n, y_n) \leq d(x_n, x_n') + d(x_n', y_n') + d(y_n, y_n')$ and $d(x_n', y_n') \leq d(x_n', x_n) + d(x_n, y_n) + d(y_n, y_n')$. Since $d(x_n, y_n)$ is a bounded sequence in the reals, $\lim_{n\to\infty} d(x_n, y_n) = d^*(x^*, y^*)$ exists. That d^* is a metric follows easily from properties of limits. For example, if $d^*(x^*, 0) = 0$ with $x^* = \{x_n\}$, then $d^*(x^*, 0) =$

$\lim_{n\to\infty} d(x_n, 0) = 0$ so that $\{x_n\} \equiv \{0\}$ i.e., $x^* = 0$ in R^*. (2) and (3) are obvious. As for (4), let $x_n^* = \{x_{n,t}\}_{t=0}^{\infty}$ be a Cauchy sequence in R^* and let $y^* = \{x_{n,n}\}$. Since x_n^* is Cauchy $d^*(x_{n+t}^*, x_n^*) \to 0$ as $n \to \infty$ i.e., $\lim_{\rho\to\infty} d(x_{n+t,\rho}, x_{n,\rho}) \to 0$ as $n \to \infty$. Since $d(y_{n+t}, y_n) = d(x_{n+t,n+t} x_{n,n}) \to 0$ as $n \to \infty$, y^* is indeed a Cauchy sequence in R i.e., $y^* \in R^*$. But

$$d^*(y^*, x_t^*) = \lim_{s\to\infty} d(y_s, x_{t,s}) = \lim_{s\to\infty} d(x_{s,s}, x_{t,s})$$

which is small for t large enough i.e., $d^*(y^*, x_t^*) \to 0$ as $t \to \infty$.

Definition A.26 R^* is the *$\mathfrak{a}$-adic completion of* R.

Since the sum and product of Cauchy sequences, $\{x_n\} + \{y_n\} = \{x_n + y_n\}$, $\{x_n\}\{y_n\} = \{x_n y_n\}$, are Cauchy, we can readily see that R^* is also a topological ring. Since $d(x-y, 0) = d(x, y)$ and $d^*(x^* - y^*, 0) = d^*(x^*, y^*)$, we can write $|x| = d(x, 0)$ and $|x^*| = d^*(x^*, 0)$. If $x^* = \lim x_n$, then $x^* = \sum_{i=1}^{\infty} y_i$ where $y_1 = x_1$, $y_2 = x_2 - x_1, \ldots$ and, conversely, $\sum_{i=1}^{\infty} y_i$ is convergent if $\{x_n\}$ with $x_n = \sum_1^n y_j$ is Cauchy.

Proposition A.27 *(1)* $|x^* \pm y^*| \leq \max\{|x^*|, |y^*|\}$; *(2) if* $|x^*| < |y^*|$, *then* $|x^* \pm y^*| = |y^*|$; *and, (3) if* $x^* = \lim x_n$, $x^* \neq 0$, *then* $|x^*| = |x_n|$ *for large* n.

Proof. (1) is a consequence of the property for R. (2) $|x^* \pm y^*| \leq |y^*|$ and $y^* = y^* \pm x^* \mp x^*$ so that $|y^*| \leq \max\{|y^* \pm x^*|, |x^*|\} = |y^* \pm x^*|$. (3) Since $x^* = x^* - x_n + x_n$, $|x_n| > |x^* - x_n|$ for n large.

Proposition A.28 *Let* $S = \{x \in R \colon x \equiv 1 (\mathrm{mod}\, \mathfrak{a})\}$ *and let* $\mathfrak{a}^* = \{x^* \in R^* \colon |x^*| < 1\}$. *Then (1)* S *is a multiplicative set in* R *and* $R_S \subset R^*$; *(2)* $\mathfrak{a}^*$ *is an ideal in* R^* *with* $R^*\mathfrak{a} \subset \mathfrak{a}^*$; *and (3) if* $x^* \equiv 1\ (\mathrm{mod}\, \mathfrak{a}^*)$, *then* x^* *is a unit in* R^*.

Proof. (1) If $x = 1 - a$, $y = 1 - a'$, $a, a' \in \mathfrak{a}$, then $xy = 1 - a - a' + aa'$ and $xy \equiv 1 (mod\, \mathfrak{a})$. Let $y^* \in R^*$ be given by $y^* = 1 + a + a^2 + \cdots$ (convergent since $|a^n| \to 0$) so that $xy^* = 1$ and $r/x = ry^* \in R^*$. (2) Since $|x^* - y^*| \leq \max\{|x^*|, |y^*|\} < 1$ for $x^*, y^* \in \mathfrak{a}^*$ and since $|z^* x^*| = |z_n x_n|$ for large n with $z_n x_n \in \mathfrak{a}$ (as $d(x_n, 0) < 1$), $\mathfrak{a}^*$ is an ideal. Clearly, $R^*\mathfrak{a} \subset \mathfrak{a}^*$. (3) Let $x^* = 1 - y^*$, $y^* \in \mathfrak{a}^*$. Since $|y^*| < 1$, $z^* = 1 + y^* + y^{*2} + \cdots \in R^*$ and $z^* x^* = z^*(1 - y^*) = 1$.

Corollary A.29 *If* $\mathfrak{a} = \mathfrak{m}$ *is maximal, then* $\mathfrak{m}^*$ *is also maximal.*

Proof. If $x^* \notin \mathfrak{m}^*$, then $|x^*| = 1$ and we may suppose that $x^* = \lim x_n$, $|x_n| = 1$ for all n. Then $|x^* - x_0| < 1$ (as $d(x^*, x_n) \to 0$) and $x^* = x_0 - y^*$ with $|y^*| < 1$, $y^* \in \mathfrak{m}^*$ and $y^*/x_0 \in \mathfrak{m}^*$. Hence $z^* = 1 - y^*/x_0$ satisfies $z^* \equiv 1 (\mathrm{mod}\, \mathfrak{m}^*)$ and $x^* = x_0 z^*$ is a unit in R^*. In other words, $\mathfrak{m}^*$ is the set of all non-units in R^*.

Corollary A.30 *If $(\mathfrak{o}, \mathfrak{m})$ is a local ring, then $(\mathfrak{o}^*, \mathfrak{m}^*)$ is also a local ring.*

Proposition A.31 *If R is Noetherian, M is a finite R-module, R^* is the $\mathfrak{a}$-adic completion of R ($\cap \mathfrak{a}^n = 0$), M^* is the $\mathfrak{a}$-adic completion of M ($\cap \mathfrak{a}^n M = 0$), then $M^* = R^* M$.*

Proof. Let $M = (m_1, \ldots, m_t)$ and let $m^* \in M^*$. $m^* = \lim \mu_n$, $\mu_n \in M$, $\{\mu_n\}$ Cauchy. Then $\mu_{n+1} - \mu_n \in \mathfrak{a}^{\sigma(n)} M$ where $\sigma(n) \to \infty$ as $n \to \infty$. Since $\mu_n \in M$, $\mu_{n+1} - \mu_n \in M$, we have

$$\mu_{n+1} - \mu_n = \sum_{j=1}^{t} \alpha_{nj} m_j$$

$$\mu_n = \sum_{j=1}^{t} \beta_{nj} m_j = \sum_{j=1}^{t} (\beta_{n-1j} + \alpha_{n-1j}) m_j$$

and we *claim* that $\{\beta_{nj}\}_n$ is Cauchy for $j = 1, \ldots, t$. Since $\mu_{n+1} - \mu_n = \sum (\beta_{n+1j} - \beta_{nj}) m_j = \sum \alpha_{nj} m_j$ and $\alpha_{nj} \in \mathfrak{a}^{\sigma(n)}$, the claim holds. Since R^* is complete, $\beta_{nj} \to \beta_j^* \in R^*$. But $m^* - \sum_{j=1}^{t} \beta_j^* m_j = m^* - \mu_n + \sum (\beta_{nj} - \beta_j^*) m_j$ and the result follows.

Corollary A.32 *Let $\mathfrak{a}^* = \{x^* \in R^* \colon |x^*| < 1\}$. Then $\mathfrak{a}^* = R^* \mathfrak{a}$, $\mathfrak{a}^{*n} = (R^* \mathfrak{a})^n = R^* \mathfrak{a}^n$, and $\mathfrak{a}^{*n} M^* = R^* \mathfrak{a}^n M$. If N is a submodule of M, then the $\mathfrak{a}$-adic closure $\overline{N} = R^* N = N^*$ (in M^*) and $R^* N \cap M = \overline{N}$ (closure in M). Moreover, $(M/N)^* \simeq R^* M / R^* N = M^*/N^*$.*

Corollary A.33 *If $(\mathfrak{o}, \mathfrak{m})$ is a Noetherian local ring, then $\mathfrak{m}^{*n} = \mathfrak{o}^* \mathfrak{m}^n$ for all n and $\mathfrak{o}/\mathfrak{m} \simeq \mathfrak{o}^*/\mathfrak{m}^*$. In particular, $G_{\mathfrak{m}}(\mathfrak{o}) = \oplus \sum \mathfrak{m}^n/\mathfrak{m}^{n+1} \simeq G_{\mathfrak{m}^*}(\mathfrak{o}^*) = \oplus \sum \mathfrak{m}^{*n}/\mathfrak{m}^{*n+1}$ and if $\ell(\mathfrak{o}/\mathfrak{m}^n) = \sum^{n} \dim_k(\mathfrak{m}^{t+1}/\mathfrak{m}^t)$, then $\ell(\mathfrak{o}/\mathfrak{m}^n) = \ell(\mathfrak{o}^*/\mathfrak{m}^{*n})$ for all $n = 1, 2, \ldots$.*

We can also view the situation of Corollary A.33 in a different way. If $u_1, \ldots, u_m$ generate $\mathfrak{m}$ and $F \in \mathfrak{o}[[X_1, \ldots, X_m]]$, then $F = \sum_{i=0}^{\infty} F_i$, F_i a form of degree i. The sequence $t_\nu = \sum_{i=0}^{\nu} F_i(u_1, \ldots, u_m)$ is Cauchy in $\mathfrak{o}$ since $t_{\nu+1} - t_\nu =$

$F_{\nu+1}(u_1,\ldots,u_m) \in \mathfrak{m}^{\nu+1}$. If $t^* = \lim\{t_\nu\}$ and if we set

$$\psi(F) = t^* \tag{A.34}$$

then $\psi\colon \mathfrak{o}[[X_1,\ldots,X_m]] \to \mathfrak{o}^*$ is a homomorphism. If $x^* = \lim x_n \in \mathfrak{o}^*$, then (as we may assume) $x_{\nu+1} - x_\nu \in \mathfrak{m}^{\nu+1}$ for $\nu \geq 0$ and there is a form $F_{\nu+1}(X_1,\ldots,X_m)$ of degree $\nu+1$ with $x_{\nu+1} - x_\nu = F_{\nu+1}(u_1,\ldots,u_m)$. Take $F_0 = x_0$. Then $\psi\left(\sum_{i=0}^{\infty} F_i\right) = x^*$ and ψ is surjective. It follows that $\mathfrak{o}^* \simeq \mathfrak{o}[[X_1,\ldots,X_m]]/\mathfrak{a}$ for some ideal $\mathfrak{a}$ and that $\psi(\mathfrak{M}) = \mathfrak{m}^* = \mathfrak{o}^*\mathfrak{m}$. Moreover, this shows that the induced maps $\psi_\nu\colon \mathfrak{o}/\mathfrak{m}^\nu \to \mathfrak{o}^*/\mathfrak{m}^{*\nu}$ are surjective. Since $k = \mathfrak{o}/\mathfrak{m} = \mathfrak{o}^*/\mathfrak{m}^*$, we can, using induction, suppose that ψ_ν is an isomorphism and then show that $\psi_{\nu+1}$ is an isomorphism. Ker $\psi_{\nu+1} \subset \mathfrak{m}^{\nu+1}/\mathfrak{m}^{\nu+2}$ is the kernel of the induced homomorphism $\tilde{\psi}_{\nu+1}\colon \mathfrak{m}^{\nu+1}/\mathfrak{m}^{\nu+2} \to \mathfrak{m}^{\nu+1}\mathfrak{o}^*/\mathfrak{m}^{\nu+2}\mathfrak{o}^*$. But $\dim_k \mathfrak{m}^{\nu+1}/\mathfrak{m}^{\nu+2} = \dim_k \mathfrak{m}^{\nu+1}\mathfrak{o}^*/\mathfrak{m}^{\nu+2}\mathfrak{o}^*$ so that Ker $\psi_{\nu+1} = 0$ and $\psi_{\nu+1}$ is bijective.

Corollary A.35 *If R is Noetherian and $0 \to L \to M \to N \to 0$ is an exact sequence of finite R-modules, then so is $0 \to L^* \to M^* \to N^* \to 0$ ($*$ for $\mathfrak{a}$-adic completion).*

Proof. Since L (may be viewed as) is a submodule of M and $\mathfrak{a}^{n+t}M \cap L = \mathfrak{a}^n(\mathfrak{a}^tM \cap L)$ for all n and some t by the Artin-Rees Theorem A.2, the $\mathfrak{a}$-adic topology on L is that induced by the $\mathfrak{a}$-adic topology on M. Thus $0 \to L^* \to M^*$ is exact and $M/L \simeq N$ gives $(M/L)^* \simeq M^*/L^* \simeq N^*$ so that $0 \to L^* \to M^* \to N^* \to 0$ is exact.

Corollary A.36 *If $\mathfrak{b}_1$ and $\mathfrak{b}_2$ are ideals in a Noetherian ring R, then*

$$(\mathfrak{b}_1 \cap \mathfrak{b}_2)^* = \mathfrak{b}_1^* \cap \mathfrak{b}_2^* \tag{A.37}$$

for the $\mathfrak{a}$-adic completions.

Proof. Since $(\mathfrak{b}_1 + \mathfrak{b}_2)^* = R^*\mathfrak{b}_1 + R^*\mathfrak{b}_2 = \mathfrak{b}_1^* + \mathfrak{b}_2^*$ and $\mathfrak{b}_2/\mathfrak{b}_1 \cap \mathfrak{b}_2 \simeq (\mathfrak{b}_1 + \mathfrak{b}_2)/\mathfrak{b}_1$ and $\mathfrak{b}_2^*/\mathfrak{b}_1^* \cap \mathfrak{b}_2^* \simeq (\mathfrak{b}_1^* + \mathfrak{b}_2^*)/\mathfrak{b}_1^*$, we deduce from the exact sequences $0 \to \mathfrak{b}_1 \cap \mathfrak{b}_2 \to \mathfrak{b}_2 \to \mathfrak{b}_2/\mathfrak{b}_1 \cap \mathfrak{b}_2 \to 0$, $0 \to \mathfrak{b}_1 \to \mathfrak{b}_1 + \mathfrak{b}_2 \to \mathfrak{b}_1 + \mathfrak{b}_2/\mathfrak{b}_1 \to 0$, that $\mathfrak{b}_2^*/(\mathfrak{b}_1 \cap \mathfrak{b}_2)^* \simeq \mathfrak{b}_2^*/\mathfrak{b}_1^* \cap \mathfrak{b}_2^*$ and hence that $(\mathfrak{b}_1 \cap \mathfrak{b}_2)^* = \mathfrak{b}_1^* \cap \mathfrak{b}_2^*$.

Corollary A.38 *If $(\mathfrak{o}, \mathfrak{m})$ is a Noetherian local ring with completion $(\mathfrak{o}^*, \mathfrak{m}^*)$ and if $\mathfrak{o}^*$ is a unique factorization domain, then $\mathfrak{o}$ is also a unique factorization domain.*

Proof. Since $\mathfrak{o} \subset \mathfrak{o}^*$, $\mathfrak{o}$ is an integral domain. To show that $\mathfrak{o}$ is a UFD it will be enough to show that $\mathfrak{o}f \cap \mathfrak{o}g = \mathfrak{o}h$ for $f, g, h \in \mathfrak{o}$ (i.e., that the intersection of principal ideals is principal). Since $\mathfrak{o}^*$ is a UFD, there is an $\tilde{h}$ in $\mathfrak{o}^*$ with $\mathfrak{o}^* f \cap \mathfrak{o}^* g = \mathfrak{o}^*(\mathfrak{o} f \cap \mathfrak{o} g) = \mathfrak{o}^* \tilde{h}$. Thus, it will be enough to prove the following: if $\mathfrak{b}$ is an ideal in $\mathfrak{o}$ and $\mathfrak{b}^* = \mathfrak{o}^* \mathfrak{b} = \mathfrak{o}^* \tilde{h}$ is principal, then $\mathfrak{b}$ itself is principal. Now $\tilde{h} = \sum r_i b_i$ with $b_i \in \mathfrak{b}$ and $r_i \in \mathfrak{o}^*$. But $b_i \in \mathfrak{b}$ implies $b_i = \alpha_i \tilde{h}$, $\alpha_i \in \mathfrak{o}^*$ and so, $\tilde{h} = \sum r_i \alpha_i \tilde{h}$ i.e., $1 = \sum r_i \alpha_i$. Not all α_i are in $\mathfrak{m}^*$ and so, suppose $\alpha_1 \notin \mathfrak{m}^*$. Then α_1 is a unit in $\mathfrak{o}^*$ and $\tilde{h} = \alpha_1^{-1} b_1$ so that $\mathfrak{o}^* \tilde{h} = \mathfrak{o}^* b_1 = \mathfrak{b}^*$. But $\mathfrak{b}^* \cap \mathfrak{o} = \mathfrak{b}$ and so $\mathfrak{b} = \mathfrak{o} b_1$. In other words, $0 \to \mathfrak{o} b_1 \to \mathfrak{b} \to \mathfrak{b}/\mathfrak{o} b_1 \to 0$ exact implies $0 \to \mathfrak{o}^* b_1 \to \mathfrak{b}^* \to \mathfrak{b}^*/\mathfrak{o}^* b_1 \to 0$ exact. But $\mathfrak{b}^*/\mathfrak{o}^* b_1 = 0$ so $\mathfrak{b}/\mathfrak{o} b_1 = 0$.

The corollary can also be proved using the following lemma ([M-5]).

Lemma A.39 *If $x, y \in \mathfrak{o}$ and x divides y in $\mathfrak{o}^*$, then x divides y in $\mathfrak{o}$.*

Proof. Let $y = z^* x$ with $z^* \in \mathfrak{o}^*$, $z^* = \lim z_n, z_n \in \mathfrak{o}$. Then $z^* - z_n \in \mathfrak{m}^{*n}$ (for n large) and $y = z_n x + (z^* - z_n)x$. But $(z^* - z_n)x \in \mathfrak{m}^{*n} \cap \mathfrak{o} = \mathfrak{m}^n$ and $y \in x\mathfrak{o} + \mathfrak{m}^n$. Therefore, $y \in \cap x\mathfrak{o} + \mathfrak{m}^n = x\mathfrak{o}$.

We now wish to prove that the formal power series ring $K[[X_1, \ldots, X_N]]$ over a field K is a UFD and that, if $(\mathfrak{o}, \mathfrak{m})$ is a Noetherian regular local ring with completion $(\mathfrak{o}^*, \mathfrak{m}^*)$ containing a field $K = \mathfrak{o}/\mathfrak{m} = \mathfrak{o}^*/\mathfrak{m}^*$ (actually isomorphic to), then $\mathfrak{o}^* = K[[X_1, \ldots, X_N]]$ where $N = K \operatorname{Dim} \mathfrak{o}$. Naturally, it will follow that the regular local rings of simple subvarieties are UFD's and consequently, are integrally closed. We begin with (Iitaka [I-1]):

Theorem A.40 *Let $(\mathfrak{o}, \mathfrak{m})$, $(\mathfrak{o}_1, \mathfrak{m}_1)$ be Noetherian local rings. Suppose that* (i) *$\mathfrak{o}$ is a subring of $\mathfrak{o}_1$;* (ii) $\mathfrak{m}\mathfrak{o}_1 \subset \mathfrak{m}_1$; *and,* (iii) *there are $u_1, \ldots, u_n \in \mathfrak{o}_1$ such that*

$$\mathfrak{o}_1 = \mathfrak{m}\mathfrak{o}_1 + \sum_{i=1}^{n} \mathfrak{o} u_i \tag{A.41}$$

or, equivalently, if $\mathfrak{o}_1/\mathfrak{m}\mathfrak{o}_1 \simeq \mathfrak{o}_1 \otimes_{\mathfrak{o}} (\mathfrak{o}/\mathfrak{m}) = \sum_{i=1}^{n} (\mathfrak{o}/\mathfrak{m})(u_i \otimes_{\mathfrak{o}} 1)$. If $\mathfrak{o}$ is complete, then

$$\mathfrak{o}_1 = \sum_{i=1}^{n} \mathfrak{o} u_i \tag{A.42}$$

is a finite $\mathfrak{o}$-module.

Proof. Let $m_1, \ldots, m_t$ generate $\mathfrak{m}$. If $x \in \mathfrak{o}_1$, then $x = \sum r_j m_j + \sum x_i u_i$ with $r_j \in \mathfrak{o}_1$, $x_i \in \mathfrak{o}$. But $r_j = \sum \alpha_{js} m_s + \sum \beta_{ji} u_i$ with $\alpha_{js} \in \mathfrak{o}_1$, $\beta_{ji} \in \mathfrak{o}$. Let $x_{i,0} = x_i$, $x_{i,1} = \sum_{j=1}^{t} \beta_{ji} m_j$ so that

$$x = \sum_{j,s} \alpha_{js} m_j m_s + \sum_{i=1}^{n} (x_{i,0} + x_{i,1}) u_i. \tag{A.43}$$

Noting that $x_{i,1} \in \mathfrak{m}$ for all i and continuing, we find that there are $x_{i,\lambda}$, $\lambda = 0, 1, \ldots$ with $x_{i,\lambda} \in \mathfrak{m}^\lambda$ and

$$x \equiv \sum_{i=1}^{n} \left(\sum_{\lambda=0}^{\rho-1} x_{i,\lambda} \right) u_i \bmod \mathfrak{m}^\rho \mathfrak{o}_1 \tag{A.44}$$

for $\rho = 1, \ldots$. But $\mathfrak{o}$ is complete and so the series $\sum_{\lambda=0}^{\infty} x_{i,\lambda}$ converge to $y_i \in \mathfrak{o}$ for $i = 1, \ldots, n$. It follows from (A.44) that

$$x - \sum_{i=1}^{n} y_i u_i \in \bigcap_{\rho=1}^{\infty} \mathfrak{m}^\rho \mathfrak{o}_1 \subset \bigcap_{1}^{\infty} \mathfrak{m}_1^\sigma = (0) \tag{A.45}$$

and the result follows.

Corollary A.46 *If $(\mathfrak{o}_1, \mathfrak{m}_1)$ is a complete regular Noetherian local ring which contains a field $K \simeq \mathfrak{o}_1/\mathfrak{m}_1$, then $\mathfrak{o}_1 \simeq K[[X_1, \ldots, X_N]]$ where $N = K\,\mathrm{Dim}\,\mathfrak{o}_1$.*

Proof. Since $\mathfrak{o}_1$ is regular of (Krull) dimension N, $\mathfrak{m}_1$ is generated by N elements $x_1, \ldots, x_N$. Let $\mathfrak{o} = K[[x_1, \ldots, x_N]]$. Then $\mathfrak{o}$ is a complete Noetherian local ring with maximal ideal $\mathfrak{m} = \sum_{i=1}^{N} \mathfrak{o} x_i$. $\mathfrak{o}$ is clearly a subring of $\mathfrak{o}_1$ and $\mathfrak{m}\mathfrak{o}_1 \subset \mathfrak{m}_1$. But $\mathfrak{o}_1 \otimes_{\mathfrak{o}} (\mathfrak{o}/\mathfrak{m}) \simeq \mathfrak{o}_1/\mathfrak{m}_1 = K$ and so, by the Theorem, $\mathfrak{o} = \mathfrak{o}_1$. If $\psi \colon K[[T_1, \ldots, T_N]] \to K[[x_1, \ldots, x_N]]$ is the natural K-homomorphism with $\psi(T_j) = x_j$, then $\mathfrak{o}_1 \simeq K[[T_1, \ldots, T_N]]/\mathrm{Ker}\ \psi$. But $K\,\mathrm{Dim}\ \mathfrak{o}_1 = N$ and $K\,\mathrm{Dim}\ K[[X_1, \ldots, X_N]] = N$ so that $\mathrm{Ker}\ \psi = 0$.

Corollary A.47 *Let $\mathfrak{o}_1 = K[[X_1, \ldots, X_N]]$ be a formal power series ring. Let $Y_1, \ldots, Y_N$ be elements of $\mathfrak{o}_1$ with $Y_i(0) = 0$, $i = 1, \ldots, N$ (i.e., the Y_i have positive order) and with*

$$\det \left[\left. \frac{\partial Y_i}{\partial X_j} \right|_0 \right] \neq 0. \tag{A.48}$$

Then $K[[Y_1, \ldots, Y_N]] = K[[X_1, \ldots, X_N]]$.

Proof. Let $\mathfrak{o} = K[[Y_1, \ldots, Y_N]]$. Then $\mathfrak{o}$ is a complete Noetherian local ring with maximal ideal $\mathfrak{m} = \sum \mathfrak{o} Y_i$. Since $\mathfrak{m}_1 = \sum \mathfrak{o}_1 X_j$, $\mathfrak{m}\mathfrak{o}_1 + \mathfrak{m}_1^2/\mathfrak{m}_1^2 = \mathfrak{m}_1/\mathfrak{m}_1^2$ by (A.48). In other words, $\mathfrak{m}\mathfrak{o}_1 + \mathfrak{m}_1^2 = \mathfrak{m}_1$ and $\mathfrak{m}\mathfrak{o}_1 = \mathfrak{m}_1$ by Nakayama's lemma. Since $K = \mathfrak{o}/\mathfrak{m} = \mathfrak{o}_1/\mathfrak{m}_1$, the result follows from the theorem.

Corollary A.49 (Weierstrass Preparation Theorem) *Let $F \in K[[X_1, \ldots, X_N]]$ which is regular in X_1 of order $m > 0$ i.e., $F(X_1, 0, \ldots, 0) = a_1 X_1^m + \cdots$*

with $a_1 \neq 0$. Then any $G \in K[[X_1, \dots, X_N]]$ can be written in the form

$$G = UF + \sum_{i=0}^{m-1} R_i(X_2, \dots, X_N) X_1^i \tag{A.50}$$

(with U, R_i unique), $R_i \in K[[X_2, \dots, X_N]]$.

Proof.
Let $\mathfrak{o} = K[[X_2, \dots, X_N]]$ and $\mathfrak{m} = \sum_{j=2}^{N} \mathfrak{o} X_j$. Let $\mathfrak{o}_1 = K[[X_1, \dots, X_N]]/(F)$. Then $\mathfrak{o}$ is (isomorphic to) a subring of $\mathfrak{o}_1$ and $\mathfrak{m}\mathfrak{o}_1 \subset \mathfrak{m}_1$. Both $\mathfrak{o}$ and $\mathfrak{o}_1$ are complete Noetherian local rings. But

$$\mathfrak{o}_1 \otimes_{\mathfrak{o}} (\mathfrak{o}/\mathfrak{m}) \simeq K[[X_1]]/F(X_1, 0, \dots, 0))$$

which is generated by $1, X_1, \dots, X_1^{m-1}$. It follows from the theorem that $\mathfrak{o}_1 = \sum_0^{m-1} \mathfrak{o} X_1^j$ and hence that (A.50) holds.

Corollary A.51 *Let $F \in K[[X_1, \dots, X_N]]$ be regular in X_1 of order $m > 0$. Then there are a (unique) unit E and (unique) $R_i(X_2, \dots, X_N)$ $i = 0, 1, \dots, m-1$ such that*

$$EF = X_1^m + R_0 X_1^{m-1} + \cdots + R_{m-1} \tag{A.52}$$

and $R_i(0, \dots, 0) = 0$ (i.e., the R_i are non-units).

Proof. Apply Corollary A.49 to X_1^m and note that $E(X_1, 0, \dots, 0)$ is of order 0 with initial term $\neq 0$ and that $R_i(0, \dots, 0) = 0$ since otherwise the order is less than m.

Corollary A.53 *$K[[X_1, \dots, X_N]]$ is a UFD.*

Proof. Let $S^* = K[[X_1, \dots, X_N]]$. It will be enough to show that if $F \in S^*$ is irreducible, then the ideal $S^*F = (F)$ is prime. Suppose that $F \mid GH$ so that $DF = GH$. We may by Corollary A.47 assume that D, F, G, H are all regular in X_1. Then

$$\begin{aligned} E_1 D = D' = X_1^r + \cdots \qquad E_2 F = F' = X_1^s + \cdots \\ E_3 G = G' = X_1^a + \cdots \qquad E_4 H = H' = X_1^b + \cdots \end{aligned} \tag{A.54}$$

so that (up to units) $D'F' = G'H'$. Let $\tilde{S} = K[[X_2, \dots, X_N]]$. Then $F' \in \tilde{S}[X_1]$. If $N = 1$, then $\tilde{S} = K$ and $\tilde{S}[X_1]$ is a UFD (Corollary I.3.8). If F' is irreducible, then $F' \mid G'H'$ in $\tilde{S}[X_1]$ implies (say) $F' \mid G'$ in $\tilde{S}[X_1]$ and hence, in $\tilde{S}[[X_1]]$.

But $G' = F'\tilde{F}$ implies $G = E_3^{-1}G' = E_3^{-1}F'\tilde{F} = (E_3^{-1}E_2\tilde{F})F$. We *claim* that F' is irreducible. If $g \mid F'$ in $\tilde{S}[X_1]$ and g is not a unit, then $g \mid F'$ in $\tilde{S}[[X_1]]$ and $E_2 g \mid F$ in $\tilde{S}[[X_1]]$. But F is irreducible in $\tilde{S}[[X_1]]$ so that either $F \mid g$ in $\tilde{S}[[X_1]]$ or g is a unit in $\tilde{S}[[X_1]]$. But $g(X_1) = \alpha X_1^t$ with $\alpha \neq 0$, $t > 0$ (lower coefficients are non-units in $\tilde{S} = K$). Thus $g(0) = 0$ and g is a non-unit in $\tilde{S}[[X_1]]$. In other words, $g = EF$ with E a unit and $g = (EE_2^{-1})F'$. Since F', g are in $\tilde{S}[X_1]$, (EE_2^{-1}) is a unit in $\tilde{S}[X_1]$. Now use induction noting that if $\tilde{S}$ is a UFD, then $\tilde{S}[X_1]$ is a UFD (Theorem I.3.7).

Definition A.55 $(\mathfrak{o}, \mathfrak{m})$ is a *geometric local ring* if $\mathfrak{o} = k[V]_\mathfrak{p}$, $\mathfrak{m} = \mathfrak{p}k[V]_\mathfrak{p}$ where $k[V]$ is an affine k-algebra and $\mathfrak{p}$ is a prime ideal in $k[V]$.

Our next goal is to show that regular geometric local rings are UFD's. We begin with a general lemma.

Lemma A.56 *Let $(\mathfrak{o}, \mathfrak{m})$, $(\tilde{\mathfrak{o}}, \tilde{\mathfrak{m}})$ be local rings and let $\psi\colon \mathfrak{o} \to \tilde{\mathfrak{o}}$ be a homomorphism with $\psi(\mathfrak{m}^n) \subset \tilde{\mathfrak{m}}^n$ for all n. Then ψ induces natural homomorphisms $\psi_G\colon G_\mathfrak{m}(\mathfrak{o}) \to G_{\tilde{\mathfrak{m}}}(\tilde{\mathfrak{o}})$ and $\psi^*\colon \mathfrak{o}^* \to \tilde{\mathfrak{o}}^*$. If ψ_G is injective, then ψ^* is injective. If ψ_G is surjective, then ψ^* is also surjective. Consequently, if ψ_G is an isomorphism, then ψ^* is an isomorphism.*

Proof. If $\bar{x} \in \mathfrak{m}^n/\mathfrak{m}^{n+1}$, then $\psi_G(\bar{x}) = \overline{\psi(x)} \in \tilde{\mathfrak{m}}^n/\tilde{\mathfrak{m}}^{n+1}$ and extend. ψ_G is well-defined as $x - y \in \mathfrak{m}^{n+1}$ implies $\psi(x-y) = \psi(x) - \psi(y) \in \tilde{\mathfrak{m}}^{n+1}$. Similarly, if $x^* = \lim x_i \in \mathfrak{o}^*$, then $\psi^*(x^*) = \lim \psi(x_i)$. ψ^* is well-defined since $x_{i+1} - x_i \in \mathfrak{m}^i$ implies $\psi(x_{i+1} - x_i) = \psi(x_{i+1}) - \psi(x_i) \in \tilde{\mathfrak{m}}^i$. Clearly, ψ_G and ψ^* are homomorphisms. Consider the exact sequences

$$\begin{array}{ccccccccc} 0 & \longrightarrow & \mathfrak{m}^n/\mathfrak{m}^{n+1} & \longrightarrow & \mathfrak{o}/\mathfrak{m}^{n+1} & \longrightarrow & \mathfrak{o}/\mathfrak{m}^n & \longrightarrow & 0 \\ & & \downarrow \psi_{G,n} & & \downarrow \psi_{n+1} & & \downarrow \psi_n & & \\ 0 & \longrightarrow & \tilde{\mathfrak{m}}^n/\tilde{\mathfrak{m}}^{n+1} & \longrightarrow & \tilde{\mathfrak{o}}/\tilde{\mathfrak{m}}^{n+1} & \longrightarrow & \tilde{\mathfrak{o}}/\tilde{\mathfrak{m}}^n & \longrightarrow & 0 \end{array} \tag{A.57}$$

Then there is an exact sequence

$$0 \to \operatorname{Ker}\psi_{G,n} \to \operatorname{Ker}\psi_{n+1} \to \operatorname{Ker}\psi_n \to \operatorname{coker}\psi_{G,n} \to \operatorname{coker}\psi_{n+1} \to \operatorname{coker}\psi_n \to 0 \tag{A.58}$$

(where the cokernel of a homomorphism $f\colon M \to N$ is $N/\operatorname{Im} f$). If ψ_G is injective, Ker $\psi_{G,n} = 0$ and Ker $\psi_n = 0$. By induction on n, all ψ_n are injective and hence, so is ψ^*. Similarly, if ψ_G is surjective, then coker $\psi_{G,n} = 0$ and coker $\psi_n = 0$. Thus, all ψ_n are surjective and hence, so is ψ^*.

If $(\mathfrak{o}, \mathfrak{m})$ is a geometric local ring, then $V(\mathfrak{p}) = W$ (and $\mathfrak{p} = \mathfrak{p}_W = I_V(W)$) is a subvariety of V. If $\dim V = r$ and $\dim W = s$, then $\operatorname{codim}_V W = r - s$ and $\mathfrak{o}$ is regular if and only if $\mathfrak{m}$ is generated by $r - s$ elements.

Proposition A.59 *If $(\mathfrak{o}, \mathfrak{m})$ is a regular geometric local ring, if $\xi_1, \ldots, \xi_{r-s}$ give a basis of $\mathfrak{m}/\mathfrak{m}^2$ over $K = \mathfrak{o}/\mathfrak{m}$ with $\xi_i \in \mathfrak{p} \subset k[V]$, if $f_\nu(X_1, \ldots, X_{r-s})$ is* *a form of degree* ν *with coefficients in* $\mathfrak{o}$ *and* $f_\nu(\xi_1, \ldots, \xi_{r-s}) \equiv 0 \pmod{\mathfrak{m}^{\nu+1}}$, *then all the coefficients of* f_ν *are in* $\mathfrak{m}$.

Proof. By making a nonsingular linear transformation $\eta_i = \sum \alpha^i_j \xi_j$, $\overline{\alpha^i_j} \in K$, $\alpha^i_j \in \mathfrak{o}$, we can assume that if some coefficient is not in $\mathfrak{m}$, then

$$f_\nu(X_1, \ldots, X_{r-s}) = X_1^\nu + R_1(X_2, \ldots, X_{r-s})X_1^{\nu-1} + \cdots + R_\nu(X_2, \ldots, X_{r-s})$$

with R_i a form of degree i and that $\eta_1, \ldots, \eta_{r-s}$ is a basis of $\mathfrak{m}/\mathfrak{m}^2$. Then $f_\nu(\eta_1, \ldots, \eta_{r-s}) = \eta_1^\nu + R_1(\eta_2, \ldots, \eta_{r-s})\eta_1^{\nu-1} + \cdots + R_\nu(\eta_2, \ldots, \eta_{r-s}) \in \mathfrak{m}^{\nu+1}$. Since $\mathfrak{m}^{\nu+1}$ is generated by the monomials of degree $\nu+1$ in the η_i, it follows that $\eta_1^\nu(1-m) \in (\eta_2, \ldots, \eta_{r-s})$ for some $m \in \mathfrak{m}$. Since $1-m$ is a unit, $\eta_1^\nu \in (\eta_2, \ldots, \eta_{r-s})$. Consider $V \cap V(\eta_2) \cap \cdots \cap V(\eta_{r-s}) \supset W$ and let W^* be an irreducible component which contains W. Then (I.16.57) $\operatorname{codim}_V W^* \leq r-s-1$ or $\dim W^* \geq s+1$. But $\eta_1^\nu \in (\eta_2, \ldots, \eta_{r-s})$ so that η_1 vanishes on W^*. In other words, $\mathfrak{p} \subset I(W^*)$ (in $k[V]$) and $W \supset W^*$. Thus, $W = W^*$ and $\dim W^* = s$. This contradiction means that the proposition holds.

Corollary A.60 $G_\mathfrak{m}(\mathfrak{o}) \simeq K[X_1, \ldots, X_{r-s}]$ *where* $K = \mathfrak{o}/\mathfrak{m}$.

Proof. Let $\mathfrak{m} = (\xi_1, \ldots, \xi_{r-s})$ and let ξ_i^* be the initial form of ξ_i i.e., $\xi_i^* \in \mathfrak{m}^{\nu_i}$, $\xi_i^* \notin \mathfrak{m}^{\nu_i+1}$, ν_i the order of ξ_i. We may assume that $\nu_i = 1$, $i = 1, \ldots, r-s$ and so, $G_\mathfrak{m}(\mathfrak{o}) = K[\xi_1^*, \ldots, \xi_{r-s}^*]$. Consider the map $\psi \colon \mathfrak{m} \to K[[X_1, \ldots, X_{r-s}]]$ with $\psi(\xi_i^*) = X_i$. If $f \in \mathfrak{o}$ with initial form f_ν^* of order ν, then $f - f_\nu^* \in \mathfrak{m}^{\nu+1}$. Similarly, there is a form $f_{\nu+1}$ with $f - f_\nu^* - f_{\nu+1} \in \mathfrak{m}^{\nu+2}$. Continuing there are forms $F_i \in K[X_1, \ldots, X_{r-s}]$, F_i of degree i or $F_i = 0$ such that $f - \sum_{i=0}^{\mu} F_i(\xi_1^*, \ldots, \xi_{r-s}^*) \in \mathfrak{m}^{\mu+1}$. Then $\psi(f) = \sum_{i=0}^{\infty} F_i$. Clearly, $\operatorname{Ker} \psi = \cap \mathfrak{m}^\mu = (0)$ so that ψ is injective. If $M = (X_1, \ldots, X_{r-s})$ in $S^* = K[[X_1, \ldots, X_{r-s}]]$, then $\psi(\mathfrak{m}) = M$, $\psi(\mathfrak{o}/\mathfrak{m}) = K$, and $\psi(\mathfrak{m}^n/\mathfrak{m}^{n+1}) = M^n/M^{n+1}$. Hence, $G_\mathfrak{m}(\mathfrak{o}) \simeq G_M(S^*) \simeq K[X_1, \ldots, X_{r-s}]$.

Corollary A.61 *If* $(\mathfrak{o}, \mathfrak{m})$ *is a regular geometric local ring, then* $\mathfrak{o}^* \simeq K[[X_1, \ldots, X_{r-s}]]$.

Proof. Since $G_\mathfrak{m}(\mathfrak{o}) \simeq G_{\mathfrak{m}^*}(\mathfrak{o}^*)$ by Corollary A.33, Lemma A.56 implies $(\mathfrak{o}^*, \mathfrak{m}^*)$ is isomorphic to the completion of (S^*, M) ($S^* = K[[X_1, \ldots, X_{r-s}]]$). But (S^*, M) is complete.

Corollary A.62 $K \operatorname{Dim} \mathfrak{o}^* = r - s$ *and* $\mathfrak{o}^*$ *is also regular.*

Corollary A.63 *If* $(\mathfrak{o}, \mathfrak{m})$ *is a regular geometric local ring, then* $\mathfrak{o}$ *is a UFD.*

Now let $(\mathfrak{o}, \mathfrak{m})$ be any Noetherian local ring and let $K = \mathfrak{o}/\mathfrak{m}$. We wish to prove the following:

Theorem A.64 *$(\mathfrak{o}, \mathfrak{m})$ is regular if and only if $G_{\mathfrak{m}}(\mathfrak{o}) \simeq K[X_1, \ldots, X_n]$ as graded rings where $n = K\,\mathrm{Dim}\,\mathfrak{o} = h(\mathfrak{m})$.*

Since $G_{\mathfrak{m}}(\mathfrak{o}) \simeq G_{\mathfrak{m}^*}(\mathfrak{o}^*)$, the theorem will imply that $\mathfrak{o}^* \simeq K[[X_1, \ldots, X_n]]$ and hence, that $\mathfrak{o}^*$ and a fortiori $\mathfrak{o}$ are UFD's. We shall prove the theorem using Hilbert polynomials.

Definition A.65 An ideal $\mathfrak{a}$ of $\mathfrak{o}$ is an *ideal of definition of* $\mathfrak{o}$ if $\sqrt{\mathfrak{a}} = \mathfrak{m}$ (or, equivalently if $\mathfrak{m}^\nu \subset \mathfrak{a} \subset \mathfrak{m}$ for some ν i.e., if $\mathfrak{a}$ is $\mathfrak{m}$-primary).

If $\mathfrak{a}$ is an ideal of definition, then $(\mathfrak{o}/\mathfrak{a}, \mathfrak{m}/\mathfrak{a})$ is a Noetherian local ring with $(\mathfrak{m}/\mathfrak{a})^\nu = 0$. Set $\overline{\mathfrak{o}} = \mathfrak{o}/\mathfrak{a}$, $\overline{\mathfrak{m}} = \mathfrak{m}/\mathfrak{a}$.

Proposition A.66 $h(\overline{\mathfrak{m}}) = K\,\mathrm{Dim}\,\overline{\mathfrak{o}} = 0$.

Proof. Suppose $(0) < \mathfrak{p} \subset \overline{\mathfrak{m}}$ with $\mathfrak{p}$ prime. If $x \in \overline{\mathfrak{m}}$, then $x^\nu = 0 \in \mathfrak{p}$. Since $\mathfrak{p}$ is prime, $x \in \mathfrak{p}$.

Definition A.67 A ring satisfies the *descending chain condition* (dcc) if any descending chain $\mathfrak{a}_1 \supset \mathfrak{a}_2 \supset \cdots \supset \mathfrak{a}_n \supset \cdots$ of ideals stabilizes i.e., $\mathfrak{a}_n = \mathfrak{a}_{n+1} = \cdots$ for some n.

Proposition A.68 $\overline{\mathfrak{o}} \supset \overline{\mathfrak{m}} \supset \overline{\mathfrak{m}}^2 \supset \cdots \supset \overline{\mathfrak{m}}^\nu = (0)$.

Proof. Consider $\overline{\mathfrak{o}} \supset \overline{\mathfrak{m}} \supset \overline{\mathfrak{m}}^2 \supset \cdots \supset \overline{\mathfrak{m}}^\nu = (0)$ and note that each $\overline{\mathfrak{m}}^j/\overline{\mathfrak{m}}^{j+1}$ is a finite dimensional K-vector space as $\overline{\mathfrak{o}}$ is Noetherian. Hence each $\overline{\mathfrak{m}}^j/\overline{\mathfrak{m}}^{j+1}$ satisfies the dcc. If $\overline{\mathfrak{m}} \supset \mathfrak{a}_1 \supset \cdots \supset \mathfrak{a}_n \supset \cdots$, then $\mathfrak{a}_1/\overline{\mathfrak{m}}^2 \supset \mathfrak{a}_2/\overline{\mathfrak{m}}^2 \supset \cdots$ and so, for large enough n, $\mathfrak{a}_n \equiv \mathfrak{a}_{n+1} \equiv \cdots (\mathrm{mod}\,\overline{\mathfrak{m}}^2)$. Then consider $\mathfrak{a}_n^1 = \mathfrak{a}_n \cap \overline{\mathfrak{m}}^2 \supset \mathfrak{a}_{n+1}^1 = \mathfrak{a}_{n+1} \cap \overline{\mathfrak{m}}^2 \supset$ and note that eventually $\mathfrak{a}_{n_1}^1 \equiv \mathfrak{a}_{n_1+1}^1 \equiv \cdots (\mathrm{mod}\,\overline{\mathfrak{m}}^3)$. Continuing we obtain the result.

The ring $(\overline{\mathfrak{o}}, \overline{\mathfrak{m}})$ is called an *Artin* (or *Artinian*) ring. If M is a finitely generated $\overline{\mathfrak{o}}$ module, then M satisfies both the acc and dcc for submodules.

Definition A.69 If M is a finitely generated $\overline{\mathfrak{o}}$ module, then the *length of M* $\ell(M)$, (or better $\ell_{\overline{\mathfrak{o}}}(M)$) is the length of a maximal chain of submodules of M.

It is easy to show that $\ell(M)$ is well-defined) and finite ([Z-3] or by analogy with subspaces of a vector space).

Proposition A.70 *Let*

$$0 \longrightarrow M_1 \xrightarrow{\varphi_1} M_2 \xrightarrow{\varphi_2} M_3 \xrightarrow{\varphi_3} 0 \tag{A.71}$$

be an exact sequence of $\bar{\mathfrak{o}}$ *modules. Then*

$$\ell(M_1) - \ell(M_2) + \ell(M_3) = 0$$

or, equivalently, $\ell(M_2) = \ell(M_1) + \ell(M_3)$.

Proof. Since $\varphi_i(M_i) \simeq M_i/\varphi_i^{-1}(0)$ for $i = 1, 2, 3$, $\ell(\varphi_i(M_i)) + \ell(\varphi_i^{-1}(0)) = \ell(M_i)$ for $i = 1, 2, 3$. Since (A.71) is exact, $\ell(\varphi_i(M_i)) = \ell(\varphi_{i+1}^{-1}(0))$. But $\ell(M_3) = \ell(\varphi_3^{-1}(0))$ and Ker $\varphi_1 = 0$ so that $\ell(\varphi_1^{-1}(0)) = 0$ and, consequently, $\ell(M_1) - \ell(M_2) + \ell(M_3) = \ell(\varphi_2^{-1}(0)) + \ell(\varphi_1^{-1}(0)) - \ell(\varphi_2^{-1}(0)) - \ell(\varphi_3^{-1}(0)) + \ell(\varphi_3^{-1}(0)) = 0$.

Corollary A.72 *If* $0 \to M_1 \to M_2 \to \cdots \to M_n \to 0$ *is exact, then*

$$\sum_{m=1}^{n} (-1)^{m-1} \ell(M_m) = 0.$$

We say that ℓ is an *additive function.* Suppose that $\mathfrak{a} = (x_1, \ldots, x_r)$ has r generators and set $\tilde{\mathfrak{o}} = \bar{\mathfrak{o}}[X_1, \ldots, X_r]$.

Remark A.73 If R is a ring, $S = R[X_1, \ldots, X_r]$ and $M = \oplus \sum M_q$ is a graded S-module, then each M_q is a R-module. If M is finitely generated, then each M_q is a finitely generated R-module.

In light of the remark, we let $M = \oplus \sum M_q$ be a finitely generated $\tilde{\mathfrak{o}}$-module. Let $\xi_1, \ldots, \xi_t$ be homogeneous generators of M with $\deg \xi_i = q_i$. Let $\tilde{\mathfrak{o}}(-q_i)$ be $\tilde{\mathfrak{o}}$ as a module but with grading $q - q_i$. Consider the map $\psi\colon \oplus \sum_1^t \tilde{\mathfrak{o}}(-q_i) \to M$ given by

$$\psi(u_1, \ldots, u_t) = \sum_{i=1}^{t} u_i \xi_i \tag{A.74}$$

where $u_i \in \tilde{\mathfrak{o}}(-q_i)$. Then ψ is surjective and preserves degrees. It follows that

$$\ell(M_q) \leq \sum_{i=1}^{t} \ell(\tilde{\mathfrak{o}}(-q_i)_q). \tag{A.75}$$

Since $\ell(\tilde{\mathfrak{o}}(-q_i)_q) = \ell(\tilde{\mathfrak{o}}_{q-q_i})$ and since the number of monomials of degree q in $X_1, \ldots, X_r$ is $(q + r - 1)!/q!(r - 1)!$, we have

$$\ell(\tilde{\mathfrak{o}}_q) \leq \binom{q + r - 1}{q} \ell(\bar{\mathfrak{o}}) < \infty \tag{A.76}$$

which implies $\ell(M_q)$ is finite (M_q is an $\bar{\mathfrak{o}}$ module).

We now come to the Hilbert-Serre Theorem.

Theorem A.77 (Hilbert-Serre) *Let* $M = \oplus \sum M_q$ *be a finitely generated graded* $\tilde{\mathfrak{o}}$ *module and let*

$$h_M(q) = \ell(M_q) = \ell_{\tilde{\mathfrak{o}}}(M_q). \tag{A.78}$$

Then, for sufficiently large q, $h_M(q)$ *is a polynomial in* q *of degree at most* $r-1$ *with integer coefficients.*

Proof. We use induction on r. If $r = 0$, then $\tilde{\mathfrak{o}} = \tilde{\mathfrak{o}}$ and so, for $q \gg 0$, $M_q = 0$ and $h_M(q) = 0$ (conventionally of degree -1). For instance, if M is generated by $\xi_1, \ldots, \xi_t$ with degree $\xi_i = d_i$, then take $q > \max d_i$. Now assume the result for $s < r$ and consider the map $\psi\colon M \to M$ given by

$$\psi(\mathbf{m}) = X_r\mathbf{m} \tag{A.79}$$

for $\mathbf{m} \in M$. ψ is a homogeneous homomorphism of degree 1. Let $N = \operatorname{Ker} \psi$ and $P = M/\psi(N) = \operatorname{coker} \psi$ so that

$$0 \longrightarrow N \longrightarrow M \xrightarrow{\psi} M \longrightarrow P \longrightarrow 0 \tag{A.80}$$

is exact. Then

$$0 \longrightarrow N_q \longrightarrow M_q \xrightarrow{\psi} M_{q+1} \longrightarrow P_{q+1} \longrightarrow 0 \tag{A.81}$$

is also exact. Since $X_rN = 0$ and $X_rP = 0$, these are submodules over $\bar{\mathfrak{o}}[X_1, \ldots, X_{r-1}]$. By additivity,

$$h_M(q+1) - h_M(q) = h_P(q+1) - h_N(q). \tag{A.82}$$

By induction, h_P and H_N are (for large q) polynomials of degree at most $r-2$ with integer coefficients. If

$$\binom{x}{d} = x(x-1)\cdots(x-d+1)/r!$$

then

$$\binom{x}{d} - \binom{x+d-1}{d} = \binom{x+d-1}{d-1}$$

and it follows that

$$h_M(q+1) - h_M(q) = c_0\binom{q}{r-2} + c_1\binom{q}{r-3} + \cdots + c_{r-2}$$

with the $c_i \in \mathbb{Z}$. Consequently,

$$h_M(q) = \alpha_0 \binom{q}{r-1} + \alpha_1 \binom{q}{r-2} + \cdots + \alpha_{r-1} \tag{A.83}$$

with $\alpha_i \in \mathbb{Z}$ for q large.

Definition A.84 $h_M(q)$ is called the *Hilbert* or *characteristic polynomial of* M.

Let $\mathfrak{a}$ be an ideal of definition of $\mathfrak{o}$ with $\mathfrak{a} = (x_1, \ldots, x_r)$. Then $G_{\mathfrak{a}}(\mathfrak{o}) = \oplus \sum \mathfrak{a}^n/\mathfrak{a}^{n+1}$ and $G_{\mathfrak{a}}(M) = \oplus \sum \mathfrak{a}^n M/\mathfrak{a}^{n+1}M$. So, $G_{\mathfrak{a}}(\mathfrak{o})$ is the image of $\bar{\mathfrak{o}}$ under a natural homomorphism and $G_{\mathfrak{a}}(M)$ is a finitely generated graded $G_{\mathfrak{a}}(\mathfrak{o})$ module. By the Theorem, $h_{G_{\mathfrak{a}}(M)}(q) = \ell(\mathfrak{a}^q M/\mathfrak{a}^{q+1}M)$ [note an $\bar{\mathfrak{o}}$ module] is a polynomial of degree at most $r-1$ with integer coefficients for q large.

Definition A.85 $H_M(\mathfrak{a}; t) = \sum_{q=0}^{t-1} h_{G_{\mathfrak{a}}(M)}(q)$ is the *Hilbert-Samuel function of* M.

We note that $H_M(\mathfrak{a}; t)$ is a polynomial of degree at most r for q large. If $\mathfrak{a}'$ is another ideal of definition with $\mathfrak{a}' = (x_1', \ldots, x_r')$, then $\mathfrak{a}'^s \subset \mathfrak{a}$ for some s and $H_M(\mathfrak{a}; t) = \ell(M/\mathfrak{a}^t M) \leq \ell(M/\mathfrak{a}'^{st}M) = H_M(\mathfrak{a}'; st)$. For large t,

$$\deg H_M(\mathfrak{a}; t) \leq \deg H_M(\mathfrak{a}'; st). \tag{A.86}$$

By symmetry, these degrees are equal. In other words, $\deg H_M$ is independent of the choice of ideal of definition. We let $d_H(M)$ denote this common degree and we observe that if $\mathfrak{a} = (x_1, \ldots, x_r)$, then $d_H(M) \leq r$. Let $d_g(\mathfrak{o})$ be the least number of generators of an ideal of definition.

Proposition A.87 $d_g(\mathfrak{o}) \geq d_H(\mathfrak{o})$.

Proof. Since there is an ideal of definition with $d_g(\mathfrak{o})$ generators, the result follows.

Proposition A.88 $d_H(\mathfrak{o}) \geq h(\mathfrak{m}) = K\,\mathrm{Dim}\ \mathfrak{o}$.

Proof. Use induction on $d_H(\mathfrak{o})$. If $d_H(\mathfrak{o}) = 0$, then $\mathfrak{m}^\nu = \mathfrak{m}^{\nu+1} = \cdots$ for some $\nu > 0$ and so, by Nakayama's lemma, $\mathfrak{m}^\nu = 0$ and $h(\mathfrak{m}) = K\,\mathrm{Dim}\ \mathfrak{o} = 0$ (by Proposition A.66). Suppose $d_H(\mathfrak{o}) > 0$ and $h(\mathfrak{m}) > 0$. Let $\mathfrak{m} \supset \mathfrak{p}_0 \supset \cdots \supset \mathfrak{p}_s \supset (0)$ be a chain of prime ideals and let $x \in \mathfrak{p}_{s-1}$, $x \notin \mathfrak{p}_s$. Then $K\,\mathrm{Dim}(\mathfrak{o}/x\mathfrak{o} + \mathfrak{p}_s) \geq s - 1$. Consider the exact sequence

$$0 \longrightarrow \mathfrak{o}/\mathfrak{p}_s \xrightarrow{x} \mathfrak{o}/\mathfrak{p}_s \longrightarrow \mathfrak{o}/x\mathfrak{o} + \mathfrak{p}_s \longrightarrow 0.$$

Then $d_H(\mathfrak{o}/x\mathfrak{o} + \mathfrak{p}_s) < d_H(\mathfrak{o}/\mathfrak{p}) \leq d_H(\mathfrak{o})$ (compute lengths). By induction $K\,\mathrm{Dim}(\mathfrak{o}/x\mathfrak{o}+\mathfrak{p}_s) \leq d_H(\mathfrak{o}/x\mathfrak{o}+\mathfrak{p}_s) < d_H(\mathfrak{o})$. In other words, $s-1 < d_H(\mathfrak{o})$ and $s \leq d_H(\mathfrak{o})$ which means that $h(\mathfrak{m}) \leq d_H(\mathfrak{o})$.

Proposition A.89 $h(\mathfrak{m}) \geq d_g(\mathfrak{o})$.

Proof. ([A-2]) Let $d = h(\mathfrak{m})$. We must show that there is an $\mathfrak{m}$-primary ideal with d generators. We will construct $u_1, \ldots, u_d$ by induction so that every prime ideal containing $(u_1, \ldots, u_j)$ has height at least j. Suppose that $u_1, \ldots, u_{j-1}$ have been chosen ($j > 0$) and let $\mathfrak{p}_\alpha$, $\alpha = 1, \ldots, t$ be the minimal prime ideals of $(u_1, \ldots, u_{j-1})$ which have height $j-1$ (there may not be any). Since $j-1 < d = h(\mathfrak{m})$, $\mathfrak{m} \neq \mathfrak{p}_\alpha$ for any α and so $\mathfrak{m} \neq \bigcup_{\alpha=1}^{t} \mathfrak{p}_\alpha$. Let $u_j \in \mathfrak{m}$, $u_j \notin \bigcup_{\alpha=1}^{t} \mathfrak{p}_\alpha$ and let $\mathfrak{p}$ be a prime with $(u_1, \ldots, u_j) \subset \mathfrak{p} \subset \mathfrak{m}$. Then $\mathfrak{p}$ contains some minimal prime ideal $\mathfrak{p}'$ of $(u_1, \ldots, u_{j-1})$. If $\mathfrak{p}' = \mathfrak{p}_\alpha$, then $u_j \in \mathfrak{p}$, $u_j \notin \bigcup_{\alpha=1}^{t} \mathfrak{p}_\alpha$ so that $\mathfrak{p} > \mathfrak{p}' = \mathfrak{p}_\alpha$ and $h(\mathfrak{p}) \geq j$. If $\mathfrak{p}' \neq \mathfrak{p}_\alpha$ for $\alpha = 1, \ldots, t$, then $h(\mathfrak{p}) \geq h(\mathfrak{p}') > j-1$. In other words, every prime ideal containing $(u_1, \ldots, u_j)$ has height $\geq j$. Every prime ideal $\mathfrak{q} \supset (u_1, \ldots, u_d)$ has $h(\mathfrak{q}) \geq d$ and $\mathfrak{q} \subset \mathfrak{m}$. But $h(\mathfrak{m}) = d$ and so $\mathfrak{q} = \mathfrak{m}$ i.e., $(u_1, \ldots, u_d)$ is $\mathfrak{m}$-primary.

Theorem A.90 *If* $(\mathfrak{o}, \mathfrak{m})$ *is a Noetherian local ring, then* $K\,\mathrm{Dim}\,\mathfrak{o} = h(\mathfrak{m}) = d_H(\mathfrak{o}) = d_G(\mathfrak{o})$.

Proof. Apply the propositions.

If $(u_1, \ldots, u_d)$ is an ideal of definition of $\mathfrak{o}$ where $d = K\,\mathrm{Dim}\,\mathfrak{o}$, then $u_1, \ldots, u_d$ is called a *system of parameters of* $\mathfrak{o}$. If $(u_1, \ldots, u_d) = \mathfrak{m}$, then $u_1, \ldots, u_d$ is a *regular system of parameters of* $\mathfrak{o}$ and $\mathfrak{o}$ is a regular local ring.

Theorem A.91 *A Noetherian local ring* $(\mathfrak{o}, \mathfrak{m})$ *is regular if and only if* $G_\mathfrak{m}(\mathfrak{o}) \simeq K[X_1, \ldots, X_d]$, *(as graded rings) where* $d = K\,\mathrm{Dim}\,\mathfrak{o}$.

Proof. If $G_\mathfrak{m}(\mathfrak{o}) \simeq K[X_1, \ldots, X_d]$, then $d_H(\mathfrak{o}) = d = K\,\mathrm{Dim}\,\mathfrak{o}$ (by Theorem A.90) and $\dim_k \mathfrak{m}/\mathfrak{m}^2 = d$ so that $\mathfrak{o}$ is regular. Conversely, if $\mathfrak{o}$ is regular, then let $u_1, \ldots, u_d$ be a regular system of parameters. If ξ_i is the initial form of u_i (i.e., $\xi_i \equiv u_i \bmod \mathfrak{m}^{\nu_i}$ where $u_i \in \mathfrak{m}^{\nu_i}$, $u_i \notin \mathfrak{m}^{\nu_i+1}$), then $G_\mathfrak{m}(\mathfrak{o}) = K[\xi_1, \ldots, \xi_d] \simeq K[X_1, \ldots, X_d]/\mathfrak{a}$ where $\mathfrak{a}$ is a homogeneous ideal. If $f \in \mathfrak{a}$, $f \neq 0$, is a form of degree q_0, then, for $q > q_0$,

$$\ell(\mathfrak{o}/\mathfrak{m}^{q+1}) \leq \binom{q+d}{d} - \binom{q+d-q_0}{d}$$

which is a polynomial of degree $d-1$ in q. But $H_{G_\mathfrak{m}(\mathfrak{o})}(\mathfrak{m}; q)$ has degree d for large q and $H_{G_\mathfrak{m}(\mathfrak{o})}(\mathfrak{m}; q) = \ell(\mathfrak{o}/\mathfrak{m}^{q+1})$. Hence $\mathfrak{a} = (0)$ and $G_\mathfrak{m}(\mathfrak{o}) \simeq K[X_1, \ldots, X_d]$.

Appendix B
Specialization, Generic Points and Spectra

Let $R = k[X_1, \ldots, X_N]$, $\mathfrak{p}$ be a prime ideal in R, and $R/\mathfrak{p} = k[x_1, \ldots, x_N] = k[x]$, $x = (x_1, \ldots, x_N)$ with $x_i = \overline{X}_i$. A polynomial $f(X_1, \ldots, X_N)$ is in $\mathfrak{p}$ if and only if $f(x_1, \ldots, x_N) = 0$ (for if $f \in \mathfrak{p}$, then $0 = \overline{f(X)} = f(\overline{X}) = f(x)$ and conversely).

Definition B.1 A prime ideal $\mathfrak{p}'$ in R is a *specialization of* $\mathfrak{p}$ if $\mathfrak{p} \subset \mathfrak{p}'$. If $R/\mathfrak{p}' = k[x'_1, \ldots, x'_N] = k[x']$, then x' is a *specialization of* x *over* k and we write $x \xrightarrow{k} x'$.

We observe that if $\mathfrak{p}'$ is a specialization of $\mathfrak{p}$, then the map $\psi \colon k[x] \to k[x']$ given by $\psi(f(x)) = f(x')$ is a surjective homomorphism with kernel $\mathfrak{p}'/\mathfrak{p}$. Conversely, if $\psi \colon S \to S'$ is a surjective homomorphism of finitely generated integral domains $S = R/\mathfrak{p}$, $S' = R/\mathfrak{p}'$, then $\mathfrak{p}'$ is a specialization of $\mathfrak{p}$.

Definition B.2 $\xi \in \mathbb{A}_k^N$ is an *algebraic specialization of* x *(or* $\mathfrak{p}$*) over* k if $x \xrightarrow{k} \xi$ i.e., if $M_\xi = (X_i - \xi_i) \supset \mathfrak{p}$ or equivalently, if $\xi \in V(\mathfrak{p}) \subset \mathbb{A}_k^N$.

Since k is algebraically closed, $V(\mathfrak{p}) = \{\xi \colon \xi$ is an algebraic specialization of $x\}$ and there are always algebraic specializations by the Nullstellensatz. Thus, $\xi \in V(\mathfrak{p})$ if and only if $f(\xi) = 0$ for all f with $f(x) = 0$. We should therefore like to call $x = (x_1, \ldots, x_N)$ a "generic point" of $V = V(\mathfrak{p})$ over k; however, x is not a point in $\mathbb{A}_k^N$. Classically, this issue was dealt with by assuming varieties had points in a so-called "Universal Domain" i.e., an algebraically closed field with infinite transcendence degree over the prime field, etc. (see [Z-3]). We shall handle things differently after a few examples.

Example B.3 Consider $\mathbb{A}_k^1$, $R = k[X_1]$. Let $\mathfrak{p} = I(\mathbb{A}_k^1) = (0)$ and $R/\mathfrak{p} =$

P. Falb, *Methods of Algebraic Geometry in Control Theory: Part II*,
Modern Birkhäuser Classics, https://doi.org/10.1007/978-3-319-96574-1

$R = k[X_1]$. Then $X_1 \xrightarrow{k} \xi$ as $(X_1 - \xi) \supset (0)$ for all $\xi \in \mathbb{A}_k^1$. Thus X_1 (an indeterminate) is a "generic point" and $\mathbb{A}_k^1$ is the "closure" of X_1 under specialization.

Example B.4 Consider $\mathbb{A}_k^2$, $R = k[X_1, X_2]$. Let $\mathfrak{p} = I(\mathbb{A}_k^2) = (0)$ and $R/\mathfrak{p} = R = k[X_1, X_2]$. Then (X_1, X_2) is a "generic point" of $\mathbb{A}_k^2$ which is the "closure" of (X_1, X_2) under specialization. The prime ideals in $k[X_1, X_2]$ are (0), $M_{(\xi_1,\xi_2)} = (X_1 - \xi_1,\, X_2 - \xi_2)$ (maximal) and (f) with f an irreducible polynomial. Suppose that $f = X_2 - \xi_2$. Then $\mathfrak{p}' = (X_2 - \xi_2) \supset (0)$ and $R/\mathfrak{p}' = k[X_1, \xi_2]$ so that $(X_1, \xi_2) \to (\xi_1, \xi_2)$ is a "specialization" for any ξ_1 on $V(\mathfrak{p}') = V(f) = V(X_2 - \xi_2) = \{(\xi_1, \xi_2)\colon X_2 = \xi_2\}$. Thus (X_1, ξ_2) is a "generic point" of $V(\mathfrak{p}')$ which is the "closure" of (X_1, ξ_2) in $\mathbb{A}_k^2$ under specialization.

Let $k[V] = R/\mathfrak{p}$ with $\mathfrak{p} = I(V)$ and let $\mathfrak{q}$ be a prime ideal in $k[V]$. If $\psi\colon R \to k[V]$ is the natural homomorphism with kernel $\mathfrak{p}$, then $\mathfrak{p}' = \psi^{-1}(\mathfrak{q})$ is a prime ideal in R with $\mathfrak{p} \subset \mathfrak{p}'$. Let $W = V(\mathfrak{q})$, $W' = V(\mathfrak{p}')$. We claim that $W = W'$ i.e., $\xi \in W$ if and only if $\mathfrak{m}_\xi = \overline{M}_\xi \supset \mathfrak{q}$. But $\psi^{-1}(\mathfrak{m}_\xi) = M_\xi \supset \mathfrak{p}'$ so that $\xi \in W$ if and only if $M_\xi \supset \mathfrak{p}'$ i.e., if and only if $\xi \in W'$. Here $\mathfrak{p}'$ is a specialization of $\mathfrak{p}$. Let $k[\alpha_1, \ldots, \alpha_N]$ be an integral domain. Then the natural homomorphism $k[X] \to k[\alpha]$ is surjective with kernel a prime ideal $\mathfrak{p}$ and $k[x] = k[X]/\mathfrak{p} \simeq k[\alpha]$. In other words, $x \xrightarrow{k} \alpha$ and $\alpha \xrightarrow{k} x$ so that both might be viewed as "generic points" of $V(\mathfrak{p})$. Suppose now that $W = V(\mathfrak{p}') \subset V = V(\mathfrak{p})$. Then $(0) \subset \mathfrak{p} \subset \mathfrak{p}'$ is a chain of specializations which translates into a chain $(X_1, \ldots, X_N) \to (x_1, \ldots, x_N) \to (x_1', \ldots, x_N')$ ($x_i = X_i \pmod{\mathfrak{p}}$, $x_i' = X_i \pmod{\mathfrak{p}'}$). In view of Lemma I.16.7, $\operatorname{tr\,deg} k[x_1', \ldots, x_N'] = s \le \operatorname{tr\,deg} k[x_1, \ldots, x_N] = r \le \operatorname{tr\,deg} k[X_1, \ldots, X_N] = N$. x is a "generic point" of V and x' is a "generic point" of W. We may assume that $x = (y_1, \ldots, y_s, v_1, \ldots, v_{N-s})$ and $x' = (y_1, \ldots, y_s, w_1, \ldots, w_{N-s})$ where $y_1, \ldots, y_s$ are algebraically independent over k. This would allow us to (at least heuristically) do certain things over the field $K^* = k(y_1, \ldots, y_s)$ making W a "point" on V over K^*. [This is sometimes referred to as "reduction to dimension 0."] We note that $\mathfrak{o}_{W,V} = \{f/g\colon f, g \in k[V],\, g \neq 0 \text{ on } W\} = \{f/g\colon f, g \in k[x],\, g(x') \neq 0\}$, that $\mathfrak{m}_{W,V} = \{h\colon h \in \mathfrak{o}_{W,V},\, h(x') = 0\}$, and that $K(\mathfrak{o}_{W,V}) = \{f/g\colon f, g \in k[x],\, f(x') \neq 0,\, g(x') \neq 0\} = K(W)$. Since $y_1, \ldots, y_s$ are transcendental over k, any (non-zero) $g(y_1, \ldots, y_s, v_1, \ldots, v_{N-s})$ which involves only $y_1, \ldots, y_s$ satisfies $g(x') \neq 0$. In other words, $\mathfrak{o}_{W,V} = \{f^*(v_1, \ldots, v_{N-s})/g^*(v_1, \ldots, v_{N-s})\colon f^*, g^* \in K^*[v_1, \ldots, v_{N-s}],\, g^*(w_1, \ldots, w_{N-s}) \neq 0\}$ and $\mathfrak{m}_{W,V} = \{h^* \in \mathfrak{o}_{W,V}\colon h^*(w_1, \ldots, w_{N-s}) = 0\}$. We shall use these ideas in Chapter 15.

Let us now indicate a more rigorous approach which will in the end be most fundamental.

Definition B.5 Let $\mathbb{A}$ be a topological space and V be a closed set in $\mathbb{A}$. A point $v \in V$ is a *generic point of* V if $\overline{\{v\}} = V$. [Note such a V is necessarily irreducible.] A point v is a *closed point* if $\overline{\{v\}} = v$.

Defintion B.6 Let R be a ring. The *spectrum of* R, $\mathrm{Spec}(R)$, is the set of all prime ideals $\mathfrak{p} < R$. If $\mathfrak{a}$ is an ideal in R, then $V(\mathfrak{a}) = \{\mathfrak{p} \in \mathrm{Spec}(R)\colon \mathfrak{p} \supset \mathfrak{a}\}$.

Proposition B.7 (i) $V(\mathfrak{a} \cap \mathfrak{b}) = V(\mathfrak{a}) \cup V(\mathfrak{b})$; (ii) $V(\sum_t \mathfrak{a}_t) = \bigcap_t V(\mathfrak{a}_t)$; (iii) $V((0)) = \mathrm{Spec}(R)$, $V(R) = \emptyset$. *Hence, the* $V(\mathfrak{a})$, *as closed sets, define a topology, called the Zariski topology on* $\mathrm{Spec}(R)$.

Proof. (i) If $\mathfrak{p} \in V(\mathfrak{a}) \cup V(\mathfrak{b})$, then, say, $\mathfrak{a} \subset \mathfrak{p}$ implies $\mathfrak{a} \cap \mathfrak{b} \subset \mathfrak{a} \subset \mathfrak{p}$. If $\mathfrak{a} \cap \mathfrak{b} \subset \mathfrak{p}$ and $\mathfrak{a} \not\subset \mathfrak{p}$, then there is an $x \in \mathfrak{a}$, $x \notin \mathfrak{p}$. But $x\mathfrak{b} \subset \mathfrak{a} \cap \mathfrak{b} \subset \mathfrak{p}$ implies $\mathfrak{b} \subset \mathfrak{p}$. (ii) If $\mathfrak{p} \in \cap V(\mathfrak{a}_t)$, then $\mathfrak{p} \supset \mathfrak{a}_t$ for all t and $\mathfrak{p} \supset \sum_t \mathfrak{a}_t$. Conversely, if $\mathfrak{p} \supset \sum_t \mathfrak{a}_t$, then $\mathfrak{p} \supset \mathfrak{a}_t$ for all t. (iii) Obvious.

The set $\mathrm{Spec}(R) - V((f)) = \{\mathfrak{p} \in \mathrm{Spec}(R)\colon f \notin \mathfrak{p}\}$ is open in $\mathrm{Spec}(R)$. We write $D(f)$ or $\mathrm{Spec}(R)_f$ for this set and call any such a *distinguished open set.* These sets form a base for the Zariski topology.

Proposition B.8 $D(f)$ *is homeomorphic to* $\mathrm{Spec}(R_f)$.

Proof. This is a simple consequence of the fact that all prime ideals $\mathfrak{q}$ in R_f are of the form $\mathfrak{p}R_f$ where $\mathfrak{p}$ is a prime ideal in R with $f \notin \mathfrak{p}$ and of the fact that $\mathfrak{p}R_f \cap R = \mathfrak{p}$.

Proposition B.9 $\overline{\{\mathfrak{p}\}} = V(\mathfrak{p})$ *and* $\mathfrak{p}$ *is a closed point if and only if* $\mathfrak{p}$ *is maximal. Moreover,* $V(\mathfrak{p})$ *is homeomorphic to* $\mathrm{Spec}(R/\mathfrak{p})$.

Proof. The first assertion is obvious and the second is a result of the fact that the prime ideals in $R/\mathfrak{p}$ are $\mathfrak{q}/\mathfrak{p}$ where $\mathfrak{q} \supset \mathfrak{p}$.

Definition B.10 If $A \subset \mathrm{Spec}(R)$, then the *ideal of* A, $I(A)$, is the set $\bigcap_{\mathfrak{p} \in A} \mathfrak{p}$.

Proposition B.11 $V(I(A)) = \overline{A}$, *the closure of* A.

Proof. If $\mathfrak{p} \in A$, then $\mathfrak{p} \supset I(A)$ and $A \subset V(I(A))$. Since $V(I(A))$ is closed, $\overline{A} \subset V(I(A))$. If $A \subset V(\mathfrak{a})$, then $\mathfrak{p} \supset \mathfrak{a}$ for all $\mathfrak{p} \in A$ and so $\mathfrak{a} \subset I(A)$ and $V(I(A)) \subset V(\mathfrak{a})$. It follows that

$$\overline{A} \subset V(I(A)) \subset \bigcap_{A \subset V(\mathfrak{a})} V(\mathfrak{a}) = \overline{A}.$$

Proposition B.12 $I(V(\mathfrak{a})) = \sqrt{\mathfrak{a}}$ *for any ideal in* R. *Thus closed subsets correspond to radical ideals.*

Proof. It is enough to show that $\sqrt{\mathfrak{a}} = \bigcap_{\mathfrak{a}\subset\mathfrak{p}} \mathfrak{p}$. By passing to $R/\mathfrak{a}$, it suffices to show that $\sqrt{(0)} = \bigcap_{\mathfrak{p}\in\mathrm{Spec}(R)} \mathfrak{p}$ which is the following lemma.

Lemma B.13 $\sqrt{(0)} = \bigcap_{\mathfrak{p}\in\mathrm{Spec}(R)} \mathfrak{p} = \{r \in R\colon r^n = 0$ *for some* $n\}$ *i.e., the set of all nilpotent elements of* R.

Proof. Clearly, $\sqrt{(0)} \subset \cap\mathfrak{p}$. Let $r \in \cap\mathfrak{p}$ and let $M = \{r^n\}_{n=1}^{\infty}$. If $r \notin \sqrt{(0)}$, then $M \cap (0) = \emptyset$ and there is a prime ideal $\mathfrak{p}$ with $\mathfrak{p} \cap M = \emptyset$ (a contradiction).

Proposition B.14 (i) $\overline{\{\mathfrak{p}\}} = V(\mathfrak{p})$ *is irreducible with* $\mathfrak{p}$ *as generic point, and,* (ii) *if* V *is closed and irreducible in* $\mathrm{Spec}(R)$, *then* $V = V(\mathfrak{p})$ *for some* $\mathfrak{p}$ *which is the unique generic point of* V.

Proof. For (i), if $V_1 \subset V(\mathfrak{p})$ is closed with $\mathfrak{p} \in V_1$, then $V(\mathfrak{p}) \subset V_1 \subset V(\mathfrak{p})$ and $V_1 = V(\mathfrak{p})$. For (ii), if V is irreducible, then $I(V)$ is prime since if $fg \in I(V)$ and $\mathfrak{p} \in V$, then $f \in \mathfrak{p}$ or $g \in \mathfrak{p}$ so that $V \subset V(f)$ or $V \subset V(g)$ (i.e., $f \in I(V)$ or $g \in I(V)$). If $I(V) = \mathfrak{p}$, then $\overline{V} = V(I(V)) = V = V(\mathfrak{p})$. As for uniqueness, if $\mathfrak{p}'$ were another generic point, then $\mathfrak{p} \supset \mathfrak{p}' \supset \mathfrak{p}$ and $\mathfrak{p}' = \mathfrak{p}$.

Proposition B.15 $\mathrm{Spec}(R)$ *is quasi-compact.*

Proof. This follows from the assertion that

$$\mathrm{Spec}(R) = \bigcup_{\alpha} \mathrm{Spec}(R_{f_\alpha})$$

if and only if $1 \in (\dots, f_\alpha, \dots)$. If $1 \in (\dots, f_\alpha, \dots)$, then no prime $\mathfrak{p}$ contains $(\dots, f_\alpha, \dots)$ and hence, if, say $f_{\alpha_0} \notin \mathfrak{p}$, then $\mathfrak{p} \in \mathrm{Spec}(R_{f_{\alpha_0}})$. Conversely, if $\mathrm{Spec}(R) = \cup\mathrm{Spec}(R_{f_\alpha})$, then, for any prime $\mathfrak{p}$, there is an f_{α_0} with $f_{\alpha_0} \notin \mathfrak{p}$. Thus, no $\mathfrak{p} \supset (\dots, f_\alpha, \dots)$ and $1 \in (\dots, f_\alpha, \dots)$.

Corollary B.16 $\mathrm{Spec}(R)$ *is compact.*

Proof.

If $\mathrm{Spec}(R) = \bigcup_{\alpha} \mathrm{Spec}(R_{f_\alpha})$, then $1 = \sum_{i=1}^{t} r_i f_{\alpha_i}$ and $\mathrm{Spec}(R) = \bigcup_{i=1}^{t} \mathrm{Spec}(R_{f_{\alpha_i}})$.

Let again $R = k[X_1, \dots, X_N]$ and consider $\mathrm{Spec}(R)$. Then $\mathfrak{p}'$ is a *specialization of* $\mathfrak{p}$ if and only if $\mathfrak{p}' \in V(\mathfrak{p}) = \overline{\{\mathfrak{p}\}}$. The *closed points* of $\mathrm{Spec}(R)$ are precisely the maximal ideals $M_\xi = (X_1 - \xi_1, \dots, X_N - \xi_N)$ and every variety has a unique generic point.

Example B.17 If $R = k$ is a field, then $\mathrm{Spec}(k) = \{(0)\}$ a single point.

Example B.18 If $R = \mathbb{Z}$, the integers, then $\text{Spec}(\mathbb{Z}) = \{\pi: \pi \text{ a prime}\} \cup \{(0)\}$. Each prime is a closed point and (0) is the generic point.

Example B.19 $\text{Spec}\, k[X] = \{f(X): f \text{ irreducible}\} \cup \{(0)\}$. If k is algebraically closed, then $f(X) = X - \xi$ and closed points are the elements of k and (0) is the unique generic point. We can write $\text{Spec}\, k[X] = k \cup (*)$, $(*)$ the generic point.

Example B.20 $\text{Spec}\, k[X,Y] = \{(0)\} \cup \{\text{maximal ideals } (X-a, Y-b)\} \cup \{f(X,Y): f \text{ irreducible}\}$ (k algebraically closed). The generic point is $\{(0)\}$, the closed points are the elements (a,b) of k^2, and, for every irreducible curve, $f(X,Y) = 0$, there is a generic point $(*)_f$. For instance, if $f(X,Y) = X - a$, then we might view $(*)_f$ as (a, Y).

Example B.21 Let $\mathfrak{o} = \mathfrak{o}_{(0,0),\mathbb{A}_k^2} = \{f/g: f,g \in k[X,Y], g(0,0) \neq 0\}$. What is $\text{Spec}(\mathfrak{o})$? $\text{Spec}(\mathfrak{o}) = \{(0)\} \cup \{(X,Y)\} \cup \{f(X,Y): f \text{ irreducible}, f(0,0) = 0\}$. For instance, if k is algebraically closed, then $\text{Spec}\, \mathfrak{o} = \{(0)\} \cup \mathfrak{m}_{(0,0)} \cup \{(aX + bY): a \text{ or } b \neq 0\}$. The generic point is (0). $\mathfrak{m}_{(0,0)} = (X,Y)$ is the unique closed point and the remaining points correspond to lines $l_{a,b}$ $(= \{aX + bY = 0\})$ through the origin. Since $V(aX + bY) = (aX + bY) \cup \mathfrak{m}_{(0,0)}$, the closure of $l_{a,b}$ is $l_{a,b} \cup \mathfrak{m}_{(0,0)}$.

Example B.22 $\text{Spec}\, \mathbb{Z}[x]$. $\mathbb{Z}[x]$ has the prime ideals (0) (unique generic point), (π) where π is a prime in $\mathbb{Z}$, (f) where f is a polynomial which is irreducible over the rationals $\mathbb{Q}$ with relatively prime coefficients, and (π, f_π) where π is a prime in $\mathbb{Z}$ and f_π is a monic polynomial which is irreducible mod π. What is $V(2) = \overline{(2)}$? $\pi = 2$ is the generic point and $(2,x)$, $(2, x+1)$ are the remaining points. Similarly, $V(3) = \overline{(3)} = \{(3), (3,x), (3,x+1), (3,x+2)\}$ and $V(5) = \overline{(5)} = \{(5), (5,x), (5,x+1), (5,x+2), (5,x+3), (5,x+4)\}$. The polynomial x^2+1 is irreducible over $\mathbb{Q}$. $V(x^2+1) = \overline{(x^2+1)} = \{(x^2+1), (2,x+1), (5,x+2), (5,x+3), \ldots\}$. Observe that $(2,x) \notin V(x^2+1)$ as $x^2+1 \not\equiv x^2 (\text{mod}\, 2)$, $(3,x) \notin V(x^2+1)$ as $x^2+1 \not\equiv x^2 \pmod 3$, $(3,x+1) \notin V(x^2+1)$ as $x^2+1 \not\equiv (x+1)^2 \pmod 3$ and $2x+1 \not\equiv 2x+2 \pmod 3$, and $(3,x+2) \notin V(x^2+1)$ as $(x^2+1) \not\equiv (x+1)(x+2) \pmod 3$. [This example indicates, in a small way, the extraordinary richness of the concept $\text{Spec}(R)$.]

Now let R be an integral domain and let $K = K(R)$ be the quotient field of R. If $\mathfrak{p}$ is an element of $\text{Spec}(R)$, then $R_\mathfrak{p}$ is a local ring contained in K. We let $V = \text{Spec}(R)$ and $\mathfrak{o}_{x,V} = R_\mathfrak{p} = R_{\mathfrak{p}_x} = R_x$ for $x \in V$ (i.e., x is the prime ideal $\mathfrak{p} = \mathfrak{p}_x$). We wish to define a sheaf of rings $\mathfrak{o}_V$ on V so that $(V, \mathfrak{o}_V)$ becomes a ringed space. In other words, we extend the notion of affine variety given in Part I Chapter 22 to arbitrary integral domains. In fact (see, for example, [M-2]), it can be extended to arbitrary rings R and, by patching, to create very general objects. We proceed essentially as in I.22.

If U is open in $V = \text{Spec}(R)$, we let

$$\mathfrak{o}_V(U) = \bigcap_{\mathfrak{p} \in U} R_{\mathfrak{p}} = \bigcap_{x \in U} \mathfrak{o}_{x,V}. \tag{B.23}$$

Clearly $\mathfrak{o}_V(U)$ is a ring containing R. If $U_1 \subset U$, then $\mathfrak{o}_V(U) \subset \mathfrak{o}_V(U_1)$ and there is a natural restriction homomorphism $\rho^U_{U_1} : \mathfrak{o}_V(U) \to \mathfrak{o}_V(U_1)$. We note that $\rho^U_U =$ identity and that if $U_2 \subset U_1 \subset U$, then $\rho^U_{U_1} \circ \rho^{U_1}_{U_2} = \rho^U_{U_2}$.

Proposition B.24 $\mathfrak{o}_V(D(f)) = R_f$ *[note that* $D(f)$ *is homeomorphic to* $\text{Spec}(R_f)$ *by Proposition B.8 so that* $\mathfrak{o}_V(\text{Spec}(R_f)) = R_f$*].*

Proof. By definition

$$\mathfrak{o}_V(D(f)) = \bigcap_{\mathfrak{p} \in D(f)} R_{\mathfrak{p}} = \bigcap_{f \notin \mathfrak{p}} R_{\mathfrak{p}}$$

and so we must show that

$$R_f = \bigcap_{f \notin \mathfrak{p}} R_{\mathfrak{p}}.$$

If $f \notin \mathfrak{p}$, then $R_f \subset R_{\mathfrak{p}}$ and so $R_f \subset \bigcap_{f \notin \mathfrak{p}} R_{\mathfrak{p}}$. Let $z \in \bigcap_{f \notin \mathfrak{p}} R_{\mathfrak{p}}$ and let $\mathfrak{a}_z = \{h \in R_f : hz \in R_f\}$. $\mathfrak{a}_z$ is an ideal in R_f. If $\mathfrak{a}_z < R_f$, then there is a prime ideal $\mathfrak{p}_z$ in R_f with $\mathfrak{a}_z \subset \mathfrak{p}_z < R_f$. Let $\mathfrak{p}'_z = \mathfrak{p}_z \cap R$. $\mathfrak{p}'_z$ is a prime ideal in R with $f \notin \mathfrak{p}'_z$. Let $z = h_1/h_2$ with $h_2 \notin \mathfrak{p}'_z$ as $z \in R_{\mathfrak{p}'_z}$. Then $h_2 z = h_1 \in R \subset R_f$ and $h_2 \in \mathfrak{a}_z$. But then $h_2 \in \mathfrak{p}_z \cap R = \mathfrak{p}'_z$ a contradiction. Therefore, $1 \in \mathfrak{a}_z$ and $z \in R_f$.

Corollary B.25 $\mathfrak{o}_V(V) = R$ *as* $V = D(1)$.

We observe that $D(f) \cap D(g) = D(fg)$, that $D(f^n) = D(f)$, and that if $U = \cup D(f_i)$, then $\mathfrak{o}_V(U) = \cap \mathfrak{o}_V(D(f_i))$ and that

$$\mathfrak{o}_V(U) = \bigcap_{D(f) \subset U} \mathfrak{o}_V(D(f)).$$

Proposition B.26 (i) *If* $U = \cup U_i$, $f_1, f_2 \in \mathfrak{o}_V(U)$, *and* $\rho^U_{U_i}(f_1) = \rho^U_{U_i}(f_2)$ *for all* i, *then* $f_1 = f_2$*; and,* (ii) *if* $U = \cup U_\alpha$ *and* $f_\alpha \in \mathfrak{o}_V(U_\alpha)$ *with* $\rho^{U_\alpha}_{U_\alpha \cap U_\beta}(f_\alpha) = \rho^{U_\beta}_{U_\beta \cap U_\alpha}(f_\beta)$ *for all* α, β, *then there is a unique (by (i))* f *in* $\mathfrak{o}_V(U)$ *with* $\rho^U_{U_\alpha}(f) = f_\alpha$. *In other words,* $\mathfrak{o}_V = \{\mathfrak{o}_V(U), \rho^U_{U_1}\}$ *is a sheaf of rings on* V *and* $(V, \mathfrak{o}_V)$ *is a ringed space.*

Proof. (i) We may suppose that $U = \cup D(h_i)$ and it will be enough to show that $\rho^U_{D(h_i)}(f) = 0$ for all i implies $f = 0$. If $\rho^U_{D(h_i)}(f) = 0$, then $h_i^{\nu_i} f = 0$ for some ν_i as $\mathfrak{o}_V(D(h_i)) = R_{h_i}$. Since $D(h_i) = D(h_i^{\nu_i})$, $\cup D(h_i^{\nu_i}) = U$ and by Proposition B.15 (proof), $1 \in (h_1^{\nu_1}, \dots)$. In other words, $1 = \sum r_i h_i^{\nu_i}$ and $f = \sum r_i(h_i^{\nu_i} f) = 0$. (ii) Suppose that $D(h_i)$ cover U, that $s_i \in \mathfrak{o}_U(D(h_i))$, and $s_i = s_j$ on $D(h_i) \cap D(h_j) = D(h_i h_j)$. Then $s_i = r_i/h_i^{m_i}$ and $\rho^{D(h_i)}_{D(h_i h_j)}(s_i) = r_i h_j^n/(h_i h_j)^n$. By hypothesis, $(h_i h_j)^{m_{ij}}(r_i h_j^n - r_j h_i^n) = 0$. It follows that $s_i = \sigma_i/h_i^\nu$ ($\sigma_i = r_i h_i^m, m+n = \nu$) and $\sigma_i h_j^\nu = \sigma_j h_i^\nu$. Since $U = \cup D(h_i^\nu)$, $1 = \sum g_i h_i^\nu$ and we let $s = \sum g_i \sigma_i$. Then $h_j^\nu s = \sum_i g_i \sigma_i h_j^\nu = \sum g_i \sigma_j h_i^\nu = \sigma_j$ so that $\rho^U_{D(h_j)}(s) = s_j$.

Let us indicate briefly how to extend these ideas to a general ring R. First note that if $D(f) \subset D(g)$, then, in $R/(g)$, $\overline{f}$ is in every prime ideal (as $\mathfrak{p} \supset (g)$ implies $\mathfrak{p} \supset (f)$). By Lemma B.13, $\overline{f^t} = (0)$ and so, $f^t = r \cdot g$ for some t and $r \in R$. We then define a map $\rho^g_f \colon R_g \to R_f$ by setting

$$\rho^g_f(h/g^s) = r^s h/f^{ts}. \tag{B.27}$$

It is easy to show that this is a well-defined homomorphism. In view of Proposition B.24, it is natural to *define* $\mathfrak{o}_V(D(f))$ as R_f. However, in contrast to the case where R is an integral domain and all the rings R_f are contained in the quotient field $K(R)$ of R, here the rings R_f are not directly comparable i.e., $\bigcap_{D(f) \subset U} \mathfrak{o}_V(D(f))$ does not make sense. The substitute is something called the *projective limit.* We have the following situation for U open in $\mathrm{Spec}(R) = V$: (i) an *ordering* of $\{f \colon D(f) \subset U\}$ by $f < g$ if $D(f) \subset D(g)$; (ii) a *family* (of rings) $\{R_f \colon D(f) \subset U\}$; and, (iii) *homomorphisms* ρ^g_f for each pair f, g with $f < g$ such that $\rho^g_f \colon R_g \to R_f$ and satisfying the condition $\rho^h_f = \rho^g_f \circ \rho^h_g$ for $f < g < h$ (this is easy to check). If we consider $\prod_f \{R_f \colon D(f) \subset U\}$ and we let π_f be the projection on the fth factor, then, the subset (actually subring) of $\prod_f R_f$ consisting of elements $\mathbf{r} = (r_f)$ such that, for *all* pairs f, g with $f < g$, $\pi_f(\mathbf{r}) = \rho^g_f \circ \pi_g(\mathbf{r})$, is called the *projective limit of* $\{R_f \colon D(f) \subset U\}$ and is denoted by $\varprojlim R_f$. Since $R_f = \mathfrak{o}_V(D(f))$, we can *define* $\mathfrak{o}_V(U)$ as $\varprojlim \mathfrak{o}_V(D(f))$ over $D(f) \subset U$. It can be shown that this gives a sheaf of rings $\mathfrak{o}_V$ on $V = \mathrm{Spec}(R)$ ([M-2], [H-3], etc.). The ringed space $(V, \mathfrak{o}_V)$ is called an *affine scheme.* Affine schemes can be "patched" together to give very, very general objects.

Appendix C
Differentials

Let k be a ring and let R and S be k-algebras and let $\psi: R \to S$ be a k-homomorphism. If N is an S-module, then N can be viewed as an R-module by defining $r \cdot n = \psi(r) \cdot n$.

Consider the map $\mu: R \otimes_k R \to R$ given by $\mu(r \otimes t) = rt$. Let $I = \text{Kernel}\ \mu$ which is an ideal in $R \otimes_k R$.

Proposition C.1 *The map $\partial: R \to R \otimes_k R$ given by $\partial(s) = s \otimes 1 - 1 \otimes s$ is a k-homomorphism such that $\partial(k \cdot 1) = 0$, $\partial(R) \subset I$, and $I = \sum_{r \in R} R\partial(r)$ (i.e., the $\partial(r)$, $r \in R$, generate I).*

Proof. Most assertions are clear and we show only that $I = \sum_{r \in R} R\partial(r)$. Let $x = \sum r_i \otimes t_i \in R \otimes_k R$. Then $x = \sum [r_i(1 \otimes t_i - t_i \otimes 1) + (r_i t_i) \otimes 1] = \sum r_i \partial(t_i) + \left[\sum r_i t_i\right] \otimes 1$. If $x \in I$, then $\sum r_i t_i = 0$ and $x \in \sum_{r \in R} R\partial(r)$.

We note that $R \otimes_k R/I \simeq R$. Since I/I^2 is an $R \otimes_k R/I$-module, it is also an R-module. We denote this R-module by $\Omega_{R/k}$ and call it the *module of (Kähler) differentials of R over k*. If $r \in R$, then the *differential of* r, dr, is $\overline{\partial_r} = \overline{r \otimes 1 - 1 \otimes r} = I^2$-residue of $r \otimes 1 - 1 \otimes r$. dr is, of course, an element of $\Omega_{R/k}$.

Corollary C.2 *$\{dr: r \in R\}$ generate $\Omega_{R/k}$ as an R-module and $d: R \to \Omega_{R/k}$ is a derivation.*

Proof. From the proposition, $I/I^2 = \sum_{r \in R} R\overline{\partial(r)} = \sum_{r \in R} R\,dr$. Now $\partial(rt) = (rt) \otimes 1 - 1 \otimes rt = r[t \otimes 1 - 1 \otimes t] + r \otimes t - [(1 \otimes r - r \otimes 1)t + r \otimes t] = r\partial(t) + t\partial(r)$ so that $d(rt) = \overline{\partial(rt)} = r\overline{\partial(t)} + t\overline{\partial(r)} = r\,dt + t\,dr$.

P. Falb, *Methods of Algebraic Geometry in Control Theory: Part II*,
Modern Birkhäuser Classics, https://doi.org/10.1007/978-3-319-96574-1

Proposition C.3 *Let M be an R-module. Then* $\mathrm{Hom}_R(\Omega_{R/k}, M)$ *is isomorphic to* $\mathrm{Der}_k(R, M)$.

Proof. If $h \in \mathrm{Hom}_R(\Omega_{R/k}, M)$, then $h \circ d$ is an element of $\mathrm{Der}_k(R, M)$ by Corollary C.2. If $h \circ d = h_1 \circ d$, then $h(dr) = h_1(dr)$ for all $r \in R$ and so $h = h_1$ since the dr generate $\Omega_{R/k}$. In other words, the map $h \to h \circ d$ is injective. Since the map $h \to h \circ d$ is clearly a homomorphism of R-modules, we need only show surjectivity. Let $D \in \mathrm{Der}_k(R, M)$ and define a map $\delta \colon R \otimes_k R \to M$ by setting

$$\delta(r \otimes t) = t\, Dr. \tag{C.4}$$

Then $\delta((r \otimes t)(r_1 \otimes t_1)) = \delta((rr_1) \otimes (tt_1)) = tt_1 D(rr_1) = tt_1 r\, Dr_1 + tt_1 r_1\, Dr = \mu(t \otimes r)\delta(r_1 \otimes t_1) + \mu(t_1 \otimes r_1) \cdot \delta(r \otimes t)$ and so $\delta(I^2) = 0$. Hence δ defines an R-module homomorphism $\overline{\delta} \colon \Omega_{R/k} \to M$ with $\overline{\delta} \circ d = D$ since $\overline{\delta}(dr) = \delta(r \otimes 1 - 1 \otimes r) = Dr$.

The pair $(\Omega_{R/k}, d)$ is universal with respect to the property described in the proposition.

If $\psi \in \mathrm{Hom}_k(R, S)$, then there is a natural homomorphism $\tilde{\psi} \colon \Omega_{R/k} \to \Omega_{S/k}$ given by

$$\tilde{\psi} \circ d_R = d_S \circ \psi \tag{C.5}$$

(note that $\Omega_{S/k}$ is an R-module via ψ). In other words, the diagram

$$\begin{array}{ccc} R & \xrightarrow{\psi} & S \\ d_R \downarrow & & d_S \downarrow \\ \Omega_{R/k} & \xrightarrow{\tilde{\psi}} & \Omega_{S/k} \end{array} \tag{C.6}$$

commutes. This is "dual" to the natural homomorphism $\hat{\psi}$ of $\mathrm{Der}_k(S, N) \to \mathrm{Der}_k(R, N)$ for (any) S-module N given by

$$\hat{\psi}(D) = D \circ \psi. \tag{C.7}$$

The kernel of $\hat{\psi}$ is $\mathrm{Der}_R(S, N)$. If N is an S-module (a fortiori an R-module), then the diagram

$$\begin{array}{ccc} \mathrm{Der}_k(S, N) & \simeq & \mathrm{Hom}_S(\Omega_{S/k}, N) \\ \hat{\psi} \downarrow & & \theta \downarrow \\ \mathrm{Der}_k(R, N) & \simeq & \mathrm{Hom}_R(\Omega_{R/k}, N) \end{array} \tag{C.8}$$

commutes where θ is induced by $\tilde{\psi}$ i.e., θ is given by

$$\theta(h) = h \circ \tilde{\psi} \tag{C.9}$$

for $h \in \mathrm{Hom}_S(\Omega_{S/k}, N)$.

Suppose that $\Omega_{R/k}$ is free over R with rank m and that $\Omega_{S/k}$ is free over S with rank n. Then $\tilde{\psi}$ induces a homomorphism $\tilde{\Psi}\colon \Omega_{R/k} \otimes_R S \to \Omega_{S/k}$ of free S-modules given by

$$\tilde{\Psi}(dr \otimes s) = s\tilde{\psi}(dr). \tag{C.10}$$

The map $\tilde{\Psi}$ may be described by an $n \times m$ matrix with entries in S. Let $d\psi\colon \mathrm{Der}_k(S,k) \to \mathrm{Der}_k(R,k)$ be given by

$$(d\psi)(D) = D \circ \psi. \tag{C.11}$$

Then:

Proposition C.12 *If $d\psi$ is surjective, then the map $h \to h \circ \tilde{\psi}$ of $\mathrm{Hom}_S(\Omega_{S/k}, k) \to \mathrm{Hom}_R(\Omega_{R/k}, k)$ is also injective.*

Proof. Apply diagram (C.8).

Corollary C.13 *If $d\psi$ is surjective, then the k-rank of $\tilde{\Psi}$, $\rho_k(\tilde{\Psi})$, is m.*

Corollary C.14 *If $d\psi$ is surjective and if $\rho_k(\tilde{\Psi}) = m$ implies $\rho(\tilde{\Psi}) = m$, then ψ is injective.*

Proof. Since $\rho(\tilde{\Psi}) = m$, $\tilde{\Psi}$ is injective. Hence $\tilde{\psi}$ is injective as a map of R-modules. But $\tilde{\psi}(\Omega_{R/k}) = \tilde{\psi}(\sum R\,dr_i)$ ($\Omega_{R/k}$ free of rank m) $= \sum \psi(R)d(r_i \circ \psi)$ and so ψ must also be injective.

Example C.15 Let $R = k[X_1, \dots, X_N]/(f_1, \dots, f_r)$ with $f_i \in k[X_1, \dots, X_N]$, $i = 1, \dots, r$. Let $x_i = \overline{X}_i$ and let $D_i = \frac{\partial}{\partial X_i}$ in $k[X_1, \dots, X_N]$. If $\psi\colon R^N \to \Omega_{R/k}$ is the map given by

$$\psi\left(\sum \alpha_i \epsilon_i\right) = \sum \alpha_i dx_i \tag{C.16}$$

(ϵ_i the standard unit vectors in R^N), then ψ is surjective and $\mathrm{Ker}\,\psi$ is generated by $\sum_{i=1}^{N} (D_i f_j)(x_1, \dots, x_N)\epsilon_i$, $j = 1, \dots, r$. Thus $R^N/\mathrm{Ker}\,\psi \simeq \Omega_{R/k}$. Moreover, if $D \in \mathrm{Der}_k(R, M)$, M an R-module, then, for $f \in k[X_1, \dots, X_N]$,

$$(Df)(x_1, \dots, x_N) = \sum_{i=1}^{N} (D_i f)(x_1, \dots, x_N) Dx_i.$$

Thus, if $f = f_j$, then $(Df_j)(x_1, \dots, x_N) = 0$. When k is a field, then $R = k[V]$ is the coordinate ring of an affine variety.

Appendix D
The Space $\mathcal{C}_m^n$

For historical reasons (see [H-5]), we develop a theory for controllable systems.*
Let $\mathbf{Y}: \mathbb{A}^{n^2+nm} \to M(n,(n+1)m)$ be given by

$$\mathbf{Y}(X,Y) = \left[Y \ XY \cdots X^n Y\right] \tag{D.1}$$

and let $\tilde{\mathbf{Y}}: \mathbb{A}^{n^2+nm} \to M(n,(n+1)m)$ be given by

$$\tilde{\mathbf{Y}}(X,Y) = \left[Y_1 \ XY_1 \cdots X^n Y_m\right]. \tag{D.2}$$

We say that $x = (A,B) \in \mathbb{A}^{n^2+nm}$ is *controllable* if $\rho(Y(A,B)) = n$ (or, equivalently, if $\rho(\tilde{\mathbf{Y}}(A,B)) = n$) and we let $\mathcal{C}_m^n = \{x = (A,B): x \text{ is controllable}\}$. $\mathcal{C}_m^n$ is open in $\mathbb{A}^{n^2+nm}$ and so is irreducible. Thus, $\mathbf{Y}(\mathcal{C}_m^n)$ and $\tilde{\mathbf{Y}}(\mathcal{C}_m^n)$ are also irreducible. Let $\mathbf{f} = (f_1, \dots, f_m)$ be an m-partition of n and let $\beta_{\mathbf{f}}$ be nice relative to f. Then:

Proposition D.3 $\tilde{\mathbf{Y}}(\mathcal{C}_m^n) \subset \cup U_{\mathbf{f}}$ *and* $\tilde{\mathbf{Y}}$ *is injective on* $\mathcal{C}_m^n$ *where* $U_{\mathbf{f}} = \{\mathbf{M} \in M_*(n,(n+1)m)$: $\det \mathbf{M}_{\beta_{\mathbf{f}}} \neq 0\}$.

If $G = \mathrm{GL}(n,k)$, then G acts on $\mathbb{A}^{n^2+nm}$ via

$$g \cdot (A,B) = (gAg^{-1}, gB) \tag{D.4}$$

and $\mathcal{C}_m^n$ is invariant under G. G also acts on $M(n,(n+1)m)$ via

$$\mathbf{M} \to g\mathbf{M} \tag{D.5}$$

---

*Of course, we could develop an entirely similar theory for observable systems. We leave this to the reader.

P. Falb, *Methods of Algebraic Geometry in Control Theory: Part II*,
Modern Birkhäuser Classics, https://doi.org/10.1007/978-3-319-96574-1

and the maps $\tilde{\mathbf{Y}}$ or $\mathbf{Y}$ naturally intertwine the action of G. There is also a natural map $\pi\colon M_*(n,(n+1)m) \to \mathrm{Gr}(n,(n+1)m)$ given by the Plücker coordinates (Chapter 8) and $\mathrm{Gr}(n,(n+1)m) \simeq M_*(n,(n+1)m)/G$ so that we see that $(\mathrm{Gr}(n,(n+1)m),\pi)$ is a natural geometric quotient. We write $\mathbf{M}_{\mathfrak{f}}$ for $\mathbf{M}_{\beta_{\mathfrak{f}}}$ and we let $V_{\mathfrak{f}} = \pi(U_{\mathfrak{f}}) = \{(\det \mathbf{M}_{(j_1,\dots,j_n)})\colon \det \mathbf{M}_{\mathfrak{f}} \neq 0\}$. Finally, we let

$$W_{\mathfrak{f}} = \left\{(\det \tilde{\mathbf{Y}}(x)_{(j_1,\dots,j_n)})\colon \det \tilde{\mathbf{Y}}(x)_{\mathfrak{f}} \neq 0\right\} = \pi(\tilde{\mathbf{Y}}(C_m^n) \cap U_{\mathfrak{f}}). \tag{D.6}$$

Let $O_G\colon C_m^n \to C_m^n/G$ be the map which carries x into its orbit $O_G(x)$. Consider the map $\tilde{y}\colon C_m^n/G \to \mathrm{Gr}(n,(n+1)m)$ given by

$$\tilde{y}(O_G(A,B)) = O_G(\tilde{\mathbf{Y}}(A,B)) = (\det \tilde{\mathbf{Y}}(A,B)_{(j_1,\dots,j_n)}). \tag{D.7}$$

Clearly, $\tilde{y}$ is well-defined. Moreover, the following diagram commutes:

$$\begin{array}{ccc} C_m^n & \xrightarrow{O_G} & C_m^n/G \\ \tilde{\mathbf{Y}}\downarrow & & \tilde{y}\downarrow \\ M_*(n,(n+1)m) & \longrightarrow & \mathrm{Gr}(n,(n+1)m) \end{array} \tag{D.8}$$

and hence,

$$W_{\mathfrak{f}} = \pi(\tilde{\mathbf{Y}}(C_m^n) \cap U_{\mathfrak{f}}) = (\tilde{y} \circ O_G)(C_m^n) \cap V_{\mathfrak{f}} = \tilde{y}(C_m^n/G) \cap V_{\mathfrak{f}}. \tag{D.9}$$

It follows that

$$(\pi \circ \tilde{\mathbf{Y}})(C_m^n) = \cup W_{\mathfrak{f}} = \tilde{y}(C_m^n/G) \cap \cup V_{\mathfrak{f}} = \tilde{y}(C_m^n/G). \tag{D.10}$$

Since $\pi \circ \tilde{\mathbf{Y}}$ is a morphism, $\tilde{y}(C_m^n/G)$ is an (irreducible) quasi-projective variety. We shall now construct a morphism $\psi_{\mathfrak{f}}\colon \mathbb{A}_k^{nm} \to C_m^n$ such that:

(1) $(\tilde{\mathbf{Y}} \circ \psi_f)(\mathbb{A}_k^{nm}) \subset U_{\mathfrak{f}}$ (D.11)

(2) $\tilde{\mathbf{Y}}(C_m^n) \cap U_{\mathfrak{f}} = G \cdot \left[(\tilde{\mathbf{Y}} \circ \psi_{\mathfrak{f}})(\mathbb{A}_k^{nm})\right]$. (D.12)

[In the language of Chapter 16, this gives a principal G-bundle.] Let $(x_1,\dots,x_m)$ be an element of $(\mathbb{A}_k^n)^m$ with $x_j \in \mathbb{A}_k^n$. Suppose that $\mathbf{f} = (f_1,\dots,f_m)$.

Case 1 $f_j \neq 0$ for all j. Then $\beta_{\mathfrak{f}} = (11\cdots 1f_1\ 21\cdots mf_m)$. Set $f_0 = 0$ and let

$$(B_{\mathfrak{f}})_j = \epsilon_{f_1+\dots+f_{j-1}+1} \tag{D.13}$$

for $j = 1,\dots,m$. Then $B_{\mathfrak{f}} = \left[(B_{\mathfrak{f}})_1 \cdots (B_{\mathfrak{f}})_m\right] = B_{\mathfrak{f}}(x_1,\dots,x_m)$ is an $n \times m$ matrix. Let

$$L_{f_j,f_j-1} = \begin{bmatrix} O_{1,f_j-1} \\ I_{f_j-1,f_j-1} \end{bmatrix} \tag{D.14}$$

and set

$$A_{\mathbf{f}}(x_1,\dots,x_m)=\begin{bmatrix} & O_{f_1,f_2-1} & & \\ L_{f_1,f_1-1} & & & \\ & x_1 & & x_2 \ \cdots \ x_m \\ O_{n-f_1,f_1-1} & & & \\ & O_{n-(f_1+f_2),f_2-1} & & \end{bmatrix} \tag{D.15}$$

We observe that

$$\begin{aligned} A_{\mathbf{f}}(x_1,\dots,x_m)^t(B_{\mathbf{f}})_j &= \epsilon_{f_1+\cdots+f_{j-1}+1+t}, \quad t=0,1,\dots,f_j-1 \\ A_{\mathbf{f}}(x_1,\dots,x_m)^{f_j}(B_{\mathbf{f}})_j &= x_j \end{aligned} \tag{D.16}$$

and we set

$$\psi_{\mathbf{f}}(x_1,\dots,x_m)=(A_{\mathbf{f}}(x_1,\dots,x_m),B_{\mathbf{f}}(x_1,\dots,x_m)). \tag{D.17}$$

Then $\psi_{\mathbf{f}}$ is a morphism. Clearly, $\tilde{\mathbf{Y}}_{\mathbf{f}}(A_{\mathbf{f}},B_{\mathbf{f}}) = [\epsilon_1\cdots\epsilon_n] = I$ and $\det(\tilde{\mathbf{Y}}\circ\psi_{\mathbf{f}})(x_1,\dots,x_m)_{\beta_{\mathbf{f}}}\neq 0$. Thus $(\tilde{\mathbf{Y}}\circ\psi_{\mathbf{f}})(\mathbb{A}^{nm})\subset U_{\mathbf{f}}$ and (D.11) holds. Since, in general, $g\cdot\tilde{\mathbf{Y}}(A,B)=\tilde{\mathbf{Y}}(gAg^{-1},B)$, we have

$$G\cdot[(\tilde{Y}\circ\psi_{\mathbf{f}})(\mathbb{A}^{nm})]\subset\tilde{\mathbf{Y}}(\mathcal{C}_m^n)\cap U_{\mathbf{f}}.$$

On the other hand, suppose that $(A,B)\in\mathcal{C}_m^n$ with $\det\tilde{\mathbf{Y}}(A,B)_{\mathbf{f}}\neq 0$. Letting $g=\tilde{\mathbf{Y}}(A,B)_{\mathbf{f}}^{-1}$, we have

$$gA^tB_j=\epsilon_{f_1+\cdots+f_{j-1}-1+t}, \quad t=0,1,\dots,f_j-1. \tag{D.18}$$

Let $x_j=gA^{f_j}B_j$ and let $\psi_{\mathbf{f}}(x_1,\dots,x_m)=(A_{\mathbf{f}}(x_1,\dots,x_m),B_{\mathbf{f}}(x_1,\dots,x_m))=(A_1,B_1)$ (for simplicity of notation). We *claim* that $gB=B_1$ and $gAg^{-1}=A_1$. But $gB_j=\epsilon_{f_1+\cdots+f_{j-1}+1}$ so that $gB=B_1$ (from (D.18)). Also, from (D.18),

$$gA^tg^{-1}(B_1)_j=\epsilon_{f_1+\cdots+f_{j-1}+1+t}=A_1^t(B_1)_j \tag{D.19}$$

for $t=0,1,\dots,f_{j-1}$. But $gA^{f_j}B_j=gA^{f_j}g^{-1}(B_1)_j=A_1^{f_j}(B_1)_j=x_j$ and so $gAg^{-1}=A_1$. It follows that

$$\tilde{\mathbf{Y}}(A,B)=g^{-1}\tilde{\mathbf{Y}}(A_1,B_1)=g^{-1}(\tilde{\mathbf{Y}}\circ\psi_{\mathbf{f}})(x_1,\dots,x_m)\in G\cdot[(\tilde{\mathbf{Y}}\circ\psi_{\mathbf{f}})(\mathbb{A}^{nm})].$$

Case 2 $f_{j_1}=f_{j_2}=\cdots=f_{j_s}=0$ for some $s\in[1,\dots,m-1]$. Let $f_1',\dots,f_{m-s}'$ be the non-zero entries in f. Set $(B_{\mathbf{f}})_{j_r}(x_1,\dots,x_m)=x_{j_r}$ for $r=1,\dots,s$ and determine the remaining columns of $B_{\mathbf{f}}(x_1,\dots,x_m)$ as well as in the matrix $A_{\mathbf{f}}(x_1,\dots,x_m)$ as in Case 1 for $m'=m-s$ and $f_1'+\cdots+f_{m-s}'=n$.

Proposition D.20 $W_{\mathbf{f}} = \{(\det(\tilde{\mathbf{Y}} \circ \psi_{\mathbf{f}})(\mathbb{A}_k^{nm})_{(j_1,\dots,j_n)})\}$.

Proof. Since $(\tilde{\mathbf{Y}} \circ \psi_{\mathbf{f}})(\mathbb{A}_k^{nm}) \subset U_{\mathbf{f}}$, $\pi((\tilde{\mathbf{Y}} \circ \psi_{\mathbf{f}})(\mathbb{A}_k^{nm})) \subset \pi(U_{\mathbf{f}})$. Since $(\tilde{\mathbf{Y}} \circ \psi_{\mathbf{f}})(\mathbb{A}_k^{nm}) \subset \tilde{\mathbf{Y}}(C_m^n)$, we then have $\pi((\tilde{\mathbf{Y}} \circ \psi_{\mathbf{f}})(\mathbb{A}_k^{nm})) \subset W_{\mathbf{f}}$. On the other hand, if $w = (\det \tilde{\mathbf{Y}}(A,B)_{(j_1,\dots,j_n)}) \in W_{\mathbf{f}}$, then $\tilde{\mathbf{Y}}(A,B) \in U_{\mathbf{f}}$ and $(A,B) \equiv_G (A_{\mathbf{f}}, B_{\mathbf{f}})$ with $(A_{\mathbf{f}}, B_{\mathbf{f}}) \in \psi_{\mathbf{f}}(\mathbb{A}_k^{nm})$. But then $w = (\det \tilde{\mathbf{Y}}(A_{\mathbf{f}}, B_{\mathbf{f}})_{(j_1,\dots,j_n)})$.

Now let $\Gamma_{\mathbf{f}} \colon \mathbb{A}_k^{nm} \to W_{\mathbf{f}}$ be given by

$$\Gamma_{\mathbf{f}}(x_1, \dots, x_m) = (\det[\tilde{\mathbf{Y}} \circ \psi_{\mathbf{f}}(x_1, \dots, x_m)]_{(j_1,\dots,j_n)}). \tag{D.21}$$

Then $\Gamma_{\mathbf{f}}$ is a surjective morphism by the Proposition. If $\Gamma_{\mathbf{f}}(x_1, \dots, x_m) = \Gamma_{\mathbf{f}}(x'_1, \dots, x'_m)$, then, clearly,

$$\tilde{\mathbf{Y}}(A_{\mathbf{f}}(x_1, \dots, x_m), B_{\mathbf{f}}(x_1, \dots, x_m)) = g\tilde{\mathbf{Y}}(A_{\mathbf{f}}(x'_1, \dots, x'_m), B_{\mathbf{f}}(x'_1, \dots, x'_m))$$

for some $g \in G$. But then $gA_{\mathbf{f}}(x')g^{-1} = A_{\mathbf{f}}(x)$, $gB_{\mathbf{f}}(x') = B_{\mathbf{f}}(x)$ and, in view of the form of $A_f(\cdot)$, $B_f(\cdot)$, we have

$$\begin{aligned} g\epsilon_{f_1+\cdots+f_{j-1}+1} &= \epsilon_{f_1+\cdots+f_{j-1}+1} \\ g\epsilon_{f_1+\cdots+f_{j-1}+1+t_j} &= gA_f^{t_j}(x')g^{-1}gB_f(x')_j \\ &= A_f^{t_j}(x)B_f(x)_j = \epsilon_{f_1+\cdots+f_{j-1}+1+t_j} \end{aligned} \tag{D.22}$$

for $t_j = 0, \dots, f_j - 1$. In other words, $g = I$. Since

$$x_j = A_{\mathbf{f}}^{f_j}(x)(B_{\mathbf{f}}(x))_j = gA_{\mathbf{f}}^{f_j}(x')(B_{\mathbf{f}}(x'))_j = I \cdot x'_j \tag{D.23}$$

we have $x_j = x'_j$ and $\Gamma_{\mathbf{f}}$ is injective. Finally, if

$$w = (\det[\tilde{\mathbf{Y}}(A_{\mathbf{f}}(x_1, \dots, x_m), B_{\mathbf{f}}(x_1, \dots, x_m))]_{(j_1,\dots,j_n)}),$$

then $\Gamma_{\mathbf{f}}^{-1}(w) = (x_1, \dots, x_m)$ where $x_j = A_{\mathbf{f}}^{f_j}(B_{\mathbf{f}})_j$ (for Case 1 and for Case 2 with $f_{j_1} = 0, \dots$ etc.). Since $x_j = A_{\mathbf{f}}^{f_j}(B_{\mathbf{f}})_j = \tilde{\mathbf{Y}}(A_{\mathbf{f}}, B_{\mathbf{f}})_{jf_j+1}$, $\Gamma_{\mathbf{f}}^{-1}$ is a morphism. In other words, we have shown that:

Proposition D.24 *$W_{\mathbf{f}}$ is isomorphic to $\mathbb{A}_k^{nm}$ and $\cup W_{\mathbf{f}} = (\pi \circ \tilde{\mathbf{Y}})(C_m^n)$ is an nm dimensional non-singular (hence normal) quasi-projective variety.*

Let $W = \cup W_{\mathbf{f}}$ and let $\psi = \pi \circ \tilde{\mathbf{Y}}$. Then $\psi(C_m^n) = W$. Assuming that G acts trivially on W, we see that ψ is a surjective G-morphism. If $x \equiv_G x_1$, then $\tilde{\mathbf{Y}}(x) \equiv_G \tilde{\mathbf{Y}}(x_1)$ and conversely. Thus, if $\psi(x) = \psi(x_1)$, then $x \equiv_G x_1$. In other words, ψ is injective on orbits. Suppose that $\eta \colon C_m^n \to X$ is a G-morphism (with G acting trivially on X) and let $\eta' \colon W \to X$ be given by $\eta'(w) = \eta(\psi^{-1}(w))$. If η' is well-defined, then $(\eta' \circ \psi)(x) = \eta(\psi^{-1}(\psi(x)) = \eta(x)$. To show that η' is well-defined, let

$$w = (\det(\tilde{\mathbf{Y}} \circ \psi_{\mathbf{f}})(x_1, \dots, x_m)_{(j_1,\dots,j_n)}) \in W_{\mathbf{f}}$$

so that $\psi^{-1}(w) = \psi_{\mathfrak{f}}^{-1}(w) = (A_{\mathfrak{f}}(x_1, \ldots, x_m), B_{\mathfrak{f}}(x_1, \ldots, x_m))$ and $\eta'(w) = \eta(A_{\mathfrak{f}}(x_1, \ldots, x_m), B_{\mathfrak{f}}(x_1, \ldots, x_m))$. Since η is a G-morphism, η' is well-defined. Also, $\psi^{-1}(w) = O_G(A_{\mathfrak{f}}(x), B_{\mathfrak{f}}(x))$ is a closed orbit with $\dim O_G(A, B) = n^2$. Since W is normal, ψ is open on invariant sets. *However*, $k[W_{\mathfrak{f}}] \simeq k[W] \simeq k[\mathbb{A}_k^{nm}]$. If $m > 1$, then $k[\mathcal{C}_m^n]^G \simeq k[\chi_i(A)]$ (see, [B-6]) and so:

Proposition D.25 *(W, ψ) is a quotient of $\mathcal{C}_m^n$ modulo G but is not a geometric quotient.*

This proposition is why the historical approach based on controllable systems was intractable.

Appendix E
Review of Affine Algebraic Geometry

We provide a brief review of the mathematical material developed in Part I. The key concepts and results are presented (without proofs). Full details can be found in either standard mathematical texts or in Part I. Here, the reader will find enough material to digest much of Part II without continually referring to Part I.

Let k be an algebraically closed field (i.e., any polynomial with coefficients in k has all its roots in k) and let $\mathbb{A}_k^N$ be affine N space over k.

Definition E.1 $V \subset \mathbb{A}_k^N$ is an *affine algebraic set* if $V = \{a = (a_1, \dots, a_N)\colon f_\alpha(a) = 0,\ f_\alpha \in k[x_1, \dots, x_N]\}$ i.e., V is the set of common zeros of a family of polynomials.

Definition E.2 An *ideal* $\mathfrak{a}$ is a subset of a ring R such that (i) if $a, b \in \mathfrak{a}$, then $a - b \in \mathfrak{a}$ and, (ii) if $a \in \mathfrak{a}$ and $r \in R$, then $ra \in \mathfrak{a}$.

If $\mathfrak{a}$ is an ideal in $k[x_1, \dots, x_N]$, then $V(\mathfrak{a}) = \{a \in \mathbb{A}_k^N\colon f(a) = 0 \text{ all } f \in \mathfrak{a}\}$ is the *zero set of* $\mathfrak{a}$. If $W \subset \mathbb{A}_k^N$, then $I(W) = \{f \in k[x_1, \dots, x_N]\colon f(W) = 0\}$ is the *ideal of* W.

Definition E.3 $W \subset \mathbb{A}_k^N$ is (*Zariski*) *closed* if W is an affine algebraic set.

We note that $\mathbb{A}_k^N$ is a topological space under Zariski closure. If $V \subset \mathbb{A}_k^N$ is an affine algebraic set and $f \notin I(V)$, then $V_f = \{v \in V\colon f(v) \neq 0\}$ is called a *principal affine open subset of* V.

Definition E.4 A ring R is *Noetherian* if every ideal $\mathfrak{a}$ in R has a finite basis i.e., if there are $f_1, \dots, f_r \in \mathfrak{a}$ such that if $f \in \mathfrak{a}$, then there are $r_i \in R$ with $f = \sum r_i f_i$.

P. Falb, *Methods of Algebraic Geometry in Control Theory: Part II*,
Modern Birkhäuser Classics, https://doi.org/10.1007/978-3-319-96574-1

Theorem E.5 (Hilbert Basis) *If R is Noetherian, then $R[x]$ is also Noetherian. In particular, $k[x_1, \dots, x_N]$ is Noetherian.*

This theorem allows one to show that the principal affine open sets form a base for the Zariski topology.

Definition E.6 If $\mathfrak{a}$ is an ideal in a ring R, then the *radical of* $\mathfrak{a}$, $\sqrt{\mathfrak{a}}$, is the ideal $\{f \colon f^m \in \mathfrak{a} \text{ for some } m\}$.

Theorem E.7 (Hilbert Nullstellensatz) *If $\mathfrak{a}$ is an ideal in $k[x_1, \dots, x_N]$, then $\sqrt{\mathfrak{a}} = I(V(\mathfrak{a}))$ (so that algebraic sets correspond to radical ideals).*

An ideal $\mathfrak{m}$ is maximal if $\mathfrak{m} \neq R$ and $\mathfrak{m} \subset \mathfrak{a} \subset R$ implies $\mathfrak{a} = \mathfrak{m}$ or $\mathfrak{a} = R$.

Corollary E.8 *If $\mathfrak{m}$ is a maximal ideal in $k[x_1, \dots, x_N]$, then $V(\mathfrak{m}) \neq \emptyset$ (k algebraically closed).*

Definition E.9 A non-empty subset V of a topological space X is *irreducible* if $V \neq V_1 \cup V_2$, V_1, V_2 closed in V and $V_1 < V$, $V_2 < V$. An ideal $\mathfrak{p}$ in a ring R is *prime* if $fg \in \mathfrak{p}$ implies that either $f \in \mathfrak{p}$ or $g \in \mathfrak{p}$.

We observe that a non-empty open subset U of an irreducible V is both irreducible and dense and, that if V is irreducible, then so is its closure $\overline{V}$.

Theorem E.10 *An algebraic set $V \subset \mathbb{A}_k^N$ is irreducible if and only if $I(V)$ is a prime ideal in $k[x_1, \dots, X_N]$.*

Thus, $V \subset \mathbb{A}_k^N$ is irreducible if and only if the ring $k[V] = k[x_1, \dots, x_N]/I(V)$ is an integral domain.

Every affine algebraic set can be represented as a finite union $V = \cup V_i$ with the V_i irreducible. The representation is unique (to within order) if no V_i is superfluous and, in that case, the V_i are called the *components of* V.

Now the key idea of affine algebraic geometry is to associate with algebraic sets the "regular" functions on these sets. If $R = k[f_1, \dots, f_n]$ has no nilpotent elements, then R is called an *affine k-algebra*. If V is an affine algebraic set, then

$$k[V] = k[x_1, \dots, x_N]/I(V) \tag{E.11}$$

is an affine k-algebra called the *affine coordinate ring of* V.

Theorem E.12 $V \simeq \mathrm{Hom}_k(k[V], k) = \{\alpha\colon$ *α is a k-homomorphism of $k[V]$ into k*$\}$.

In other words, V can be recovered from $k[V]$ and the elements of $k[V]$ can be viewed as regular functions on V.

Corollary E.13 $V \simeq \mathrm{spm}(k[V]) = \{\mathfrak{m}\colon \mathfrak{m}$ *a maximal ideal in* $k[V]\}$.

Definition E.14 If $V \subset \mathbb{A}_k^N$ and $W \subset \mathbb{A}_k^M$ are affine algebraic sets, then a mapping $\varphi\colon V \to W$ is a *morphism* if $\varphi^*(g) = g \circ \varphi \in k[V]$ for all $g \in k[W]$ (φ^* is called the *comorphism*). A mapping $\psi\colon V \to W$ is *regular* if ψ is the restriction of a polynomial mapping.

If $\mathrm{Hom}_k(V,W) = \{\varphi\colon \varphi$ a morphism of V into $W\}$, $\mathrm{Reg}_k(V,W) = \{\psi\colon \psi$ is a regular map of V into $W\}$, and $\mathrm{Hom}_k(k[W],k[V]) = \{\lambda\colon \lambda$ is a k-homomorphism of $k[W]$ into $k[V]\}$, then we have:

Theorem E.15 $\mathrm{Hom}_k(V,W) \simeq \mathrm{Reg}_k(V,W) \simeq \mathrm{Hom}_k(k[W],k[V])$.

A bijective morphism φ is an *isomorphism* if φ^{-1} is also a morphism.

Corollary E.16 *A morphism* $\varphi\colon V \to W$ *is an isomorphism if and only if the comorphism* $\varphi^*\colon k[W] \to k[V]$ *is an isomorphism.*

If $V \subset \mathbb{A}_k^N$ is irreducible, then $I(V) = \mathfrak{p}_V$ is prime and $k[V]$ is an integral domain. The quotient field $k(V)$ of $k[V]$ is then called the *function field of* V.

Definition E.17 The set $\mathfrak{o}_{\xi,V} = \{f/g\colon f,g \in k[V],\ g(\xi) \neq 0\}$ is called the *local ring of* V *at* ξ.

A subset M of a ring R is *multiplicatively closed* if $0 \notin M$ and $m, m' \in M$ implies $mm' \in M$. If $M_{\mathfrak{p}} = R - \mathfrak{p} = \{r\colon r \notin \mathfrak{p}\}$ for $\mathfrak{p}$ a prime ideal, then $M_{\mathfrak{p}}$ is multiplicatively closed and we can define the *quotient ring of* R *with respect to* $M_{\mathfrak{p}}$, $R_{\mathfrak{p}}$ $(= \{r/m\colon m \in M_{\mathfrak{p}}\})$. Then, we have:

$$\mathfrak{o}_{\xi,V} = k[V]_{\mathfrak{m}_\xi} = \left(k[x_1,\dots,x_N]_{M_\xi}/I(V)k[x_1,\dots,x_N]_{M_\xi}\right) \tag{E.18}$$

where $\mathfrak{m}_\xi = \{f \in k[V]\colon f(\xi) = 0\}$ and $M_\xi = \{g \in k[x_1,\dots,x_N]\colon g(\xi) = 0\}$. The elements of $\mathfrak{o}_{\xi,V}$ are regular functions at ξ.

Definition E.19 If U is open in V, then

$$\mathfrak{o}_V(U) = \bigcap_{\xi \in U} \mathfrak{o}_{\xi,V}$$

is the *ring of regular functions on* U.

We note that $\mathfrak{o}_V(V) = k[V]$ and that $\mathfrak{o}_V(V_f) = k[V]_f = \{h/f^s\colon h \in k[V],\ s \in \mathbb{Z}\}$.

Definition E.20 If $V \subset \mathbb{A}_k^N$, $W \subset \mathbb{A}_k^M$ are affine algebraic sets and if $\mathfrak{a}$ is the ideal in $k[X_1, \ldots, X_N, Y_1, \ldots, Y_M]$ generated by $I(V) \cup I(W)$, then $V(\mathfrak{a}) = V \times W$ is called the *product of* V *and* W.

The map $\pi_V \colon V \times W \to V$ given by $\pi_V(\xi, \eta) = \xi$ is called the *projection of* $V \times W$ *on* V. π_V is a morphism and also is an open map. The notion of a product allows us to develop various properties of morphisms.

Proposition E.21 (a) *Let* $\varphi \colon V \to W$ *be a morphism and let* $\mathrm{gr}(\varphi) = \{(\xi, \eta) \in V \times W \colon \eta = \varphi(\xi)\}$ *be the graph of* φ. *Then* $\mathrm{gr}(\varphi)$ *is closed and isomorphic to* V. (b) *If* $\varphi \colon V \to W$, $\psi \colon V \to W$ *are morphisms and* $\varphi = \psi$ *on a dense subset, then* $\varphi = \psi$.

In order to deal with the Geometric Quotient Theorem (and a number of other results), we require the notion of dimension.

Definition E.22 If X is a topological space, then the *dimension of* X, $\dim X$, is the maximum integer n such that there is a chain $V_0 < V_1 < \cdots < V_n$ of distinct closed irreducible subsets of X so that $\dim_k V$ is the dimension of V as a topological space. If V is an affine variety (irreducible), then the *algebraic dimension of* V *over* k, $\mathrm{Dim}_k V$, is the transcendence degree of $k(V)$ over k.

We can show (I.16) that the two notions are equal. Several concepts of general utility are required.

Definition E.23 Let $R \subset S$ be rings. An element u of S is *integral over* R if u satisfies a monic equation

$$u^n + r_1 u^{n-1} + \cdots + r_n = 0$$

with coefficients r_i in R. The set $\{u \in S \colon u \text{ is integral over } R\}$ is called the *integral closure of* R *in* S. An integral domain R is *integrally closed* if the integral closure of R in $K(R)$ (the quotient field of R) is R itself.

Definition E.24 Let R be an integral domain and let $\mathfrak{p}$ be a prime ideal in R. Then $h = h(\mathfrak{p})$ is the *height of* $\mathfrak{p}$ if there is a (strict) chain $(0) = \mathfrak{p}_0 < \mathfrak{p}_1 < \cdots < \mathfrak{p}_h = \mathfrak{p}$ of prime ideals and there does not exist such a chain with more elements. Also, $d = d(\mathfrak{p})$ is the *depth of* $\mathfrak{p}$ if there is a (strict) chain $\mathfrak{p} = \mathfrak{p}_d < \mathfrak{p}_{d-1} < \cdots < \mathfrak{p}_0 < R$ and there does not exist such a chain with more elements.

The key result which makes things work is the following:

Theorem E.25 *Let* R *be a finitely generated integral domain of transcendence degree* ν *over* k *and let* $\mathfrak{p}$ *be a prime ideal in* R. *If* $\dim_k \mathfrak{p}$ $(= \mathrm{tr}\deg K(R/\mathfrak{p})/k) = s$, *then*

$$h(\mathfrak{p}) = \nu - s, \quad d(\mathfrak{p}) = s, \quad h(\mathfrak{p}) + d(\mathfrak{p}) = \nu$$

so that $\dim_k \mathfrak{p} = d(\mathfrak{p})$.

This result depends on the so-called "Going Down" theorem ([A-2] or I.16.40) and leads to some interesting properties of dimension.

Proposition E.26 (a) *An affine variety $V \subset \mathbb{A}_k^N$ has dimension $N-1$ if and only if $V = V(f)$ where f is a prime in $k[x_1, \ldots, x_N]$*; (b) *If $f_1, \ldots, f_r \in k[W]$ and Z is an irreducible component of $V(f_1, \ldots, f_r)$, then* $\text{codim}_W Z \leq r$; *and,* (c) *If W is an affine variety and Z is a closed irreducible subset with* $\text{codim}_W Z = r \geq 1$, *then there are $f_1, \ldots, f_r$ in $k[V]$ such that Z is a component of $V(f_1, \ldots, f_r)$.*

Now, let us suppose that $\psi \colon X \to Y$ is a morphism. If $y \in Y$, the closed set $\psi^{-1}(y) = \{x \in X \colon \psi(x) = y\}$ is the *fiber of ψ over y*. If W is a closed irreducible subset of Y, then the closed set $\psi^{-1}(W) = \{x \in X \colon \psi(x) \in W\}$ is the *fiber of ψ over W*.

Definition E.27 A morphism $\psi \colon X \to Y$ of varieties is *dominant* if $\psi(X)$ is dense in Y. A morphism is *finite* if $k[X]$ is integral over $\psi^*(k[Y])$.

Proposition E.28 *Let $\psi \colon X \to Y$ be a dominant morphism and let W be a closed irreducible subset of Y. If Z is a component of $\psi^{-1}(W)$ which dominates W, then* $\dim Z \geq \dim W + \dim X - \dim Y$. *In particular,* $\dim \psi^{-1}(y) \geq \dim X - \dim Y$ *for $y \in Y$.*

Proposition E.29 *A dominant finite morphism is a surjective, closed map. If, in addition, Y is normal (i.e., $k[Y]$ is integrally closed), then $\psi(Z) = W$ for all components Z of $\psi^{-1}(W)$.*

Theorem E.30 *Let $\psi \colon X \to Y$ be a dominant morphism. Then there is an open set $U \subset Y$ such that* (a) *$U \subset \psi(X)$ and,* (b) *if W is a closed irreducible subset of Y with $W \cap U \neq \emptyset$ and Z is a component of $\psi^{-1}(W)$ with $Z \cap \psi^{-1}(U) \neq \emptyset$, then* $\dim Z = \dim W + \dim X - \dim Y$.

These results lead to an important theorem of Chevalley ([D-2]) which is the key to proving a part of the Geometric Quotient Theorem.

Theorem E.31 (Chevalley) *Let $\psi \colon X \to Y$ be a dominant morphism and let* $r = \dim X - \dim Y$. *Suppose that Y is normal and that all the components of $\psi^{-1}(y)$, $y \in Y$, are r-dimensional. Then ψ is an open map.*

An important characteristic of the space of linear systems and its quotient under coordinate change is *nonsingularity*.

Definition E.32 Let V be an affine variety so that $\mathfrak{p} = I(V)$ is prime and let $\xi = (\xi_1, \ldots, \xi_N) \in V$. Then the (Zariski) *tangent space to* V *at* ξ, $T_{V,\xi}$, is the linear variety defined by

$$\sum_{i=1}^{N} \frac{\partial f}{\partial X_i}(\xi)(X_i - \xi_i) = 0 \tag{E.33}$$

for all $f \in \mathfrak{p}$.

The tangent space can also be characterized by $T_{V,\xi} = \mathrm{Der}_k(k[V], k_\xi)$ (the space of k-derivations of $k[V]$ into k_ξ) or by $T_{V,\xi} = \mathrm{Der}_k(\mathfrak{o}_{\xi,V}, k_\xi)$ or by $T_{V,\xi} = (\mathfrak{m}_\xi/\mathfrak{m}_\xi^2)^*$ where $\mathfrak{m}_\xi = \{f \in k[V]\colon f(\xi) = 0\}$.

Proposition E.34 *If* ν *is an integer, then* $\{\xi \in V\colon \dim_k T_{V,\xi} \geq \nu\}$ *is closed.*

Definition E.35 A point $\xi \in V$ is a *simple point of* V if $\dim_k T_{V,\xi} = \dim_k V$ and V is *nonsingular* or *smooth* if all points of V are simple.

If $\mathrm{Sing}(V) = \{\xi \in V\colon \xi \text{ is singular}\}$, then $\mathrm{Sing}(V)$ is a proper closed subset of V.

Since $T_{V,\xi} = (\mathfrak{m}_\xi/\mathfrak{m}_\xi^2)^*$, ξ is a simple point if and only if $\mathfrak{o}_{\xi,V}$ is a regular local ring.

Finally, there is the notion of an *abstract affine algebraic variety.* Let R be an affine k-algebra and let $X = \mathrm{spm}(R) = \{\mathfrak{m}\colon \mathfrak{m} \text{ a maximal ideal in } R\}$. If E is a subset of R, then $V(E) = \{\mathfrak{m} \in X\colon \mathfrak{m} \supset E\}$ is the *closure of* E.

Proposition E.36 (i) $V(E) = V(\mathfrak{a}) = V(\sqrt{\mathfrak{a}})$ *where* $\mathfrak{a} = (E)$ *is the ideal generated by* E*;* (ii) $V(0) = X$, $V(R) = \emptyset$*;* (iii) $V(\cup E_i) = \cap V(E_i)$*; and,* (iv) $V(\mathfrak{a} \cap \mathfrak{b}) = V(\mathfrak{a}) \cup V(\mathfrak{b})$.

Thus, the sets $V(\mathfrak{a})$ define a topology, the *Zariski topology,* on $X = \mathrm{Spm}(R)$. The *principal open sets* $X_f = X - V(f) = \{\mathfrak{m}\colon f \notin \mathfrak{m}\}$ form a base for this topology. A closed set $V(\mathfrak{a})$ is irreducible if and only if $\mathfrak{a}$ is a prime ideal and so X is irreducible if and only if R is an integral domain.

If $\mathfrak{m}$ is an element of X, then $R_\mathfrak{m}$ is a local ring and we write $x = \mathfrak{m}_x$ so that $R_x = R_{\mathfrak{m}_x}$. The elements of R_x are called regular functions and we write $\mathfrak{o}_{x,X} = R_x$. $\mathfrak{o}_{x,X}$ is the *local ring of* X *at* x. If U is open in X, then we set

$$\mathfrak{o}_X(U) = \bigcap_{x \in U} R_x$$

and call $\mathfrak{o}_X(U)$ the k-algebra of *regular functions on* U.

Proposition E.37 *Suppose that* X *is irreducible. Then* (i) $\mathfrak{o}_X(X_f) = R_f$ *and* $\mathfrak{o}_X(X) = R$*;* (ii) *if* $V \subset U$ *are open sets and* $\rho_{U,V}\colon \mathfrak{o}_X(U) \to \mathfrak{o}_X(V)$ *is* *the restriction map i.e.,* $\rho_{U,V}(f)(\mathfrak{m}) = f(\mathfrak{m})$ *for* $\mathfrak{m} \in V$ *and* $f \in \mathfrak{o}_X(U)$*, then* $\rho_{U,V} \in \mathrm{Hom}_k(\mathfrak{o}_X(U), \mathfrak{o}_X(V))$*,* $\rho_{U,U} =$ *identity and if* $W \subset V \subset U$*, then* $\rho_{U,V} \circ \rho_{V,W} = \rho_{U,W}$*;* (iii) *if* $U = \cup U_i$*,* $f_1, f_2 \in \mathfrak{o}_X(U)$ *and* $\rho_{U,U_i}(f_1) = \rho_{U,U_i}(f_2)$ *for all* i*, then* $f_1 = f_2$*; and,* (iv) *if* $U = \cup U_\alpha$ *and* $f_\alpha \in \mathfrak{o}_X(U_\alpha)$ *with* $\rho_{U_\alpha, U_\alpha \cap U_\beta}(f_\alpha) = \rho_{U_\beta, U_\beta \cap U_\alpha}(f_\beta)$ *for all* α, β*, then there is a (unique by (iii))* f *in* $\mathfrak{o}_X(U)$ *such that* $\rho_{U,U_\alpha}(f) = f_\alpha$*.*

So we have a topological space X and a family of k-algebras $\mathfrak{o}_X(U)$, U open in X, and, a family of k-homomorphisms $\rho_{U,V}\colon \mathfrak{o}_X(U) \to \mathfrak{o}_X(V)$ for $U \supset V$, such that (ii), (iii) and (iv) are satisfied. The family $\mathfrak{o}$ of $\mathfrak{o}_X(U)$ and maps $\rho_{U,V}$ is called a *sheaf* and the pair $(X, \mathfrak{o})$ is called a *ringed space.*

Definition E.38 If R is an irreducible affine k-algebra, then the ringed space $(X, \mathfrak{o}_X)$ where $X = \mathrm{Spm}(R)$ and $\mathfrak{o}_X$ is the sheaf $\{\mathfrak{o}_X(U), \rho_{U,V}\}$ is an *abstract affine algebraic variety over* k*.*

This notion is ultimately the most fundamental.

References

[A-1] Athans, M. and Falb, P.L., *Optimal Control: An Introduction to the Theory and its Applications*, McGraw-Hill, New York, 1966.

[A-2] Atiyah, M.F. and MacDonald, I.G., *Introduction to Commutative Algebra*, Addison-Wesley, Reading, MA, 1969.

[B-1] Birkhoff, G. and Maclane, S., *A Survey of Modern Algebra*, Revised Edition, Macmillan, New York, 1953.

[B-2] Borel, A., *Linear Algebraic Groups*, Benjamin, New York, 1969.

[B-3] Bourbaki, N., *Algèbre Commutative*, I–VII, Hermann, Paris, 1961–1965.

[B-4] Brockett, R. and Byrnes, C.I., Multivariable Nyquist criteria root loci, and pole placement: A geometric viewpoint, *IEEE Trans. Aut. Cont.*, **AC-26** (1981).

[B-5] Brockett, R., *Finite-Dimensional Linear Systems*, Wiley-Interscience, New York, 1970.

[B-6] Byrnes, C., The moduli space for a linear dynamical system, in *Geometric Control Theory* (C. Martin, R. Hermann, Eds.), Math. Sci. Press, Brookline, MA, 1977.

P. Falb, *Methods of Algebraic Geometry in Control Theory: Part II*,
Modern Birkhäuser Classics, https://doi.org/10.1007/978-3-319-96574-1

[B-7] Byrnes, C.I., Pole assignment by output feedback, in *Lecture Notes in Control and Information Sciences*, **135**, Springer-Verlag, Berlin, Heidelberg, New York, 1989.

[B-8] Byrnes, C. and Falb, P.L., Applications of algebraic geometry in system theory, *Am. J. Math.*, **101** (1979).

[B-9] Byrnes, C. and Hurt, N., On the moduli of linear dynamical systems, *Adv. in Math: Studies in Analysis*, **4** (1978).

[C-1] Clark, J.M.C., The consistent selection of local coordinates in linear system identification, *Proc. JACC*, Purdue, 1976.

[D-1] Davison, E. and Wang, S., Properties of linear time-invariant multivariable systems subject to arbitrary output and state feedback, *IEEE Trans. Aut. Cont.*, **AC-18** (1973).

[D-2] Dieudonné, J., *Cours de Géométrie Algébrique*, 1, 2, Presses Universitaires de France, 1974.

[D-3] Dieudonné, J. and Carrell, J., Invariant theory, old and new, *Advances in Math.*, **4** (1970).

[F-1] Falb, P.L., *Linear Systems and Invariants*, Lectures Notes, Control Group, Lund University, Sweden, 1974.

[F-2] Fogarty, J., *Invariant Theory*, Benjamin, New York, 1969.

[F-3] Fuhrmann, P., Linear algebra and finite dimensional systems, Math. Report 143, Ben Gurion University, 1978.

[F-4] Fulton, W., *Algebraic Curves*, Benjamin, New York, 1969.

[F-5] Fulton, W., *Intersection Theory*, Springer-Verlag, Berlin, Heidelberg, 1984.

[G-1] Gantmacher, F.R., *Theory of Matrices* I, II, Chelsea, New York, 1959.

[H-1] Haboush, W.J., Reductive groups are geometrically reductive, *Ann. of Math.*, **102** (1975).

[H-2] Harris, J., *Algebraic Geometry*, Springer-Verlag, Berlin, Heidelberg, New York, 1992.

[H-3] Hartshorne, R., *Algebraic Geometry*, Springer-Verlag, Berlin, Heidelberg, New York, 1977.

[H-4] Hazewinkel, M., Moduli and canonical forms for linear dynamical systems III: The algebraic-geometric case, in *Geometric Control Theory* (C. Martin, R. Hermann, eds.), Math. Sci. Press, Brookline, MA, 1977.

[H-5] Hazelwinkel, M. and Kalman, R.E., On invariants, canonical forms and moduli for linear constant finite-dimensional, dynamical systems, in *Lecture Notes Economics-Math System Theory*, Springer-Verlag, Berlin, Heidelberg, New York, 1976.

[H-6] Hermann, R., *Linear Systems Theory and Introductory Algebraic Geometry*, Math. Sci. Press, Brookline, MA, 1974.

[H-7] Hermann, R. and Martin, C., Applications of algebraic geometry to system theory—Part I, *IEEE Trans. Aut. Cont.*, **AC-22** (1977).

[H-8] Hochschild, G., *Basic Theory of Algebraic Groups and Lie Algebras*, Springer-Verlag, Berlin, Heidelberg, New York, 1981.

[H-9] Hodge, W.V.D. and Pedoe, D., *Methods of Algebraic Geometry*, Vols. I,II,III, Cambridge University Press, 1952.

[H-10] Humphreys, J., *Linear Algebraic Groups*, Springer-Verlag, Berlin, Heidelberg, New York, 1975.

[I-1] Iitaka, Shigeru, *Algebraic Geometry; An Introduction to Birational Geometry of Algebraic Varieties*, Springer-Verlag, New York, 1982.

[J-1] Jacobson, N., *Lectures in Abstract Algebra* I, II, III, Van Nostrand, Princeton, NJ, 1953.

[K-1] Kalman, R.E., Global structure of classes of linear dynamical systems, in *Lectures*, NATO Adv. Study Inst., London, 1971.

[K-2] Kalman, R.E., Falb, P.L., and Arbib, M.A., *Topics in Mathematical System Theory*, McGraw-Hill, New York, 1969.

[K-3] Kendig, K., *Elementary Algebraic Geometry*, Springer-Verlag, Berlin, Heidelberg, New York, 1977.

[K-4] Kimura, H., Pole assignment by gain output feedback, *IEEE Trans. Aut. Cont.*, **AC-20** (1975).

[K-5] Kunz, E., *Introduction to Commutative Algebra and Algebraic Geometry*, Birkhäuser Boston, Cambridge, MA, 1985.

[L-1] Lang, S., *Algebra*, Addison-Wesley, Reading, MA, 1971.

[M-1] Matsumura, H., *Commutative Algebra*, 2nd ed., Benjamin, New York, 1980.

[M-2] Mumford, D.B., *Introduction to Algebraic Geometry* (Preliminary Version of First 3 Chapters), Lecture Notes, Harvard University, 1965.

[M-3] Mumford, D.B., *Geometric Invariant Theory*, Springer-Verlag, Berlin, Heidelberg, New York, 1965. [2nd Enlarged Edition, with J. Fogarty, 1982] .

[M-4] Mumford, D. and Suominen, K., Introduction to the theory of moduli, in Proc. Fifth Summer School in Math., Oslo, 1970.

[M-5] Mumford, D., *Algebraic Geometry I: Complex Projective Varieties*, Springer-Verlag, Berlin, Heidelberg, New York, 1976.

[N-1] Nagata, M., Invariants of a group in an affine ring, *J. Math. Kyoto* (1964).

[N-2] Neeman, A., Topics in Algebraic Geometry, Thesis, Harvard University, 1983.

[N-3] Northcott, D.G., *Affine Sets and Affine Groups*, Cambridge University Press, Cambridge, U.K., 1980.

[R-1] Rissanen, J., Basis of invariants and canonical forms for linear dynamic systems, *Automatica*, **10** (1974).

[R-2] Rosenbrock, H., *State-space and Multivariable Theory*, Wiley-Interscience, New York, 1970.

[S-1] Samuel, P., *Méthodes d'Algèbre Abstraite en Géométrie Algébrique*, Springer-Verlag, Berlin, Heidelberg, New York, 1967.

[S-2] Shafarevich, I.R., *Basic Algebraic Geometry*, Springer-Verlag, Berlin, Heidelberg, New York, 1977.

[S-3] Sontag, E., Linear systems over commutative rings: A survey, *Richerche di Automatica*, **7** (1976).

[S-4] Springer, T.A., *Linear Algebraic Groups*, Birkhäuser Boston, Cambridge, MA, 1981.

[W-1] Wang, S. and Davison, E., Canonical forms of linear multivariable systems, Cont. Sys. Report 7203, Dept. of Elec. Eng., Toronto, 1972.

[W-2] Wang, X., Pole placement by static output feedback, *J. Math. Sys. Est. Cont.*, **2** (1992).

[W-3] Wang, X., Grassmannian, central projection, and output feedback pole assignment of linear systems, *IEEE Trans. Aut. Cont.*, **AC-41** (1996).

[W-4] Wolovich, W., *Linear Multivariable Systems*, Springer-Verlag, Berlin, Heidelberg, New York, 1974.

[W-5] Wonham, W.M., On pole assignment in multi-input controllable linear systems, *IEEE Trans. Aut. Cont.*, **AC-12** (1967).

[W-6] Wonham, W.M. and Morse, A.S., Feedback invariants of linear multivariable systems, *Automatica*, **8** (1972).

[Z-1] Zariski, O., A new proof of the Hilbert Nullstellensatz, *Bull. Am. Math. Soc.*, **53** (1947).

[Z-2] Zariski, O., The concept of a simple point of an abstract algebraic variety, *Trans. Am. Math. Soc.*, **62** (1947).

[Z-3] Zariski, O. and Samuel, P., *Commutative Algebra* I, II, Van Nostrand, Princeton, NJ, 1958, 1960.

Glossary of Notations

P. Falb, *Methods of Algebraic Geometry in Control Theory: Part II*,
Modern Birkhäuser Classics, https://doi.org/10.1007/978-3-319-96574-1

Rat, 13, 70
Rat(2, 2, 2), 46
Rat(3, 2, 2), 40
Rat(n, 2, 2), 2, 42
Rat(n, 3, 2), 2
Rat(n, m, p), 3, 166, 167
Rat($n; p, 1$), 13, 14, 15
Res(p, q), 16

$\sigma_k(Z)$, 196
$\Sigma(n, m, p)$, 219

$\hat{S}_+$, 8
$S^n_{m,p}$, 219

$\tau(n, m, p)$, 181
$\Theta(p, N)$, 194

$V_{\mathrm{Gr}}(n, m, p)$, 184
$\mathbf{V}(n, m, p)$, 3, 165, 166, 170, 181
$\mathbf{V}_A(n, m, p)$, 4, 211, 244

Index

P. Falb, *Methods of Algebraic Geometry in Control Theory: Part II*,
Modern Birkhäuser Classics, https://doi.org/10.1007/978-3-319-96574-1

Printed in the United States
By Bookmasters